SIGNAL ANALYSIS

SIGNAL ANALYSIS

ATHANASIOS PAPOULIS

POLYTECHNIC INSTITUTE OF NEW YORK

McGRAW-HILL BOOK COMPANY

New York St. Louis San Francisco Auckland Bogotá Düsseldorf
Johannesburg London Madrid Mexico Montreal New Delhi Panama Paris
São Paulo Singapore Sydney Tokyo Toronto

SIGNAL ANALYSIS

67890 DODO 83

This book was set in Times New Roman.
The editors were Peter D. Nalle and Michael Gardner;
the cover was designed by Scott Chelius;
the production supervisor was Leroy A. Young.
The drawings were done by J & R Services, Inc.
R. R. Donnelley & Sons Company was printer and binder.

Library of Congress Cataloging in Publication Data

Papoulis, Athanasios, date
Signal analysis.

Includes bibliographies and index.
1. Signal processing. I. Title.
TK5102.5.P35 621.38'043 76-54353
ISBN 0-07-048460-0

CONTENTS

PREFACE

Educators are in general agreement that engineering students are best prepared for future challenges if they are instructed not only in their specialty but also in disciplines of broad scope. One such discipline, rich in theory and applications, is signal analysis. It is used in astronomy, oceanography, crystallography, bioengineering, antenna design, communications technology, system theory, computer sciences, and many other fields.

I have selected from this elegant subject several areas related to Fourier transforms and linear systems, and have sought to develop them with clarity, perspective, and economy. This book is the result.

Most derivations follow rigorously from a small number of clearly defined concepts. However, in certain cases, I give only plausibility arguments relating the conclusions to familiar concepts. This occasional sacrifice of rigor for the sake of brevity is not uncommon, even in pure mathematics where, for example, Jordan's curve theorem is usually accepted as self-evident and the axiom of choice is ignored.

The material is divided into three self-contained but closely related parts. The level of difficulty of each part is fairly uniform, but the degree of elaboration varies depending on the importance or novelty of the topic. I have used in several instances the word *elaborate* in parenthesis to indicate that the explanation is brief.

The book includes several applied topics, some of them new. However, I am primarily an educator, and I concentrated on what I thought I could do best—concepts of general interest rather than detailed coverage of special topics.

During the preparation of the manuscript, I had the benefit of lengthy discussions with many students and colleagues. I am indebted in particular to Professors R. Haddad, A. E. Laemmel, and D. C. Youla. I also wish to express my appreciation to Professor G. Temes of U.C.L.A. for his critical comments and constructive suggestions.

Athanasios Papoulis

SIGNAL ANALYSIS

PART ONE

SIGNALS, SYSTEMS, AND TRANSFORMS

In this part, we develop the theory of continuous and discrete systems, Fourier transforms, and z transforms, stressing discrete-continuous parallels and the role of digital filters in analog signal processing.

We define a system as a mapping of a set F (input) into a set G (output). The system is *continuous* or *analog* if the elements of F and G are functions of a continuous variable. The system is *discrete* or *digital* if the elements of F and G are sequences of numbers. Fourier transforms and z transforms are introduced as eigenvalues of such systems (system functions).

This material is covered at the Polytechnic Institute of New York in a one-semester course for beginning graduate students. The required background is general calculus and some knowledge of circuit theory.

CHAPTER

ONE

INTRODUCTION

In this chapter, we present briefly the basic concepts of the theory of discrete and continuous systems.

In Sec. 1-1, we develop the discrete form of linearity, establish the equivalence between convolution and terminal characterization of linear systems, and introduce the z transform as system function.[1] *The discussion is repeated in Sec. 1-2 for continuous systems and Fourier transforms.*[2] *In Sec. 1-3, we explain the principle of digital simulation of analog systems.*

These topics are developed with greater elaboration in subsequent chapters.

---

[1] B. Gold and C. Rader, "Digital Processing of Signals," McGraw-Hill Book Company, New York, 1969.

[2] L. E. Franks, "Signal Theory," Prentice-Hall, Inc., Englewood Cliffs, N.J., 1969.

1-1 DISCRETE SIGNALS AND SYSTEMS

The notation $f[n]$ will mean a sequence of numbers, real or complex, defined for every integer n. The sequence $f[n]$ will be called a *discrete* or *digital signal*, and the index n *discrete time*.

The following special cases will be used often (Fig. 1-1):

Step sequence *Delta sequence*

$$U[n] = \begin{cases} 1 & n \geq 0 \\ 0 & n < 0 \end{cases} \qquad \delta[n] = \begin{cases} 1 & n = 0 \\ 0 & n \neq 0 \end{cases}$$

We note that $\delta[n-3]$ equals 1 for $n = 3$, and 0 for $n \neq 3$. For any k,

$$\delta[n-k] = \begin{cases} 1 & n = k \\ 0 & n \neq k \end{cases} \tag{1-1}$$

In the above, it is understood that n is the discrete time and k is a constant parameter.

From Eq. (1-1) it follows that an arbitrary sequence $f[n]$ can be written as a weighted sum of delta sequences:

$$f[n] = \sum_{k=-\infty}^{\infty} f[k]\,\delta[n-k] \tag{1-2}$$

This is illustrated in Fig. 1-2.

Discrete Systems

A discrete system is a rule for assigning to a sequence $f[n]$ another sequence $g[n]$. Thus, a discrete system is a mapping (transformation) of the sequence $f[n]$ into the sequence $g[n]$. We shall use the notation

$$g[n] = L\{f[n]\}$$

for this mapping. The sequence $f[n]$ will be called the *input*, and the sequence $g[n]$ the *output*, or *response* (Fig. 1-3).

In general, to determine the value of the output $g[n]$ for a specific n, we must know the input $f[n]$ for every n, past and future. However, as we see in the following illustrations, this is not always necessary.

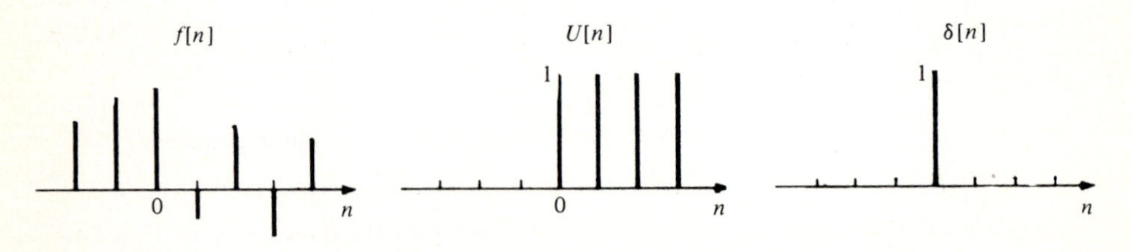

Figure 1-1

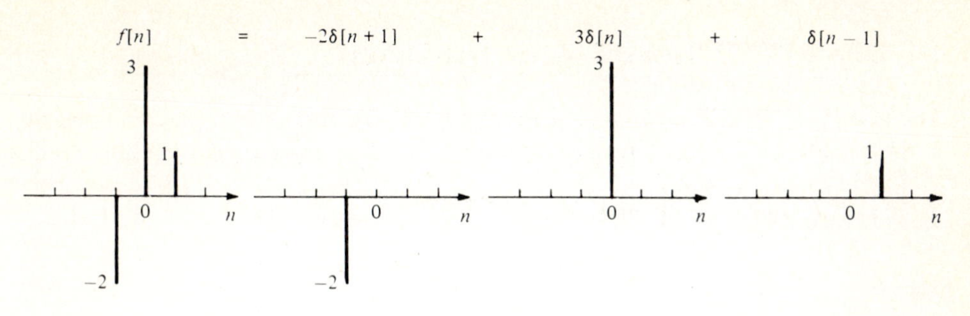

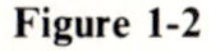
Figure 1-2

Example 1-1

(*a*) $g[n] = f^2[n]$. This system is nonlinear, and the present value $g[n]$ of the output depends only on $f[n]$ (memoryless system).

(*b*) $g[n] = nf[n]$. This is a linear, memoryless, time-varying system.

(*c*) $g[n] = 2f[n] + 3f[n-1]$. The present value $g[n]$ depends on $f[n]$ and the preceding value $f[n-1]$. The system has finite memory.

In the systems (nonrecursive) of Example 1-1, $g[n]$ was expressed directly in terms of $f[n]$.

Example 1-2

$$g[n] + 2g[n-1] = f[n]$$

In this example, to find $g[n]$ we need to know not only $f[n]$ but also $g[n-1]$. Thus, $g[n]$ is obtained by solving a *recursion equation*. We have, in fact, an infinite set of equations, one for each n. As we shall show, under certain conditions (causality), these equations have a unique solution; therefore, they define a system (*recursive*).

The following simple systems are of particular interest:

Delay element	*Multiplier*
$g[n] = f[n-1]$	$g[n] = af[n]$

These systems will be represented by the block diagrams of Fig. 1-4. The letter a in the triangle representing the multiplier is its *gain*. The significance of the letter z^{-1} (the system function) in the block representing the delay element will be discussed presently.

We show later that an arbitrary linear system can be realized by a combination of delay elements and multipliers. As an illustration, Fig. 1-5 gives the realization of the system $g[n] = 2f[n] + 3f[n-1]$.

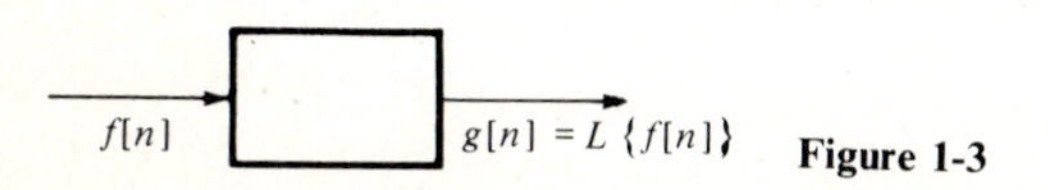

Figure 1-3

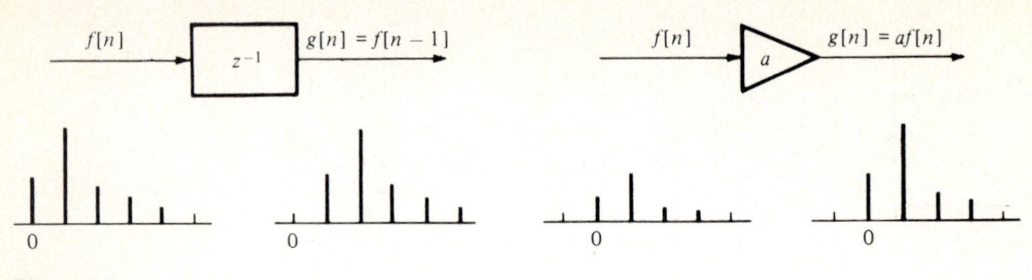

Figure 1-4

Linearity A system L is linear if

$$L\{a_1 f_1[n] + a_2 f_2[n]\} = a_1 L\{f_1[n]\} + a_2 L\{f_2[n]\} \qquad (1\text{-}3)$$

for any a_1, a_2, $f_1[n]$, and $f_2[n]$.

From the definition it follows that the response to $af[n]$ equals $ag[n]$. Furthermore, if $g_1[n]$ and $g_2[n]$ are the responses to $f_1[n]$ and $f_2[n]$, respectively, then the response to $f_1[n] + f_2[n]$ equals $g_1[n] + g_2[n]$.

Time-invariance A system L is *time-invariant* if

$$L\{f[n-k]\} = g[n-k] \qquad (1\text{-}4)$$

for any k. In words: A shift of the input results in an equal shift of the output.

Example 1-3
(*a*) The system $g[n] = |f[n]|$ is nonlinear (explain) but time-invariant.
(*b*) The system $g[n] = nf[n]$ is linear, but time-varying, because the response to $f[n-k]$ equals $nf[n-k]$, whereas $g[n-k] = (n-k)f[n-k]$.
(*c*) The system $g[n] = 2f[n] + 3f[n-1]$ is linear and time-invariant.

In the following, the term "linear system," or simply "system," will mean linear and time-invariant.

The delta response We shall denote by $h[n]$ the response of a system to the delta sequence $\delta[n]$ (Fig. 1-6):

$$L\{\delta[n]\} = h[n] \qquad (1\text{-}5)$$

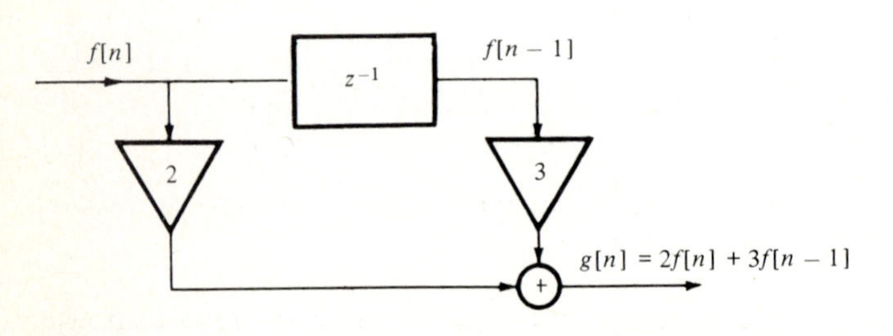

Figure 1-5

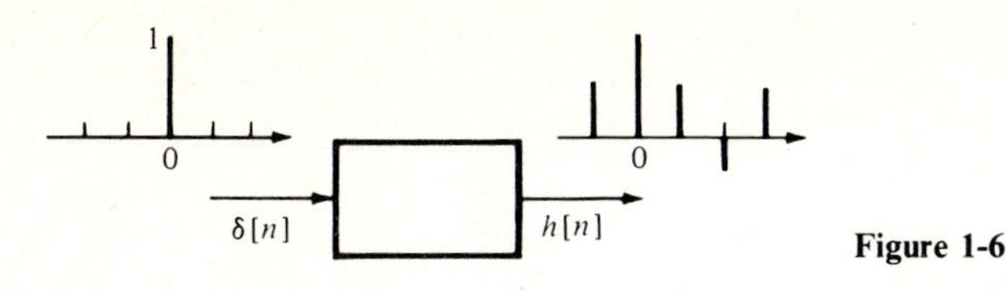

Figure 1-6

We note that the sequence $h[n]$ is not necessarily zero for $n < 0$. If

$$h[n] = 0 \qquad \text{for} \qquad n < 0 \tag{1-6}$$

then the system is called *causal*.

Example 1-4

$$g[n] = 2f[n] + 3f[n-1]$$

In this example, $g[n]$ is expressed directly in terms of $f[n]$ (nonrecursive system). Setting $f[n] = \delta[n]$, we can, therefore, find the delta response $h[n]$:

$$h[n] = 2\,\delta[n] + 3\,\delta[n-1]$$

Example 1-5

$$g[n] = f[n] + \tfrac{1}{2}f[n-1] + \cdots + (\tfrac{1}{2})^k f[n-k] + \cdots \tag{1-7}$$

As in Example 1-4,

$$h[n] = \delta[n] + \tfrac{1}{2}\,\delta[n-1] + \cdots + (\tfrac{1}{2})^k\,\delta[n-k] + \cdots = \begin{cases} (\tfrac{1}{2})^n & n \ge 0 \\ 0 & n < 0 \end{cases} = (\tfrac{1}{2})^n U[n]$$

Example 1-6 We wish to find the delta response $h[n]$ of a causal system such that

$$g[n] - \tfrac{1}{2}g[n-1] = f[n] \tag{1-8}$$

From Eqs. (1-5) and (1-6) it follows that

$$h[n] - \tfrac{1}{2}h[n-1] = \delta[n] \qquad \text{for all } n$$

and

$$h[n] = 0 \qquad \text{for } n < 0$$

Setting $n = 0, 1, \ldots$, and noting that $h[-1] = 0$, we obtain

$n = 0$: $\quad h[0] = 1$

$n = 1$: $\quad h[1] - \tfrac{1}{2}h[0] = 0 \qquad h[1] = \tfrac{1}{2}$

$n = 2$: $\quad h[2] - \tfrac{1}{2}h[1] = 0 \qquad h[2] = (\tfrac{1}{2})^2$

In general, $h[n] = \tfrac{1}{2}h[n-1]$ for $n > 0$. By an easy induction, we find

$$h[n] = (\tfrac{1}{2})^n U[n] \tag{1-9}$$

In Chap. 2 we develop simpler methods for determining $h[n]$. In Fig. 1-7 we show a block-diagram realization of the above system.

We note that the two systems in (1-7) and (1-8) have the same delta response $h[n]$; hence, they are equivalent. That is, they yield the same response to the same

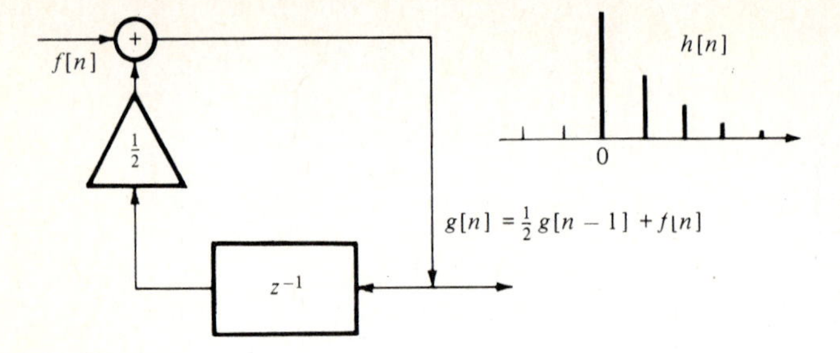

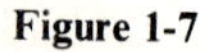

Figure 1-7

input [see (1-11)]. The system in (1-7) is nonrecursive, but an infinite number of delay elements is needed for its realization. The system in Eq. (1-8) (recursive) is realized with one delay element only.

Discrete convolution We shall express the response $g[n]$ of a linear system to an arbitrary input $f[n]$ in terms of $h[n]$ and $f[n]$.

We note that the response to $\delta[n-k]$ equals $h[n-k]$ for any k (time-invariance):

$$L\{\delta[n-k]\} = h[n-k] \tag{1-10}$$

Hence, the response to $f[k]\,\delta[n-k]$ equals $f[k]h[n-k]$ (linearity). From the above and (1-2) it follows that

$$g[n] = L\{f[n]\} = \sum_{k=-\infty}^{\infty} f[k]L\{\delta[n-k]\} = \sum_{k=-\infty}^{\infty} f[k]h[n-k]$$

The last sum is the *discrete convolution* of $f[n]$ with $h[n]$. This operation will be denoted by $f[n] * g[n]$. As it is easy to see, it is commutative. We have, thus, reached the important conclusion that

$$g[n] = f[n] * h[n] = \sum_{k=-\infty}^{\infty} f[k]h[n-k] = \sum_{k=-\infty}^{\infty} f[n-k]h[k] \tag{1-11}$$

Example 1-7

$$h[n] = (\tfrac{1}{2})^n U[n] \qquad f[n] = U[n] - U[n-4] = \begin{cases} 1 & 0 \le n < 4 \\ 0 & \text{otherwise} \end{cases}$$

Find $$g[n] = f[n] * h[n] \text{ for } n = 2 \text{ and } n = 5.$$

To find $g[2]$, we multiply $f[k]$ by $h[2-k]$ and sum for all k. As we see from Fig. 1-8a,

$$g[2] = f[2]h[0] + f[1]h[1] + f[0]h[2] = 1 + \tfrac{1}{2} + \tfrac{1}{4}$$

Similarly,

$$g[5] = f[3]h[2] + f[2]h[3] + f[1]h[4] + f[0]h[5] = (\tfrac{1}{2})^2 + (\tfrac{1}{2})^3 + (\tfrac{1}{2})^4 + (\tfrac{1}{2})^5$$

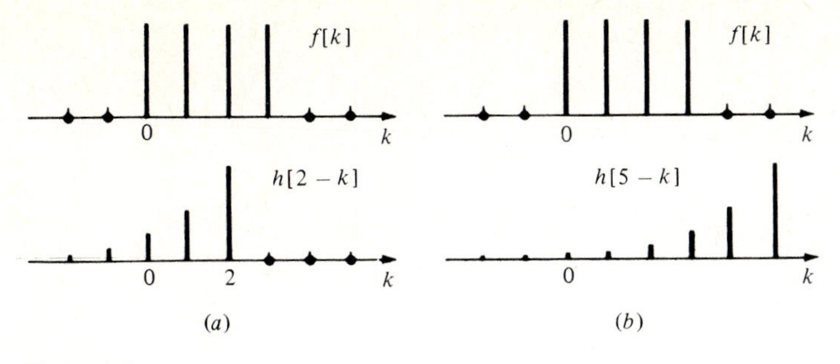

Figure 1-8

We note that, if $h[n] = 0$ for $n < 0$, then

$$g[n] = \sum_{k=-\infty}^{n} f[k]h[n-k] = \sum_{k=0}^{\infty} f[n-k]h[k] \tag{1-12}$$

If, also, $f[n] = 0$ for $n < 0$, then $g[n] = 0$ for $n < 0$; for $n \geq 0$ it is given by

$$g[n] = \sum_{k=0}^{n} f[k]h[n-k] = \sum_{k=0}^{n} f[n-k]h[k] \tag{1-13}$$

In particular,

$$g[0] = f[0]h[0] \qquad g[1] = f[0]h[1] + f[1]h[0]$$

$$g[2] = f[0]h[2] + f[1]h[1] + f[2]h[0]$$

Example 1-8

$$f[n] = U[n] - U[n-3] = h[n]$$

Find $g[n]$. It is easy to see that $g[n] = 0$ for $n < 0$ and $n > 4$.

For $0 \leq n \leq 2$: $\quad g[n] = \sum_{k=0}^{n} 1 \cdot 1 = n + 1$

For $2 < n \leq 4$: $\quad g[n] = \sum_{k=n-2}^{2} 1 \cdot 1 = 5 - n$

as in Fig. 1-9.

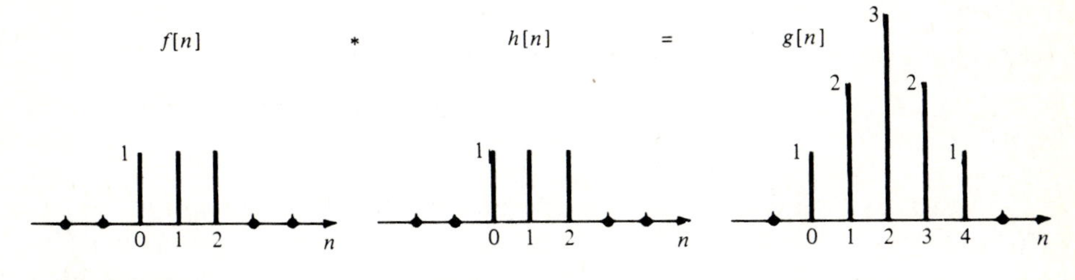

Figure 1-9

Example 1-9

$$f[n] = 2^n \qquad h[n] = (\tfrac{1}{3})^n U[n]$$

From (1-12) it follows that

$$g[n] = \sum_{k=0}^{\infty} 2^{n-k}(\tfrac{1}{3})^k = \tfrac{6}{5} \times 2^n$$

The System Function

Suppose that the input to a linear system is a geometric progression

$$f[n] = r^n$$

As we see from (1-11), the resulting response

$$g[n] = \sum_{k=-\infty}^{\infty} r^{n-k}h[k] = r^n \sum_{k=-\infty}^{\infty} h[k]r^{-k} \tag{1-14}$$

is also a geometric progression multiplied by the value $H(r)$ of the z *transform*

$$H(z) = \sum_{n=-\infty}^{\infty} h[n]z^{-n} \tag{1-15}$$

of the sequence $h[n]$.

Reasoning as in (1-14), we conclude that

$$L\{z^n\} = H(z)z^n \tag{1-16}$$

for any z, real or complex, for which the series of Eq. (1-15) converges.

We shall call the factor $H(z)$ the *system function.* This function can be determined either from (1-15) or from (1-16). To determine $H(z)$ from (1-16), we apply the input $f[n] = z^n$. The coefficient of z^n of the resulting response $H(z)z^n$ equals $H(z)$.

For the delay element, $H(z) = z^{-1}$. Indeed, if $f[n] = z^n$, then $g[n] = f[n-1] = z^{n-1} = z^{-1}z^n$.

For the multiplier, $H(z) = a$. Indeed, if $f[n] = z^n$, then $g[n] = af[n] = az^n$.

Example 1-10 The recursion equation

$$6g[n] + 5g[n-1] + g[n-2] = f[n]$$

defines a system with input $f[n]$ and output $g[n]$. If $f[n] = z^n$, then $g[n] = H(z)z^n$. Substituting, we obtain

$$6H(z)z^n + 5H(z)z^{n-1} + H(z)z^{n-2} = z^n$$

Hence,

$$H(z) = \frac{1}{6 + 5z^{-1} + z^{-2}}$$

In Fig. 1-10, we have four systems specified by their block diagrams. In part (a) (nonrecursive), $H(z)$ is found directly by tracing z^n. In part (b) (recursive), $H(z)$ is found by solving an equation. The third example is a combination of the first two (cascade). For part (d) it is necessary to introduce an auxiliary output with

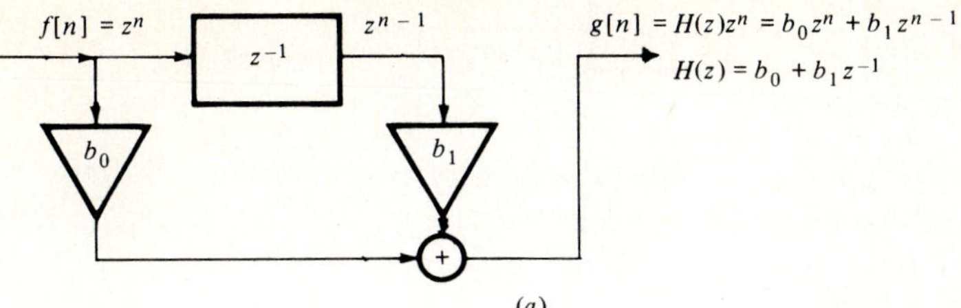

(*a*)

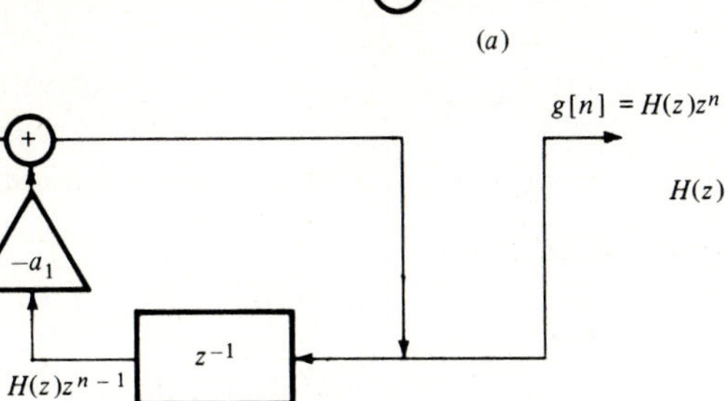

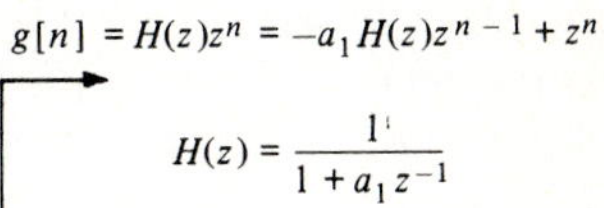

(*b*)

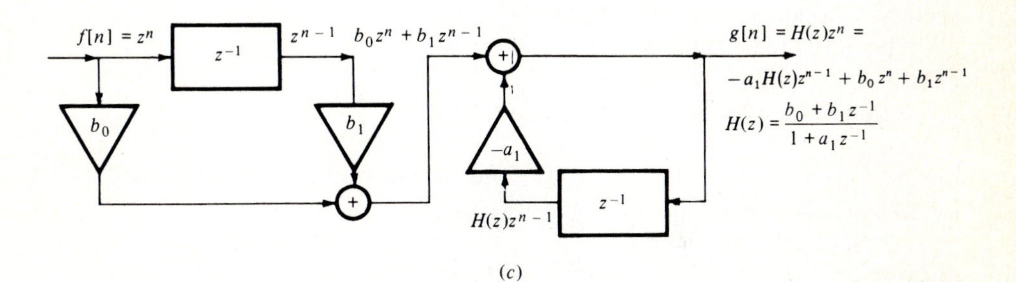

(*c*)

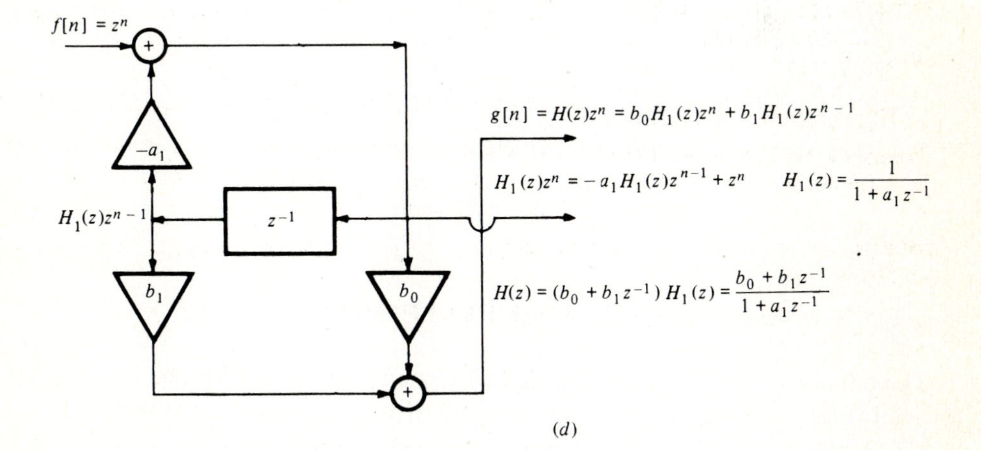

(*d*)

Figure 1-10

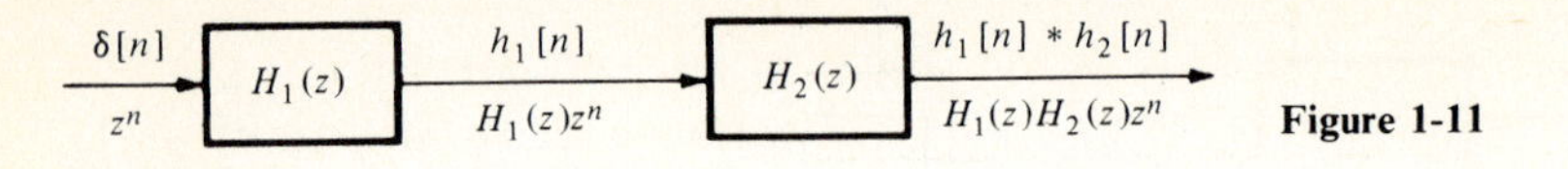

Figure 1-11

system function $H_1(z)$ and to solve a system of two equations. Systems (c) and (d) have the same system function; hence they are terminally equivalent. However, whereas (c) contains two delay elements, (d) contains only one.

Systems in cascade Two systems are connected *in cascade* if the output of the first is the input to the second. Denoting by $h[n]$ and $H(z)$ the delta response and the system function, respectively, of the system so formed, we conclude, with the explanation in Fig. 1-11, that (elaborate)

$$h[n] = h_1[n] * h_2[n] \qquad H(z) = H_1(z)H_2(z) \tag{1-17}$$

Convolution theorem By definition, $H(z)$, $H_1(z)$, and $H_2(z)$ are the z transforms of $h[n]$, $h_1[n]$, and $h_2[n]$, respectively. And since the sequences $h_1[n]$ and $h_2[n]$ are arbitrary, it follows from (1-17) that the z transform of the convolution of two sequences equals the product of their z transforms.

This leads to the conclusion that if

$$F(z) = \sum_{n=-\infty}^{\infty} f[n]z^{-n} \qquad G(z) = \sum_{n=-\infty}^{\infty} g[n]z^{-n}$$

are the z transforms of the input $f[n]$ and output $g[n]$ of the system $H(z)$, then

$$G(z) = F(z)H(z) \tag{1-18}$$

because

$$g[n] = f[n] * h[n]$$

We have, thus, the following three equivalent definitions of the system function $H(z)$.

1. $H(z)$ is the z transform of $h[n]$.
2. If $f[n] = z^n$, then $H(z)$ is the coefficient of the resulting response $g[n] = H(z)z^n$.
3. $H(z)$ equals the ratio $G(z)/F(z)$.

1-2 ANALOG SIGNALS AND SYSTEMS

The term *continuous*, or *analog signal*, will mean a function $f(t)$, real or complex, defined for every real t. The following signals will be used often (Fig. 1-12):

Step function $\quad U(t) = \begin{cases} 1 & t > 0 \\ 0 & t < 0 \end{cases}$

Sign function $\quad \operatorname{sgn} t = \begin{cases} 1 & t > 0 \\ -1 & t < 0 \end{cases}$

Rectangular pulse $\quad p_a(t) = \begin{cases} 1 & |t| < a \\ 0 & |t| > a \end{cases}$

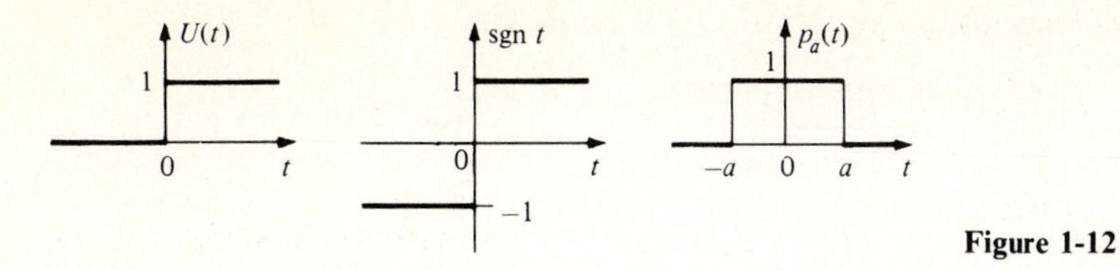

Figure 1-12

The delta function $\delta(t)$ This important concept will be discussed in Appendix 3-C. We note here only that $\delta(t)$ can be viewed as the limit of a family of functions $f_c(t)$ such that

$$\int_{-\infty}^{\infty} f_c(t)\,dt = 1 \qquad \int_{-\infty}^{\infty} f_c(t)\varphi(t)\,dt \underset{c\to 0}{\to} \varphi(0) \tag{1-19}$$

for any function $\varphi(t)$ continuous at the origin. This interpretation of $\delta(t)$ leads to the identity

$$\int_{-\infty}^{\infty} \delta(t)\,\varphi(t)\,dt = \varphi(0) \tag{1-20}$$

from which all formal properties of $\delta(t)$ can be deduced.

Analog Systems

An analog system is a rule for assigning to a function $f(t)$ another function $g(t)$. A system is thus a transformation mapping the *input* $f(t)$ to the *output* or *response:*

$$g(t) = L[f(t)]$$

Linearity A system L is linear if

$$L[a_1 f_1(t) + a_2 f_2(t)] = a_1 L[f_1(t)] + a_2 L[f_2(t)] \tag{1-21}$$

for any a_1, a_2, $f_1(t)$, and $f_2(t)$.

Time-invariance A system L is *time-invariant* if

$$L[f(t - t_0)] = g(t - t_0) \tag{1-22}$$

for any real t_0.

Example 1-11
(*a*) $g(t) = |f(t)|$ (rectifier): nonlinear, time-invariant
(*b*) $g(t) = t^2 f(t)$: linear, time-varying
(*c*) $g(t) = f(t - a)$ (delay line): linear, time-invariant

Note: If a system is linear and $f(t) \equiv 0$, then $g(t) \equiv 0$, because then $f(t) = 2f(t)$; hence, $g(t) = 2g(t)$. If, however, $f(t) = 0$ for $t \le t_0$ only, it does not follow that $g(t_0) = 0$ because $g(t_0)$ depends on all values of $f(t)$, past and future.

Causality We shall say that a function $f(t)$ is *causal* if

$$f(t) = 0 \qquad \text{for} \qquad t < 0$$

We shall say that a *system* is causal if *a causal input yields a causal output*. Thus, a causal system has the following property [see Eq. (1-22)]:

$$\text{If} \qquad f(t) = 0 \qquad \text{for} \qquad t \leq t_0 \qquad \text{then} \qquad g(t) = 0 \text{ for } t \leq t_0 \tag{1-23}$$

Corollary If a system is causal and $f_1(t) = f_2(t)$ for $t \leq t_0$, then $g_1(t) = g_2(t)$ for $t \leq t_0$ because the response to $f(t) = f_1(t) - f_2(t)$ equals $g_1(t) - g_2(t)$, and $f(t) = 0$ for $t \leq t_0$.

A physical system is always causal if t is real time; however, not all physical systems are causal. If, for example, L is an optical system and its input $f(x)$ is an object such that $f(x) = 0$ on the left half-plane $x \leq 0$, the resulting output (image) $g(x)$ might not be zero for $x \leq 0$.

Reality We shall say that a system is *real* if the response to a *real* input $f(t)$ is a real function $g(t)$. From the definition and the linearity of the system it follows that if $f_1(t)$ and $f_2(t)$ are two real functions and

$$L\{f_1(t) + jf_2(t)\} = g_1(t) + jg_2(t)$$

then

$$g_1(t) = L\{f_1(t)\} \qquad g_2(t) = L\{f_2(t)\} \tag{1-24}$$

Differential equations An important special case of a linear system is a differential equation with constant coefficients and *zero initial conditions*. Suppose, for example, that

$$g'(t) + \alpha g(t) = f(t) \tag{1-25}$$

If this equation holds *for all t* and is so interpreted that $g(t) = 0$ for $t \leq t_0$ when $f(t) = 0$ for $t \leq t_0$, then it defines a linear, causal system with input $f(t)$ and output $g(t)$. If Eq. (1-25) holds for $t \geq 0$ only and $g(0) = 0$ (zero initial conditions), then, setting $f(t) = 0$ for $t < 0$, we can assume that it holds for all t. The system so defined is real if α is a real number.

In Fig. 1-13 we show the solution of (1-25) for two special forms of $f(t)$ of unit area:

$$f_1(t) = \frac{1}{c}[U(t) - U(t-c)] \qquad g_1(t) = \begin{cases} \dfrac{1}{\alpha c}(1 - e^{-\alpha t}) & 0 \leq t < c \\ \dfrac{1}{\alpha c}(e^{\alpha c} - 1)e^{-\alpha t} & t > c \end{cases}$$

$$f_2(t) = \frac{1}{c} e^{-t/c} U(t) \qquad g_2(t) = \frac{1}{1 - \alpha c}(e^{-\alpha t} - e^{-t/c}) \qquad t > 0$$

It is of interest to observe that, if c is small, then

$$g_1(t) \simeq g_2(t) \simeq e^{-\alpha t} U(t) \qquad \text{for} \qquad t \gg c$$

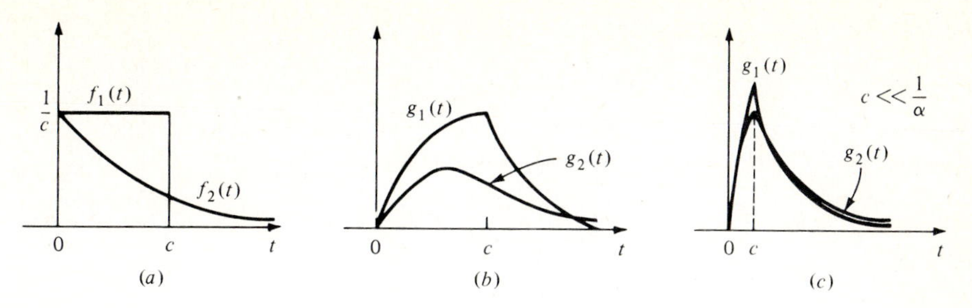

Figure 1-13

Thus, as $c \to 0$, the response reaches a limiting value $e^{-\alpha t}U(t)$ that is the same for both inputs. This important observation leads to the concept of the impulse response.

The impulse response To an arbitrary linear system L we apply a sequence of inputs

$$f_c(t) = \frac{1}{c} f_0\left(\frac{t}{c}\right) \tag{1-26}$$

of unit area as in Eq. (1-19). The resulting response $g_c(t)$ is a signal that depends on the system L, the form $f_0(t)$ of the input, and the scaling factor c. It can be shown [see (4-9)] that, as $c \to 0$, $g_c(t)$ tends to a limit:

$$L[f_c(t)] = g_c(t) \to h(t) \qquad c \to 0$$

The limit $h(t)$ depends on the system L but *it is independent of the form of the input*, as long as its area equals 1. Since $f_c(t)$ tends to $\delta(t)$ [see Eq. (1-19)] we shall say that $h(t)$ is the response of L to the delta function $\delta(t)$:

$$h(t) = L[\delta(t)] \tag{1-27}$$

The function $h(t)$ we shall call the *impulse response* of the system.

From the above it follows that if the input to a system is a signal $f(t)$ of arbitrary form but of sufficiently short duration[1] relative to $h(t)$, then the resulting response is approximately equal to $Ah(t)$, where A is the area of $f(t)$.

If $f(t)$ takes significant values not near the origin but near the point t_0 as in Fig. 1-14, then (time-invariance) the approximate response will equal $Ah(t - t_0)$.

Example 1-12 The equation $g'(t) + \alpha g(t) = f(t)$ defines a linear, causal system with impulse response $e^{-\alpha t}U(t)$. If $f(t)$ is a triangle of area $A = B\varepsilon$ (as in Fig. 1-14) and $\varepsilon \ll 1/\alpha$, then

$$g(t) \simeq Ah(t - t_0) = B\varepsilon e^{-\alpha(t-t_0)} \qquad t > t_0 + \varepsilon$$

[1] The expression "$f(t)$ is of short duration relative to $h(t)$" will mean loosely that $f(t)$ is negligible for $|t| > \varepsilon$ and $h(t)$ is approximately constant in any interval of length 2ε.

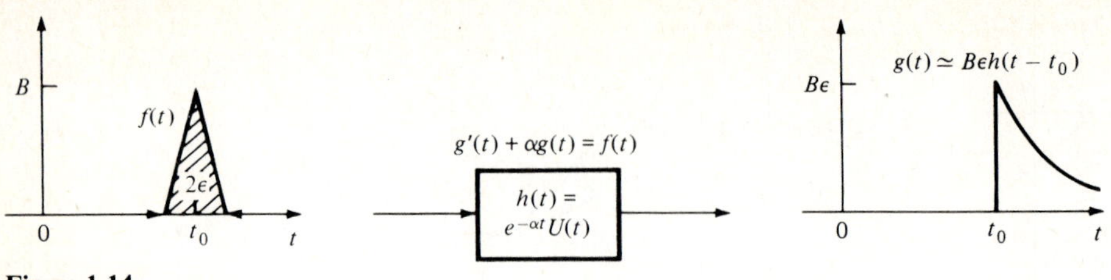

Figure 1-14

Convolution We shall show that the response $g(t)$ of a linear system L to an arbitrary input $f(t)$ is given by

$$g(t) = \int_{-\infty}^{\infty} f(\tau)h(t-\tau)\,d\tau \tag{1-28}$$

Proof The function $f(t)$ can be written as a sum of elementary functions $f_i(t)$ as in Fig. 1-15. Denoting by $g_i(t)$ the response of L to $f_i(t)$, we have (linearity)

$$f(t) = \sum_i f_i(t) \qquad g(t) = \sum_i g_i(t) \tag{1-29}$$

If $\Delta\tau$ is sufficiently small, then the area of $f_i(t)$ equals $f(\tau_i)\,\Delta\tau$. Hence, the resulting response is approximately $f(\tau_i)\,\Delta\tau\; h(t-\tau_i)$ because $f_i(t)$ is concentrated near the point τ_i. With $\Delta\tau \to 0$, we thus conclude that

$$\sum_i g_i(t) \simeq \sum_i f(\tau_i)h(t-\tau_i)\,\Delta\tau \to \int_{-\infty}^{\infty} f(\tau)h(t-\tau)\,d\tau$$

and (1-28) follows.

The integral in Eq. (1-28) is the *convolution* of the input $f(t)$ with the impulse response $h(t)$:

$$g(t) = f(t) * h(t) = \int_{-\infty}^{\infty} f(\tau)h(t-\tau)\,d\tau = \int_{-\infty}^{\infty} f(t-\tau)h(\tau)\,d\tau \tag{1-30}$$

If the system L is causal, then

$$h(t) = 0 \qquad \text{for} \qquad t < 0 \tag{1-31}$$

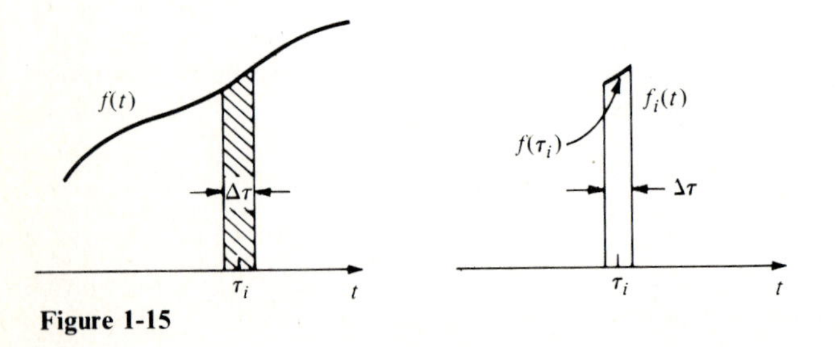

Figure 1-15

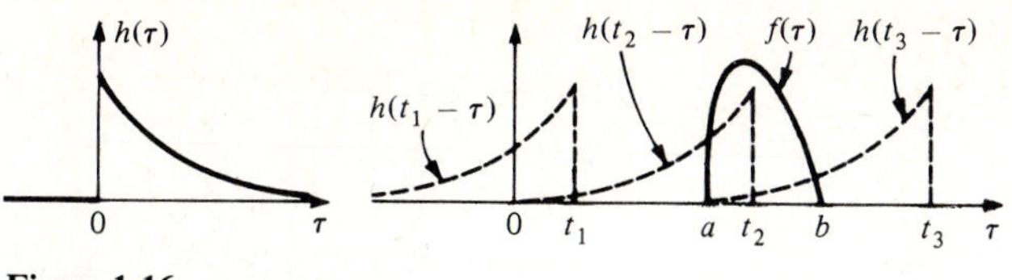

Figure 1-16

because $\delta(t) = 0$ for $t < 0$. In this case, (1-30) yields

$$g(t) = \int_{-\infty}^{t} f(\tau)h(t-\tau)\,d\tau = \int_{0}^{\infty} f(t-\tau)h(\tau)\,d\tau \tag{1-32}$$

If also $f(t) = 0$ for $t < 0$, then $g(t) = 0$ for $t < 0$; for $t > 0$ it is given by

$$g(t) = \int_{0}^{t} f(\tau)h(t-\tau)\,d\tau = \int_{0}^{t} f(t-\tau)h(\tau)\,d\tau \tag{1-33}$$

The evaluation of the convolution integral is often facilitated by a semi-graphical method: To find $g(t_0)$ for some $t = t_0$, we form the function $h(-\tau)$ and its displacement $h(t_0 - \tau)$. The area of the product $f(\tau)h(t_0 - \tau)$ yields $g(t_0)$.

Example 1-13 We wish to find the convolution $g(t)$ of the two functions $f(t)$ and $h(t)$ of Fig. 1-16. As we see, $h(t)$ is causal and $f(t)$ is zero outside the interval (a, b). From the figure it follows that $g(t) = 0$ for $t = t_1 < a$, and

$$g(t_2) = \int_{a}^{t_2} f(\tau)h(t_2-\tau)\,d\tau \qquad a < t_2 < b$$

$$g(t_3) = \int_{a}^{b} f(\tau)h(t_3-\tau)\,d\tau \qquad t_3 > b$$

We note that, if $f(t) = 0$ outside the interval (a_1, b_1) and $h(t) = 0$ outside the interval (a_2, b_2), then $g(t) = 0$ outside the interval $(a_1 + a_2, b_1 + b_2)$.

Example 1-14 If $f(t) = U(t-1) - U(t-3)$ is a pulse as in Fig. 1-17, and $h(t) = f(t)$, then $g(t)$ is a triangle.

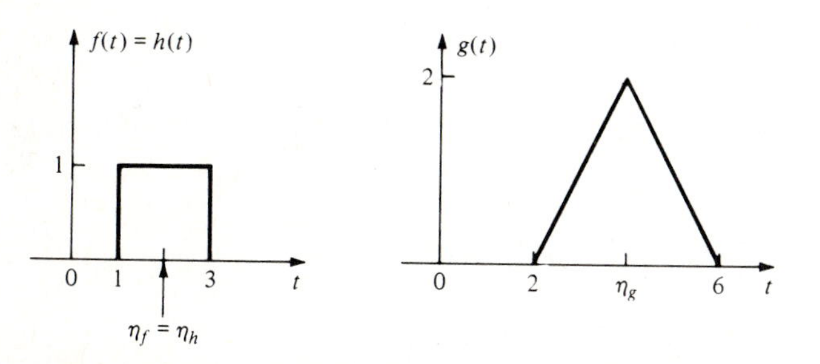

Figure 1-17

Output center of gravity We denote by

$$A_f = \int_{-\infty}^{\infty} f(t)\,dt \qquad m_f = \int_{-\infty}^{\infty} tf(t)\,dt \qquad \eta_f = \frac{m_f}{A_f} \tag{1-34}$$

the area, first moment, and center of gravity, respectively, of $f(t)$. We define the corresponding quantities for $h(t)$ and $g(t)$ similarly.

Theorem If $g(t) = f(t) * h(t)$, then

$$A_g = A_f A_h \qquad \eta_g = \eta_f + \eta_h \tag{1-35}$$

Proof Integrating both sides of Eq. (1-28) and changing the order of integrations on the right side, we have, with $t - \tau = x$,

$$\int_{-\infty}^{\infty} g(t)\,dt = \int_{-\infty}^{\infty} f(\tau) \int_{-\infty}^{\infty} h(t-\tau)\,dt\,d\tau = \int_{-\infty}^{\infty} f(\tau)\,d\tau \int_{-\infty}^{\infty} h(x)\,dx$$

Hence, $A_g = A_f A_h$. Similarly,

$$\int_{-\infty}^{\infty} tg(t)\,dt = \int_{-\infty}^{\infty} f(\tau) \int_{-\infty}^{\infty} th(t-\tau)\,dt\,d\tau = \int_{-\infty}^{\infty} f(\tau) \int_{-\infty}^{\infty} (x+\tau)h(x)\,dx\,d\tau$$

$$= \int_{-\infty}^{\infty} f(\tau)\,d\tau \int_{-\infty}^{\infty} xh(x)\,dx + \int_{-\infty}^{\infty} \tau f(\tau)\,d\tau \int_{-\infty}^{\infty} h(x)\,dx$$

Thus,
$$m_g = A_f m_h + A_h m_f$$

Dividing both sides by $A_g = A_f A_h$, we conclude that $\eta_g = \eta_f + \eta_h$.

In Example 1-14, $A_g = 4$, $A_f = 2$, and $A_h = 2$; also, $\eta_g = 4$, $\eta_f = 2$, and $\eta_h = 2$.

The System Function

Suppose that the input to a linear system is an exponential

$$f(t) = e^{j\omega_0 t}$$

As we see from (1-30), the resulting response is given by

$$g(t) = \int_{-\infty}^{\infty} e^{j\omega_0 (t-\tau)} h(\tau)\,d\tau = e^{j\omega_0 t} \int_{-\infty}^{\infty} e^{-j\omega_0 \tau} h(\tau)\,d\tau \tag{1-36}$$

Thus, $g(t)$ is also an exponential multiplied by the value $H(\omega_0)$ of the *Fourier transform*

$$H(\omega) = \int_{-\infty}^{\infty} h(t) e^{-j\omega t}\,dt \tag{1-37}$$

of $h(t)$. Since this is true for every ω_0, we conclude that

$$L\{e^{j\omega t}\} = H(\omega) e^{j\omega t} \tag{1-38}$$

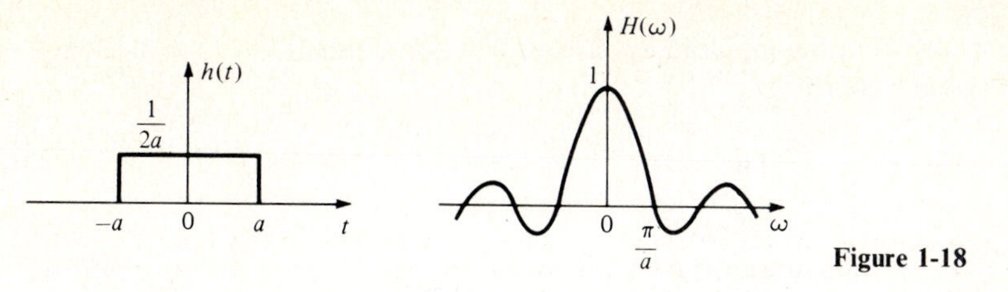

Figure 1-18

We shall call the factor $H(\omega)$ the *system function.* It can be determined either from (1-37) or from (1-38). To determine $H(\omega)$ from (1-38), we apply the input $f(t) = e^{j\omega t}$. The function $H(\omega)$ equals the coefficient of the resulting response $H(\omega)e^{j\omega t}$.

In general, $H(\omega)$ is complex with amplitude $A(\omega)$ and phase $\varphi(\omega)$:

$$H(\omega) = A(\omega)e^{j\varphi(\omega)} \tag{1-39}$$

Example 1-15 (Smoothing) Consider a system whose impulse response $h(t)$ is a pulse as in Fig. 1-18. From Eq. (1-37) it follows that the corresponding system function is given by

$$H(\omega) = \frac{1}{2a}\int_{-a}^{a} e^{-j\omega t}\,dt = \frac{\sin a\omega}{a\omega}$$

It is easy to see that

$$g(t) = f(t) * \frac{1}{2a}p_a(t) = \frac{1}{2a}\int_{t-a}^{t+a} f(\tau)\,d\tau = \frac{1}{2a}\int_{-a}^{a} f(t-\tau)\,d\tau \tag{1-40}$$

Example 1-16 From (1-37) it follows that if $h(t) = e^{-\alpha t}U(t)$, for $\alpha > 0$, then

$$H(\omega) = \int_{0}^{\infty} e^{-\alpha t}e^{-j\omega t}\,dt = \frac{1}{\alpha + j\omega} \tag{1-41}$$

Example 1-17 (Differentiator) Consider a system whose output is the derivative $f'(t)$ of the input. If $f(t) = e^{j\omega t}$, then

$$g(t) = f'(t) = j\omega e^{j\omega t}$$

Hence, $H(\omega) = j\omega$.

Example 1-18 The differential equation

$$g'(t) + \alpha g(t) = f(t) \qquad \text{all } t \tag{1-42}$$

specifies a causal system with input $f(t)$ and output $g(t)$. As we see from (1-38), if $f(t) = e^{j\omega t}$, then

$$g(t) = H(\omega)e^{j\omega t}$$

Inserting into (1-42), we obtain

$$j\omega H(\omega)e^{j\omega t} + \alpha H(\omega)e^{j\omega t} = e^{j\omega t}$$

Hence,

$$H(\omega) = \frac{1}{\alpha + j\omega}$$

From (1-41) it follows that the inpulse response of the system equals $e^{-\alpha t}U(t)$. Therefore, the solution of (1-42) is given by

$$g(t) = e^{-\alpha t} \int_{-\infty}^{t} f(\tau)e^{\alpha\tau}\, d\tau$$

Real systems The system function of a linear system is, in general, an arbitrary function $H(\omega)$ of the real variable (frequency) ω. However, if $h(t)$ is real, then

$$H^*(\omega) = \int_{-\infty}^{\infty} h(t)e^{j\omega t}\, dt = A(\omega)e^{-j\varphi(\omega)}$$

[see Eq. (1-37)]. But the above integral equals $H(-\omega)$; hence,

$$H^*(\omega) = H(-\omega) \qquad A(-\omega) = A(\omega) \qquad \varphi(-\omega) = -\varphi(\omega) \tag{1-43}$$

Thus, the system function of a real system has even amplitude and odd phase; therefore, if it is known for $\omega \geq 0$, then it is known also for $\omega < 0$.

The response of an arbitrary system, real or complex, to

$$f(t) = \cos \omega t = \frac{e^{j\omega t} + e^{-j\omega t}}{2}$$

equals

$$g(t) = \frac{H(\omega)}{2} e^{j\omega t} + \frac{H(-\omega)}{2} e^{-j\omega t}$$

From (1-43) it follows that, if $h(t)$ is real, then

$$g(t) = \text{Re}\,[H(\omega)e^{j\omega t}]$$

This follows also from Eq. (1-24):

$$\cos \omega t = \text{Re}\, e^{j\omega t}$$

Hence,

$$g(t) = \text{Re}\,[H(\omega)e^{j\omega t}] = \text{Re}\,[A(\omega)e^{j\varphi(\omega)}e^{j\omega t}]$$

We thus conclude that

$$L\{\cos \omega t\} = A(\omega) \cos\,[\omega t + \varphi(\omega)] \tag{1-44}$$

Example 1-19 We wish to solve the equation

$$g'(t) + 2g(t) = \cos 3t$$

As we have seen, $g(t)$ can be considered as the response of a system with input $\cos 3t$ and system function $1/(2 + j\omega)$. Since

$$H(3) = \frac{1}{2 + j3} = \frac{1}{\sqrt{13}} e^{-j\tan^{-1} 3/2}$$

it follows from (1-44) that

$$g(t) = \frac{1}{\sqrt{13}} \cos\left(3t - \tan^{-1}\frac{3}{2}\right)$$

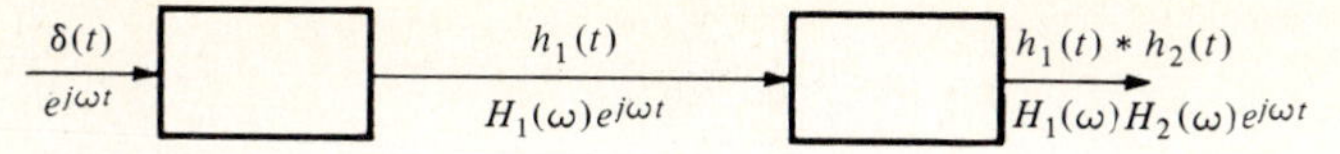

Figure 1-19

Delay time We shall show that the center of gravity η_h of $h(t)$ [see Eq. (1-34)] equals the slope at the origin of the phase lag, $-\varphi(\omega)$:

$$\eta_h = -\varphi'(0) \tag{1-45}$$

Proof From (1-43) we conclude that $\varphi(0) = 0$ and $A'(0) = 0$. Hence, $H(0) = A(0)$ and $H'(0) = j\varphi'(0)A(0)$. Furthermore [see Eq. (1-37)],

$$H(0) = \int_{-\infty}^{\infty} h(t)\,dt = A_h$$

and $$H'(\omega) = \int_{-\infty}^{\infty} (-jt)h(t)e^{-j\omega t}\,dt \qquad H'(0) = -j\int_{-\infty}^{\infty} th(t)\,dt = -jm_h$$

Thus, $j\varphi'(0)A_h = -jm_h$, and (1-45) results because $m_h = A_h\eta_h$.

From Eqs. (1-45) and (1-35) it follows that if a signal $f(t)$ passes through a real system, its center of gravity is delayed by an amount equal to $-\varphi'(0)$.

Systems in cascade Two systems are connected in cascade as in Fig. 1-19. Denoting by $h(t)$ and $H(\omega)$ the impulse response and the system function, respectively, of the combined system, we conclude from the figure that

$$h(t) = h_1(t) * h_2(t) \qquad H(\omega) = H_1(\omega)H_2(\omega) \tag{1-46}$$

Convolution theorem By definition, $H(\omega)$, $H_1(\omega)$, and $H_2(\omega)$ are the Fourier transforms of $h(t)$, $h_1(t)$, and $h_2(t)$, respectively. And, since the functions $h_1(t)$ and $h_2(t)$ are arbitrary, it follows from (1-46) that the Fourier transform of the convolution of two functions equals the product of their Fourier transforms.

This leads to the conclusion that if

$$F(\omega) = \int_{-\infty}^{\infty} f(t)e^{-j\omega t}\,dt \qquad \text{and} \qquad G(\omega) = \int_{-\infty}^{\infty} g(t)e^{-j\omega t}\,dt$$

are the Fourier transforms of $f(t)$ and $g(t)$, respectively, then

$$G(\omega) = F(\omega)H(\omega) \tag{1-47}$$

because $g(t) = f(t) * h(t)$.

We thus have the following three equivalent definitions of the system function $H(\omega)$:

1. $H(\omega)$ is the Fourier transform of $h(t)$.
2. If $f(t) = e^{j\omega t}$, then $H(\omega)$ is the coefficient of the resulting response $g(t) = H(\omega)e^{j\omega t}$.
3. $H(\omega)$ equals the ratio $G(\omega)/F(\omega)$.

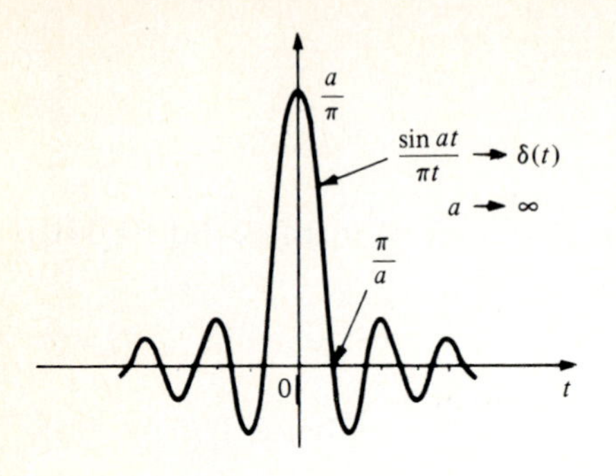

Figure 1-20

The Inversion Formula

We shall express the impulse response $h(t)$ of a system in terms of its system function $H(\omega)$. For this purpose, we shall use the identity

$$\delta(t) = \frac{1}{2\pi} \int_{-\infty}^{\infty} e^{j\omega t}\, d\omega \tag{1-48}$$

This identity can be justified as follows: It is well known that, if $a > 0$, then

$$\int_{-\infty}^{\infty} \frac{\sin at}{\pi t}\, dt = \int_{-\infty}^{\infty} \frac{\sin t}{\pi t}\, dt = 1 \tag{1-49}$$

This leads to the conclusion that [see Fig. 1-20 and (3-C-21)]

$$\lim_{a\to\infty} \int_{-\infty}^{\infty} \frac{\sin at}{\pi t} \varphi(t)\, dt = \varphi(0) \tag{1-50}$$

from which it follows as in Eq. (1-19) that

$$\delta(t) = \lim_{a\to\infty} \frac{\sin at}{\pi t} \tag{1-51}$$

and (1-48) results because

$$\frac{\sin at}{\pi t} = \frac{1}{2\pi} \int_{-a}^{a} e^{j\omega t}\, d\omega \tag{1-52}$$

Theorem

$$h(t) = \frac{1}{2\pi} \int_{-\infty}^{\infty} H(\omega) e^{j\omega t}\, d\omega \tag{1-53}$$

Proof The response of our system to $\delta(t)$ equals $h(t)$; the response to the right side of (1-48) equals the right side of (1-53). Therefore, (1-53) is a consequence of (1-48) (Fig. 1-21).

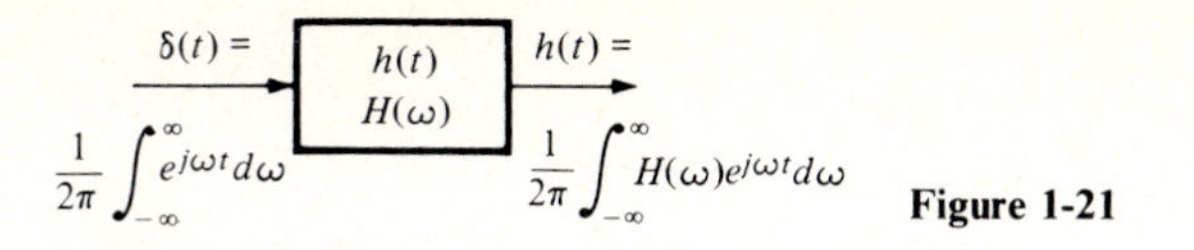

Figure 1-21

Example 1-20 An *ideal low-pass filter* is a system with constant amplitude $A(\omega)$ and linear phase $\varphi(\omega)$ in the interval $(-\omega_c, \omega_c)$:

$$H(\omega) = p_{\omega_c}(\omega)e^{-jt_0\omega}$$

From (1-53) it follows that its impulse response is given by

$$h(t) = \frac{1}{2\pi}\int_{-\omega_c}^{\omega_c} e^{-jt_0\omega}e^{j\omega t}\,d\omega = \frac{\sin \omega_c(t-t_0)}{\pi(t-t_0)}$$

as in Fig. 1-22.

If the input is a step function $f(t) = U(t)$, then the resulting output, denoted by $a(t)$, is the *step response* of the system. As we see from (1-30),

$$a(t) = \int_{-\infty}^{t} h(\tau)\,d\tau = \int_{-\infty}^{t} \frac{\sin \omega_c(\tau - t_0)}{\pi(\tau - t_0)}\,d\tau = \frac{1}{2} + \frac{1}{\pi}\int_{0}^{\omega_c(t-t_0)} \frac{\sin x}{x}\,dx$$

We note that, if $f(t) = \cos \omega_0 t$, then $g(t) = A(\omega_0)\cos \omega_0(t - t_0)$ [see Eq. (1-44)]. Therefore, if $\omega_0 > \omega_c$, then $g(t) = 0$; if $\omega_0 < \omega_c$, then $g(t) = \cos \omega_0(t - t_0)$.

The preceding theorem expresses an arbitrary function $h(t)$ in terms of its Fourier transform $H(\omega)$ defined by (1-37). The assumption that $h(t)$ is the impulse response of a system was introduced merely to derive (1-53) from (1-48) in a simple way. Applying the theorem to the output $g(t)$ of a system and its transform $G(\omega) = F(\omega)H(\omega)$, we obtain

$$g(t) = \frac{1}{2\pi}\int_{-\infty}^{\infty} G(\omega)e^{j\omega t}\,d\omega = \frac{1}{2\pi}\int_{-\infty}^{\infty} F(\omega)H(\omega)e^{j\omega t}\,d\omega \tag{1-54}$$

Periodic inputs It is well known that a periodic function with period T can be written as a sum of exponentials (see Fourier series, Sec. 3-2):

$$f(t) = \sum_{n=-\infty}^{\infty} a_n e^{jn\omega_0 t} \qquad \omega_0 = \frac{2\pi}{T} \tag{1-55}$$

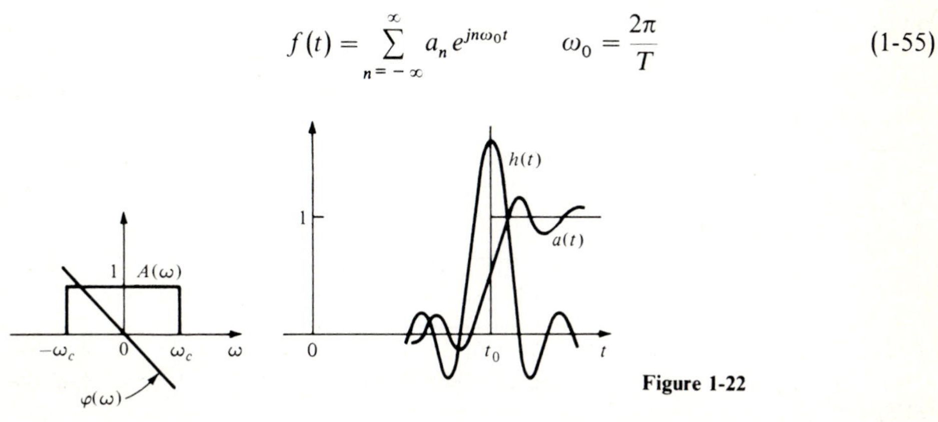

Figure 1-22

where
$$a_n = \frac{1}{T}\int_{-T/2}^{T/2} f(t)e^{-jn\omega_0 t}\,dt \tag{1-56}$$

From (1-38) it follows that the response of a linear system to a periodic input $f(t)$ is given by

$$g(t) = \sum_{n=-\infty}^{\infty} a_n H(n\omega_0)e^{jn\omega_0 t} \tag{1-57}$$

This formula is useful only if a small number of terms in the above sum are significant. Suppose, for example, that the system is a low-pass filter, as in Example 1-20. If $\omega_0 < \omega_c < 2\omega_0$, then (1-57) yields

$$g(t) = a_0 + a_1 e^{j\omega_0(t-t_0)} + a_{-1}e^{-j\omega_0(t-t_0)}$$

If $H(\omega)$ is not narrow band, or, equivalently, if the duration of $h(t)$ is of the order of T (see Sec. 8-2), then $g(t)$ is best determined from the convolution integral, Eq. (1-28).

The Poisson sum formula We have used the notion of a system to prove the convolution theorem (1-47) and the inversion formula (1-53). Proceeding similarly, we shall prove the Poisson sum formula (1-59). This important identity will be used repeatedly.

The impulse train

$$\sum_{n=-\infty}^{\infty} \delta(t + nT)$$

(Fig. 1-23) is a periodic function with Fourier-series coefficients

$$a_n = \frac{1}{T}\int_{-T/2}^{T/2} \delta(t)\, e^{-jn\omega_0 t}\,dt = \frac{1}{T}$$

This follows from Eqs. (1-20) and (1-56) and from the fact that, in the interval $(-T/2, T/2)$, the above sum equals $\delta(t)$. Inserting into (1-55), we obtain

$$\sum_{n=-\infty}^{\infty} \delta(t + nT) = \frac{1}{T}\sum_{n=-\infty}^{\infty} e^{jn\omega_0 t} \tag{1-58}$$

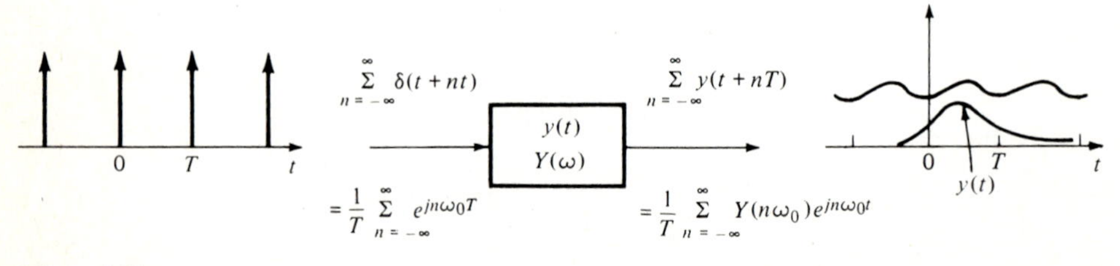

Figure 1-23

Consider an arbitrary function $y(t)$ with Fourier transform

$$Y(\omega) = \int_{-\infty}^{\infty} y(t)e^{-j\omega t}\,dt$$

Using (1-58), we shall show that

$$\sum_{n=-\infty}^{\infty} y(t + nT) = \frac{1}{T}\sum_{n=-\infty}^{\infty} Y(n\omega_0)e^{jn\omega_0 t} \tag{1-59}$$

where T is an arbitrary constant and $\omega_0 = 2\pi/T$.

Proof We form a system (Fig. 1-23) with impulse response $y(t)$ and system function $Y(\omega)$. The response of this system to the left side of (1-58) equals the left side of (1-59), and the response to the right side of (1-58) equals the right side of (1-59). Hence, (1-59) is a consequence of (1-58).

We note that the left side of (1-59) is a periodic function and its Fourier-series coefficients equal the sample values $Y(n\omega_0)/T$ of $Y(\omega)/T$.

1-3 DIGITAL SIMULATION OF ANALOG SYSTEMS

Consider an analog system $H_a(\omega)$ with input $f(t)$ and output $g(t)$. We wish to find a discrete system $H(z)$ such that, if its input $f[n]$ equals the sample values $f(nT)$ of $f(t)$, then the resulting output $g[n]$ equals the sample values $g(nT)$ of $g(t)$:

$$\text{If} \qquad f[n] = f(nT) \qquad \text{then} \qquad g[n] = g(nT) \tag{1-60}$$

If such a system exists, then we shall say that it is a *digital simulator* of $H_a(\omega)$. As we shall see, (1-60) cannot, in general, be true for any $f(t)$. The simulation is possible only if the class of inputs is restricted.

Indeed, suppose first that $f(t)$ is an exponential. As we know [see (1-36)], $g(t)$ is, then, also an exponential (Fig. 1-24):

$$f(t) = e^{j\omega_0 t} \qquad g(t) = H_a(\omega_0)e^{j\omega_0 t} \tag{1-61}$$

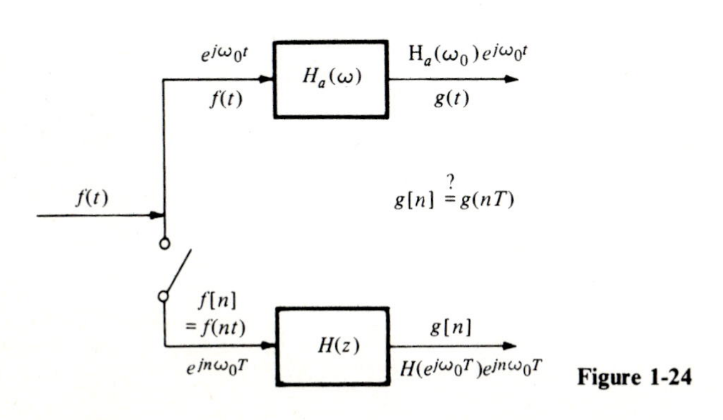

Figure 1-24

The sample values $f(nT)$ of $f(t)$ form a geometric progression with ratio $e^{j\omega_0 T}$

$$f[n] = f(nT) = e^{jn\omega_0 T}$$

Hence, the response $g[n]$ of the discrete system $H(z)$ is given by

$$g[n] = H(e^{j\omega_0 T})e^{jn\omega_0 T} \tag{1-62}$$

This follows from (1-14) with $r = e^{j\omega_0 T}$

Since $g(nT) = H_a(\omega_0)e^{jn\omega_0 T}$, to satisfy (1-60) we must choose $H(z)$ such that

$$H(e^{j\omega_0 T}) = H_a(\omega_0) \tag{1-63}$$

Consider next an arbitrary signal with Fourier transform $F(\omega)$. From the inversion formula it follows that

$$f(t) = \frac{1}{2\pi}\int_{-\infty}^{\infty} F(\omega)e^{j\omega t}\, d\omega \qquad g(t) = \frac{1}{2\pi}\int_{-\infty}^{\infty} F(\omega)H_a(\omega)e^{j\omega t}\, d\omega \tag{1-64}$$

Since
$$f[n] = f(nT) = \frac{1}{2\pi}\int_{-\infty}^{\infty} F(\omega)e^{jn\omega T}\, d\omega$$

we conclude as in (1-62) that the response $g[n]$ of $H(z)$ is given by

$$g[n] = \frac{1}{2\pi}\int_{-\infty}^{\infty} F(\omega)H(e^{j\omega T})e^{jn\omega T}\, d\omega \tag{1-65}$$

But [see (1-64)]

$$g(nT) = \frac{1}{2\pi}\int_{-\infty}^{\infty} F(\omega)H_a(\omega)e^{jn\omega T}\, d\omega \tag{1-66}$$

Hence, for (1-60) to be true, the integrals in (1-65) and (1-66) must be equal for every n. This is the case if

$$H(e^{j\omega T}) = H_a(\omega) \tag{1-67}$$

for every ω for which $F(\omega) \neq 0$. If, therefore, (1-60) holds for every $f(t)$, (1-67) must hold for every ω. However, this is not always possible because $H(e^{j\omega T})$ is a periodic function of ω whereas, in general, $H_a(\omega)$ is not. To satisfy the simulation condition (1-60), we must restrict the class of inputs.

Simulation theorem If $f(t)$ is *bandlimited*, i.e., if

$$F(\omega) = 0 \qquad \text{for} \qquad |\omega| > \sigma \qquad \sigma = \frac{\pi}{T} \tag{1-68}$$

and $H(z)$ is such that

$$H(e^{j\omega T}) = H_a(\omega) \qquad \text{for} \qquad |\omega| < \sigma \tag{1-69}$$

(Fig. 1-25), then the simulation condition (1-60) is satisfied.

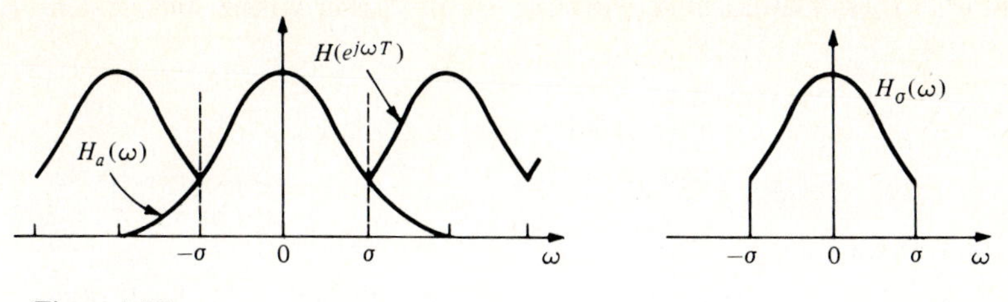

Figure 1-25

Proof From Eqs. (1-68) and (1-69) it follows that

$$F(\omega)H(e^{j\omega T}) = F(\omega)H_a(\omega) \qquad \text{for all } \omega$$

Hence, the two integrals in (1-65) and (1-66) are equal for every n.

Note: As ω increases from $-\sigma$ to σ, the number $z = e^{j\omega T}$ covers the unit circle of the z plane. Hence, (1-69) yields all values of $H(z)$ on this circle. We maintain that $H(z)$ is uniquely determined from these values.

Delta response of simulator Denoting by $h[n]$ the delta response of the discrete system, we conclude from (1-15) that

$$H(e^{j\omega T}) = \sum_{n=-\infty}^{\infty} h[n]e^{-jn\omega T} \tag{1-70}$$

Thus, the coefficients of the Fourier-series expansion of the periodic function $H(e^{j\omega T})$ equal $h[n]$. Hence,

$$h[n] = \frac{1}{2\sigma}\int_{-\sigma}^{\sigma} H(e^{j\omega T})e^{jn\omega T}\, d\omega = \frac{1}{2\sigma}\int_{-\sigma}^{\sigma} H_a(\omega)e^{jn\omega T}\, d\omega \tag{1-71}$$

Therefore, the discrete system is uniquely determined from (1-69).

We introduce the analog system function $H_\sigma(\omega) = H_a(\omega)p_\sigma(\omega)$ obtained by truncating $H_a(\omega)$ for $|\omega| > \sigma$ (Fig. 1-25). Denoting by $h_\sigma(t)$ its impulse response, we have

$$h_\sigma(t) = \frac{1}{2\pi}\int_{-\sigma}^{\sigma} H_a(\omega)e^{j\omega t}\, d\omega \tag{1-72}$$

and Eq. (1-71) yields

$$h[n] = Th_\sigma(nT) \tag{1-73}$$

From the preceding discussion it follows that to determine the discrete simulator $H(z)$ of the analog system $H_a(\omega)$, we proceed as follows:

1. We evaluate $h[n]$ by sampling the impulse response $h_\sigma(t)$ of the truncated system $H_\sigma(\omega)$.
2. Or, we specify $H(z)$ in terms of its values on the unit circle [see (1-69)].

Example 1-21 We wish to simulate a differentiator, i.e., an analog system with system function

$$H_a(\omega) = j\omega$$

Since

$$h_\sigma(t) = \frac{1}{2\pi}\int_{-\sigma}^{\sigma} j\omega e^{j\omega t}\, d\omega = \frac{\sigma t \cos \sigma t - \sin \sigma t}{\pi t^2} \tag{1-74}$$

(1-73) yields

$$h[n] = Th_\sigma(nT) = \frac{(-1)^n}{nT} \qquad \text{for } n \neq 0 \text{ and } h[0] = 0$$

Note: As we have shown, if $F(\omega) = 0$ for $|\omega| > \sigma$, then $g(nT) = f'(nT)$. This leads to the identity

$$f'(nT) = \sum_{\substack{k=-\infty \\ k \neq 0}}^{\infty} \frac{(-1)^k}{kT} f(nT - kT) \tag{1-75}$$

expressing the derivative of a bandlimited function in terms of its sample values.

Example 1-22 We wish to simulate the ideal low-pass filter $H_a(\omega) = p_{\omega_c}(\omega)$ with cut-off frequency $\omega_c < \sigma$ (Fig. 1-26). In this case, $H_\sigma(\omega) = H_a(\omega)$ because $H_a(\omega) = 0$ for $|\omega| > \sigma$. Hence,

$$h_\sigma(t) = h_a(t) = \frac{\sin \omega_c t}{\pi t} \qquad h[n] = Th_a(nT) = \frac{\sin \omega_c Tn}{\pi n}$$

Comb filters We shall say that an analog system is a *comb filter* if its system function $H_a(\omega)$ is periodic:

$$H_a(\omega + c) = H_a(\omega) \tag{1-76}$$

Such a system can be simulated digitally for any input provided that $T = 2\pi/c$, that is, $\sigma = c/2$. Indeed, if $H(z)$ is determined from Eq. (1-69), then $H(e^{j\omega T}) = H_a(\omega)$ for all ω.

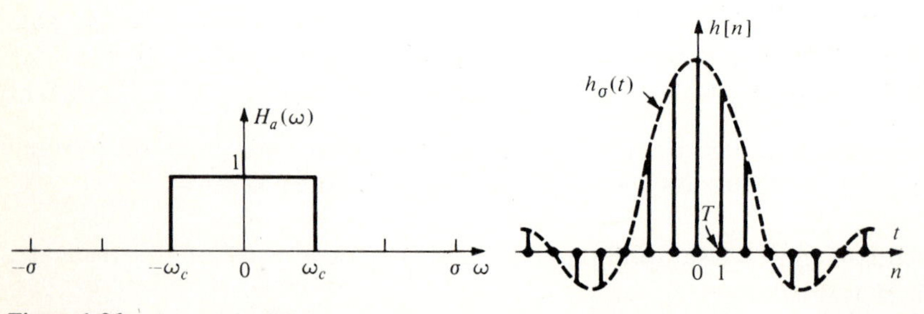

Figure 1-26

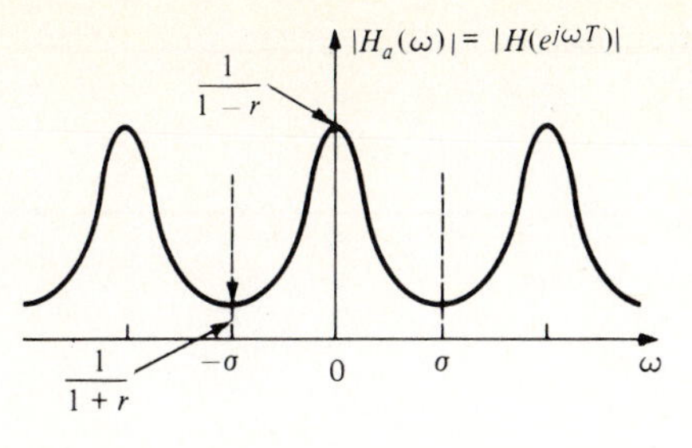

Figure 1-27

Example 1-23 The equation

$$\frac{1}{T}[g(t) - g(t - T)] + \alpha g(t) = f(t) \tag{1-77}$$

defines a system with input $f(t)$ and output $g(t)$. If $f(t) = e^{j\omega t}$, then $g(t) = H_a(\omega)e^{j\omega t}$. Inserting into (1-77), we have

$$\frac{1}{T} H_a(\omega)e^{j\omega t} - \frac{1}{T} H_a(\omega)e^{j\omega(t-T)} + \alpha H_a(\omega)e^{j\omega t} = e^{j\omega t}$$

Hence (Fig. 1-27),

$$H_a(\omega) = \frac{A}{1 - re^{-j\omega T}}$$

where

$$r = \frac{1}{\alpha T + 1} \qquad \text{and} \qquad A = \frac{T}{\alpha T + 1}$$

Clearly, $H_a(\omega)$ is a comb filter with period $2\pi/T$ and its digital simulator is given by

$$H(e^{j\omega T}) = \frac{A}{1 - re^{-j\omega T}} \qquad H(z) = \frac{A}{1 - rz^{-1}}$$

Example 1-24 If $H_a(\omega) = (1 + e^{-j\omega T})^2$, then $H(z) = (1 + z^{-1})^2$ and $h[n] = \delta[n] + 2\delta[n-1] + \delta[n-2]$.

In many applications, it is desirable to have a system whose system function $H_a(\omega)$ meets certain design requirements. The system is realized either by a combination of various elements, lumped or distributed, or by a computer. The first approach involves network synthesis techniques; the second approach is based on the digital simulation theorem. In Sec. 5-3 we rederive this theorem and discuss various methods for approximating the discrete system $H(z)$ with a system that can be realized with a finite number of delay elements.

CHAPTER

TWO

DISCRETE SYSTEMS

Continuing the discussion of Sec 1-1, we develop in this chapter the properties of discrete systems, stressing clarification of the main concepts rather than detailed coverage of specific applications.[1, 2]

In Sec. 2-1, we give a self-contained and, for our purposes, complete presentation of the theory of z transforms. To avoid any reliance on the theory of complex variables, we derive the inversion formula in terms of Fourier series, thus demonstrating the connection between z transforms and Fourier series.

In Sec. 2-2, we show that a system that can be realized with a finite number of delay elements and multipliers is equivalent to a recursion equation of finite order with constant coefficients. We discuss the solutions of such equations.

In Sec. 2-3, we cover analysis and synthesis of finite-order systems with a brief discussion of the properties of their frequency response. The approximation problem is considered in Chap. 5.

---

[1] A. V. Oppenheim and R. W. Schafer, "Digital Signal Processing," Prentice-Hall, Inc., Englewood Cliffs, N.J., 1975.

[2] L. R. Rabiner and B. Gold, "Theory and Applications of Digital Signal Processing," Prentice-Hall, Inc., Englewood Cliffs, N.J., 1975.

2-1 *z* TRANSFORMS

The series

$$F(z) = \sum_{n=-\infty}^{\infty} f[n]z^{-n} \tag{2-1}$$

establishes a correspondence between the sequence $f[n]$ and the function $F(z)$. We shall use the notation

$$f[n] \leftrightarrow F(z)$$

for this correspondence. The function $F(z)$ will be called the *z transform* of $f[n]$.

Example 2-1
(*a*) If $f[n] = \delta[n-k]$, then $F(z) = z^{-k}$.
(*b*) If $f[n] = 3\,\delta[n-2] + 2\,\delta[n-5]$ as in Fig. 2-1, then $F(z) = 3z^{-2} + 2z^{-5}$.

The z transform $F(z)$ is defined only for the values of z, real or complex, for which the series (2-1) converges. As is known from the theory of functions of complex variables, the region R of convergences of the Laurent series (2-1) is a ring $r_1 < |z| < r_2$ (Fig. 2-2*a*) whose inner and outer radii r_i and r_2 depend on the behavior of $f[n]$ as $n \to +\infty$ and $-\infty$, respectively. In this ring, $F(z)$ is an analytic function of z, that is, it has derivatives of any order. The poles or any other singularities of $F(z)$ are outside the region R.

If $f[n] = 0$ for $n < 0$, then $r_2 = \infty$ because (2-1) has only negative powers of z. In this case, $R = R_1$ is the exterior $|z| > r_1$ of the circle of Fig. 2-2*b*. If $f[n] = 0$ for $n > 0$, then $r_1 = 0$ because (2-1) has only positive powers of z. In this case, $R = R_2$ is the interior $|z| < r_2$ of the circle of Fig. 2-2*c*.

Example 2-2

(*a*)
$$f[n] = U[n] \qquad F(z) = \sum_{n=0}^{\infty} z^{-n} = \frac{1}{1-z^{-1}} = \frac{z}{z-1}$$

The above sum is a geometric progression, and it converges for $|z^{-1}| < 1$; hence, the region R_1 of existence of $F(z)$ is the exterior $|z| > 1$ of the unit circle (Fig. 2-3).

(*b*)
$$f[n] = -U[-n-1] = \begin{cases} -1 & n < 0 \\ 0 & n \ge 0 \end{cases} \qquad F(z) = -\sum_{n=-\infty}^{-1} z^{-n} = -\sum_{n=1}^{\infty} z^{n} = \frac{z}{z-1}$$

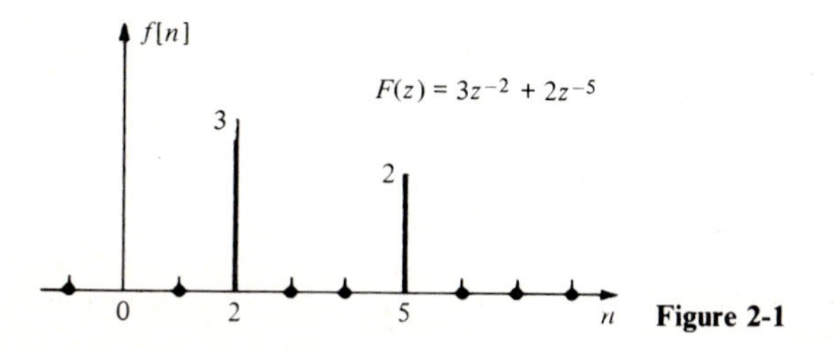

Figure 2-1

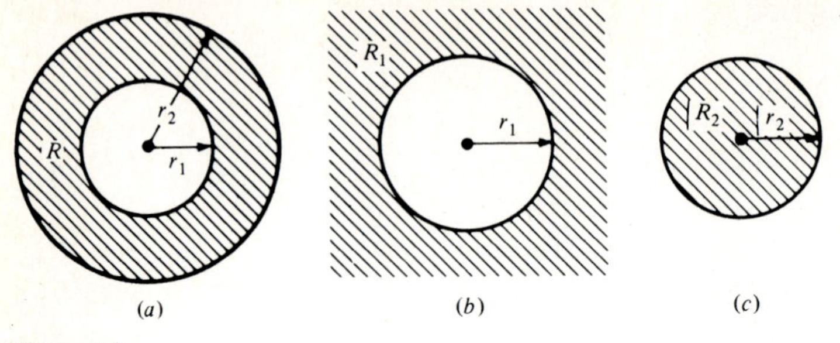

Figure 2-2

The sum converges for $|z| < 1$; hence, the region R_2 of existence of $F(z)$ is the interior $|z| < 1$ of the unit circle.

> **Note:** The z transforms of the above two sequences $U[n]$ and $-U[-n-1]$ have the *same* algebraic expression $z/(z-1)$. However, the corresponding regions R_1 and R_2 of their existence are different.

Example 2-3

(*a*) From the identity

$$\sum_{n=0}^{\infty} a^n z^{-n} = \frac{z}{z-a} \qquad |z| > a \tag{2-2}$$

it follows that

$$a^n U[n] \leftrightarrow \frac{z}{z-a} \qquad |z| > a \tag{2-3}$$

(*b*) Differentiating Eq. (2-2) with respect to a, we obtain

$$\sum_{n=0}^{\infty} n a^{n-1} z^{-n} = \frac{z}{(z-a)^2}$$

Hence,

$$n a^{n-1} U[n] \leftrightarrow \frac{z}{(z-a)^2} \qquad |z| > a$$

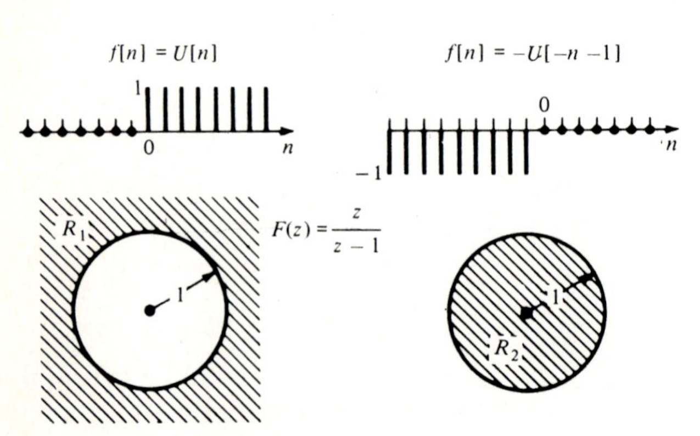

Figure 2-3

(*c*) Differentiating (2-2) m times with respect to a, we find

$$\sum_{n=0}^{\infty} n(n-1)\cdots(n-m+1)a^{n-m}z^{-n} = \frac{m!\,z}{(z-a)^{m+1}}$$

Therefore,
$$\binom{n}{m}a^{n-m}U[n] \leftrightarrow \frac{z}{(z-a)^{m+1}} \qquad |z| > a \tag{2-4}$$

Inversion of the z Transform

We wish to find a sequence $f[n]$ whose z transform equals a given function $F(z)$. If $F(z)$ is specified by a series expansion

$$F(z) = \sum_{n=-\infty}^{\infty} c[n]z^{-n} \qquad z \in R \tag{2-5}$$

then $f[n]$ is unique and it equals $c[n]$.

Example 2-4 If $F(z) = 3z + 4z^{-2} - 5z^{-10}$, then $f[-1] = 3$, $f[2] = 4$, $f[10] = -5$, and $f[n] = 0$ otherwise.

If $F(z)$ is specified by an algebraic expression, its inverse is not, in general, unique. Indeed, as we have seen in Example 2-2, the function $z/(z-1)$ is the z transform of two different sequences. To have a unique inverse, we must know not only $F(z)$, but also the region R of convergence of the series (2-1). If it is known that $F(z)$ is the transform of a causal sequence, then, its region of convergence is the exterior of a circle containing all its poles.

We shall discuss three methods for determining $f[n]$ from $F(z)$.

1. **Series expansion** We expand $F(z)$ into a series as in (2-5). The unknown numbers $f[n]$ equal the coefficients $c[n]$ of this expansion. Since $F(z)$ has a unique expansion in a ring R, its inverse $f[n]$ is uniquely determined if R is known.

Example 2-5

$$F(z) = \frac{z^{-5}}{z+2}$$

It is easy to see that

$$\frac{z^{-5}}{z+2} = z^{-6}\sum_{k=0}^{\infty}(-2z^{-1})^k \qquad \text{or} \qquad \frac{z^{-5}}{z+2} = \frac{z^{-5}}{2}\sum_{k=0}^{\infty}\left(-\frac{z}{2}\right)^k$$

The first expansion is valid for $|z| > 2$, and the second for $|z| < 2$. Hence,

$$f[n] = f_1[n] = \begin{cases} (-2)^{n-6} & n \ge 6 \\ 0 & n < 6 \end{cases} \qquad \text{or} \qquad f[n] = f_2[n] = \begin{cases} \frac{1}{2}(-\frac{1}{2})^{5-n} & n \le 5 \\ 0 & n > 5 \end{cases}$$

The inverse of $F(z)$ is $f_1[n]$ if R is the region $|z| > 2$; the inverse of $F(z)$ is $f_2[n]$ if R is the region $0 < |z| < 2$ (interior of the circle $|z| = 2$ excluding the origin).

2. Tables If the function $F(z)$ can be written as a sum

$$F(z) = \sum_i a_i F_i(z) \tag{2-6}$$

where $F_i(z)$ are functions with known inverses $f_i[n]$, then (uniqueness)

$$f[n] = \sum_i a_i f_i[n]$$

Example 2-6 From the identity

$$\frac{z^3}{z-1} = z^2 + z + \frac{z}{z-1}$$

we conclude that the inverse $f[n]$ of the function $z^3/(z-1)$ in the region $R = R_1$ is given by (see Example 2-2)

$$f[n] = \delta[n+2] + \delta[n+1] + U[n] = U[n+2]$$

The above method is of particular interest if $F(z)$ is rational.

Rational transforms A rational function is the ratio of two polynomials of z:

$$F(z) = \frac{N(z)}{D(z)} \tag{2-7}$$

or, equivalently, of z^{-1}. We prefer to use polynomials of z. We can assume, if necessary removing a polynomial from $F(z)$ by synthetic division, that the degree of $N(z)$ does not exceed the degree of $D(z)$.

We shall first determine the *causal inverse* $f[n]$ of $F(z)$ using the pair

$$z_i^n U[n] \leftrightarrow \frac{z}{z - z_i}$$

for the simple roots, and the pair (2-4) for the multiple roots.

Suppose that all the roots z_i of the denominator of $F(z)$ are simple. As is well known, the proper fraction $F(z)/z$ can be written as a sum

$$\frac{F(z)}{z} = \sum_i \frac{A_i}{z - z_i} \qquad \text{where} \qquad A_i = \left.\frac{F(z)(z - z_i)}{z}\right|_{z = z_i}$$

Hence,

$$F(z) = \sum_i \frac{A_i z}{z - z_i} \qquad f[n] = U[n] \sum_i A_i z_i^n \tag{2-8}$$

Example 2-7

$$F(z) = \frac{z+2}{2z^2 - 7z + 3} \qquad z_1 = \frac{1}{2} \qquad z_2 = 3$$

$$\frac{F(z)}{z} = \frac{2/3}{z} - \frac{1}{z - 1/2} + \frac{1/3}{z-3} \qquad f[n] = \frac{2}{3}\delta[n] - \left(\frac{1}{2}\right)^n U[n] + \frac{1}{3} \times 3^n U[n]$$

Multiple roots can be handled similarly. We give only an illustration.

Example 2-8

$$F(z) = \frac{z}{(z - 1/2)(z - 1)^2}$$

$$\frac{F(z)}{z} = \frac{4}{z - 1/2} + \frac{2}{(z - 1)^2} - \frac{4}{z - 1}$$

As we see from (2-4), the inverse of $z/(z - 1)^2$ equals $nU[n]$. Hence,

$$f[n] = [4(\tfrac{1}{2})^n + 2n - 4]U[n]$$

Noncausal inverse The rational function $F(z)$ has a causal inverse $f[n]$ given by Eq. (2-8) if the region of existence of $F(z)$ is the exterior R_1 of the circle $|z| = r_1 = \max |z_i|$ containing all poles z_i of $F(z)$. We shall now determine the inverse of $F(z)$ assuming that R is an arbitrary ring as in Fig. 2-4.

Example 2-9
(*a*) From the identity

$$\sum_{n=-\infty}^{-1} a^n z^{-n} = \sum_{n=1}^{\infty} \left(\frac{z}{a}\right)^n = \frac{-z}{z - a} \qquad |z| < a \tag{2-9}$$

it follows that

$$a^n U[-n - 1] \leftrightarrow \frac{-z}{z - a} \qquad |z| < a \tag{2-10}$$

(*b*) Differentiating (2-9) m times with respect to a, we obtain, as in (2-4),

$$\binom{n}{m} a^{n-m} U[-n - 1] \leftrightarrow \frac{-z}{(z - a)^{m+1}} \qquad |z| < a \tag{2-11}$$

We denote by p_i the *interior* poles of $F(z)$, that is, the roots of $D(z)$ that are in the interior of the inner boundary $|z| = r_1$ of R; and by s_i the *exterior* poles, that is, the roots of $D(z)$ that are outside the outer boundary $|z| = r_2$ of R as in Fig. 2-4. We separate the expansion (2-8) of $F(z)$ into two groups:

$$F(z) = \sum_i \frac{B_i z}{z - p_i} + \sum_i \frac{C_i z}{z - s_i}$$

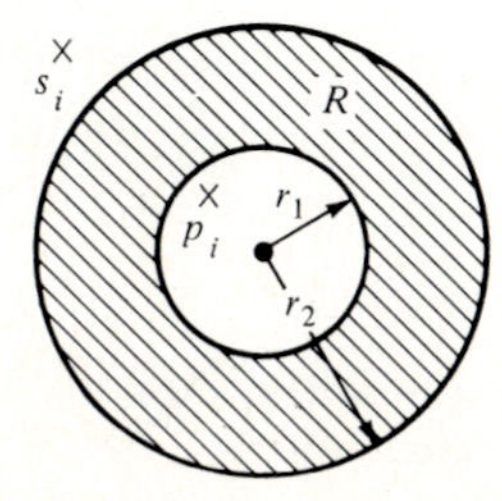

Figure 2-4

From (2-3) and (2-10) it follows that

$$f[n] = U[n] \sum_i B_i p_i^n - U[-n-1] \sum_i C_i s_i^n$$

Multiple roots can be treated similarly [see (2-11)].

Example 2-10

$$F(z) = \frac{z+2}{2z^2 - 7z + 3} = \frac{2}{3} - \frac{z}{z - 1/2} + \frac{z/3}{z-3}$$

as in Example 2-7. This function has two poles, $z_1 = 1/2$ and $z_2 = 3$, and the following three inverse transforms:

(*a*) In the region $|z| > 3$, all poles are *interior*; hence,

$$f[n] = \tfrac{2}{3}\,\delta[n] - (\tfrac{1}{2})^n U[n] + \tfrac{1}{3} \times 3^n U[n]$$

(*b*) In the ring $\frac{1}{2} < |z| < 3$, the pole $p_1 = \frac{1}{2}$ is *interior*, and the pole $s_1 = 3$ is *exterior;* hence,

$$f[n] = \tfrac{2}{3}\,\delta[n] - (\tfrac{1}{2})^n U[n] - \tfrac{1}{3} \times 3^n U[-n-1]$$

(*c*) In the region $|z| < \frac{1}{2}$, both poles are *exterior;* hence,

$$f[n] = \tfrac{2}{3}\,\delta[n] + (\tfrac{1}{2})^n U[-n-1] - \tfrac{1}{3} \times 3^n U[-n-1]$$

3. The inversion formula With $z = e^{j\omega T}$, it follows from (2-1) that

$$F(e^{j\omega T}) = \sum_{n=-\infty}^{\infty} f[n] e^{-jn\omega T}$$

Thus, on the unit circle, $F(z)$ is a periodic function of ω with period $2\pi/T$ and Fourier-series coefficients $f[n]$. Hence,

$$f[n] = \frac{1}{2\sigma} \int_{-\sigma}^{\sigma} F(e^{j\omega T}) e^{jn\omega T}\, d\omega \qquad \sigma = \frac{\pi}{T} \tag{2-12}$$

Note: For arbitrary sequences, the constant T has no particular significance and can be assumed to equal 1. In this case, ω is the phase of z and $\sigma = \pi$. If $f[n]$ is obtained by sampling an analog signal as in Eq. (1-71), then T equals the sampling interval.

Complex form If $z = e^{j\omega T}$, then

$$dz = jTe^{j\omega T}\, d\omega = jTz\, d\omega$$

Hence, (2-12) can be written as a contour integral

$$f[n] = \frac{1}{2\pi j} \oint_C F(z) z^{n-1}\, dz \tag{2-13}$$

where C is the unit circle. Thus, to compute $f[n]$, it suffices to evaluate the above integral. This can be done with the calculus of *residues*, which for rational functions involves the fraction expansion (2-8).

Theorems

1. Shifting. If $f[n] \leftrightarrow F(z)$, then for any integer m,

$$f[n-m] \leftrightarrow z^{-m}F(z) \tag{2-14}$$

Proof

$$\sum_{n=-\infty}^{\infty} f[n-m]z^{-n} = \sum_{k=-\infty}^{\infty} f[k]z^{-(m+k)} = z^{-m}F(z)$$

Example 2-11 Since $z^3/(z-1) = z^2 \cdot z/(z-1)$ and the inverse of $z/(z-1)$ equals $U[n]$, it follows from (2-14) with $m = -2$ that

$$U[n+2] \leftrightarrow \frac{z^3}{z-1}$$

2. Convolution. If $f_1[n] \leftrightarrow F_1(z)$ and $f_2[n] \leftrightarrow F_z(z)$, then

$$\sum_{k=-\infty}^{\infty} f_1[k]f_2[n-k] \leftrightarrow F_1(z)F_2(z) \tag{2-15}$$

Proof The above was derived in (1-18). It can also be proved by multiplying the series expansions of $F_1(z)$ and $F_2(z)$. We shall carry out the details for causal sequences:

$$F_1(z) = f_1[0] + f_1[1]z^{-1} + \cdots + f_1[n]z^{-n} + \cdots$$

$$F_2(z) = f_2[0] + f_2[1]z^{-1} + \cdots + f_2[n]z^{-n} + \cdots$$

$$F_1(z)F_2(z) = f_1[0]f_2[0] + (f_1[0]f_2[1] + f_1[1]f_2[0])z^{-1} + \cdots$$

$$+ \left(\sum_{k=0}^{n} f_1[k]f_2[n-k]\right)z^{-n} + \cdots$$

The right side is the series expansion of $F_1(z)F_2(z)$; hence, its inverse is the coefficient of z^{-n}.

Example 2-12 Using z transforms, we shall find the sum of the integers from 0 to n and the sum of their squares.

(*a*)

$$f_1[n] = \sum_{k=0}^{n} k = (nU[n]) * U[n] \leftrightarrow \frac{z}{(z-1)^2}\frac{z}{z-1} = z\frac{z}{(z-1)^3}$$

But

$$\frac{n(n-1)}{2}U[n] \leftrightarrow \frac{z}{(z-1)^3}$$

Hence (shifting),

$$f_1[n] = \frac{n(n+1)}{2}$$

(*b*) $$f_2[n] = \sum_{k=0}^{n} k^2 = (n^2 U[n]) * U[n] \leftrightarrow \frac{z^3 + z^2}{(z-1)^4}$$

because $$n^2 U[n] = \left[2\frac{n(n-1)}{2} + n\right] U[n] \leftrightarrow \frac{2z}{(z-1)^3} + \frac{z}{(z-1)^2}$$

But $$\binom{n}{3} U[n] \leftrightarrow \frac{z}{(z-1)^4}$$

Hence, $$f_2[n] = \frac{(n+2)(n+1)n}{6} + \frac{(n+1)n(n-1)}{6} = \frac{(2n+1)(n+1)n}{6}$$

3. Conjugate sequences. If $f[n] \leftrightarrow F(z)$, then

$$f^*[-n] \leftrightarrow F^*\left(\frac{1}{z^*}\right) \tag{2-16}$$

Proof

$$\sum_{n=-\infty}^{\infty} f^*[-n] z^{-n} = \sum_{k=-\infty}^{\infty} f^*[k] z^k$$

Notes: If the sequence $f[n]$ is real, then $F^*(1/z^*) = F(1/z)$. If $F(z)$ converges in the ring $r_1 < |z| < r_2$, then $F^*(1/z^*)$ converges in the ring $1/r_2 < |z| < 1/r_1$.

Example 2-13 From (2-16) and the pair (2-4) it follows that

$$\binom{-n}{m} a^{-n-m} U[-n] \leftrightarrow \frac{z^{-1}}{(z^{-1} - a)^{m+1}} \qquad |z| < \frac{1}{|a|} \tag{2-17}$$

This result agrees with (2-11) if a is replaced with $1/a$ (elaborate).

4. Parseval's formula. If $x[n] \leftrightarrow X(z)$ and $y[n] \leftrightarrow Y(z)$, then for any $T = \pi/\sigma$,

$$\sum_{n=-\infty}^{\infty} x[n] y^*[n] = \frac{1}{2\sigma} \int_{-\sigma}^{\sigma} X(e^{j\omega T}) Y^*(e^{j\omega T})\, d\omega \tag{2-18}$$

Proof From (2-15) and (2-12) it follows that

$$\sum_{k=-\infty}^{\infty} f_1[k] f_2[n-k] = \frac{1}{2\sigma} \int_{-\sigma}^{\sigma} F_1(e^{j\omega T}) F_2(e^{j\omega T}) e^{jn\omega T}\, d\omega \tag{2-19}$$

If $f_1[n] = x[n]$ and $f_2[-n] = y^*[n]$, then $F_1(z) = X(z)$ and $F_2(z) = Y^*(1/z^*)$, and (2-18) follows from Eq. (2-19) with $n = 0$.

Corollary From (2-18) it follows, with $x[n] = y[n]$, that

$$\sum_{n=-\infty}^{\infty} |x[n]|^2 = \frac{1}{2\sigma}\int_{-\sigma}^{\sigma} |X(e^{j\omega T})|^2 \, d\omega \tag{2-20}$$

5. Initial and final value. If the sequence $f[n]$ is causal, then $F(z)$ exists for $|z| > r_1$ and, as we see from (2-1), $f[0] = F(\infty)$. The following is a generalized converse of the above:

If $F(z)$ exists for $|z| > r_1$ and if for some integer m, positive or negative, $\lim_{z \to \infty} z^m F(z) = A < \infty$, then

$$f[m] = A \qquad \text{and} \qquad f[n] = 0 \qquad \text{for } n < m \tag{2-21}$$

For example, if $F(z) = z^3/(z-1)$, then $z^{-2}F(z) \to 1$. Hence, $f[-2] = 1$ and $f[n] = 0$ for $n < -2$.

The above theorem relates the behavior of $F(z)$ for large z to the "initial value" $f[m]$ of $f[n]$. The following result relates the behavior of $f[n]$ for large n to the radius r_1 of the region R_1.

If $F(z)$ exists for $|z| > r_1$ and $r_1 < 1$, then

$$f[n] \to 0 \qquad \text{for } n \to \infty \tag{2-22}$$

This theorem is a consequence of the following property of Laurent series: If the series (2-1) converges for $r_1 < |z| < r_2$, then

$$r_1 = \overline{\lim} \sqrt[n]{|f[n]|} \qquad r_2 = \underline{\lim} \sqrt[n]{|f[-n]|} \qquad \text{as } n \to \infty \tag{2-23}$$

Clearly, if a rational function $F(z)$ is the z transform of a causal sequence $f[n]$, then (2-22) holds if all the poles z_i of $F(z)$ are inside the unit circle. In this case, $f[n] \to 0$ with $n \to \infty$.

6. Unilateral *z* transform. The unilateral z transform of a sequence $f[n]$ is the sum

$$F_1(z) = \sum_{n=0}^{\infty} f[n] z^{-n} \tag{2-24}$$

Clearly, if $f[n] = 0$ for $n < 0$, then $F_1(z) = F(z)$. If $f[n]$ is an arbitrary sequence, then $F_1(z)$ is the z transform of $f[n]U[n]$. Hence, the theorems given above, properly modified, hold also for unilateral transforms. The shifting theorem (2-14) merits special consideration:

If $F_1(z)$ is the unilateral transform of $f[n]$ and $m > 0$, then the unilateral transform of $f[n-m]$ equals

$$z^{-m}F_1(z) + f[-1]z^{-m+1} + f[-2]z^{-m+2} + \cdots + f[-m] \tag{2-25}$$

Proof With $n - m = k$, we have

$$\sum_{n=0}^{\infty} f[n-m]z^{-N} = \sum_{k=-m}^{\infty} f[k]z^{-(m+k)} = f[-m] + \cdots + f[-1]z^{-m+1} + z^{-m}\sum_{k=0}^{\infty} f[k]z^{-k} \tag{2-26}$$

and (2-25) results because the sum on the left is the unilateral transform of the sequence $f[n-m]$ and the last sum on the right equals $F_1(z)$.

2-2 RECURSION EQUATIONS

A recursion equation of order m is a relationship between two sequences $f[n]$ and $g[n]$ of the form

$$g[n] + a_1 g[n-1] + \cdots + a_m g[n-m] = b_0 f[n] + \cdots + b_m f[n-m] \tag{2-27}$$

We shall assume that, if $f[n] = 0$ for $n < n_0$, then $g[n] = 0$ for $n < n_0$ (causality). With this assumption, if Eq. (2-27) holds for all n, then $g[n]$ is uniquely determined in terms of $f[n]$.

Example 2-14

$$g[n] + 2g[n-1] = U[n] \qquad \text{for all } n$$

Clearly, $g[n] = 0$ for $n < 0$ because $f[n] = U[n] = 0$ for $n < 0$. Setting $n = 0, 1, \ldots$ in the given system of equations, we obtain

$$g[0] + 2g[-1] = 1 \qquad g[1] + 2g[0] = 1 \qquad g[2] + 2g[1] = 1 \qquad \cdots$$

Hence,

$$g[0] = 1 \qquad g[1] = -1 \qquad g[2] = 3 \qquad \cdots$$

The determination of $g[n]$ is simplified with the use of z transforms. Denoting by $F(z)$ and $G(z)$ the z transforms of $f[n]$ and $g[n]$, respectively, we conclude from (2-27) and the shifting theorem (2-14) that

$$G(z) + a_1 z^{-1}G(z) + \cdots + a_m z^{-m}G(z) = b_0 F(z) + \cdots + b_m z^{-m}F(z) \tag{2-28}$$

because the two sides of (2-27) are equal for all n, and hence their z transforms are equal.

Solving, we obtain

$$G(z) = F(z)H(z) \tag{2-29}$$

where

$$H(z) = \frac{b_0 + b_1 z^{-1} + \cdots + b_m z^{-m}}{1 + a_1 z^{-1} + \cdots + a_m z^{-m}} \tag{2-30}$$

The sequence $g[n]$ can be evaluated from (2-29) in one of two ways:

1. We determine $F(z)$ from $f[n]$ and compute the inverse z transform of the product $F(z)H(z)$. This approach is used mainly if $F(z)$ is rational.
2. We compute the inverse $h[n]$ of $H(z)$ and convolve $h[n]$ with the given $f[n]$.

Example 2-15

$$g[n] + 2g[n-1] = f[n] \qquad \text{for all } n$$

From Eq. (2-30) it follows that

$$H(z) = \frac{1}{1 + 2z^{-1}} = \frac{z}{z+2}$$

Hence,

$$h[n] = (-2)^n U[n]$$

(*a*) If $f[n] = U[n]$, then $F(z) = z/(z-1)$. Hence,

$$G(z) = \frac{z}{z-1}\frac{z}{z+2} = \frac{z/3}{z-1} + \frac{2z/3}{z+2}$$

The inverse of the above yields $g[n] = \frac{1}{3} + \frac{2}{3}(-2)^n$ for $n \geq 0$.

(*b*) $\quad f[n] = U[n] - U[n-3] = \delta[n] + \delta[n-1] + \delta[n-2]$

$$g[n] = f[n] * h[n] = (-2)^n U[n] + (-2)^{n-1} U[n-1] + (-2)^{n-2} U[n-2]$$

Hence, $g[0] = 1$ and $g[1] = -1$, and $g[n] = \frac{3}{4}(-2)^n$ for $n \geq 2$.

The solution (2-29) was obtained under the assumption that (2-27) was true for all n. Suppose now that (2-27) is true for $n \geq 0$ only. To determine $g[n]$ for $n \geq 0$, we need to know not only $f[n]$ but also the values (initial conditions)

$$g[-m] \qquad g[-m+1] \qquad \cdots \qquad g[-1] \tag{2-31}$$

of $g[n]$. If these values and the corresponding values $f[-m], \ldots, f[-1]$ of $f[n]$ are zero, then the solution $g[n]$ of (2-27) is given by (2-29). Indeed, setting $g[n] = 0$ and $f[n] = 0$ for $n < 0$, we conclude that (2-27) is true for all n; hence, (2-29) holds.

Example 2-16 The sequence $f[n]$ and $g[n]$ in Fig. 2-5 satisfy the recursion equation

$$g[n] - \tfrac{1}{2}g[n-1] = f[n] \tag{2-32}$$

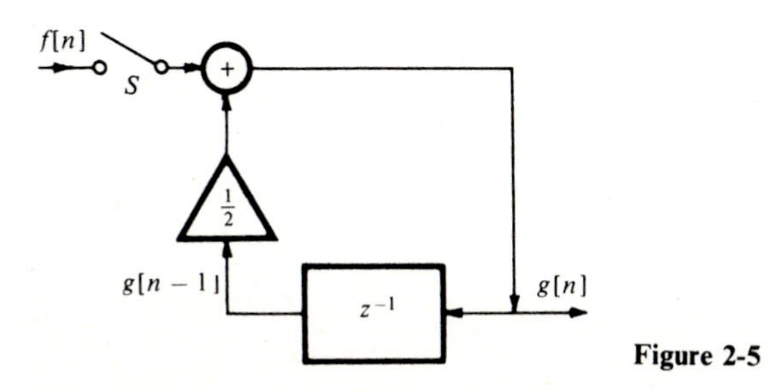

Figure 2-5

The switch S closes at $n = 0$, and the input $f[n] = 5$ for $n \geq 0$. We wish to find $g[n]$ under the assumption that $g[-1] = 0$. We know that Eq. (2-32) holds for $n \geq 0$; setting $f[n] = 5U[n]$ and $g[n] = 0$ for $n < 0$, we can assume that it holds for all n. Hence, we can use (2-29):

$$G(z) = \frac{F(z)}{1 - z^{-1}/2} = \frac{5z^2}{(z - \frac{1}{2})(z - 1)} = \frac{-5z}{z - \frac{1}{2}} + \frac{10z}{z - 1}$$

From the above we conclude that $g[n] = -5(\frac{1}{2})^n + 10$ for $n \geq 0$.

If (2-27) holds for $n \geq 0$ only and the initial conditions are not zero, then we can no longer use (2-29). To determine $g[n]$, we must apply unilateral z transforms and the modified form (2-25) of the shifting theorem. We illustrate with two examples.

Example 2-17

$$g[n] - 5g[n-1] + 6g[n-2] = 1 \qquad n \geq 0 \qquad g[-1] = 3 \qquad g[-2] = 2$$

We take unilateral transforms of both sides of the equation. Since the unilateral transform of $f[n] = 1$ equals $z/(z - 1)$, we conclude from (2-25) that

$$G_1(z) - 5\{z^{-1}G_1(z) + g[-1]\} + 6\{z^{-2}G_1(z) + z^{-1}g[-1] + g[-2]\} = \frac{z}{z-1}$$

Hence,
$$G_1(z) = \frac{3 - 18z^{-1}}{1 - 5z^{-1} + 6z^{-2}} + \frac{z/(z-1)}{1 - 5z^{-1} + 6z^{-2}}$$

The first fraction is due to the initial conditions, and the second to the input $f[n]$. Expanding into partial fractions, we find

$$G_1(z) = \frac{z/2}{z-1} + \frac{8z}{z-2} - \frac{9z/2}{z-3} \qquad g[n] = \frac{1}{2} + 8 \times 2^n - \frac{9}{2} \times 3^n \qquad n \geq 0$$

Example 2-18

(*a*) Show that if $g[n] + a_1 g[n-1] + a_2 g[n-2] = 0$ for $n \geq 0$, then

$$g[n] = Az_1{}^n + Bz_2{}^n$$

where z_1 and z_2 are the roots of the equation $z^2 + a_1 z + a_2 = 0$.

Proof Taking unilateral transforms and expanding into partial fractions, we find

$$G_1(z) = \frac{-g[-1](a_1 z^2 + a_2 z) - a_2 z^2 g[-2]}{z^2 + a_1 z + a_2} = \frac{Az}{z - z_1} + \frac{Bz}{z - z_2}$$

The constants A and B can be expressed in terms of $g[-1]$ and $g[-2]$.

(*b*) Show that if $g[n] - 2ag[n-1] + g[n-2] = 0$ for $n \geq 0$ and $a > 1$, then

$$g[n] = C \cosh \alpha n + D \sinh \alpha n \qquad \text{where} \qquad \cosh \alpha = a$$

Proof Denoting by z_1 and z_2 the roots of the equation $z^2 - 2az + 1 = 0$, we have $z_1 + z_2 = 2a$ and $z_1 z_2 = 1$. With $z_1 = e^{\alpha}$, it follows that $z_2 = 1/z_1 = e^{-\alpha}$ and $e^{\alpha} + e^{-\alpha} = 2a$. Hence,

$$g[n] = Ae^{\alpha n} + Be^{-\alpha n} = \frac{A+B}{2}\cosh \alpha n + \frac{A-B}{2}\sinh \alpha n$$

The Recursion Equation as a Discrete System

If (2-27) holds for all n, or if it holds for $n \geq 0$ but its initial conditions (2-31) are zero, then it defines a digital, causal system with input $f[n]$, output $g[n]$, and system functions $H(z)$ as in (2-30). If $f[n] = \delta[n]$, then $g[n] = h[n]$; hence, the delta response of the system is the solution of the recursion equation

$$h[n] + a_1 h[n-1] + \cdots + a_m h[n-m] = b_0\,\delta[n] + \cdots + b_m\,\delta[n-m] \tag{2-33}$$

such that

$$h[-1] = h[-2] = \cdots = h[-m] = 0 \tag{2-34}$$

Setting $n = 0, 1, \ldots, m$ into (2-33), we obtain

$$\begin{aligned} h[0] &= b_0 \\ h[1] + a_1 h[0] &= b_1 \\ &\cdots\cdots \\ h[m] + a_1 h[m-1] + \cdots + a_m h[0] &= b_m \end{aligned} \tag{2-35}$$

For $n > m$, $h[n]$ satisfies the homogeneous equation

$$h[n] + a_1 h[n-1] + \cdots + a_m h[n-m] = 0 \tag{2-36}$$

Example 2-19 We shall find the delta response of the system specified by the equation

$$g[n] - \tfrac{5}{6}g[n-1] + \tfrac{1}{6}g[n-2] = f[n] - f[n-1]$$

Clearly,

$$H(z) = \frac{1 - z^{-1}}{1 - \frac{5}{6}z^{-1} + \frac{1}{6}z^{-2}} = \frac{4z}{z - \frac{1}{3}} - \frac{3z}{z - \frac{1}{2}}$$

Hence, $h[n] = [4(\frac{1}{3})^n - 3(\frac{1}{2})^n]U[n]$.

2-3 FINITE-ORDER SYSTEMS

We shall say that a discrete system is *realizable* if it is equivalent to a system consisting of a finite number of delay elements and multipliers. A realizable system is, thus, a causal system with rational system function

$$H(z) = \frac{b_0 + b_1 z^{-1} + \cdots + b_m z^{-m}}{1 + a_1 z^{-1} + \cdots + a_m z^{-m}} = \frac{N(z)}{D(z)} \tag{2-37}$$

and its output $g[n]$ satisfies the recursion equation (2-27).

From the assumed causality it follows that the region of existence of $H(z)$ is the exterior of the circle $|z| = r_1$ containing all roots of $D(z)$. Hence, $h[n]$ is given by [see (2-8)]

$$h[n] = A_0\,\delta[n] + A_1 z_1{}^n U[n] + \cdots + A_m z_m{}^n U[n] \tag{2-38}$$

if all poles z_i are simple. If $H(z)$ has multiple poles, then $h[n]$ contains terms of the form (2-4).

If all poles z_i are inside the unit circle, then the system is *stable*. The *order* m of $H(z)$ is the largest exponent of z in any nonzero term of $N(z)$ or $D(z)$.

Realization of a Finite-Order Causal System

We shall show that a system of order m can be synthesized with m delay elements. This can be done in many ways. We give below canonical realizations for the following three cases: constant denominator (nonrecursive); constant numerator; general.

1. Figure 2-6 is the block diagram realization of the nonrecursive system

$$N(z) = b_0 + b_1 z^{-1} + \cdots + b_m z^{-m} \tag{2-39}$$

2. In Fig. 2-7, we show the system

$$\frac{1}{D(z)} = \frac{1}{1 + a_1 z^{-1} + \cdots + a_m z^{-m}} \tag{2-40}$$

where $f[n]$ is the input and the output $g[n]$ is taken at the terminal T_0. Note that the z transform of $g[n-k]$ equals

$$z^{-k}G(z) = \frac{z^{-k}}{D(z)}F(z)$$

Hence, the diagram of Fig. 2-7 also realizes the system $z^{-k}/D(z)$ if the output is taken at the terminal T_k.

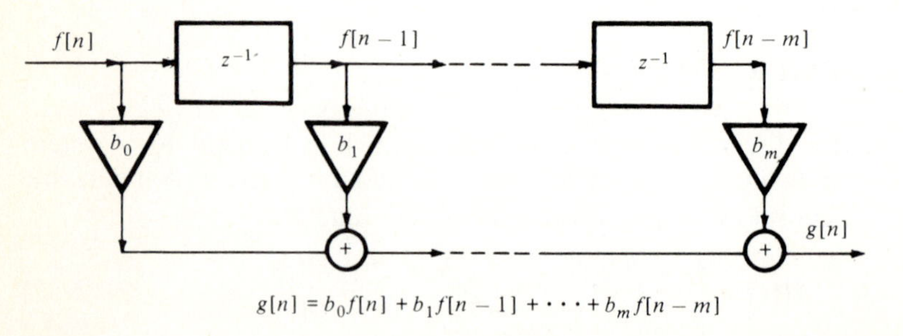

Figure 2-6

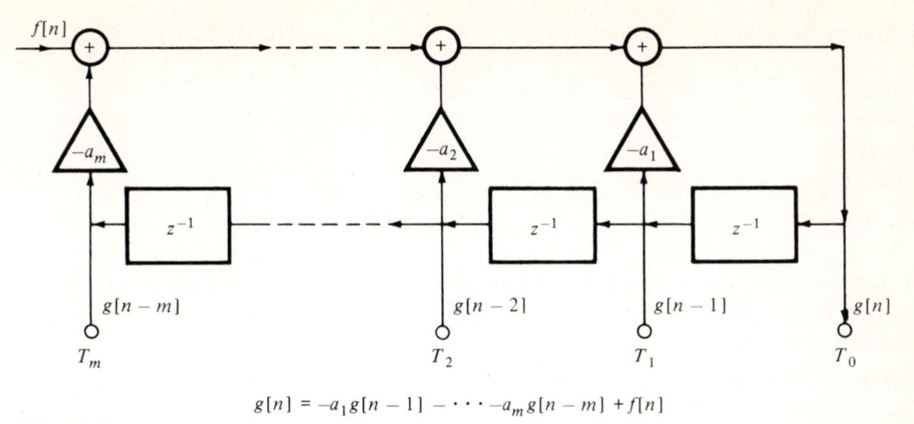

Figure 2-7

3. If we multiply each output $g[n-k]$ of the system of Fig. 2-7 by b_k and add the products as in Fig. 2-8, we obtain a system with system function

$$\frac{b_0}{D(z)} + b_1 \frac{z^{-1}}{D(z)} + \cdots + b_m \frac{z^{-m}}{D(z)} = \frac{N(z)}{D(z)}$$

Thus, the system of Fig. 2-8 is a realization of $H(z)$.

There are, of course, many other ways of realizing a digital system. For example, writing $H(z)$ as a product $H(z) = H_1(z) \cdots H_r(z)$ or as a sum $H(z) = H_1(z) + \cdots + H_r(z)$ of rational functions, we can realize it in terms of low-order systems connected in cascade or in parallel. If we allow complex factors in the multipliers, then all components $H_i(z)$ can be of order 1. Otherwise, we must include second-order systems to realize the complex zeros or poles of $H(z)$ (see also Prob. 26).

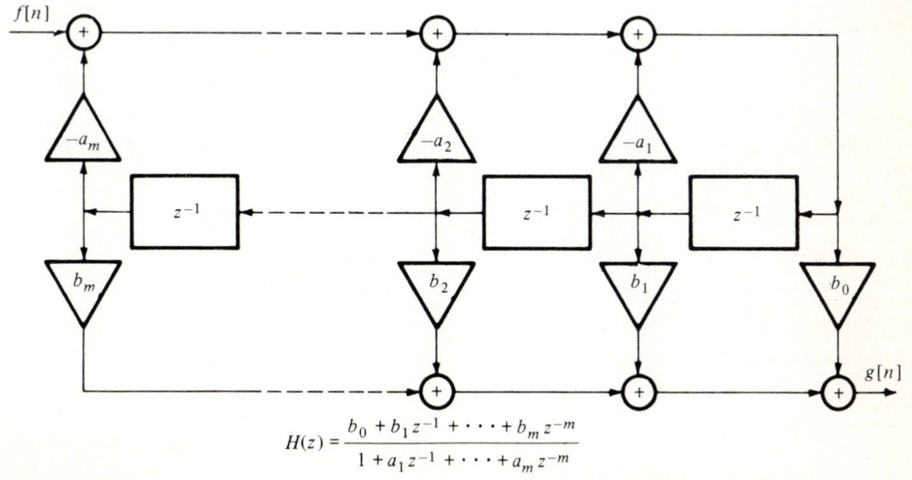

Figure 2-8

Frequency Response

If a system is stable, then the unit circle $|z| = 1$ is in the region of existence of $H(z)$. The behavior of $H(z)$ on this circle we shall call the *frequency response* of the system:

$$H(e^{j\omega T}) = \frac{b_0 + b_1 e^{-j\omega T} + \cdots + b_m e^{-jm\omega T}}{1 + a_1 e^{-j\omega T} + \cdots + a_m e^{-jm\omega T}} = A(\omega)e^{j\varphi(\omega)} \tag{2-41}$$

The function $H(e^{j\omega T})$ is of particular interest if the system is used as a digital simulator of an analog system $H_a(\omega)$. In this case [see Eq. (1-69)], $H(e^{j\omega T})$ equals $H_a(\omega)$ in the band $|\omega| < \sigma = \pi/T$ of the input of $H_a(\omega)$. This is possible only if $H_a(\omega)$ is a rational function of $e^{j\omega T}$. Otherwise, the simulation of $H_a(\omega)$ by a digital system can only be approximate. In Chap. 5, we consider the problem of approximating an arbitrary $H_a(\omega)$ with a rational function.

We shall call $A(\omega)$ the *amplitude* and $\varphi(\omega)$ the *phase* of $H(e^{j\omega T})$. The function $H(e^{j\omega T})$ is periodic with period $2\pi/T = 2\sigma$, and its Fourier-series expansion

$$H(e^{j\omega T}) = \sum_{n=0}^{\infty} h[n]e^{-jn\omega T} \tag{2-42}$$

contains only negative powers of $e^{j\omega T}$ with coefficients the values $h[n]$ of the delta response of the filter.

Example 2-20
(*a*) If $H(z) = 1 - z^{-1}$, then

$$|H(e^{j\omega T})| = 2\left|\sin\frac{\omega T}{2}\right|$$

(*b*) If (Fig. 2-9) $H(z) = (1 - z^{-1})^m$, then

$$|H(e^{j\omega T})| = 2^m\left|\sin\frac{\omega T}{2}\right|^m$$

mth difference The output of the system $1 - z^{-1}$ is denoted by $\Delta f[n]$ and is known as the *first difference* of $f[n]$. The output of the filter $(1 - z^{-1})^m$ is denoted by $\Delta^m f[n]$ and is known as the *mth difference* of $f[n]$:

$$\Delta f[n] = f[n] - f[n-1] \qquad \Delta^m f[n] = \Delta\{\Delta^{(m-1)} f[n]\}$$

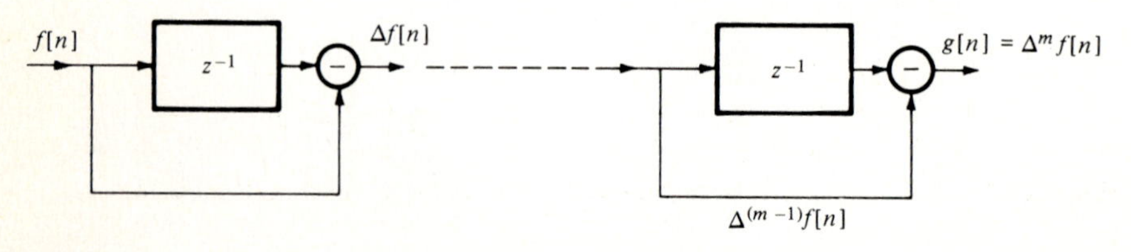

Figure 2-9

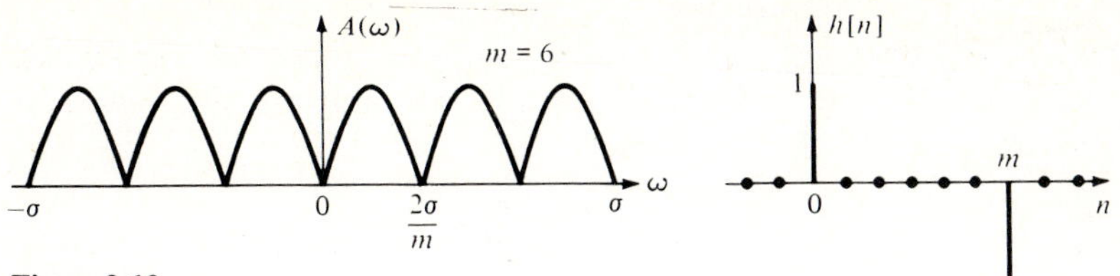

Figure 2-10

Since
$$G(z) = (1 - z^{-1})^m F(z) = \sum_{k=0}^{m} \binom{m}{k} (-z^{-1})^k F(z)$$

it follows from (2-14) that

$$\Delta^m f[n] = \sum_{k=0}^{m} (-1)^k \binom{m}{k} f[n-k]$$

Example 2-21
(*a*) If $H(z) = 1 - z^{-m}$, then (Fig. 2-10)

$$|H(e^{j\omega T})| = 2\left|\sin \frac{m\omega T}{2}\right| \qquad h[n] = \delta[n] - \delta[n-m]$$

(*b*) If

$$H(z) = 1 + z^{-1} + \cdots + z^{-(m-1)} = \frac{1 - z^{-m}}{1 - z^{-1}}$$

then (Fig. 2-11)
$$H(e^{j\omega T}) = \left|\frac{\sin(m\omega T/2)}{\sin(\omega T/2)}\right|$$

and $h[n] = U[n] - U[n-m]$.

The evaluation of the frequency response of the system is often facilitated if $H(e^{j\omega T})$ is expressed as a product of linear factors. Denoting by ζ_i the zeros and by z_i the poles of $H(z)$, we conclude from Eq. (2-41) that (Fig. 2-12)

$$H(e^{j\omega T}) = b_0 \frac{(PM_1)\cdots(PM_m)}{(PN_1)\cdots(PN_m)} \qquad PM_i = e^{j\omega T} - \zeta_i \qquad PN_i = e^{j\omega T} - z_i \quad (2\text{-}43)$$

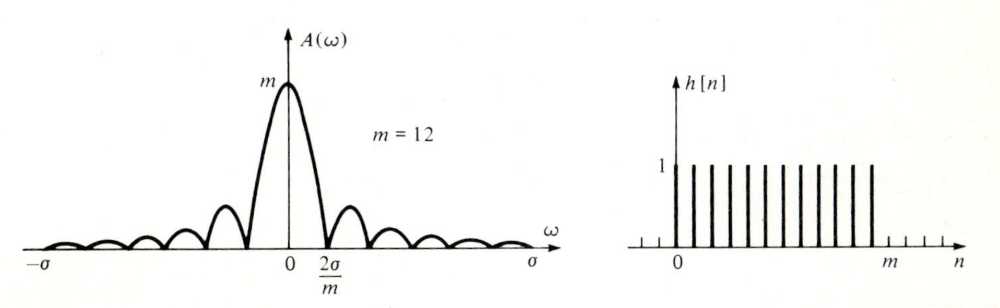

Figure 2-11

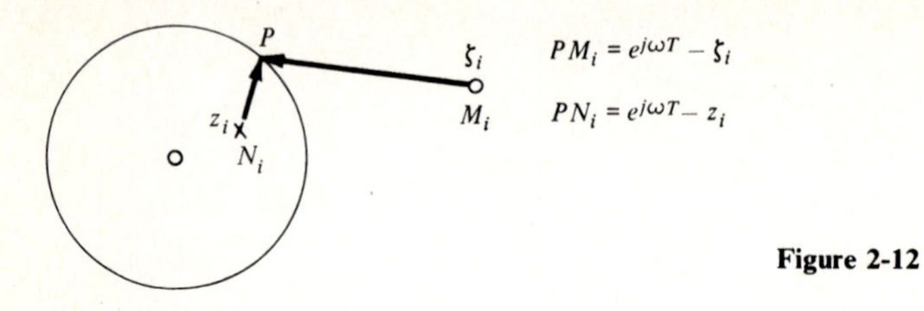

Figure 2-12

The above is particularly useful in the determination of the response of a narrow-band system. In this case, only the singularities that are near the unit circle are of interest; all other factors can be assumed to be constant.

Example 2-22 If $H(z) = 1/(z - r)$, then (Fig. 2-13)

$$H(e^{j\omega T}) = \frac{1}{e^{j\omega T} - r}$$

We note that, if $1 - r \ll 1$, then, in the interval $(-\sigma, \sigma)$, $A(\omega)$ takes significant values only if $|\omega T|$ is of the order of $1 - r$. In this region, $e^{j\omega T} - r \simeq 1 - r + j\omega T$. Hence,

$$A(\omega) \simeq \frac{1}{\sqrt{(1-r)^2 + \omega^2 T^2}}$$

Example 2-23 (Resonance)

$$H(z) = \frac{1}{z^2 - 2rz\cos\alpha + r^2} \qquad z_{1,2} = re^{\pm j\alpha}$$

We shall assume that $1 - r \ll 1$. In this case, $A(\omega)$ takes significant values only when ωT is near $\pm\alpha$ (Fig. 2-14). For ωT near α, we have the approximations

$$PN_2 \simeq 2\sin\alpha \qquad PN_1 = e^{j\omega T} - re^{j\alpha} \simeq e^{j\alpha}[1 - r + j(\omega T - \alpha)]$$

Hence,

$$A(\omega) \simeq \frac{1/2\sin\alpha}{\sqrt{(1-r)^2 + (\omega T - \alpha)^2}} \qquad \text{for } 0 < \omega < \sigma$$

Clearly, $A(\omega)$ is maximum for $\omega = \alpha/T = \omega_r$. For $\omega = \omega_r \pm (1 - r)/T$, it equals $A_{max}/\sqrt{2}$ (half-power points).

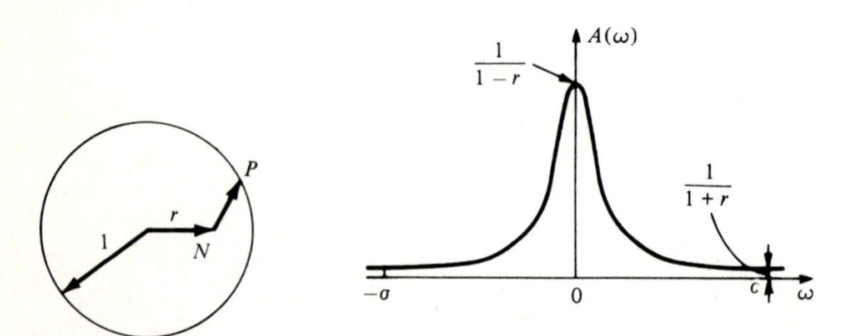

Figure 2-13

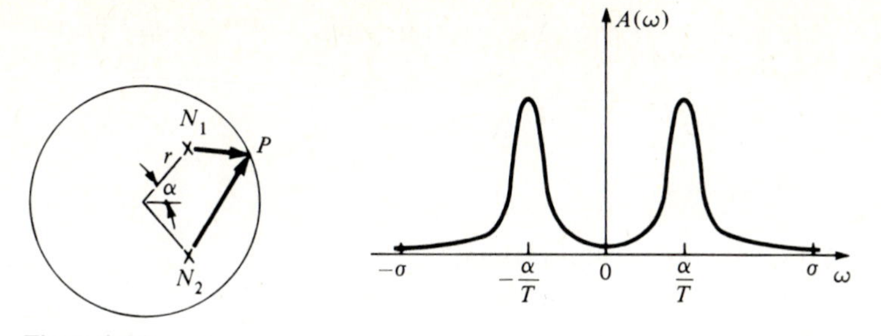

Figure 2-14

Spectrum We shall examine the properties of the *energy spectrum* $A^2(\omega)$ of a *real* system. Since the coefficients a_i and b_i of $H(z)$ are real, it follows from Eq. (2-41) that $H^*(e^{j\omega T}) = H(e^{-j\omega T})$. Hence,

$$A^2(\omega) = |H(e^{j\omega T})|^2 = H(e^{j\omega T})H(e^{-j\omega T}) = H(z)H\left(\frac{1}{z}\right)\bigg|_{z=e^{j\omega T}} \tag{2-44}$$

Suppose that z_i is a root of the numerator or the denominator of $H(z)$. The function $H(z)$ will then contain the factor $z - z_i$, and the function $H(1/z)$ the factor $z^{-1} - z_i$. Therefore, the product $H(z)H(1/z)$ will consist of factors of the form

$$(z - z_i)\left(\frac{1}{z} - z_i\right) = 1 - z_i\left(z + \frac{1}{z}\right) + z_i^2$$

As we see, the above product is a function of $z + 1/z$; and, since this is true for every z_i, we conclude that

$$H(z)H\left(\frac{1}{z}\right) = V(w) \qquad w = \frac{1}{2}\left(z + \frac{1}{z}\right) \tag{2-45}$$

where $V(w)$ is a rational function of w. This leads to the following results:

Theorem The spectrum $A^2(\omega)$ of a finite-order system is a nonnegative rational function of $\cos \omega T$.

Proof If $z = e^{j\omega T}$, then $\frac{1}{2}(z + 1/z) = w = \cos \omega T$. Hence,

$$A^2(\omega) = V(\cos \omega T) \tag{2-46}$$

as we can see from (2-44) and (2-45).

Example 2-24

$$H(z) = \frac{z - 2}{3z - 1} \qquad H(z)H\left(\frac{1}{z}\right) = \frac{1 - 2(z + 1/z) + 4}{9 - 3(z + 1/z) + 1}$$

$$V(w) = \frac{5 - 4w}{10 - 6w} \qquad A^2(\omega) = \left|\frac{e^{j\omega T} - 2}{3e^{j\omega T} - 1}\right|^2 = \frac{5 - 4\cos \omega T}{10 - 6\cos \omega T}$$

Conversely, given a nonnegative rational function $V(\cos \omega T)$, we can find a system $H(z)$ such that

$$|H(e^{j\omega T})|^2 = V(\cos \omega T) \tag{2-47}$$

The function $H(z)$ is determined as follows:

1. Replacing $\cos \omega T$ by w, we obtain the rational function $V(w)$.
2. We find all roots w_i of the numerator and denominator of $V(w)$.
3. We form the equation

$$\frac{1}{2}\left(z + \frac{1}{z}\right) = w_i \tag{2-48}$$

   for each w_i. This equation has two roots, z_i and $1/z_i$, where by z_i we mean the root that is inside the unit circle.

4. The zeros and poles of the unknown $H(z)$ equal the numbers z_i so obtained. If w_i is a zero (pole) of $V(w)$, then the corresponding z_i is a zero (pole) of $H(z)$. The constant factor needed to complete the determination of $H(z)$ is such that $H(z)H(1/z) = V(w)$ as in Eq. (2-45).

Example 2-25 Find $H(z)$ such that

$$|H(e^{j\omega T})|^2 = \frac{5 - 4\cos \omega T}{10 - 6\cos \omega T}$$

1. $V(w) = \dfrac{5 - 4w}{10 - 6w}$

2. One zero $w_1 = \frac{5}{4}$; one pole $w_2 = \frac{10}{6}$
3. The roots of $z + 1/z = \frac{5}{2}$ and $z + 1/z = \frac{10}{3}$, inside the unit circle, are $z_1 = \frac{1}{2}$ and $z_2 = \frac{1}{3}$, respectively

4. $H(z) = A\dfrac{z - \frac{1}{2}}{z - \frac{1}{3}} = \dfrac{2z - 1}{3z - 1}$

$$A = \tfrac{2}{3} \qquad \text{because} \qquad H^2(1) = V(1) = \tfrac{1}{4}$$

Note: To obtain a stable system, the poles that we assign to $H(z)$ must be inside the unit circle. However, its zeros can be anywhere. Hence, $H(z)$ is not uniquely determined from (2-47). For example, the functions

$$H_m(z) = \frac{2z - 1}{3z - 1} \qquad \text{and} \qquad H(z) = \frac{z - 2}{3z - 1} = \frac{2z - 1}{3z - 1}\frac{z - 2}{2z - 1}$$

have the same amplitude because

$$\left|\frac{e^{j\omega T} - 2}{2e^{j\omega T} - 1}\right| = \left|\frac{e^{j\omega T} - 2}{2 - e^{-j\omega T}}\right| = 1$$

The zero $\zeta = \frac{1}{2}$ of $H_m(z)$ is smaller than 1, whereas the zero $\zeta = 2$ of $H(z)$ is greater than 1. If we impose the condition that $H(z)$ is *minimum phase*, i.e., its zeros also are inside the unit circle, then $H(z)$ is uniquely determined from its amplitude $A(\omega)$.

Bilinear transformations We shall discuss a second method for determining $H(z)$ from $|H(e^{j\omega T})|$. For this purpose, we introduce the variable

$$Z = \frac{z-1}{z+1} \tag{2-49}$$

Clearly, if $z = e^{j\omega T}$, then

$$Z = j \tan \frac{\omega T}{2} \tag{2-50}$$

Hence, the circle $|z| = 1$ of the z plane is mapped into the imaginary axis of the Z plane (Fig. 2-15), and, as is easily seen, the interior of the unit circle is mapped into the left-hand part of the Z plane:

$$\text{If} \qquad |z| \le 1 \qquad \text{then} \qquad \operatorname{Re} Z \le 0 \tag{2-51}$$

Suppose that

$$w = \frac{1}{2}\left(z + \frac{1}{z}\right) \qquad W = \frac{w-1}{w+1} \tag{2-52}$$

It then follows that

$$\frac{w-1}{w+1} = \left(\frac{z-1}{z+1}\right)^2 \qquad W = Z^2 \tag{2-53}$$

We introduce the functions

$$H_1(Z) = H\left(\frac{1+Z}{1-Z}\right) \qquad \text{and} \qquad V_1(W) = V\left(\frac{1+W}{1-W}\right) \tag{2-54}$$

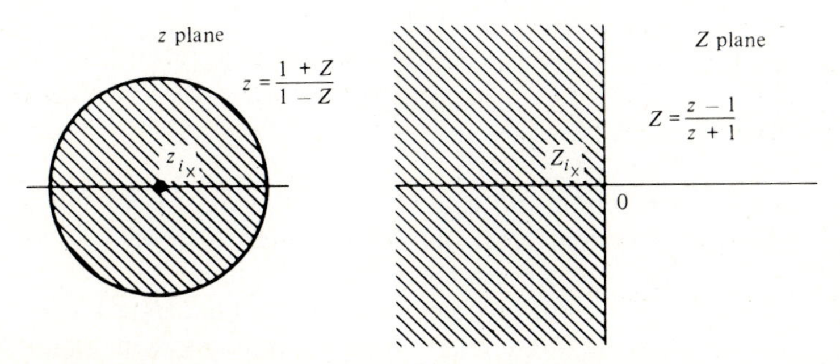

Figure 2-15

Clearly,
$$V_1(W) = H\left(\frac{1+Z}{1-Z}\right)H\left(\frac{1-Z}{1+Z}\right)$$

because $V(w) = H(z)H(1/z)$ [see Eq. (2-45)]. Therefore,

$$V_1(Z^2) = H_1(Z)H_1(-Z) \tag{2-55}$$

This equation is equivalent to (2-45) and can be used to determine $H_1(Z)$ from $V(\cos \omega T)$. This is done as follows:

1. Replacing $\cos \omega T$ with $(1 + W)/(1 - W)$, we obtain from $V(\cos \omega T)$ the rational function $V_1(W)$.
2. We find all roots W_i of $V_1(W)$.
3. We form the equation $Z^2 = W_i$ and find its roots $Z_i = \sqrt{W_i}$ and $-Z_i$, where by Z_i we mean the root with negative real part.
4. The zeros and poles of $H_1(Z)$ equal the numbers Z_i so obtained.
5. The unknown $H(z)$ is given by
$$H(z) = H_1\left(\frac{z-1}{z+1}\right)$$

Example 2-26 Find $H(z)$ such that

$$|H(e^{j\omega T})|^2 = \frac{5 - 4\cos \omega T}{10 - 6\cos \omega T} = V(\cos \omega T)$$

1. $V_1(W) = \dfrac{1 - 9W}{4 - 16W}$
2. One zero $W_1 = \frac{1}{9}$; one pole $W_2 = \frac{1}{4}$
3. $Z_1 = -\frac{1}{3} \qquad Z_2 = -\frac{1}{2}$
4. $H_1(Z) = A\dfrac{Z + \frac{1}{3}}{Z + \frac{1}{2}} = \dfrac{1 + 3Z}{2 + 4Z}$

   $A = \frac{3}{4} \qquad$ because $\qquad H_1{}^2(0) = V_1(0) = \frac{1}{4}$
5. $H(z) = \dfrac{1 + 3(z-1)/(z+1)}{2 + 4(z-1)/(z+1)} = \dfrac{2z - 1}{3z - 1}$

Notes:
1. Since $\operatorname{Re} Z_i < 0$ by construction and $z_i = (1 + Z_i)/(1 - Z_i)$, we conclude that $|z_i| < 1$. Hence, the system $H(z)$ obtained in step 5 of the above construction has all its zeros and poles inside the unit circle.
2. As we shall note in Sec. 4-3, the problem of determining $H_1(Z)$ from $V_1(Z^2)$ is the same as the problem of determining an analog system $H_a(\omega)$ from its amplitude $|H_a(\omega)|$.

All-pass systems We shall say that a finite-order system $H_0(z)$ is *all-pass* if it is stable and

$$|H_0(e^{j\omega T})| = 1 \qquad \text{for all } \omega \tag{2-56}$$

A real all-pass system has the following properties.
Property 1: Its zeros ζ_i and poles z_i are symmetrical with respect to the unit circle; i.e.,

$$\zeta_i = \frac{1}{z_i^*} \tag{2-57}$$

Proof If

$$H_0(z) = \frac{(zz_1^* - 1)\cdots(zz_m^* - 1)}{(z - z_1)\cdots(z - z_m)} \tag{2-58}$$

then $\qquad |H_0(e^{j\omega T})| = 1$

because

$$\left|\frac{e^{j\omega T}z_i^* - 1}{e^{j\omega T} - z_i}\right| = \left|\frac{z_i^* - e^{-j\omega T}}{z_i - e^{j\omega T}}\right| = 1 \tag{2-59}$$

Conversely, if $H_0(z)$ is all-pass, then it must be of the form (2-58). This follows by determining $H_0(z)$ from its amplitude $V(\cos \omega T) = 1$ [see Eqs. (2-47) and (2-48)] because the numerator of $V(w)$ equals its denominator (elaborate). Details are given in Sec. 4-3 for the analog case.

Circle of Apollonius Equation (2-59) has the following geometric interpretation: The unit circle $|z| = 1$ is the locus of points P such that the ratio $|PM_i|/|PN_i|$ of the distances from the points $1/z_i^*$ and z_i is constant (Fig. 2-16).
Property 2: The phase angle $\varphi(\omega)$ of an all-pass system is monotone decreasing from $\varphi(-\sigma)$ to $\varphi(-\sigma) - 2m\pi$.

Proof The angle α_i of the ratio PM_i/PN_i decreases monotonically from α_{i0} to $\alpha_{i0} - 2\pi$ as ω increases from $-\sigma$ to σ (see Fig. 2-16). The above follows because $\varphi(\omega) = \alpha_1 + \alpha_2 + \cdots + \alpha_m$.

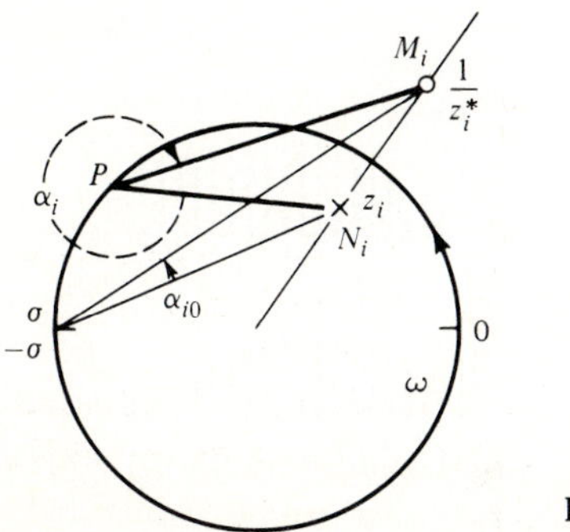

Figure 2-16

Property 3: Suppose that the input to an all-pass system $H_0(z)$ is a causal sequence $y[n]$. Denoting by $g[n]$ the resulting response, we conclude from the energy theorem (2-20) that

$$\sum_{n=0}^{\infty} g^2[n] = \frac{1}{2\sigma}\int_{-\sigma}^{\sigma} |Y(e^{j\omega T})H_0(e^{j\omega T})|^2\, d\omega = \frac{1}{2\sigma}\int_{-\sigma}^{\sigma} |Y(e^{j\omega T})|^2\, d\omega = \sum_{n=0}^{\infty} y^2[n] \tag{2-60}$$

because the z transform of $g[n]$ equals $Y(z)H_0(z)$.

We have, thus, shown that the energy of the output $g[n]$ of an all-pass system equals the energy of the input $y[n]$. The following is a stronger result:

Corollary For any n_0,

$$\sum_{n=0}^{n_0} y^2[n] \geq \sum_{n=0}^{n_0} g^2[n] \tag{2-61}$$

Proof We form the sequence

$$y_1[n] = \begin{cases} y[n] & n \leq n_0 \\ 0 & n > n_0 \end{cases}$$

(Fig. 2-17). Denoting by $g_1[n]$ the response of the all-pass system $H_0(z)$ to $y_1[n]$, we conclude from the causality of the system that

$$g_1[n] = g[n] \qquad \text{for } n \leq n_0$$

Hence [see (2-60)],

$$\sum_{n=0}^{n_0} y^2[n] = \sum_{n=0}^{\infty} y_1{}^2[n] = \sum_{n=0}^{\infty} g_1{}^2[n] = \sum_{n=0}^{n_0} g^2[n] + \sum_{n=n_0+1}^{\infty} g_1{}^2[n]$$

and (2-61) results.

Minimum-phase filters We shall say that a system $H_m(z)$ is *minimum phase* if all its zeros ζ_i and poles z_i are inside the unit circle:

$$|\zeta_i| < 1 \qquad |z_i| < 1 \tag{2-62}$$

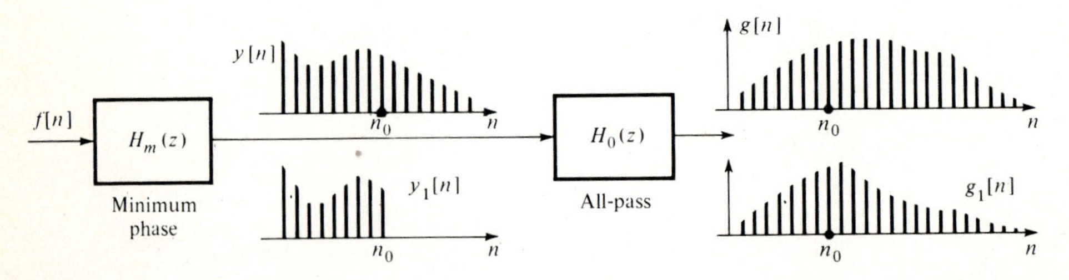

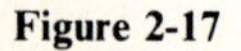
Figure 2-17

Minimum-phase systems have the following properties:

Property 1: If $H_m(z)$ is a minimum-phase system with phase $\varphi_m(\omega)$, and $H(\omega)$ is an arbitrary stable system with phase $\varphi(\omega)$ such that

$$|H(e^{j\omega T})| = |H_m(e^{j\omega T})| \tag{2-63}$$

then $$\varphi'(\omega) < \varphi'_m(\omega)$$

Proof With

$$\frac{H(z)}{H_m(z)} = H_0(z)$$

it follows from our assumption that $|H_0(e^{j\omega T})| = 1$. Furthermore, $H_0(z)$ has all its poles inside the unit circle because they are either poles of $H(z)$ or zeros of $H_m(z)$. Hence, $H_0(z)$ is all-pass and, as we have shown, its phase $\varphi_0(\omega)$ is monotone decreasing; that is, $\varphi'_0(\omega) < 0$. This completes the proof because $\varphi(\omega) = \varphi_m(\omega) + \varphi_0(\omega)$.

Property 2 (energy concentration): If the systems $H(z)$ and $H_m(z)$ have the same amplitude as in (2-63) and their responses to the same input $f[n]$ are $g[n]$ and $y[n]$, respectively, then for any n_0,

$$\sum_{n=0}^{n_0} y^2[n] \geq \sum_{n=0}^{n_0} g^2[n] \tag{2-64}$$

Proof The system $H(z)$ can be obtained by connecting $H_m(z)$ and $H_0(z)$ in cascade (Fig. 2-17). Thus, $g[n]$ is the output of the all-pass system $H_0(z)$ with input $y[n]$. Hence, (2-64) is a consequence of Eq. (2-61).

CHAPTER

THREE

FOURIER ANALYSIS

In this chapter, we introduce the basic properties of Fourier integrals[1,2] and Fourier series, including numerical techniques for their evaluation. Other aspects of the theory of Fourier transforms are developed throughout the book. The material is essentially self-contained; however, since it is presented with economy, prior familiarity with this subject, although not essential, is desirable.

In Secs. 3-1 and 3-2 we cover simple theorems and examples.

In Secs. 3-3 and 3-4 we deal with fast Fourier transforms (abbreviated FFT). In the literature of the past decade, FFT theory is often invoked in a variety of problems. In our view, there is no such theory. FFT is merely a fast algorithm for the numerical evaluation of Fourier integrals. We discuss the underlying concepts, stressing the relationship between Fourier integrals, Fourier series, and discrete Fourier series.

In Appendix 3-A, we present, for easy reference, various forms of the mean-value theorem.

In Appendix 3-B we prove Riemann's lemma and use it to relate the continuity properties of a signal to the asymptotic properties of its transform.

In Appendix 3-C we introduce the delta function as an integrand or as a limit and discuss its formal properties.

---

[1] E. C. Titchmarsh, "Introduction to the Theory of Fourier Integrals," Oxford University Press, New York, 1948.

[2] A. Papoulis, "The Fourier Integral and Its Applications," McGraw-Hill Book Company, New York, 1962.

3-1 FOURIER TRANSFORMS

The notation $f(t) \leftrightarrow F(\omega)$ will mean that $F(\omega)$ is the Fourier transform of $f(t)$:

$$F(\omega) = \int_{-\infty}^{\infty} f(t)e^{-j\omega t}\,dt \tag{3-1}$$

In this notation, it will be assumed that the above integral exists as a Cauchy principal value[1] for every real ω.

Example 3-1 If $f(t) = e^{-\alpha t}U(t)$, then (Fig. 3-1)

$$F(\omega) = \int_{0}^{\infty} e^{-\alpha t}e^{-j\omega t}\,dt = \frac{1}{\alpha + j\omega} \qquad \alpha > 0$$

The inversion formula The function $f(t)$ can be expressed in terms of $F(\omega)$:

$$f(t) = \frac{1}{2\pi}\int_{-\infty}^{\infty} F(\omega)e^{j\omega t}\,d\omega \tag{3-2}$$

This result was derived in Sec. 1-2 as a corollary of the identity (1-48). It can be also justified as follows:

We form the function

$$f_\sigma(t) = \frac{1}{2\pi}\int_{-\sigma}^{\sigma} F(\omega)e^{j\omega t}\,d\omega \tag{3-3}$$

We wish to show that if $F(\omega)$ is given by Eq. (3-1), then

$$f_\sigma(t) \underset{\sigma\to\infty}{\rightrightarrows} f(t) \tag{3-4}$$

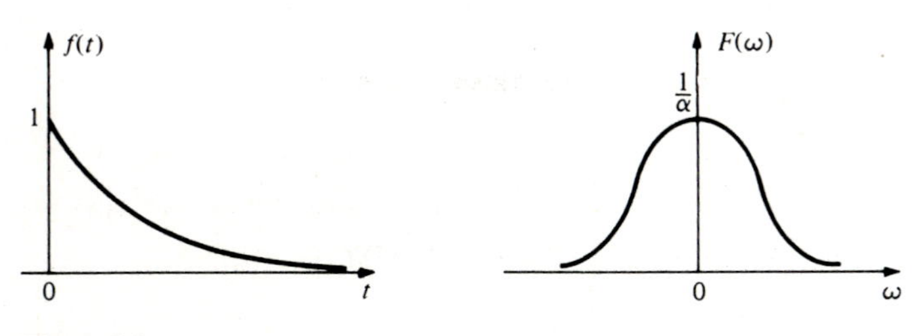

Figure 3-1

[1] The Cauchy principal value of an integral from $-\infty$ to ∞ is the limit

$$\int_{-\infty}^{\infty} x(t)\,dt = \lim_{A\to\infty}\int_{-A}^{A} x(t)\,dt$$

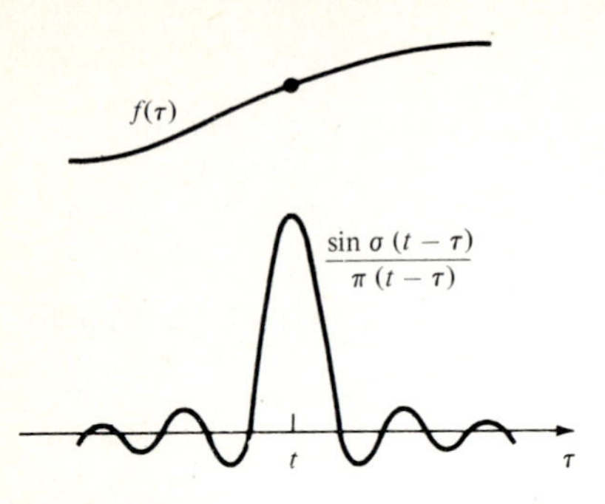

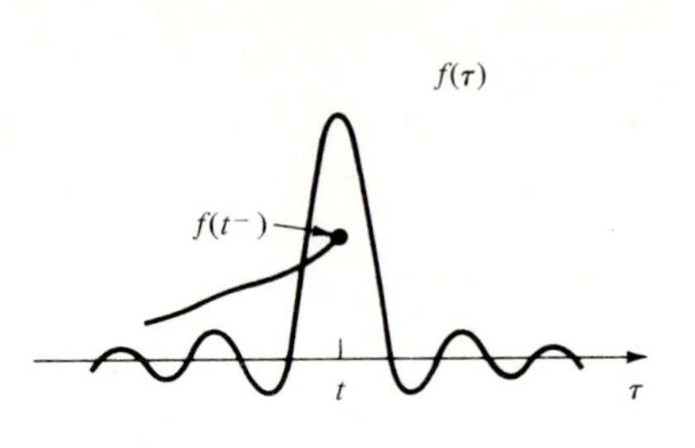

Figure 3-2

Inserting (3-1) into (3-3), we have

$$f_\sigma(t) = \frac{1}{2\pi}\int_{-\sigma}^{\sigma} e^{j\omega t}\int_{-\infty}^{\infty} f(\tau)e^{-j\omega\tau}\,d\tau\,d\omega = \frac{1}{2\pi}\int_{-\infty}^{\infty} f(\tau)\int_{-\sigma}^{\sigma} e^{j\omega(t-\tau)}\,d\omega\,d\tau$$

Thus,
$$f_\sigma(t) = \int_{-\infty}^{\infty} f(\tau)\frac{\sin\sigma(t-\tau)}{\pi(t-\tau)}\,d\tau \tag{3-5}$$

The above integral equals the convolution of $f(t)$ with the *Fourier-integral* kernal $\sin \sigma t/\pi t$, and it tends to $f(t)$ as $\sigma \to \infty$ at every point of continuity of $f(t)$ [see Eq. (3-C-15)]. If $f(t)$ is discontinuous at a point t, then (Fig. 3-2)

$$f_\sigma(t) \underset{\sigma\to\infty}{\to} \frac{f(t^+) + f(t^-)}{2} \tag{3-6}$$

Note: If $f(t)$ is discontinuous at $t = t_0$, then, in the vicinity of t_0, the function $f_\sigma(t)$ is not close to $f(t)$ no matter how large σ is. As t approaches t_0, $f_\sigma(t)$ oscillates rapidly (*Gibbs' phenomenon*). However, for large σ, the ripple is concentrated near t_0, passing any point $t \neq t_0$.

Example 3-2 If $f(t)$ is a pulse $p_a(t)$ as in Fig. 3-3, then

$$F(\omega) = \int_{-a}^{a} e^{j\omega t}\,dt = \frac{2\sin a\omega}{\omega}$$

$$f_\sigma(t) = \int_{-a}^{a} \frac{\sin\sigma(t-\tau)}{\pi(t-\tau)}\,d\tau = \frac{1}{\pi}\{\mathrm{Si}[\sigma(t+a)] - \mathrm{Si}[\sigma(t-a)]\}$$

where
$$\mathrm{Si}(t) = \int_0^t \frac{\sin\tau}{\tau}\,d\tau$$

is the *sine integral.*

In general, the functions $f(t)$ and $F(\omega)$ are complex:

$$f(t) = f_1(t) + jf_2(t) \qquad F(\omega) = R(\omega) + jX(\omega)$$

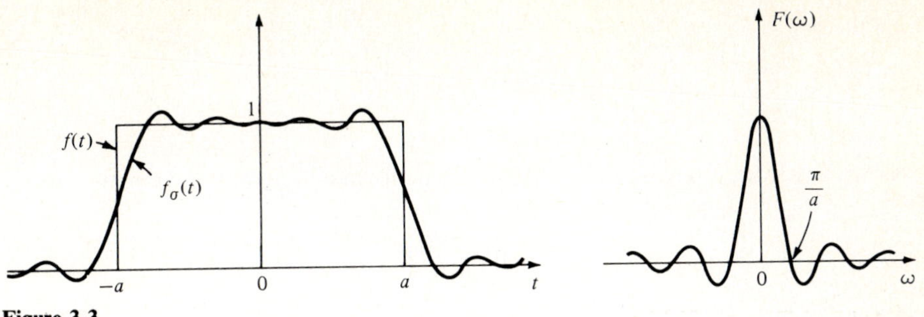

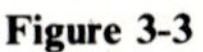

Figure 3-3

Since $e^{j\omega t} = \cos \omega t + j \sin \omega t$, we conclude from Eqs. (3-1) and (3-2), equating real and imaginary parts, that

$$R(\omega) = \int_{-\infty}^{\infty} [f_1(t) \cos \omega t + f_2(t) \sin \omega t]\, dt$$
$$X(\omega) = \int_{-\infty}^{\infty} [f_2(t) \cos \omega t - f_1(t) \sin \omega t]\, dt \tag{3-7}$$

$$f_1(t) = \frac{1}{2\pi} \int_{-\infty}^{\infty} [R(\omega) \cos \omega t - X(\omega) \sin \omega t]\, d\omega$$
$$f_2(t) = \frac{1}{2\pi} \int_{-\infty}^{\infty} [R(\omega) \sin \omega t + X(\omega) \cos \omega t]\, d\omega \tag{3-8}$$

Real signals If $f(t)$ is real, then (3-7) yields

$$R(\omega) = \int_{-\infty}^{\infty} f(t) \cos \omega t\, dt \qquad X(\omega) = -\int_{-\infty}^{\infty} f(t) \sin \omega t\, dt \tag{3-9}$$

From the above, it follows that $R(-\omega) = R(\omega)$ and $X(-\omega) = -X(\omega)$. Hence,

$$F^*(\omega) = F(-\omega) \tag{3-10}$$

If $f(-t) = f(t)$, then $X(\omega) = 0$. Hence, the transform of a real, even function is real:

$$F(\omega) = R(\omega) = \int_{-\infty}^{\infty} f(t) \cos \omega t\, dt = 2 \int_{0}^{\infty} f(t) \cos \omega t\, dt \tag{3-11}$$

If $f(-t) = -f(t)$, then $R(\omega) = 0$. Hence, $F(\omega)$ is purely imaginary and is given by

$$F(\omega) = jX(\omega) = -j \int_{-\infty}^{\infty} f(t) \sin \omega t\, dt = -2j \int_{0}^{\infty} f(t) \sin \omega t\, dt$$

Example 3-3 The function $1/t$ is odd; hence,

$$F(\omega) = -j \int_{-\infty}^{\infty} \frac{\sin \omega t}{t}\, dt = \begin{cases} -j\pi & \omega > 0 \\ j\pi & \omega < 0 \end{cases}$$

yielding the pair

$$\frac{1}{\pi t} \leftrightarrow -j \operatorname{sgn} \omega \tag{3-12}$$

As we see from Eq. (3-1), the Fourier transform of $f(-t)$ is given by

$$\int_{-\infty}^{\infty} f(-t)e^{-j\omega t}\,dt = \int_{-\infty}^{\infty} f(t)e^{j\omega t}\,dt = F(-\omega)$$

But $F(-\omega) = R(\omega) - jX(\omega)$ because $f(t)$ is real; hence,

$$f(t) \leftrightarrow R(\omega) + jX(\omega) \qquad f(-t) \leftrightarrow R(\omega) - jX(\omega) \tag{3-13}$$

With

$$f_e(t) = \frac{f(t) + f(-t)}{2} \qquad f_o(t) = \frac{f(t) - f(-t)}{2}$$

the even and odd parts of $f(t)$, it follows from (3-13) that

$$f_e(t) \leftrightarrow R(\omega) \qquad f_o(t) \leftrightarrow jX(\omega) \tag{3-14}$$

If $f(t) = 0$ for $t < 0$, then (Fig. 3-4)

$$f(t) = 2f_e(t) = 2f_o(t) \qquad \text{for } t > 0$$

From the above and (3-14), it follows that a real, causal function is uniquely determined in terms of $R(\omega)$ or $X(\omega)$:

$$f(t) = \frac{2}{\pi}\int_0^{\infty} R(\omega)\cos\omega t\,d\omega = -\frac{2}{\pi}\int_0^{\infty} X(\omega)\sin\omega t\,d\omega \qquad t > 0 \tag{3-15}$$

Example 3-4 From the pair (see Example 3-1)

$$e^{-\alpha t}U(t) \leftrightarrow \frac{1}{\alpha + j\omega} = \frac{\alpha}{\alpha^2 + \omega^2} - j\frac{\omega}{\alpha^2 + \omega^2}$$

and (3-14) it follows that

$$e^{-\alpha|t|} \leftrightarrow \frac{2\alpha}{\alpha^2 + \omega^2} \tag{3-16}$$

because $e^{-\alpha|t|}$ is the even part of $2e^{-\alpha t}U(t)$.

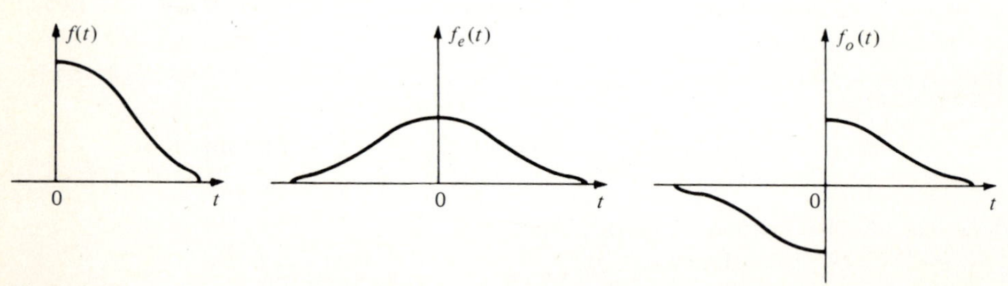

Figure 3-4

Theorems

If $f(t) \leftrightarrow F(\omega)$, then

1. Symmetry

$$F(t) \leftrightarrow 2\pi f(-\omega) \tag{3-17}$$

Proof From Eq. (3-2) it follows that

$$2\pi f(-\omega) = \int_{-\infty}^{\infty} F(x) e^{-j\omega x}\, dx$$

Example 3-5 Applying (3-17) to the pair in (3-16), we obtain

$$\frac{2\alpha}{\alpha^2 + t^2} \leftrightarrow 2\pi e^{-\alpha|\omega|}$$

2. Conjugate functions

$$f^*(t) \leftrightarrow F^*(-\omega) \tag{3-18}$$

Corollary If $f(t)$ is real, then $f(t) = f^*(t)$. Hence, $F(\omega) = F^*(-\omega)$.

3. Scaling For any real a,

$$f(at) \leftrightarrow \frac{1}{|a|} F\left(\frac{\omega}{a}\right) \tag{3-19}$$

Proof

$$\int_{-\infty}^{\infty} f(at) e^{-j\omega t}\, dt = \frac{1}{|a|} \int_{-\infty}^{\infty} f(x) e^{-j\omega x/a}\, dx = \frac{1}{|a|} F\left(\frac{\omega}{a}\right)$$

4. Shifting For any real a,

$$f(t-a) \leftrightarrow e^{-ja\omega} F(\omega) \qquad e^{jat} f(t) \leftrightarrow F(\omega - a) \tag{3-20}$$

Proof

$$\int_{-\infty}^{\infty} f(t-a) e^{-j\omega t}\, dt = \int_{-\infty}^{\infty} f(x) e^{-j\omega(x+a)}\, dx = e^{-ja\omega} F(\omega)$$

5. Modulation

$$f(t) \cos \omega_0 t \leftrightarrow \frac{1}{2}[F(\omega + \omega_0) + F(\omega - \omega_0)] \tag{3-21}$$

Proof It follows from (3-20) because

$$\cos \omega_0 t = \tfrac{1}{2}(e^{j\omega_0 t} + e^{-j\omega_0 t})$$

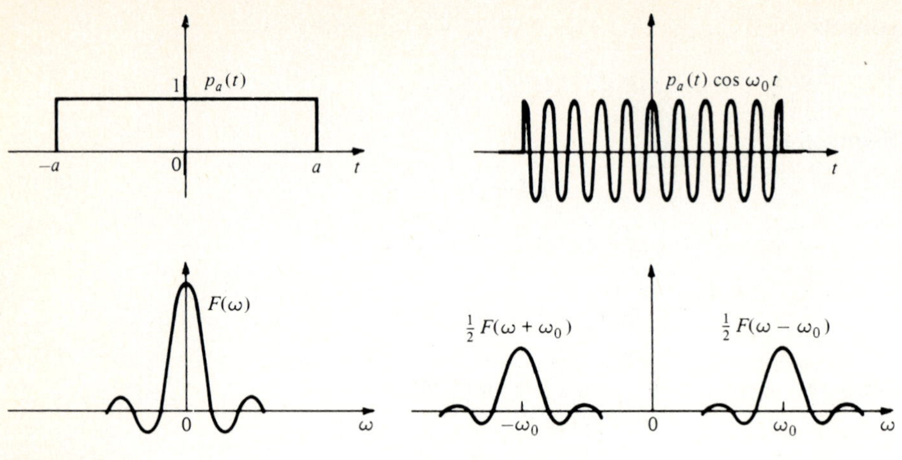

Figure 3-5

Example 3-6 From the pair (Fig. 3-5)

$$p_a(t) \leftrightarrow \frac{2 \sin a\omega}{\omega} \tag{3-22}$$

and (3-21) it follows that

$$p_a(t) \cos \omega_0 t \leftrightarrow \frac{\sin a(\omega - \omega_0)}{\omega - \omega_0} + \frac{\sin a(\omega + \omega_0)}{\omega + \omega_0} \tag{3-23}$$

Example 3-7 The pair

$$p_a(t)\left(1 + \cos \frac{\pi}{a} t\right) \leftrightarrow \frac{2\pi^2 \sin a\omega}{\omega(\pi^2 - a^2\omega^2)} \tag{3-24}$$

of Fig. 3-6 is the sum of (3-22) and (3-23) if $\omega_0 = \pi/a$.

6. Derivatives

$$(-jt)^n f(t) \leftrightarrow F^{(n)}(\omega) \qquad f^{(n)}(t) \leftrightarrow (j\omega)^n F(\omega) \tag{3-25}$$

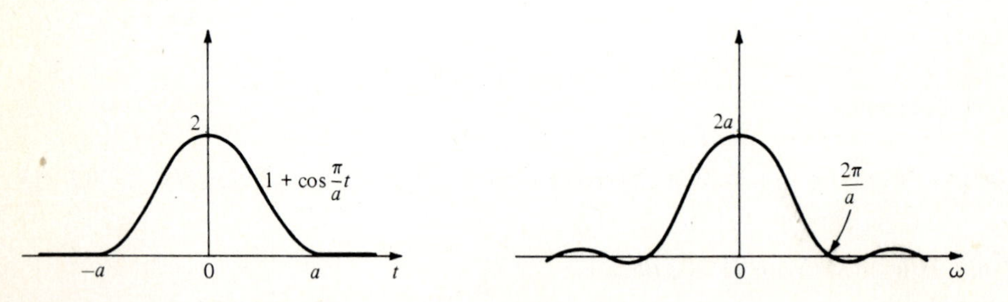

Figure 3-6

Proof Differentiating Eqs. (3-1) and (3-2), we obtain

$$F^{(n)}(\omega) = \int_{-\infty}^{\infty} (-jt)^n f(t) e^{-j\omega t}\, dt \qquad f^{(n)}(t) = \frac{1}{2\pi}\int_{-\infty}^{\infty} (j\omega)^n F(\omega) e^{j\omega t}\, d\omega$$

Example 3-8 From the pair $e^{-\alpha t}U(t) \leftrightarrow 1/(\alpha + j\omega)$ and (3-25), it follows that

$$t^n e^{-\alpha t} U(t) \leftrightarrow \frac{n!}{(\alpha + j\omega)^{n+1}} \tag{3-26}$$

and since $|t|e^{-\alpha|t|}$ is the even part of $2te^{-\alpha t}U(t)$, we conclude from (3-14) that

$$|t|e^{-\alpha|t|} \leftrightarrow \text{Re}\,\frac{2}{(\alpha + j\omega)^2} = \frac{2(\alpha^2 - \omega^2)}{(\alpha^2 + \omega^2)^2} \tag{3-27}$$

7. Convolution theorem If $f_1(t) \leftrightarrow F_1(\omega)$ and $f_2(t) \leftrightarrow F_2(\omega)$, then

$$f_1(t) * f_2(t) \leftrightarrow F_1(\omega) F_2(\omega) \tag{3-28}$$

Proof This important theorem was derived in (1-46) in the context of linear systems. We give below a direct proof: From the definition of convolution and (3-1), it follows that the Fourier transform of $f_1(t) * f_2(t)$ equals

$$\int_{-\infty}^{\infty} e^{-j\omega t}\left[\int_{-\infty}^{\infty} f_1(\tau) f_2(t - \tau)\, d\tau\right] dt$$

Setting $t = \tau + x$ and changing the order of integration, we obtain

$$\int_{-\infty}^{\infty} f_1(\tau) \int_{-\infty}^{\infty} e^{-j\omega(\tau + x)} f_2(x)\, dx\, d\tau = \int_{-\infty}^{\infty} f_1(\tau) e^{-j\omega\tau}\, d\tau \int_{-\infty}^{\infty} f_2(x) e^{-j\omega x}\, dx$$

and (3-28) results.

Note: Applying the inversion formula (3-2) to the pair in (3-28), we obtain

$$\int_{-\infty}^{\infty} f_1(\tau) f_2(t - \tau)\, d\tau = \frac{1}{2\pi}\int_{-\infty}^{\infty} F_1(\omega) F_2(\omega) e^{j\omega t}\, d\omega \tag{3-29}$$

Frequency convolution Reasoning as in the proof of (3-28), we can show that (elaborate)

$$f_1(t) f_2(t) \leftrightarrow \frac{1}{2\pi} F_1(\omega) * F_2(\omega) \tag{3-30}$$

Example 3-9 If $p_a(t)$ is a rectangular pulse, then $p_a(t) * p_a(t) = 2aq_{2a}(t)$, where

$$q_c(t) = \begin{cases} 1 - \dfrac{|t|}{c} & |t| < c \\ 0 & |t| > c \end{cases}$$

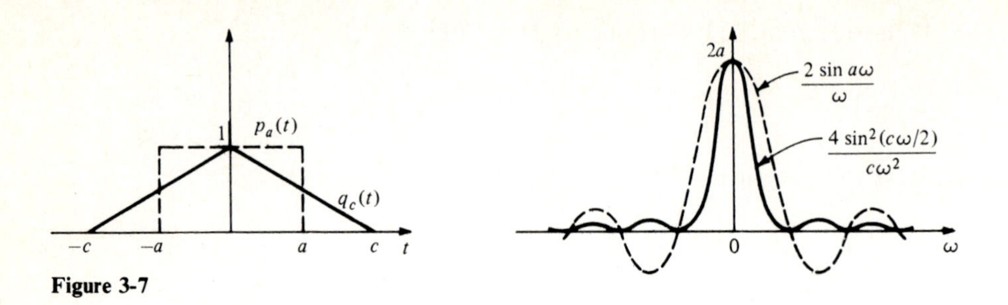

Figure 3-7

is a triangle as in Fig. 3-7. Since $p_a(t) \leftrightarrow 2 \sin a\omega/\omega$, it follows from (3-28) that $2aq_{2a}(t) \leftrightarrow 4 \sin^2 a\omega/\omega^2$. With $2a = c$, the above yields the pair

$$q_c(t) \leftrightarrow \frac{4 \sin^2 (c\omega/2)}{c\omega^2} \tag{3-31}$$

and, with the symmetry theorem (3-17), the pair

$$\frac{2 \sin^2 (ct/2)}{\pi c t^2} \leftrightarrow q_c(\omega) \tag{3-32}$$

Example 3-10 Find the inverse transform $\rho(t)$ of the *energy spectrum* $|F(\omega)|^2$ of $f(t)$:

$$\rho(t) \leftrightarrow |F(\omega)|^2$$

Since $|F(\omega)|^2 = F(\omega)F^*(\omega)$ and the inverse of $F^*(\omega)$ equals $f^*(-t)$, it follows from (3-28) that

$$\rho(t) = f(t) * f^*(-t) = \int_{-\infty}^{\infty} f(t-\tau)f^*(-\tau)\, d\tau = \int_{-\infty}^{\infty} f(t+\alpha)f^*(\alpha)\, d\alpha$$

The function $\rho(t)$ is known as the *autocorrelation* of the signal $f(t)$.

Windows If $w(t)$ is a signal such that $w(t) = 0$ for $|t| > T$, then the integral

$$F_w(\omega) = \int_{-T}^{T} f(t)w(t)e^{-j\omega t}\, dt \tag{3-33}$$

is the Fourier transform of the product $f_w(t) = f(t)w(t)$. With

$$W(\omega) = \int_{-T}^{T} w(t)e^{-j\omega t}\, dt$$

it follows from (3-30) that

$$F_w(\omega) = \frac{1}{2\pi}\int_{-\infty}^{\infty} F(\omega - y)W(y)\, dy \tag{3-34}$$

Example 3-11
(*a*) If $w(t) = p_T(t)$ is a pulse, then $W(\omega) = 2 \sin T\omega/\omega$. Hence,

$$F_w(\omega) = \int_{-T}^{T} f(t)e^{-j\omega t}\, dt = \int_{-\infty}^{\infty} F(\omega - y)\frac{\sin Ty}{\pi y}\, dy$$

(*b*) If $w(t)$ is a triangle, then [see (3-31)]

$$F_w(\omega) = \int_{-T}^{T} f(t)\left(1 - \frac{|t|}{T}\right)e^{-j\omega t}\,dt = \int_{-\infty}^{\infty} F(\omega - y)\,\frac{2\sin^2(Ty/2)}{\pi T y^2}\,dy$$

8. Parseval's formula If $y_1(t) \leftrightarrow Y_1(\omega)$ and $y_2(t) \leftrightarrow Y_2(\omega)$, then

$$\int_{-\infty}^{\infty} y_1(t)y_2^*(t)\,dt = \frac{1}{2\pi}\int_{-\infty}^{\infty} Y_1(\omega)Y_2^*(\omega)\,d\omega \tag{3-35}$$

Proof Setting $t = 0$ in (3-29), we obtain

$$\int_{-\infty}^{\infty} f_1(\tau)f_2(-\tau)\,d\tau = \frac{1}{2\pi}\int_{-\infty}^{\infty} F_1(\omega)F_2(\omega)\,d\omega \tag{3-36}$$

With $y_1(t) = f_1(t)$ and $y_2^*(t) = f_2(-t)$, we have $Y_1(\omega) = F_1(\omega)$ and $Y_2(\omega) = F_2^*(\omega)$. Inserting into (3-36), we obtain Eq. (3-35).

Energy theorem

$$\int_{-\infty}^{\infty} |y(t)|^2\,dt = \frac{1}{2\pi}\int_{-\infty}^{\infty} |Y(\omega)|^2\,d\omega \tag{3-37}$$

Proof It follows from (3-35) with $y_1(t) = y_2(t) = y(t)$.

Note: The Fourier integral is an operator mapping the function $f(t)$ into the function $F(\omega)$. This operator is obviously linear and, as we see from Parseval's theorem, it preserves inner products if they are defined as in Eq. (3-35). (See also Sec. 5-2.)

Example 3-12 From the pair

$$\frac{\sin at}{t} \leftrightarrow \pi p_a(\omega)$$

and (3-35) it follows that

$$\int_{-\infty}^{\infty} \frac{\sin^2 at}{t^2}\,dt = \frac{1}{2\pi}\int_{-a}^{a} \pi^2\,d\omega = a\pi \tag{3-38}$$

Example 3-13 If $f(t) \leftrightarrow F(\omega) = A(\omega)e^{j\varphi(\omega)}$, then

$$\int_{-\infty}^{\infty} t^2|f(t)|^2\,dt = \frac{1}{2\pi}\int_{-\infty}^{\infty} \{[A'(\omega)]^2 + A^2(\omega)[\varphi'(\omega)]^2\}\,d\omega \tag{3-39}$$

$$\int_{-\infty}^{\infty} t|f(t)|^2\,dt = \frac{-1}{2\pi}\int_{-\infty}^{\infty} A^2(\omega)\varphi'(\omega)\,d\omega \tag{3-40}$$

Proof Since [see (3-25)]

$$tf(t) \leftrightarrow jF'(\omega) = j[A'(\omega) + jA(\omega)\varphi'(\omega)]e^{j\varphi(\omega)} \tag{3-41}$$

Eq. (3-39) follows from (3-37) with $y(t) = tf(t)$. With $y_1(t) = tf(t)$ and $y_2(t) = f(t)$, (3-35) yields

$$\int_{-\infty}^{\infty} tf(t)f^*(t)\,dt = \frac{1}{2\pi}\int_{-\infty}^{\infty} [-A(\omega)\varphi'(\omega) + jA'(\omega)]e^{j\varphi(\omega)}A(\omega)e^{-j\varphi(\omega)}\,d\omega$$

and (3-40) results because the left side is real.

9. Moment theorem With

$$m_n = \int_{-\infty}^{\infty} t^n f(t)\,dt \tag{3-42}$$

the moments of $f(t)$, and $F^{(n)}(0)$ the derivatives of its transform at the origin, we maintain that

$$F^{(n)}(0) = (-j)^n m_n \tag{3-43}$$

Proof Expanding the exponential in (3-1) and integrating termwise, we have

$$F(\omega) = \int_{-\infty}^{\infty} f(t)\left[\sum_{n=0}^{\infty} \frac{(-j\omega t)^n}{n!}\right] dt = \sum_{n=0}^{\infty} (-j)^n m_n \frac{\omega^n}{n!}$$

and (3-43) results because the last sum is the Maclaurin expansion of $F(\omega)$. Hence, the coefficient $(-j)^n m_n$ of $\omega^n/n!$ equals $F^{(n)}(0)$.

Center of gravity Clearly, $F(0) = A(0) = m_0$ is the area of $f(t)$. The ratio $\eta = m_1/m_0$ is the center of gravity of $f(t)$. From Eq. (3-43), it follows that, if the signal $f(t)$ is real, then

$$\eta = -\varphi'(0) \tag{3-44}$$

Proof For real signals, $A(\omega)$ is even and $\varphi(\omega)$ is odd [see (3-10)]. Hence, $A'(0) = 0$ and $\varphi(0) = 0$, and (3-41) yields $F'(0) = jA(0)\varphi'(0)$ from which (3-44) follows because $F'(0) = -jm_1$.

3-2 LINE SPECTRA AND FOURIER SERIES

The properties of the delta function $\delta(t)$ are based on the identity

$$\int_{-\infty}^{\infty} \delta(t)\varphi(t)\,dt = \varphi(0) \tag{3-45}$$

valid for any $\varphi(t)$ continuous at $t = 0$. From the above, it follows that

$$\int_{-\infty}^{\infty} \delta(t)e^{-j\omega t}\,dt = 1$$

Hence, the Fourier transform of $\delta(t)$ equals 1 (Fig. 3-8):

$$\delta(t) \leftrightarrow 1 \tag{3-46}$$

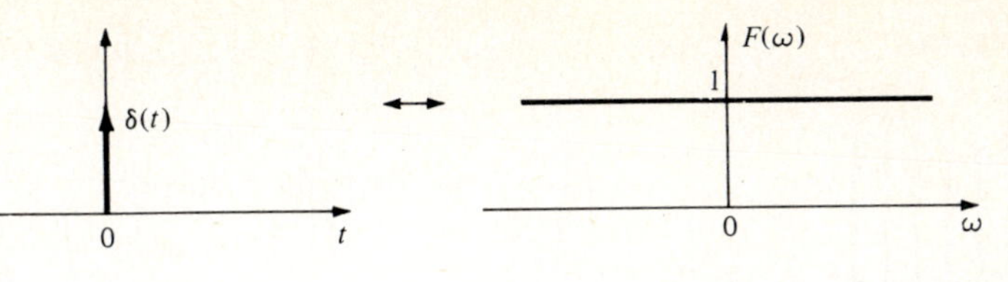

Figure 3-8

Inserting the above into the inversion formula (3-2), we obtain the identity

$$\delta(t) = \frac{1}{2\pi}\int_{-\infty}^{\infty} e^{j\omega t}\,d\omega \tag{3-47}$$

If $f(t) = 1$, then

$$F(\omega) = \int_{-\infty}^{\infty} e^{-j\omega t}\,dt = 2\pi\,\delta(\omega)$$

This follows from Eq. (3-47) with t and ω interchanged. Hence,

$$1 \leftrightarrow 2\pi\,\delta(\omega) \tag{3-48}$$

Example 3-14 The step function $U(t)$ can be written as a sum

$$U(t) = \tfrac{1}{2} + \tfrac{1}{2}\,\text{sgn}\,t$$

From (3-12) and the symmetry theorem (3-17), it follows that

$$\text{sgn}\,t \leftrightarrow \frac{2}{j\omega} \tag{3-49}$$

Hence [see (3-48)],

$$U(t) \leftrightarrow \pi\,\delta(\omega) + \frac{1}{j\omega} \tag{3-50}$$

Corollary If $f(t) \leftrightarrow F(\omega)$, then

$$\int_{-\infty}^{t} f(\tau)\,d\tau = f(t) * U(t) \leftrightarrow \pi F(0)\,\delta(\omega) + \frac{F(\omega)}{j\omega} \tag{3-51}$$

Proof $f(t) * U(t) \leftrightarrow F(\omega)[\pi\,\delta(\omega) + 1/j\omega]$ [see (3-50) and (3-28)], and (3-51) results because $F(\omega)\,\delta(\omega) = F(0)\,\delta(\omega)$.

From Eqs. (3-45) and (3-48) and the shifting theorem, it follows that

$$\delta(t - a) \leftrightarrow e^{-ja\omega} \qquad e^{jat} \leftrightarrow 2\pi\,\delta(\omega - a) \tag{3-52}$$

Hence, $$\cos\omega_0 t = \tfrac{1}{2}(e^{j\omega_0 t} + e^{-j\omega_0 t}) \leftrightarrow \pi\,\delta(\omega - \omega_0) + \pi\delta(\omega + \omega_0) \tag{3-53}$$

The first pair in (3-52) leads to the useful identity

$$\delta(t - a) * f(t) = f(t - a) \tag{3-54}$$

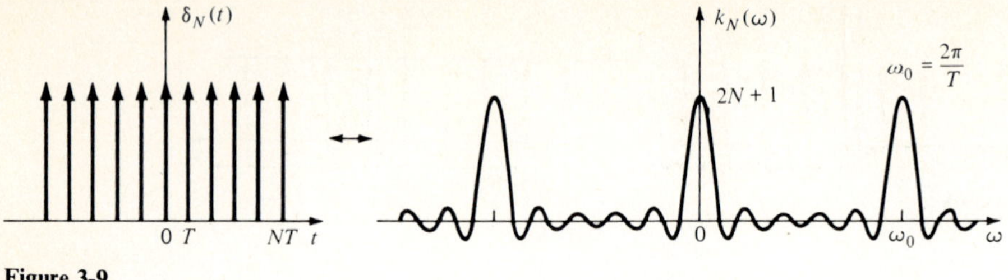

Figure 3-9

Indeed, the transform of the left side equals $e^{-ja\omega}F(\omega)$ [see (3-28)], and (3-54) follows from (3-20).

Example 3-15 Consider the sum

$$\delta_N(t) = \sum_{n=-N}^{N} \delta(t + nT) \tag{3-55}$$

consisting of $2N + 1$ impulses as in Fig. 3-9. Denoting by $k_N(\omega)$ its Fourier transform, we conclude from (3-52) that

$$k_N(\omega) = \sum_{n=-N}^{N} e^{jnT\omega} = \frac{\sin (N + \frac{1}{2})T\omega}{\sin (T\omega/2)} \tag{3-56}$$

Clearly, $k_N(\omega)$ is a periodic function of ω with period $\omega_0 = 2\pi/T$, and

$$\int_{-\omega_0/2}^{\omega_0/2} k_N(\omega)\, d\omega = \omega_0 \qquad \text{because} \qquad \int_{-\omega_0/2}^{\omega_0/2} e^{jnT\omega}\, d\omega = \begin{cases} 0 & n \neq 0 \\ \omega_0 & n = 0 \end{cases} \tag{3-57}$$

Example 3-16 Given a function $f_0(t)$ with Fourier transform $F_0(\omega)$, we form the sum (Fig. 3-10)

$$f_N(t) = \sum_{n=-N}^{N} f_0(t + nT) \tag{3-58}$$

consisting of $f_0(t)$ and its displacements. We maintain that the Fourier transform $F_N(\omega)$ of $f_N(t)$ is given by

$$F_N(\omega) = F_0(\omega)k_N(\omega) \tag{3-59}$$

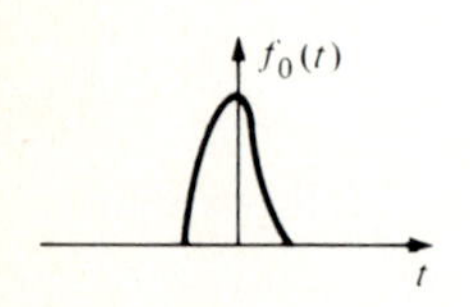

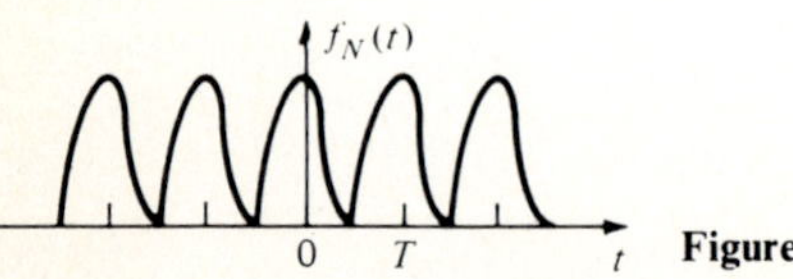

Figure 3-10

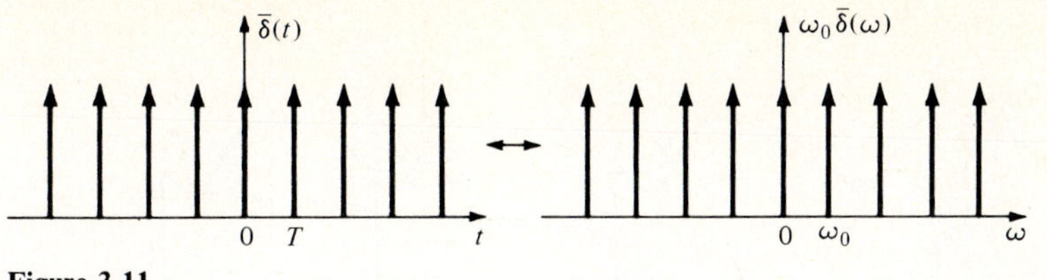

Figure 3-11

Indeed, since $\delta(t + nT) * f_0(t) = f_0(t + nT)$, we conclude that

$$f_N(t) = \delta_N(t) * f_0(t)$$

and Eq. (3-59) follows from the convolution theorem (3-28) because the transform of $\delta_N(t)$ equals $k_N(\omega)$.

Note: From the pair $\delta_N(t) \leftrightarrow k_N(\omega)$ of Example 3-15, it follows with $N \to \infty$ that

$$\bar{\delta}(t) \leftrightarrow \omega_0 \, \bar{\delta}(\omega) \tag{3-60}$$

where $\bar{\delta}(t)$ and $\bar{\delta}(\omega)$ are two impulse trains as in Fig. 3-11:

$$\bar{\delta}(t) = \sum_{n=-\infty}^{\infty} \delta(t + nT) \qquad \bar{\delta}(\omega) = \sum_{n=-\infty}^{\infty} \delta(\omega - n\omega_0) \tag{3-61}$$

Indeed, $\bar{\delta}(t)$ is obviously the limit of $\delta_N(t)$. Furthermore,

$$k_N(\omega) \to \omega_0 \, \bar{\delta}(\omega) \qquad N \to \infty \tag{3-62}$$

because $k_N(\omega)$ is a periodic function with period ω_0 and, in the interval $(-\omega_0/2, \omega_0/2)$, it tends to $\omega_0 \, \delta(\omega)$; or, equivalently, it satisfies the equation [see (3-C-11)]

$$\lim_{N \to \infty} \int_{-\omega_0/2}^{\omega_0/2} k_N(\omega)\varphi(\omega) \, d\omega = \omega_0 \, \varphi(0) \tag{3-63}$$

This follows by reasoning as in the proof of the Riemann lemma (3-B-4).

Fourier Series

Using the pair (3-60), we shall express the Fourier transform $F(\omega)$ of a periodic function $f(t)$ in terms of the Fourier transform

$$F_0(\omega) = \int_{-T/2}^{T/2} f(t)e^{-j\omega t} \, dt \tag{3-64}$$

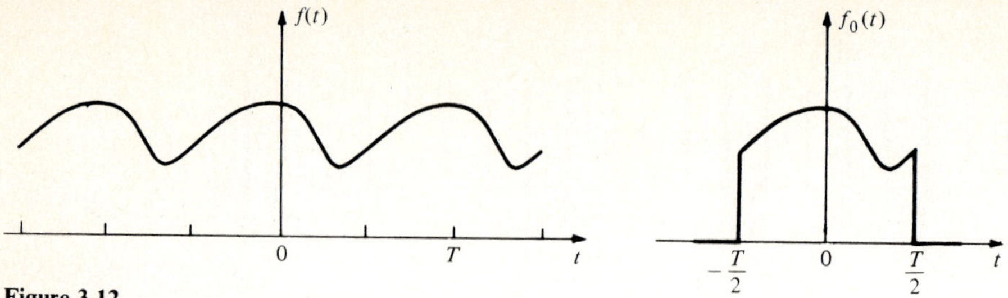

Figure 3-12

of the segment (Fig. 3-12)

$$f_0(t) = \begin{cases} f(t) & |t| < \dfrac{T}{2} \\ 0 & |t| > \dfrac{T}{2} \end{cases} \tag{3-65}$$

of $f(t)$.

It is easy to see that

$$f(t) = \sum_{n=-\infty}^{\infty} f_0(t + nT) = \bar{\delta}(t) * f_0(t) \tag{3-66}$$

Taking transforms of both sides, we conclude with (3-60) and the convolution theorem that

$$F(\omega) = \omega_0 \, \bar{\delta}(\omega) \, F_0(\omega) = \omega_0 \, F_0(\omega) \sum_{n=-\infty}^{\infty} \delta(\omega - n\omega_0) \tag{3-67}$$

Hence,

$$F(\omega) = \omega_0 \sum_{n=-\infty}^{\infty} F_0(n\omega_0) \, \delta(\omega - n\omega_0) \tag{3-68}$$

because [see Eq. (3-C-6)] $F_0(\omega) \, \delta(\omega - n\omega_0) = F_0(n\omega_0) \, \delta(\omega - n\omega_0)$.

We have, thus, shown that the Fourier transform $F(\omega)$ of a periodic function $f(t)$ with period T consists of a sequence of impulses at a distance $\omega_0 = 2\pi/T$ apart. This result leads to the familiar Fourier-series expansion of $f(t)$. Indeed, with

$$a_n = \frac{1}{T} F_0(n\omega_0) = \frac{1}{T} \int_{-T/2}^{T/2} f(t) e^{-jn\omega_0 t} \, dt \tag{3-69}$$

we obtain, taking the inverse transforms of both sides of Eq. (3-68),

$$f(t) = \sum_{n=-\infty}^{\infty} a_n e^{jn\omega_0 t} \qquad \omega_0 = \frac{2\pi}{T} \tag{3-70}$$

because the inverse transform of $\omega_0 \, \delta(\omega - n\omega_0)$ equals $e^{jn\omega_0 t}/T$.

As we show in the next example, to find a_n, it is sometimes simplest to determine $F_0(\omega)$ and use its sample values.

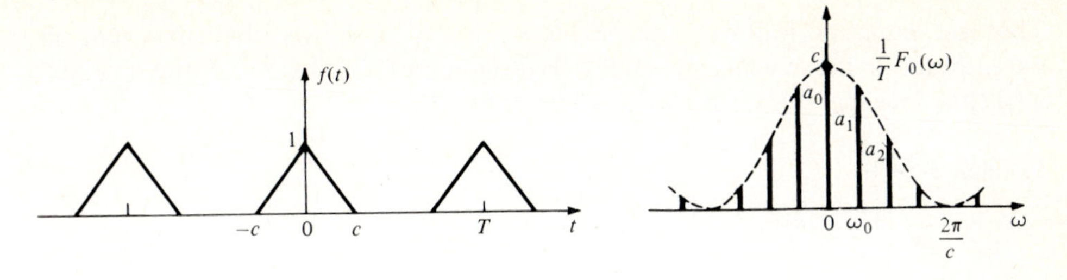

Figure 3-13

Example 3-17 The function $f(t)$ in Fig. 3-13 consists of a train of triangles. Clearly [see (3-31)],

$$f_0(t) = q_c(t) \qquad F_0(\omega) = \frac{4 \sin^2 (c\omega/2)}{c\omega^2}$$

Hence,

$$a_n = \frac{\omega_0}{2\pi} F_0(n\omega_0) = \frac{2 \sin^2 (n\omega_0 c/2)}{\pi\omega_0 cn^2}$$

Convolution theorem If

$$f_1(t) = \sum_{n=-\infty}^{\infty} a_n e^{jn\omega_0 t} \qquad f_2(t) = \sum_{n=-\infty}^{\infty} b_n e^{jn\omega_0 t} \tag{3-71}$$

then

$$\frac{1}{T}\int_{-T/2}^{T/2} f_1(t-\tau)f_2(\tau)\,d\tau = \sum_{n=-\infty}^{\infty} a_n b_n e^{jn\omega_0 t} \tag{3-72}$$

and

$$\frac{1}{T}\int_{-T/2}^{T/2} f_1(t)f_2(t)e^{-jn\omega_0 t}\,dt = \sum_{m=-\infty}^{\infty} a_m b_{n-m} \tag{3-73}$$

Proof Inserting the Fourier expansion of $f_1(t)$ into Eq. (3-72), we have

$$\frac{1}{T}\int_{-T/2}^{T/2} f_1(t-\tau)f_2(\tau)\,d\tau = \frac{1}{T}\int_{-T/2}^{T/2} \sum_{n=-\infty}^{\infty} a_n e^{jn\omega_0(t-\tau)} f_2(\tau)\,d\tau$$

$$= \sum_{n=-\infty}^{\infty} a_n e^{jn\omega_0 t} \frac{1}{T}\int_{-T/2}^{T/2} f_2(\tau)e^{-jn\omega_0\tau}\,d\tau$$

and (3-72) results.

Multiplying the two equations in (3-71), we obtain

$$f_1(t)f_2(t) = \sum_{m=-\infty}^{\infty} a_m e^{jm\omega_0 t} \sum_{k=-\infty}^{\infty} b_k e^{jk\omega_0 t}$$

$$= \sum_{m=-\infty}^{\infty} \sum_{k=-\infty}^{\infty} a_m b_k e^{j(m+k)\omega_0 t} = \sum_{n=-\infty}^{\infty} \left(\sum_{k=-\infty}^{\infty} a_{n-k} b_k \right) e^{jn\omega_0 t}$$

The last equality follows with $m + k = n$, and it shows that the sum in parenthesis is the coefficient of the Fourier-series expansion of the product $f_1(t)f_2(t)$. Hence, (3-73) is a consequence of (3-69).

Example 3-18 If

$$a_n = b_n = \begin{cases} 1 & |n| \le N \\ 0 & |n| > N \end{cases} \qquad \text{then} \qquad \sum_{k=-\infty}^{\infty} a_{n-k} b_k = \begin{cases} 2N + 1 - |n| & |n| \le 2N \\ 0 & |n| > 2N \end{cases}$$

With

$$f_1(t) = f_2(t) = \sum_{n=-N}^{N} e^{jn\omega_0 t} = \frac{\sin (N + \frac{1}{2})\omega_0 t}{\sin (\omega_0 t/2)} = k_N(t)$$

the *Fourier-series kernel*, it follows from (3-73) that

$$\frac{\sin^2 (N + \frac{1}{2})\omega_0 t}{(2N + 1) \sin^2 (\omega_0 t/2)} = \sum_{n=-2N}^{2N} \left(1 - \frac{|n|}{2N + 1}\right) e^{jn\omega_0 t} \tag{3-74}$$

(elaborate). The above function is known as the *Fejér kernel.*

Windows If w_n is a sequence of numbers such that $w_n = 0$ for $|n| > N$, then the sum

$$f_w(t) = \sum_{n=-N}^{N} a_n w_n e^{jn\omega_0 t}$$

is a periodic function with Fourier-series coefficients $a_n w_n$. With

$$w(t) = \sum_{n=-N}^{N} w_n e^{jn\omega_0 t}$$

it follows from Eq. (3-72), with $f_1(t) = f(t)$ and $f_2(t) = w(t)$, that

$$f_w(t) = \frac{1}{T} \int_{-T/2}^{T/2} f(t - \tau) w(\tau)\, d\tau \tag{3-75}$$

Example 3-19

(*a*) If $w_n = 1$, then $w(t) = k_N(t)$ as in Example 3-18. Hence,

$$f_w(t) = \sum_{n=-N}^{N} a_n e^{jn\omega_0 t} = \frac{1}{T} \int_{-T/2}^{T/2} f(t - \tau) \frac{\sin (N + \frac{1}{2})\omega_0 \tau}{\sin (\omega_0 \tau/2)}\, d\tau$$

(*b*) If $w_n = 1 - |n|/(2N + 1)$, then [see (3-74)]

$$f_w(t) = \sum_{n=-2N}^{2N} \left(1 - \frac{|n|}{2N + 1}\right) a_n e^{jn\omega_0 t} = \frac{1}{T} \int_{-T/2}^{T/2} f(t - \tau) \frac{\sin^2 (N + \frac{1}{2})\omega_0 \tau}{(2N + 1) \sin^2 (\omega_0 \tau/2)}\, d\tau$$

Parseval's formula If $f_1(t)$ and $f_2(t)$ are two periodic functions as in (3-71), then

$$\frac{1}{T} \int_{-T/2}^{T/2} f_1(t) f_2^*(t)\, dt = \sum_{m=-\infty}^{\infty} a_m b_m^* \tag{3-76}$$

Proof It is easy to see that the Fourier-series coefficients of $f_2^*(t)$ equal b_{-m}^*. Replacing $f_2(t)$ in Eq. (3-73) with $f_2^*(t)$ and setting $n = 0$, we obtain (3-76).

Energy theorem If $f(t) = \sum_{n=-\infty}^{\infty} a_n e^{jn\omega_0 t}$, then

$$\frac{1}{T}\int_{-T/2}^{T/2} |f(t)|^2 \, dt = \sum_{n=-\infty}^{\infty} |a_n|^2 \tag{3-77}$$

Proof It follows from (3-76) with $f_1(t) = f_2(t) = f(t)$.

Real functions, cosine and sine series Consider a real, periodic function $f(t)$ with Fourier-series coefficients

$$a_n = \beta_n + j\gamma_n$$

It is easy to see from (3-69) that

$$a_{-n} = a_n^* \qquad \beta_{-n} = \beta_n \qquad \gamma_{-n} = -\gamma_n$$

$$\beta_n = \frac{1}{T}\int_{-T/2}^{T/2} f(t) \cos n\omega_0 t \, dt \qquad \gamma_n = -\frac{1}{T}\int_{-T/2}^{T/2} f(t) \sin n\omega_0 t \, dt \tag{3-78}$$

Inserting into (3-70), we obtain (elaborate)

$$f(t) = \beta_0 + 2\sum_{n=1}^{\infty} (\beta_n \cos n\omega_0 t - \gamma_n \sin n\omega_0 t) \tag{3-79}$$

Note: An *arbitrary* real function $f(t)$ can be expanded into a series of the form (3-79). The expansion holds for $|t| < T/2$ only. The right side of (3-79) is periodic, and for $|t| = T/2$ it equals the average of $f(T/2)$ and $f(-T/2)$.

Even functions It is easy to see from Eq. (3-78) that, if $f(-t) = f(t)$, then

$$\beta_n = \frac{2}{T}\int_{0}^{T/2} f(t) \cos n\omega_0 t \, dt \qquad \gamma_n = 0 \tag{3-80}$$

and (3-79) yields

$$f(t) = \beta_0 + 2\sum_{n=1}^{\infty} \beta_n \cos n\omega_0 t \tag{3-81}$$

Note: An *arbitrary* real function $f(t)$ can be expanded into a series of the form (3-81). The expansion holds for $0 \le t \le T/2$ only. The right side is an even, periodic function.

Odd functions If $f(-t) = -f(t)$, then

$$\beta_n = 0 \qquad \gamma_n = -\frac{2}{T}\int_{0}^{T/2} f(t) \sin n\omega_0 t \, dt \tag{3-82}$$

and (3-79) yields

$$f(t) = -2 \sum_{n=1}^{\infty} \gamma_n \sin n\omega_0 t \tag{3-83}$$

Note: An *arbitrary* real function $f(t)$ can be expanded into a series of the form (3-83). The expansion holds for $0 < t < T/2$ only. The right side is an odd, periodic function, and for $t = 0$ and $t = T/2$ it equals zero.

3-3 FROM FOURIER INTEGRALS TO DISCRETE FOURIER SERIES[1]

We shall consider the problem of evaluating numerically the integral

$$F(\omega) = \int_{-\infty}^{\infty} f(t)e^{-j\omega t}\, dt \tag{3-84}$$

and the related integral (3-69). The function $f(t)$ is specified either graphically (or by a table) or analytically.

To compute $F(\omega)$, we must replace the infinite limits in (3-84) with finite limits, approximate the resulting integral with a sum, and evaluate the sum for a discrete set of values of ω. In this section, we present the underlying theory.

The Poisson Sum Formula

We introduce the functions

$$\bar{f}(t) = \sum_{n=-\infty}^{\infty} f(t + nT) \qquad \bar{F}(\omega) = \sum_{n=-\infty}^{\infty} F(\omega + n\omega_1) \tag{3-85}$$

Clearly, $\bar{f}(t)$ is periodic with period T, and it equals the sum of the function $f(t)$ and all its displacements (Fig. 3-14). Similarly, $\bar{F}(\omega)$ is periodic with period ω_1, and it equals the sum of $F(\omega)$ and all its displacements. The constants T and ω_1 are arbitrary.

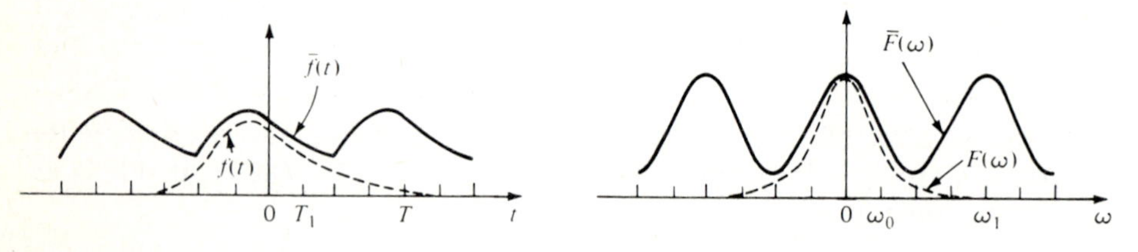

Figure 3-14

[1] J. W. Cooley, P. A. W. Lewis, and P. D. Welch, The Finite Fourier Transform, *IEEE Trans. Audio Electroacoust.*, vol. AU-17, no. 2, pp. 77–95, June, 1969.

We have shown in Sec. 1-2 that [see Eq. (1-59)]

$$\bar{f}(t) = \frac{1}{T} \sum_{n=-\infty}^{\infty} F(n\omega_0)e^{jn\omega_0 t} \qquad \omega_0 = \frac{2\pi}{T} \tag{3-86}$$

This important result, known as the *Poisson sum formula*, forms the basis for the transformation of the improper integral (3-84) to an integral with finite limits [see (3-88)].

We note that a similar formula holds also for the function $\bar{F}(\omega)$:

$$\bar{F}(\omega) = \frac{2\pi}{\omega_1} \sum_{n=-\infty}^{\infty} f(nT_1)e^{-jnT_1\omega} \qquad T_1 = \frac{2\pi}{\omega_1} \tag{3-87}$$

This can be deduced from (3-86) and the symmetry theorem (3-17). It can also be established directly as a consequence of the Fourier expansion of a frequency-domain impulse train:

$$\sum_{n=-\infty}^{\infty} \delta(\omega + n\omega_1) = \frac{1}{\omega_1} \sum_{n=-\infty}^{\infty} e^{-jnT_1\omega}$$

Indeed, convolving both sides of the above identity with $F(\omega)$, we obtain Eq. (3-87) because [see (3-C-4)]

$$\delta(\omega + n\omega_1) * F(\omega) = \int_{-\infty}^{\infty} \delta(\omega - y + n\omega_1)\, F(y)\, dy = F(\omega + n\omega_1)$$

and [see (3-2)]

$$e^{-jnT_1\omega} * F(\omega) = \int_{-\infty}^{\infty} e^{-jnT_1(\omega - y)}F(y)\, dy = 2\pi f(nT_1)e^{-jnT_1\omega}$$

Notes:

1. It is obvious that $\bar{F}(\omega)$ is not the Fourier transform of $\bar{f}(t)$.
2. If the function $f(t)$ is discontinuous at the sampling points, then (3-87) is still valid, provided that

$$f(nT_1) = \tfrac{1}{2}[f(nT_1{}^+) + f(nT_1{}^-)]$$

because (3-87) is based on the inversion formula (3-2) which holds only if $f(t)$ equals the average of $f(t^+)$ and $f(t^-)$. Suppose, for example, that $f(t) = 0$ for $t < 0$ and that for $t > 0$ it is continuous. If $f(0^+) \neq 0$, then $f(t)$ is discontinuous at $t = 0$ because $f(0^-) = 0$. This leads to the conclusion that, for causal functions,

$$\sum_{n=-\infty}^{\infty} F(n\omega_1) = T_1 \frac{f(0^+)}{2} + T_1 \sum_{n=1}^{\infty} f(nT_1)$$

as we can see from (3-87) with $\omega = 0$.

3. As we show in the next example, the Poisson sum formula can be used to derive a variety of interesting identities.

Example 3-20 From the pair

$$e^{-\alpha|t|} \leftrightarrow \frac{2\alpha}{\alpha^2 + \omega^2}$$

and (3-87) it follows, with $\omega = 0$ and $T_1 = 1$, that

$$\sum_{n=-\infty}^{\infty} \frac{2\alpha}{\alpha^2 + (2\pi n)^2} = \sum_{n=-\infty}^{\infty} e^{-\alpha|n|}$$

From Fourier Integrals to Fourier Series

From Poisson's formula (3-86), it follows that the sample values $F(n\omega_0)$ of $F(\omega)$ equal the Fourier-series coefficients of the periodic function $T\bar{f}(t)$. Hence [see Eq. (3-69)],

$$F(n\omega_0) = \int_{-T/2}^{T/2} \bar{f}(t) e^{-jn\omega_0 t}\, dt \tag{3-88}$$

Thus, for $\omega = n\omega_0$ the infinite integral (3-84) is reduced to the finite integral (3-88) involving the function $\bar{f}(t)$. This function can be determined exactly if $f(t)$ is known for all t.

If $f(t)$ is given over a finite interval $(-a/2, a/2)$, and $f(t) = 0$ for $|t| > a/2$, then $\bar{f}(t)$ is again known exactly; but now the defining sum in (3-85) contains only m terms if $a = mT$. In this case, we could replace T with a. However, as we shall see, the choice of T is dictated by other considerations.

> **Note:** If we wish to find the transform $G(\omega)$ of a signal $g(t)$ but know only the segment $f(t) = g(t)p_{a/2}(t)$, then (3-88) yields the samples of the transform [see Example 3-11]:
>
> $$F(\omega) = \int_{-a/2}^{a/2} f(t) e^{-j\omega t}\, dt = \int_{-\infty}^{\infty} G(\omega - \gamma) \frac{\sin a\gamma/2}{\pi\gamma}\, d\gamma$$
>
> The resulting error $G(n\omega_0) - F(n\omega_0)$ can be reduced if $f(t)$ is replaced with the product $f_w(t) = w(t)f(t)$, where $w(t)$ is a suitable factor (window). In this section we shall not be concerned with the truncation problem. We shall determine only the transform $F(\omega)$ of the given $f(t)$.

From Fourier Series to Discrete Fourier Series

As we have seen, the samples $F(n\omega_0)$ of $F(\omega)$ are the Fourier-series coefficients of $T\bar{f}(t)$. For their numerical evaluation, it suffices to turn to the Fourier-series problem. The following considerations will permit us to *approximate* the integral (3-88) with a finite sum.

Consider the periodic function

$$y(t) = \sum_{k=-\infty}^{\infty} c_k e^{jk\omega_0 t} \tag{3-89}$$

With N an arbitrary integer and $T_1 = T/N$, the sample values $y(mT_1)$ of $y(t)$ are given by

$$y(mT_1) = \sum_{k=-\infty}^{\infty} c_k e^{jk\omega_0 m T_1} = \sum_{k=-\infty}^{\infty} c_k w_N{}^{km} \qquad w_N = e^{j2\pi/N} \tag{3-90}$$

because $\omega_0 T_1 = \omega_0 T/N = 2\pi/N$. The number w_N is the Nth primitive root of 1.

The integer k can be written as a sum

$$k = n + rN \qquad n = 0, \ldots, N-1 \qquad r = \ldots, -1, 0, 1, \ldots \tag{3-91}$$

and since $\qquad w_N{}^N = 1 \qquad w_N{}^{km} = w_N{}^{(n+rN)m} = w_N{}^{mn}$

Eq. (3-90) yields

$$y(mT_1) = \sum_{n=0}^{N-1} \sum_{r=-\infty}^{\infty} c_{n+rN} w_N{}^{(n+rN)m} = \sum_{n=0}^{N-1} w_N{}^{mn} \sum_{r=-\infty}^{\infty} c_{n+rN}$$

Defining the "aliased" coefficients $\bar{c}_n$ by

$$\bar{c}_n = \sum_{r=-\infty}^{\infty} c_{n+rN} \tag{3-92}$$

we obtain $\qquad y(mT_1) = \sum_{n=0}^{N-1} \bar{c}_n w_N{}^{mn} \qquad m = 0, \ldots, N-1 \qquad$ (3-93)

Thus, the sample values $y(mT_1)$ of a periodic function $y(t)$ and the aliased coefficients $\bar{c}_n$ are related by the above system of N equations.

Note: Equation (3-92) defines $\bar{c}_n$ for all n, and the sequence $\bar{c}_n$ so obtained is periodic:

$$\bar{c}_{n+N} = \bar{c}_n \tag{3-94}$$

Example 3-21 Suppose that

$$y(t) = \sum_{n=-10}^{10} c_n e^{jn\omega_0 t} \qquad N = 9$$

Since $c_n = 0$ for $|n| > 10$, it follows from (3-92) that

$$\bar{c}_0 = c_{-9} + c_0 + c_9 \qquad \bar{c}_1 = c_{-8} + c_1 + c_{10} \qquad \bar{c}_2 = c_{-7} + c_2 \qquad \cdots$$

and $\qquad \bar{c}_7 = c_{-2} + c_7 \qquad \bar{c}_8 = c_{-10} + c_{-1} + c_8$

In general, the coefficients c_n cannot be determined in terms of $\bar{c}_n$ unless at most N of them are different from zero. This is the case for the following class of periodic functions.

Trigonometric polynomials A trigonometric polynomial is a Fourier series with finitely many terms:

$$y(t) = \sum_{n=-M}^{M} c_n e^{jn\omega_0 t} \tag{3-95}$$

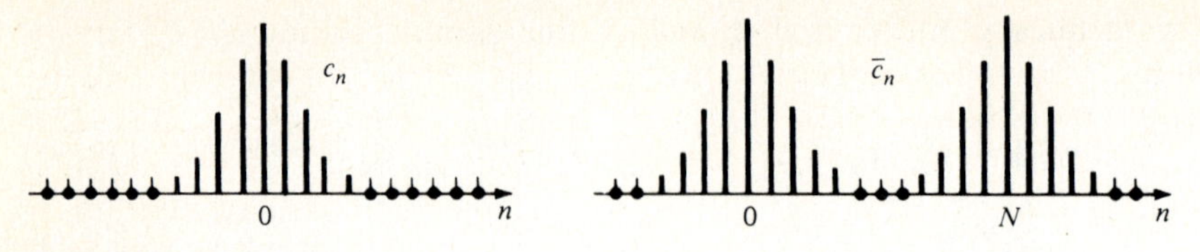

Figure 3-15

Suppose that

$$N > 2M + 1 \tag{3-96}$$

It then follows from (3-92) that (Fig. 3-15)

$$c_n = \begin{cases} \bar{c}_n & |n| \le M \\ 0 & |n| > M \end{cases} \tag{3-97}$$

Hence, if we know $\bar{c}_n$, we can find c_n. And since $\bar{c}_n$ is uniquely determined from $y(mT_1)$, we conclude that the coefficients c_n of a trigonometric polynomial can be expressed in terms of its sample values $y(mT_1)$.

Fundamental theorem The numerical evaluation of the Fourier integral (3-84) is based on the following results:

If T is an arbitrary constant, N an arbitrary integer, and

$$T_1 = \frac{T}{N} \qquad \omega_0 = \frac{2\pi}{T} \qquad \omega_1 = \frac{2\pi}{T_1} = N\omega_0 \tag{3-98}$$

then for any m,

$$\bar{f}(mT_1) = \frac{1}{T} \sum_{n=0}^{N-1} \bar{F}(n\omega_0) w_N{}^{mn} \qquad w_N = e^{j2\pi/N} \tag{3-99}$$

where $\bar{f}(t)$ and $\bar{F}(\omega)$ are the functions defined in (3-85) (Fig. 3-14).

Proof The above is a consequence of Eqs. (3-86) and (3-93). Indeed, $\bar{f}(t)$ is a periodic function with Fourier-series coefficients $c_n = F(n\omega_0)/T$; hence,

$$\bar{c}_n = \frac{1}{T} \sum_{r=-\infty}^{\infty} F[(n + rN)\omega_0] = \frac{1}{T} \sum_{r=-\infty}^{\infty} F(n\omega_0 + r\omega_1) = \frac{1}{T} \bar{F}(n\omega_0) \tag{3-100}$$

and (3-99) follows from (3-93). The last equality in (3-100) is a consequence of (3-85).

Setting $m = 0, \ldots, N - 1$ in Eq. (3-99), we obtain a system of N equations whose solution yields the samples $\bar{F}(n\omega_0)$ of $\bar{F}(\omega)$ in terms of the samples $\bar{f}(mT_1)$ of $\bar{f}(t)$. We note that, in general, $F(n\omega_0)$ cannot be determined in terms of $\bar{F}(n\omega_0)$. If, however,

$$F(\omega) = 0 \quad \text{for} \quad |\omega| > \sigma \quad \text{and} \quad \omega_1 > 2\sigma \tag{3-101}$$

then

$$F(\omega) = \bar{F}(\omega) \quad \text{for} \quad |\omega| < \sigma \tag{3-102}$$

Hence, the solution of (3-99) yields $F(n\omega_0)$. If the function $f(t)$ is not bandlimited as in (3-101) but ω_1 is sufficiently large so that $F(\omega)$ can be neglected for $|\omega| > \omega_1/2$, then $F(n\omega_0)$ is approximately equal to $\bar{F}(n\omega_0)$ for $|n| < \omega_1/2\omega_0$; hence, it can again be determined from (3-99). The difference $F(n\omega_0) - \bar{F}(n\omega_0)$ is the aliasing error. In the next section we discuss the numerical solution of (3-99).

3-4 DISCRETE FOURIER SERIES AND FAST FOURIER TRANSFORMS

As we have seen, the evaluation of Fourier integrals and Fourier series is reduced to the solution of a system of N equations of the form [see Eqs. (3-93) and (3-99)]

$$A_m = \sum_{n=0}^{N-1} a_n w_N^{mn} \qquad m = 0, \ldots, N-1 \qquad w_N = e^{j2\pi/N} \tag{3-103}$$

We maintain that the solution of this system is given by

$$a_n = \frac{1}{N} \sum_{m=0}^{N-1} A_m w_N^{-mn} \qquad n = 0, \ldots, N-1 \tag{3-104}$$

Proof To prove (3-104) for a specific n, we change the summation index in (3-103) from n to k, multiply the mth equation by w_N^{-mn}, and sum from 0 to $N-1$:

$$\sum_{m=0}^{N-1} A_m w_N^{-mn} = \sum_{m=0}^{N-1} w_N^{-mn} \sum_{k=0}^{N-1} a_k w_N^{mk} = \sum_{k=0}^{N-1} a_k \sum_{m=0}^{N-1} w_N^{m(k-n)} \tag{3-105}$$

The last sum is a geometric progression with ratio w_N^{k-n}; hence,

$$\sum_{m=0}^{N-1} w_N^{m(k-n)} = \frac{w_N^{N(k-n)} - 1}{w_N^{k-n} - 1} = \begin{cases} 0 & k \neq n \\ N & k = n \end{cases}$$

because $w_N^N = e^{j2\pi} = 1$. Thus, the factor of a_k on the right side of (3-105) equals zero if $k \neq n$, and equals N if $k = n$. Therefore, the right side equals Na_n and (3-104) results.

Definition Equations (3-103) and (3-104) establish a one-to-one correspondence between the N numbers a_n and the N numbers A_m. We shall use the notation

$$a_n \underset{N}{\leftrightarrow} A_m$$

for this correspondence, and we shall say that the numbers a_n and A_m form a *discrete Fourier series* (abbreviated DFS) pair of order N.

If we assume that (3-103) and (3-104) hold for every m and n, then the sequences A_m and a_n must be periodic:

$$A_{m+N} = A_m \qquad a_{n+N} = a_n \tag{3-106}$$

because

$$w_N^{\pm(m+N)n} = w_N^{\pm mn}$$

Thus, a DFS pair can be viewed as a correspondence between two periodic sequences a_n and A_m of period N satisfying Eqs. (3-103) and (3-104).

Notes:

1. With a_n and A_m so interpreted, it follows that the summation limits 0 and $N - 1$ in (3-103) and (3-104) can be replaced with n_1 and $n_1 + N - 1$, respectively, where n_1 is any integer. The following special case is of interest: If $N = 2M + 1$, then with $n_1 = -M$, we obtain

$$A_m = \sum_{n=-M}^{M} a_n w_N^{mn} \qquad a_n = \frac{1}{N} \sum_{m=-M}^{M} A_m w_N^{-mn} \tag{3-107}$$

2. The recursion equation

$$g[n] - w_N^{-m} g[n-1] = x[n] \qquad x[n] = \begin{cases} a_n & 0 \le n \le N-1 \\ 0 & \text{otherwise} \end{cases}$$

defines a discrete system with delta response $h[n] = w_N^{-mn} U[n]$. Its output at $n = N$ is given by [see Eq. (1-13)]

$$g[N] = \sum_{k=0}^{N-1} w_N^{-(N-k)m} a_k = A_m$$

i.e., it equals the DFS A_m of the input. This idea is further developed in Sec. 5-6.

DFS and interpolation Consider the polynomial

$$F(z) = \sum_{n=0}^{N-1} a_n z^n$$

From (3-103) it follows that

$$F(w_N^m) = A_m$$

Clearly, the numbers w_N^m are the Nth roots of 1 (Fig. 3-16). Hence, the coefficients a_n of a polynomial of degree N and its values $F(w_N^m)$ on the N points w_N^m of the unit circle form a DFS pair. This leads to a simple solution of the following interpolation problem:

Given a function $y(\theta)$, find a polynomial $F(z)$ of degree $N - 1$ such that (Fig. 3-16)

$$F(e^{jm\theta_0}) = y(m\theta_0) \qquad \theta_0 = \frac{2\pi}{N} \qquad m = 0, 1, \ldots, N-1$$

Since $w_N = e^{j\theta_0}$, we conclude that $F(w_N^m) = y(m\theta_0)$. Hence,

$$a_n = \frac{1}{N} \sum_{m=0}^{N-1} y(m\theta_0) w_N^{-mn}$$

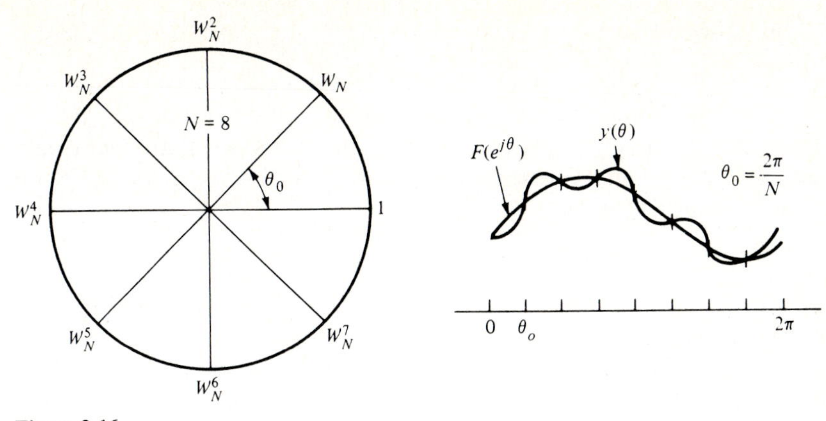

Figure 3-16

Example 3-22 If a_n is a periodic sequence with period N (Fig. 3-17), and if $a_n \underset{N}{\leftrightarrow} A_m$ and $a_n \underset{2N}{\leftrightarrow} \bar{A}_m$, then

$$\bar{A}_m = \begin{cases} 2A_r & \text{for } m = 2r \\ 0 & \text{for } m = 2r + 1 \end{cases}$$

Proof From the definition (3-103), it follows that

$$\bar{A}_m = \sum_{n=0}^{N-1} a_n w_{2N}{}^{mn} + \sum_{n=N}^{2N-1} a_n w_{2N}{}^{mn} = \sum_{n=0}^{N-1} a_n w_{2N}{}^{mn} + \sum_{k=0}^{N-1} a_{k+N} w_{2N}{}^{m(k+N)}$$

But $\qquad a_{k+N} = a_k \qquad w_{2N}{}^{2rm} = w_N{}^{rm} \qquad w_{2N}{}^{mN} = (-1)^m$

Hence, if $s = 2r$, then each of the last two sums equals A_r; and if $m = 2r + 1$, then they are of opposite signs. Therefore, their sum is zero.

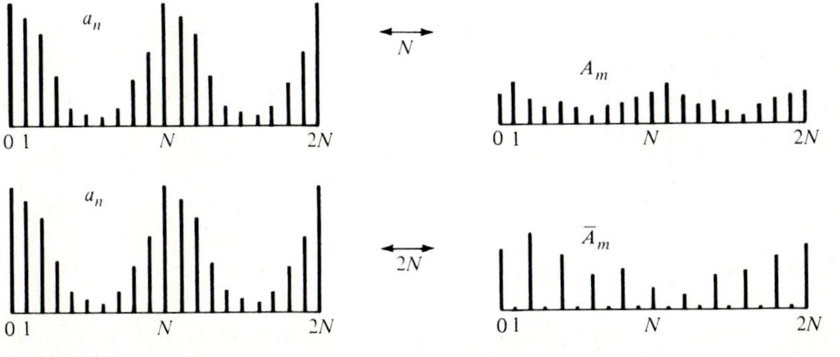

Figure 3-17

The FFT Algorithm

We consider next the computational problem in the determination of the N numbers A_m from Eq. (3-103). The same considerations hold also for Eq. (3-104). Since additions are usually much faster than multiplications, we shall count only the required multiplications. Clearly, $A_0 = a_0 + a_1 + \cdots + a_{N-1}$ involves no multiplications. To find A_m from

$$A_m = a_0 + a_1 w_N{}^m + a_2 w_N{}^{2m} + \cdots + a_{N-1} a_N{}^{m(N-1)}$$

we must perform $N - 1$ multiplications for each $m > 0$. Hence, the total number of required multiplications equals $(N - 1)^2$. This number can be reduced considerably by the following technique, known as fast Fourier transform.

We shall show that for the determination of the $2N$ coefficients A_m of a DFS of order $2N$,

$$a_n \underset{2N}{\leftrightarrow} A_m \tag{3-108}$$

it suffices to compute two sets B_m and C_m of DFS coefficients of order N:

$$b_n \underset{N}{\leftrightarrow} B_m \qquad c_n \underset{N}{\leftrightarrow} C_m$$

Fundamental theorem If

$$b_n = a_{2n} \qquad c_n = a_{2n+1} \tag{3-109}$$

are the even and odd components, respectively, of the given sequence a_n (Fig. 3-18), then for any m,

$$A_m = B_m + w_{2N}{}^m C_m \tag{3-110}$$

Proof Replacing N with $2N$ in (3-103) and summing over the even and odd values of n, we obtain

$$A_m = \sum_{n=0}^{2N-1} a_n w_{2N}{}^{mn} = \sum_{k=0}^{N-1} a_{2k} w_{2N}{}^{2km} + \sum_{k=0}^{N-1} a_{2k+1} w_{2N}{}^{(2k+1)m}$$

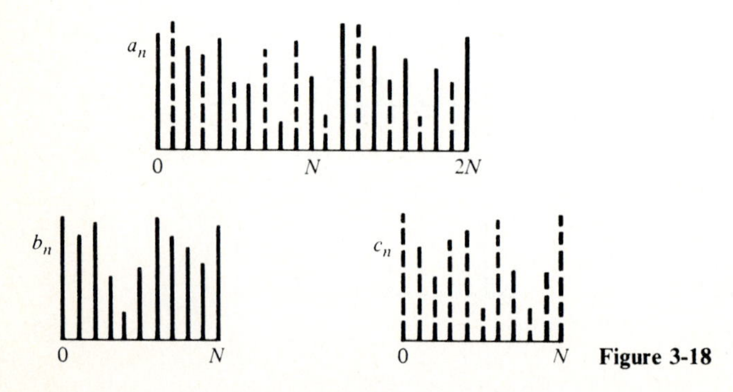

Figure 3-18

But
$$w_{2N}^{2km} = w_N^{km} \qquad w_{2N}^{(2k+1)m} = w_N^{km} w_{2N}^{m}$$

Hence
$$\sum_{k=0}^{N-1} a_{2k} w_{2N}^{2km} = \sum_{k=0}^{N-1} b_k w_N^{km} = B_m$$

$$\sum_{k=0}^{N-1} a_{2k+1} w_{2N}^{(2k+1)m} = w_{2N}^{m} \sum_{k=0}^{N-1} c_k w_N^{km} = w_{2N}^{m} C_m$$

and Eq. (3-110) results.

Notes:

1. For the evaluation of B_m and C_m, the total number of required multiplications equals $2(N-1)^2$. Hence, to determine A_m from (3-110) we need $2(N-1)^2 + 2N - 1 = 2N^2 - 2N + 1$ multiplications because (3-110) involves one multiplication $w_{2N}^{m} C_m$ for each $m \neq 0$. If A_m is determined directly from (3-108), then the required number of multiplications equals $(2N-1)^2 = 4N^2 - 4N + 1$. Thus, the use of (3-110) results in a reduction of the required number of multiplications by a factor of about 2. An additional minor reduction results if we use the fact that

$$B_{m+N} = B_m \qquad C_{m+N} = C_m \qquad w_{2N}^{m+N} = -w_{2N}^{m} \tag{3-111}$$

Indeed, replacing m in (3-110) with $m + N$ and using (3-111), we conclude that

$$A_{m+N} = B_m - w_{2N}^{m} C_m \tag{3-112}$$

Thus, we compute A_m from (3-110) for $m = 0, \ldots, N-1$, and from (3-112) for $m = N, \ldots, 2N-1$.

2. A DFS of order $N = m \times r$ can be reduced to m DFS of order r (see Prob. 50).

FFT for $N = 2^s$ If N is even, then the preceding result can be repeated; that is, B_m and C_m can be determined in terms of four DFS of order $N/2$. If $N = 2^s$, then the process can start with $N = 2$. To determine the resulting computational economy, we denote by $F(N)$ the required number of multiplications for a DFS of order $N = 2^s$ obtained by repeated use of (3-110). From the fundamental theorem, it follows that

$$F(2N) = 2F(N) + N \tag{3-113}$$

Indeed, we need $2F(N)$ multiplications for the evaluation of B_m and C_m, and $N - 1$ multiplications for the evaluation of A_m from (3-110). No new multiplications are involved in Eq. (3-112). However, we have added 1 to account for the required subtractions.

Since $F(2) = 1$, it follows from (3-113) that $F(4) = 4$ and $F(8) = 12$. We shall show by induction that

$$F(N) = \frac{N}{2} \log_2 N \tag{3-114}$$

Proof The above is true for $N = 2$. Suppose that it is true for some N. Then, using (3-113), we conclude

$$F(2N) = N \log_2 N + N = N \log_2 2N$$

Hence, (3-114) is true for $2N$.

Properties of DFS

In the numerical evaluation of DFS, we compute N numbers A_m in terms of N given numbers a_n. We shall, however, interpret a_n and A_m as two periodic sequences. In particular, we shall assume that

$$a_{-n} = a_{N-n} \qquad A_{-m} = A_{N-m} \tag{3-115}$$

From the definition (3-103) it follows, with $n = N - k$, that

$$A_m^* = \sum_{n=1}^{N} a_n^* w_N^{-mn} = \sum_{k=0}^{N-1} a_{N-k}^* w_N^{-m(N-k)} = \sum_{k=0}^{N-1} a_{-k}^* w_N^{mk}$$

because $w_N^* = w_N^{-1}$. But the last sum is the DFS of a_{-n}^*. Hence,

$$a_{-n}^* \underset{N}{\leftrightarrow} A_m^* \tag{3-116}$$

Similarly, since

$$A_{-m} = \sum_{n=1}^{N} a_n w_N^{-mn} = \sum_{k=0}^{N-1} a_{N-k} w_N^{-m(N-k)} = \sum_{k=0}^{N-1} a_{-k} w_N^{mk}$$

we conclude that

$$a_{-n} \underset{N}{\leftrightarrow} A_{-m} \tag{3-117}$$

If a_n is a sequence of *real* numbers, then $a_{-n}^* = a_{-n}$; hence,

$$A_m^* = A_{-m} \tag{3-118}$$

and since $A_{-m} = A_{N-m}$, if we know A_m for $m = 0, \ldots, N/2$, then we can find A_m for every m.

For real sequences, the following computational saving is possible: If a_n and b_n are real and

$$a_n \underset{N}{\leftrightarrow} A_m \qquad b_n \underset{N}{\leftrightarrow} B_m \qquad a_n + jb_n \underset{N}{\leftrightarrow} C_m$$

then

$$A_m = \frac{1}{2}(C_m + C_{N-m}^*) \qquad B_m = \frac{1}{2j}(C_m - C_{N-m}^*) \tag{3-119}$$

Hence, to determine the DFS of two real sequences, it suffices to find the DFS of a single complex sequence.

Proof From the linearity of DFS it follows that the DFS of $a_n \pm jb_n$ equals $A_m \pm jB_m$. But $a_n - jb_n$ is the conjugate of $a_n + jb_n$; hence [see Eqs. (3-116) and (3-115)], its DFS equals $C^*_{-m} = C^*_{N-m}$. Thus,

$$A_m + jB_m = C_m \qquad A_m - jB_m = C^*_{N-m}$$

and (3-119) results. We note that the sequences A_m and B_m are not, in general, real.

Convolution theorem If $a_n \underset{N}{\leftrightarrow} A_m$ and $b_n \underset{N}{\leftrightarrow} B_m$, then

$$\sum_{k=0}^{N-1} a_k b_{n-k} = \frac{1}{N} \sum_{m=0}^{N-1} A_m B_m w_N^{-mn} \tag{3-120}$$

Proof As we see from (3-103), the DFS of the left side of Eq. (3-120) is given by

$$\sum_{n=0}^{N-1} \left(\sum_{k=0}^{N-1} a_k b_{n-k} \right) w_N^{mn} = \sum_{k=0}^{N-1} a_k \sum_{n=0}^{N-1} b_{n-k} w_N^{mn} \tag{3-121}$$

With $n - k = r$, we have

$$\sum_{n=0}^{N-1} b_{n-k} w_N^{mn} = w_N^{mk} \sum_{r=-k}^{N-1-k} b_r w_N^{mr} \tag{3-122}$$

But b_r and w_N^m are two periodic sequences with period N. Hence, the sum from $-k$ to $N - 1 - k$ equals the sum from 0 to $N - 1$. Thus, the last sum equals B_m, and the right side of (3-121) equals

$$\sum_{k=0}^{N-1} a_k B_m w_N^{mk} = A_m B_m$$

Shifting From (3-122) it follows that

$$\text{If} \qquad b_n \underset{N}{\leftrightarrow} B_m \qquad \text{then} \qquad b_{n-k} \underset{N}{\leftrightarrow} w_N^{mk} B_m \tag{3-123}$$

Parseval's formula We set $n = 0$ in (3-120) and, in the result, replace b_{-k} with b_k^*. From (3-116) it follows that B_m must be replaced with B_m^*. Hence,

$$\sum_{k=0}^{N-1} a_k b_k^* = \frac{1}{N} \sum_{m=0}^{N-1} A_m B_m^* \tag{3-124}$$

Energy theorem With $a_k = b_k$, the above yields

$$\sum_{k=0}^{N-1} |a_k|^2 = \frac{1}{N} \sum_{m=0}^{N-1} |A_m|^2 \tag{3-125}$$

DFS as eigenvalues of circulant matrices Consider the circulant matrix

$$L = \begin{pmatrix} a_0 & a_1 & \cdots & a_{N-1} \\ a_{N-1} & a_0 & \cdots & a_{N-2} \\ \cdots & \cdots & \cdots & \cdots \\ a_1 & a_2 & \cdots & a_0 \end{pmatrix} \tag{3-126}$$

where a_n are N given numbers. This matrix has N eigenvalues $z_1, \ldots, z_N$ that can be found by solving the equation $|L - zI| = 0$, and N eigenvectors $v_1, \ldots, v_N$, each of which is a solution of the system

$$Lv_m = z_m v_m \tag{3-127}$$

We maintain that the eigenvalues of L equal the DFS of a_n, and the components of its eigenvectors are powers of the Nth root of unity:

$$z_m = A_m \qquad v_m = [1, w_N{}^m, w_N{}^{2m}, \ldots, w_N{}^{(N-1)m}] \tag{3-128}$$

Proof Multiplying both sides of Eq. (3-103) by $w_N{}^{mk}$, we obtain

$$a_0 w_N{}^{mk} + a_1 w_N{}^{m(k+1)} + \cdots + a_{N-1} w_N{}^{m(k+N-1)} = w_N{}^{mk} A_m$$

Since $w_N{}^{mN} = 1$, we obtain, on rearranging the terms,

$$a_{N-k} + a_{N-k+1} w_N{}^m + \cdots + a_0 w_N{}^{mk} + \cdots + a_{N-k-1} w_N{}^{m(N-1)} = w_N{}^{mk} A_m$$

But this is the $(k+1)$th equation of the system (3-127). Hence, (3-128) follows.

Evaluation of Fourier Integrals by DFS

We have shown in (3-99) that the sample values

$$\bar{f}(mT_1) = \sum_{k=-\infty}^{\infty} f(mT_1 + kT) \qquad \bar{F}(n\omega_0) = \sum_{r=-\infty}^{\infty} F(n\omega_0 + r\omega_1) \tag{3-129}$$

of the functions $\bar{f}(t)$ and $\bar{F}(\omega)$ form a DFS pair

$$\frac{1}{T} \bar{F}(n\omega_0) \underset{N}{\leftrightarrow} \bar{f}(mT_1) \tag{3-130}$$

Hence [see (3-104)],

$$\bar{F}(n\omega_0) = T_1 \sum_{m=0}^{N-1} \bar{f}(mT_1) w_N{}^{-mn} \tag{3-131}$$

This result will be used to determine the values $F(n\omega_0)$ of the transform $F(\omega)$ of a signal $f(t)$. It will be assumed that $f(t)$ is specified in the interval $(0,a)$ and equals zero outside this interval. From the five parameters T, N, T_1, ω_0, and ω_1, we need to determine T and N.

1. *Data-interval sampling.* We choose $T = a$ and $N = N_0$, where N_0 is the available DFS order. With this choice of T, $\bar{f}(t)$ equals $f(t)$ in the $(0,T)$ interval (Fig.

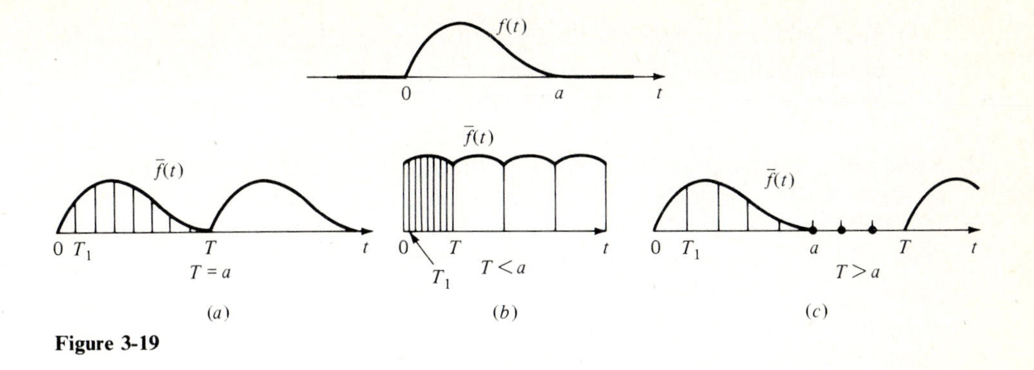

Figure 3-19

3-19a). Hence, to determine $\bar{F}(n\omega_0)$ it suffices to use for $\bar{f}(mT_1)$ the samples $f(mT_1)$ of the given signal $f(t)$. The resulting DFS in (3-131) yields *exactly* the samples $\bar{F}(n\omega_0)$ of $\bar{F}(\omega)$. From these samples we can determine *approximately* the samples $F(n\omega_0)$ of $F(\omega)$, provided that the band of $F(\omega)$ is sufficiently narrow. Indeed, suppose that $F(\omega)$ is negligible for $|\omega| > \sigma$ and that $\sigma < \omega_1/2$. It then follows from Eq. (3-129) that (Fig. 3-20)

$$F(n\omega_0) \simeq \begin{cases} \bar{F}(n\omega_0) & \text{for } |n| \leq \dfrac{N}{2} \\ 0 & \text{for } |n| > \dfrac{N}{2} \end{cases} \tag{3-132}$$

Notes:

(a) We assumed that the interval $(-\sigma, \sigma)$ in which $F(\omega)$ takes significant values is symmetrical about the origin. We did so because $F(-\omega) = F^*(\omega)$ if $f(t)$ is real.

(b) In the numerical evaluation of the DFS in (3-131), $\bar{F}(n\omega_0)$ is computed for $n = 0, \ldots, N-1$. The resulting numbers $\bar{F}(n\omega_0)$ for $n > N/2$ are used to find $F(n\omega_0)$ for $n < 0$ [see (3-106)]:

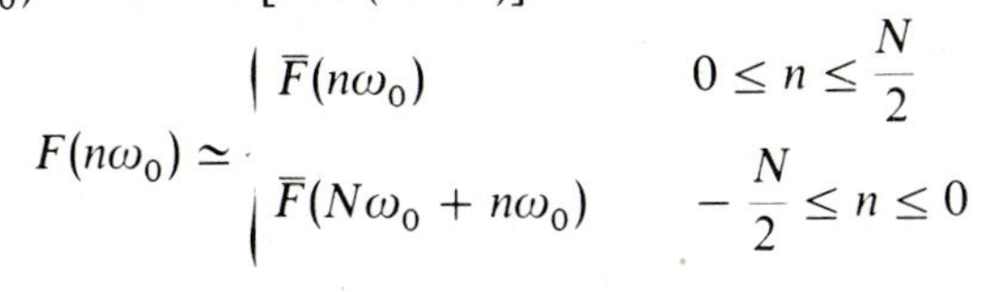

$$F(n\omega_0) \simeq \begin{cases} \bar{F}(n\omega_0) & 0 \leq n \leq \dfrac{N}{2} \\ \bar{F}(N\omega_0 + n\omega_0) & -\dfrac{N}{2} \leq n \leq 0 \end{cases}$$

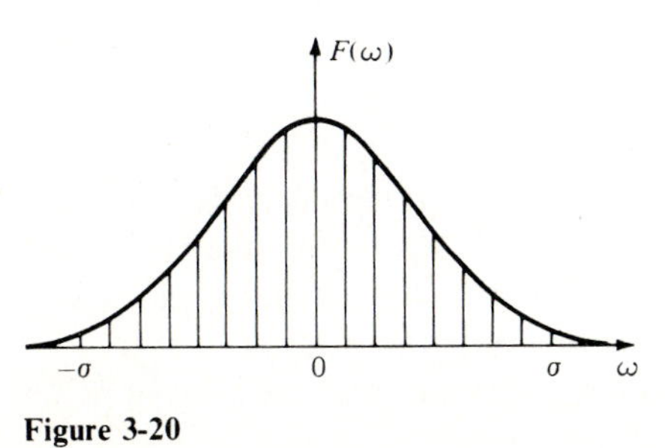

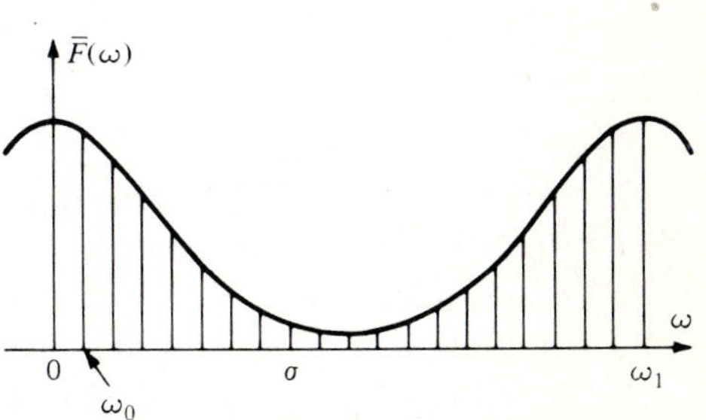

Figure 3-20

(c) Since $f(t) = 0$ outside the interval $(0,a)$ by assumption, $F(\omega)$ cannot be bandlimited; hence the aliasing error $F(n\omega_0) - \bar{F}(n\omega_0)$ cannot be eliminated. It can, however, be reduced for sufficiently large ω_1.

2. *Reduction of aliasing error.* To reduce the error $F(n\omega_0) - \bar{F}(n\omega_0)$, we must increase ω_1. This is done by choosing for T a value smaller than the length a of the data interval (Fig. 3-19*b*) because $\omega_1 = N\omega_0 = 2\pi N/T$. The samples $\bar{f}(mT_1)$ of $\bar{f}(t)$ are then no longer equal to $f(mT_1)$ but are obtained by summing $f(mT_1)$ as in Eq. (3-129). If $T = a/M$, then the sum in (3-129) contains M terms that are different from zero. Again, $\bar{f}(mT_1)$ is computed exactly, and $F(n\omega_0)$ is determined approximately from (3-132).
3. *Improvement in resolution.* The choice $T = a$ leads to the value $\omega_0 = 2\pi/a$ for the sampling interval ω_0. If it is desirable to have a smaller ω_0, then T must be larger than a. Since $f(t) = 0$ for $t > a$, this leads to a function $\bar{f}(t)$ that equals the given $f(t)$ for $0 < t < a$ and that equals zero for $a < t < T$ (Fig. 3-19*c*). The samples $\bar{f}(mT_1)$ are, therefore, equal to $f(mT_1)$ for m such that $mT_1 < a$; for $mT_1 > a$, they are zero.

   Since $\omega_1 = N\omega_0$, the decrease in ω_0 results also in a decrease in ω_1. This method can, therefore, be used only if ω_1 is larger than 2σ. Otherwise, the aliasing error will be unacceptable.

   A decrease of ω_0 with fixed ω_1 is possible if the order N of the DFS is increased. Using the available program for a DFS of order N_0, we evaluate two sequences B_m and C_m. The DFS of order $2N_0$ is then determined from (3-110). In the context of Fourier transforms, this method leads to the following scheme:

   We evaluate the samples $F(n\omega_0)$ of $f(t)$ and the samples $F_1(n\omega_0)$ of the Fourier transform $F_1(\omega)$ of

$$f_1(t) = e^{-j\omega_0 t/2} f(t)$$

But $\quad F_1(n\omega_0) = F\left(n\omega_0 + \frac{\omega_0}{2}\right) \quad$ because $\quad F_1(\omega) = F\left(\omega + \frac{\omega_0}{2}\right)$

Hence, the above yields the samples of $F(\omega)$ with $\omega_0/2$ as sampling interval. Thus, by computing two DFS, we double the resolution without increasing the aliasing error. The preceding scheme also provides a proof of theorem (3-110), as it is easy to show (elaborate).

Notes:

1. The evaluation of $f(t)$ from $F(\omega)$ can be carried out similarly. In this case, $\bar{F}(n\omega_0)$ is determined exactly from (3-129) in terms of the available data $F(n\omega_0)$, and $\bar{f}(mT_1)$ is found from (3-99). The unknown samples $f(mT_1)$ are thus specified within the aliasing error $f(mT_1) - \bar{f}(mT_1)$.
2. As we illustrate in the following examples, DFS can also be used in other areas of numerical analysis.

Example 3-23 Find the values $f'(mT_1)$ of the derivative $f'(t)$ of a signal $f(t)$.
Since

$$f(t) \leftrightarrow F(\omega) \qquad f'(t) \leftrightarrow j\omega F(\omega)$$

it follows from (3-131) and (3-99) that

$$\bar{F}(n\omega_0) = T_1 \sum_{m=0}^{N-1} \bar{f}(mT_1) w_N^{-mn} \qquad \bar{f}'(mT_1) = \frac{1}{T} \sum_{n=0}^{N-1} jn\omega_0 \bar{F}(n\omega_0) w_N^{mn} \qquad (3\text{-}133)$$

Thus, to find $f'(mT_1)$, we compute $\bar{f}(mT_1)$ from $f(t)$ using Eq. (3-129) and $\bar{F}(n\omega_0)$ from the first sum in (3-133). Inserting the result in the second sum, we obtain $f'(mT_1) \simeq \bar{f}'(mT_1)$.

Example 3-24 (Interpolation) Given $f(mT_1)$ for every m, find $f(mT_1 + T_1/2)$.
Clearly,

$$f\left(t + \frac{T_1}{2}\right) \leftrightarrow e^{j\omega T_1/2} F(\omega)$$

and (3-99) yields

$$f\left(mT_1 + \frac{T_1}{2}\right) \simeq \bar{f}\left(mT_1 + \frac{T_1}{2}\right) = \frac{1}{T} \sum_{n=0}^{N-1} w_N^{n/2} \bar{F}(n\omega_0) w_N^{mn}$$

because $e^{jn\omega_0 T_1/2} = w_N^{n/2}$.

Example 3-25 The evaluation of the convolution $g(t) = f(t) * h(t)$ of two given functions $f(t)$ and $h(t)$ can be carried out with three DFS. If ω_1 is sufficiently large, then $\bar{G}(n\omega_0) \simeq G(n\omega_0) = F(n\omega_0)H(n\omega_0)$ because $G(\omega) = F(\omega)H(\omega)$. Hence,

$$\bar{g}(mT_1) = \frac{1}{T} \sum_{n=0}^{N-1} F(n\omega_0)H(n\omega_0) w_N^{mn} \, .$$

If the durations of $f(t)$ and $h(t)$ are a and b, respectively, then the duration of $g(t)$ is $a + b$. Therefore, to conclude that $g(mT_1)$ equals $\bar{g}(mT_1)$, we must select for T a value at least equal to $a + b$. This requirement leads to large errors if $b \ll a$, because then few samples of $h(t)$ are used. If $f(t)$ is smooth relative to b, then $g(t)$ can be computed from the moment expansion (4-9). Otherwise, convolution with recursive DFS is recommended (Sec. 5-6).

The evaluation of Fourier integrals by DFS is, in general, approximate. If a function $f(t)$ is bandlimited, that is, if $F(\omega) = 0$ for $|\omega| > \sigma$ and $\omega_1 > 2\sigma$, then $F(n\omega_0) = \bar{F}(n\omega_0)$ in the band of $f(t)$; that is, the aliasing error is zero. However, in this case, $f(t)$ cannot be time-limited (see Sec. 6-1); to compute $F(n\omega_0)$ from real data, we must introduce truncation errors. If $f(t)$ is time-limited, then it cannot be bandlimited.

It is possible, however, that $f(t)$ is time-limited and $F(n\omega_0) = 0$ for $|n| > N/2$ (Fig. 3-21). For this class of time-limited functions, $F(n\omega_0)$ can be found exactly from Eq. (3-129). It follows from (3-69) that the Fourier-series expansion of such a function has finitely many terms; that is, $f(t)$ is a trigonometric polynomial as in (3-95).

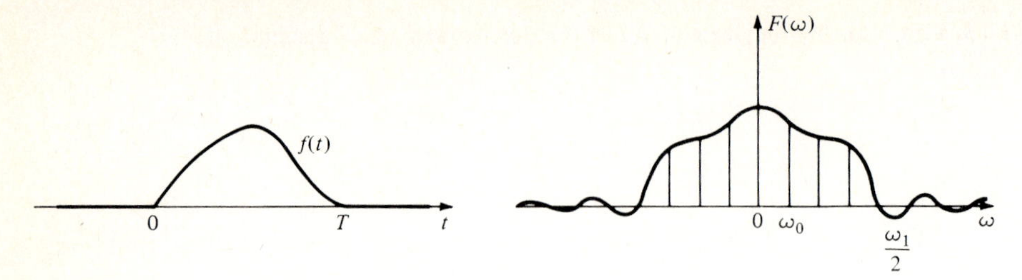

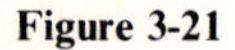
Figure 3-21

Fourier Series

We have shown in Eq. (3-93) that if

$$y(t) = \sum_{n=-\infty}^{\infty} c_n e^{jn\omega_0 t} \qquad \text{and} \qquad \bar{c}_n = \sum_{r=-\infty}^{\infty} c_{n+rN} \tag{3-134}$$

then the samples $y(mT_1)$ of $y(t)$ and the aliased coefficients $\bar{c}_n$ form a DFS pair:

$$\bar{c}_n \leftrightarrow y(mT_1) \tag{3-135}$$

Hence [see (3-104)],

$$\bar{c}_n = \frac{1}{N} \sum_{m=0}^{N-1} y(mT_1) w_N^{-mn} \qquad T_1 = \frac{T}{N} \tag{3-136}$$

This result is the basis for using DFS in the study of periodic functions.

Evaluation of c_n We compute $\bar{c}_n$ from Eq. (3-136) using for N the order N_0 of the available FFT algorithm. If c_n is negligible for $|n| > M$ and $N > 2M$, then it follows from (3-134) that (Fig. 3-22)

$$c_n \simeq \begin{cases} \bar{c}_n & \text{for } |n| \leq \dfrac{N}{2} \\ 0 & \text{for } |n| > \dfrac{N}{2} \end{cases} \tag{3-137}$$

The difference $c_n - \bar{c}_n$ is the aliasing error. This error is zero if $c_n = 0$ for $|n| > M$, that is, if $y(t)$ is a trigonometric polynomial.

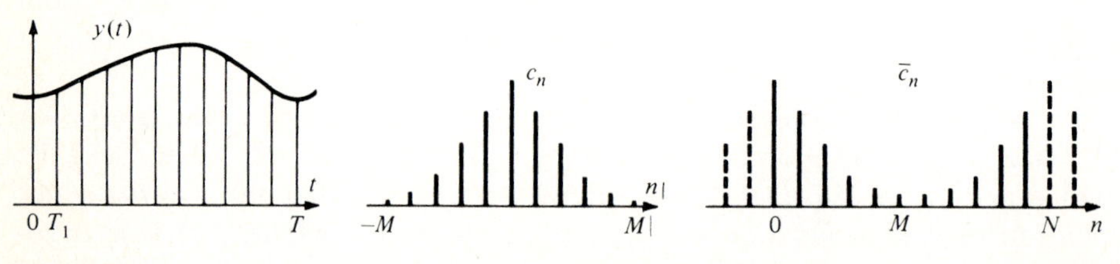

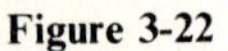
Figure 3-22

From the above it follows that, if

$$y(t) = \sum_{n=-M}^{M} c_n e^{jn\omega_0 t}$$

then, for any $N > 2M$ and $T_1 = T/N$,

$$c_n = \frac{1}{T}\int_0^T y(t)e^{-jn\omega_0 t}\,dt = \frac{1}{N}\sum_{m=0}^{N-1} y(mT_1)w_N^{-mn} \tag{3-138}$$

Notes:

1. We assumed that the range $(-M, M)$ in which c_n takes significant values is symmetric about the point $n = 0$ because if $y(t)$ is real, then $c_{-n} = c_n^*$.
2. The DFS (3-136) yields $\bar{c}_n$ for $n = 0, \ldots, N-1$. The required values in (3-137) for $n < 0$ are obtained from $\bar{c}_n$ for $n > M/2$ because $\bar{c}_n = \bar{c}_{n+N}$.
3. If c_n is not negligible for $|n| > N/2$, then in order to reduce the aliasing error $c_n - \bar{c}_n$, we must increase the DFS order N.

Evaluation of $y(mT_1)$ We assume that the coefficients c_n of a periodic function $y(t)$ are specified for every n from $-M$ to M and that they are zero for $|n| > M$. The truncation problem will not be considered here. We wish to evaluate $y(mT_1)$ for a specified value $T_1 = T/N$ of the sampling interval (resolution). If the required N does not exceed the order N_0 of the available algorithm, then we choose $N = N_0$. If, however, $N > N_0$, then we must increase the available order. We assume then that N is specified.

If $N > 2M$, then $\bar{c}_n = c_n$ for $0 \le n \le M$ and $\bar{c}_n = c_{n-N}$ for $M < n < N$. If $N < 2M$, then $\bar{c}_n$ is no longer equal to c_n but is obtained from the last sum in (3-134).

APPENDIX 3-A

The Mean-Value Theorems

We present for easy reference the mean-value theorems of differential and integral calculus.

Integral Calculus

As is well known, if a function $F(x)$ is continuous in an interval (a,b), and μ is a number between the maximum M and minimum m of $F(x)$, then $F(\xi) = \mu$ for some ξ between a and b. This leads to the following:

Theorem 1 If $F(x)$ is continuous and $p(x) \geq 0$, then

$$\int_a^b F(x)p(x)\,dx = F(\xi)\int_a^b p(x)\,dx \qquad a \leq \xi \leq b \tag{3-A-1}$$

Proof Since $m \leq F(x) \leq M$, and $p(x) \geq 0$, it follows that

$$m\int_a^b p(x)\,dx \leq \int_a^b F(x)p(x)\,dx \leq M\int_a^b p(x)\,dx$$

Hence,
$$\int_a^b F(x)p(x)\,dx = \mu\int_a^b p(x)\,dx \qquad \text{where} \qquad m \leq \mu \leq M$$

and Eq. (3-A-1) results.

Theorem 2 If the function $q(x)$ is monotone increasing or decreasing and $q(b) = 0$, then for any $f(x)$,

$$\int_a^b f(x)q(x)\,dx = q(a)\int_a^\xi f(x)\,dx \qquad a \leq \xi \leq b \tag{3-A-2}$$

Proof If the theorem is true for $q(x)$, it is also true for $-q(x)$; it suffices, therefore, to assume that $q(x)$ is decreasing. We form the function

$$F(x) = \int_a^x f(\alpha)\,d\alpha$$

Clearly, $F(x)$ is continuous and $F'(x) = f(x)$. Hence (integration by parts),

$$\int_a^b f(x)q(x)\,dx = q(x)F(x)\Big|_a^b - \int_a^b q'(x)F(x)\,dx = -\int_a^b q'(x)F(x)\,dx$$

because $F(a) = 0$ and $q(b) = 0$ by assumption. From the monotonicity of $q(x)$, it follows that $-q'(x) \geq 0$. Therefore [see (3-A-1)],

$$-\int_a^b F(x)q'(x)\,dx = -F(\xi)\int_a^b q'(x)\,dx = F(\xi)q(a)$$

yielding (3-A-2).

Corollary If the function $g(x)$ is monotone increasing or decreasing, then for any $f(x)$,

$$\int_a^b f(x)g(x)\,dx = g(a)\int_a^\xi f(x)\,dx + g(b)\int_\xi^b f(x)\,dx \qquad a \leq \xi \leq b \quad \text{(3-A-3)}$$

Proof It follows from (3-A-2) with $q(x) = g(x) - g(b)$.

Differential Calculus

Using (3-A-1), we shall derive the mean-value theorem for derivatives.

Theorem 3 If the derivative $y'(x)$ of a function $y(x)$ is continuous, then

$$y(b) = y(a) + y'(\xi)(b - a) \qquad a \leq \xi \leq b \quad \text{(3-A-4)}$$

Proof

$$y(b) - y(a) = \int_a^b y'(x)\,dx = y'(\xi)\int_a^b dx = y'(\xi)(b - a) \quad \text{(3-A-5)}$$

The second equality follows from (3-A-1) with $p(x) = 1$ and $F(x) = y'(x)$.

Theorem 4 If the second derivative $y''(x)$ of $y(x)$ is continuous, then

$$y(b) = y(a) + y'(a)(b - a) + y''(\xi_1)\frac{(b - a)^2}{2} \qquad a \leq \xi_1 \leq b \quad \text{(3-A-6)}$$

Proof Since $dx = d(x - b)$, we have

$$y(b) - y(a) = \int_a^b y'(x)\,d(x - b) = y'(x)(x - b)\Big|_a^b - \int_a^b y''(x)(x - b)\,dx$$

$$= -y'(a)(a - b) - y''(\xi_1)\int_a^b (x - b)\,dx$$

and (3-A-6) results. The last equality follows from Eq. (3-A-1) with $p(x) = -(x - b)$ and $F(x) = y''(x)$.

Reasoning similarly, we can show that

$$y(b) = y(a) + \cdots + y^{(n-1)}(a)\frac{(b - a)^{n-1}}{(n - 1)!} + y^{(n)}(\xi_i)\frac{(b - a)^n}{n!} \qquad a \leq \xi_i \leq b \quad \text{(3-A-7)}$$

APPENDIX 3-B

Asymptotic Properties of Fourier Transforms

We shall relate the behavior of the Fourier transform $F(\omega)$ of a function $f(t)$ for large ω to the continuity properties of $f(t)$ and its derivatives.

Definitions A function $f(t)$ is *bounded* in an interval (a,b) if there exists a constant M such that $|f(t)| < M$ for every t in this interval.

An arbitrary function $f(t)$ can be written as a difference of two increasing functions $f_1(t)$ and $f_2(t)$ constructed as in Fig. 3-23:

$$f(t) = f_1(t) - f_2(t) \tag{3-B-1}$$

The quantity

$$V = f_1(b) - f_1(a) + f_2(b) - f_2(a) \tag{3-B-2}$$

is the *total variation* of $f(t)$. If $V < \infty$, then $f(t)$ is of *bounded variation* (abbreviated BV) in the interval (a,b).

If a function $f(t)$ is not BV, then it is either unbounded or it oscillates rapidly. The functions $1/t$ and $\sin(1/t)$ are not BV in any interval including the origin. The first is unbounded, the second oscillatory.

If $f(t)$ is bounded and has finitely many extrema, then it is BV; it might be BV even if it has infinitely many extrema. The function $\sin \omega_0 t$ is BV in any finite interval; it is not BV in the interval $(-\infty,\infty)$. The function $\sin \omega_0 t/(1 + t^2)$ is BV in the interval $(-\infty,\infty)$ although it has infinitely many extrema.

Notation The expression

$$f(x) = o\left(\frac{1}{x^n}\right) \qquad x \to \infty$$

means that $x^n f(x) \to 0$ as $x \to \infty$. The expression

$$f(x) = O\left(\frac{1}{x^n}\right) \qquad |x| \to \infty$$

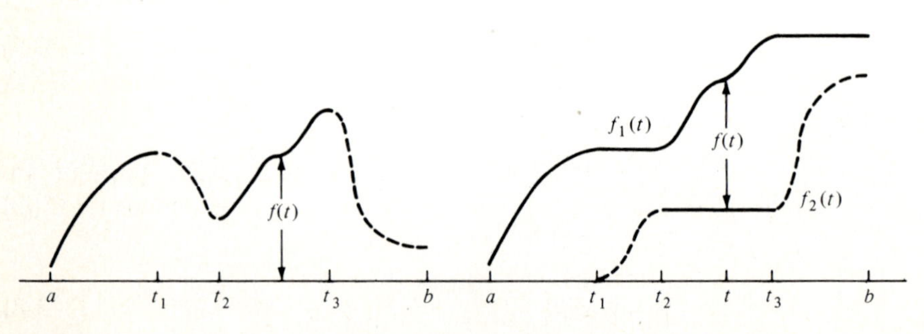

Figure 3-23

means that $x^n f(x)$ is bounded; that is, $|x^n f(x)| < c < \infty$ for every $|x| > x_0$, where x_0 is some finite constant.

The asymptotic properties of Fourier transforms are based on the following basic theorem.

The Riemann lemma If a function $f(t)$ of bounded variation is Riemann integrable in an interval (a,b), finite or infinite, and

$$F(\omega) = \int_a^b f(t)e^{-j\omega t}\,dt \tag{3-B-3}$$

then

$$F(\omega) = O\left(\frac{1}{\omega}\right) \qquad |\omega| \to \infty \tag{3-B-4}$$

That is, $F(\omega)$ goes to zero at least as fast as $1/\omega$ for $|\omega| \to \infty$.

Proof It is sufficient to assume that $f(t)$ is real. Suppose, first, that $f(t)$ is increasing. From Eq. (3-A-3), it follows that

$$\int_a^b f(t) \cos \omega t\,dt = f(a) \int_a^\tau \cos \omega t\,dt + f(b) \int_\tau^b \cos \omega t\,dt \tag{3-B-5}$$

But

$$\left|\int_\alpha^\beta \cos \omega t\,dt\right| = \left|\frac{\sin \alpha\omega - \sin \beta\omega}{\omega}\right| \le \frac{2}{|\omega|}$$

Hence,

$$\left|\int_a^b f(t) \cos \omega t\,dt\right| \le \frac{4M}{|\omega|} \tag{3-B-6}$$

where M is the largest absolute value of the two numbers $f(a)$ and $f(b)$.

Consider next an arbitrary $f(t)$. As we know [see Eq. (3-B-1)], $f(t) = f_1(t) - f_2(t)$, where $f_1(t)$ and $f_2(t)$ are two increasing functions. Applying (3-B-6) to $f_1(t)$ and $f_2(t)$, we conclude that $f(t)$ satisfies (3-B-6) where, now, M equals twice the maximum absolute value of $f_1(a)$, $f_1(b)$, $f_2(a)$, and $f_2(b)$.

The same bound holds for the integral of $f(t) \sin \omega t$, and (3-B-4) results because M is finite (bounded-variation assumption).

Suppose now that $f(t)$ is continuous with bounded derivative $f'(t)$. Since the transform of $f'(t)$ equals $j\omega F(\omega)$, we conclude from (3-B-4) that $j\omega F(\omega)$ is of the order of $1/\omega$ as $\omega \to \infty$. Hence,

$$F(\omega) = O\left(\frac{1}{\omega^2}\right) \qquad |\omega| \to \infty \tag{3-B-7}$$

Repeated application of the above leads to the following:

Theorem 1 If the function $f(t)$ and all its derivatives of order up to n exist and are of bounded variation in the interval $(-\infty, \infty)$, then its Fourier transform $F(\omega)$ tends to zero at least as fast as $1/\omega^{n+1}$ as $|\omega| \to \infty$:

$$F(\omega) = O\left(\frac{1}{\omega^{n+1}}\right) \qquad |\omega| \to \infty \tag{3-B-8}$$

Conversely, if (3-B-8) holds, then $f(t)$ and its derivatives of order up to $n - 1$ are continuous. The nth derivative is bounded but might be discontinuous.

Example 3-26

$$F(\omega) = \frac{1}{1 + \omega^4} = O\left(\frac{1}{\omega^4}\right) \qquad |\omega| \to \infty$$

Because $f(t)$, $f'(t)$, and $f''(t)$ are continuous, and $f'''(t)$ is bounded.

APPENDIX 3-C

Singularity Functions

Singularity functions can be defined in two equivalent ways.[1]

The delta function as distribution A distribution $f(t)$ is a process of assigning a number (functional) $N_f(\varphi)$ to a function $\varphi(t)$ of a certain class. If $f(t)$ is an ordinary function, then it will be considered as a distribution, and $N_f(\varphi)$ will be the integral

$$N_f(\varphi) = \int_{-\infty}^{\infty} \varphi(t) f(t)\, dt \tag{3-C-1}$$

We shall write the number $N_f(\varphi)$ as an integral even if $f(t)$ is not an ordinary function because we shall require that its properties be, formally, the properties of integrals. For example, the linearity

$$N_f(\varphi_1 + \varphi_2) = N_f(\varphi_1) + N_f(\varphi_2)$$

is a consequence of Eq. (3-C-1). We emphasize, however, that for arbitrary distributions, the right side of (3-C-1) has no independent meaning as an integral. It is merely the number $N_f(\varphi)$.

The delta function $\delta(t)$ is a functional assigning to a function $\varphi(t)$, continuous at the origin, the value $\varphi(0)$:

$$\int_{-\infty}^{\infty} \varphi(t)\, \delta(t)\, dt = \varphi(0) \tag{3-C-2}$$

The following properties of $\delta(t)$ are consistent with the above representation:

Property 1: $\qquad \delta(at) = \dfrac{1}{|a|}\delta(t) \qquad$ (3-C-3)

[1] M. J. Lighthill, "An Introduction to Fourier Analysis and Generalized Functions," Cambridge University Press, New York, 1959.

Proof With $at = x$, we have

$$\int_{-\infty}^{\infty} \varphi(t)\,\delta(at)\,dt = \frac{1}{|a|}\int_{-\infty}^{\infty} \varphi\left(\frac{x}{a}\right)\delta(x)\,dx = \frac{\varphi(0)}{|a|}$$

$$\frac{1}{|a|}\int_{-\infty}^{\infty} \varphi(t)\,\delta(t)\,dt = \frac{\varphi(0)}{|a|}$$

Property 2: $$\int_{-\infty}^{\infty} \varphi(\tau)\,\delta(t-\tau)\,d\tau = \int_{-\infty}^{\infty} \varphi(\tau)\,\delta(\tau - t)\,d\tau = \varphi(t) \tag{3-C-4}$$

Proof The first equality is a consequence of the evenness of $\delta(t)$. With $\tau - t = x$, the second integral yields

$$\int_{-\infty}^{\infty} \varphi(\tau)\,\delta(\tau - t)\,d\tau = \int_{-\infty}^{\infty} \varphi(t+x)\,\delta(x)\,dx = \varphi(t)$$

Note: From (3-C-4) it follows that

$$\varphi(t) * \delta(t) = \varphi(t)$$

Thus, $\delta(t)$ is the *identity element* in the operation of convolution.

Property 3: $$\int_{-\infty}^{\infty} \varphi(\tau)\,\delta^{(n)}(\tau)\,d\tau = (-1)^n \varphi^{(n)}(0) \tag{3-C-5}$$

Proof Differentiating the second equality in (3-C-4) n times with respect to t, we obtain

$$(-1)^n \int_{-\infty}^{\infty} \varphi(\tau)\,\delta^{(n)}(\tau - t)\,d\tau = \varphi^{(n)}(t)$$

and (3-C-5) results with $t = 0$.
Property 4: If $g(t)$ is an ordinary function continuous at $t = a$, then

$$g(t)\,\delta(t-a) = g(a)\,\delta(t-a) \tag{3-C-6}$$

Proof From (3-C-4) it follows that

$$\int_{-\infty}^{\infty} \varphi(t)g(t)\,\delta(t-a)\,dt = \varphi(a)g(a) \qquad \int_{-\infty}^{\infty} \varphi(t)g(a)\,\delta(t-a)\,dt = \varphi(a)g(a)$$

Property 5: $$\frac{dU(t)}{dt} = \delta(t) \tag{3-C-7}$$

where $U(t)$ is the step function and both sides are interpreted as distributions.

Proof

$$\int_{-\infty}^{\infty} \varphi(t)\frac{dU(t)}{dt}\,dt = \varphi(t)U(t)\Big|_{-\infty}^{\infty} - \int_{-\infty}^{\infty} U(t)\varphi'(t)\,dt = \varphi(\infty) - \int_{0}^{\infty} \varphi'(t)\,dt$$
$$= \varphi(\infty) - [\varphi(\infty) - \varphi(0)] = \varphi(0)$$

Notes:

1. The above properties *do not follow* from Eq. (3-C-2); they are assumptions *consistent with* the formal properties of integrals.
2. Integrals with finite limits can be considered as special cases of Eq. (3-C-4). Indeed, assuming that $\varphi(t) = 0$ for t outside the interval (a, b), we obtain

$$\int_a^b \varphi(t)\,\delta(t-c)\,dt = \int_{-\infty}^{\infty} \varphi(t)\,\delta(t-c)\,dt = \begin{cases} \varphi(c) & a < c < b \\ 0 & \text{otherwise} \end{cases} \tag{3-C-8}$$

3. Although $\delta(t)$ is not a function defined for every t, the statement "$\delta(t)$ is zero in any interval not containing the origin" has a meaning consistent with (3-C-8).

The delta function as limit Consider a family of functions $f_c(t)$ such that the limit

$$\lim_{c \to 0} \int_{-\infty}^{\infty} \varphi(t) f_c(t)\,dt = I_f(\varphi) \tag{3-C-9}$$

exists for every $\varphi(t)$ of a certain class. Clearly, this limit is a number depending on $\varphi(t)$; hence, it defines a distribution function $f(t)$. The function $f(t)$ so formed is called the *limit* of $f_c(t)$:

$$f(t) = \lim_{c \to 0} f_c(t) \tag{3-C-10}$$

The above limit need not exist in the ordinary sense.

If the given family $f_c(t)$ is such that the resulting $I_f(\varphi)$ equals $\varphi(0)$, then its limit $f(t)$ is the delta function $\delta(t)$:

$$\delta(t) = \lim_{c \to 0} f_c(t) \qquad \lim_{c \to 0} \int_{-\infty}^{\infty} \varphi(t) f_c(t)\,dt = \varphi(0) \tag{3-C-11}$$

The above is true if

$$\int_{-\infty}^{\infty} f_c(t)\,dt = 1 \tag{3-C-12}$$

and

$$\lim_{c \to 0} \int_{|t| > \epsilon} \varphi(t) f_c(t)\,dt = 0 \tag{3-C-13}$$

for any $\epsilon > 0$ (elaborate). It is not necessary, however, that $f_c(t) \to 0$ with $c \to 0$ for $t \neq 0$.

The following special cases are of particular interest (Fig. 3-24):

1. $$\delta(t) = \lim_{c \to 0} \frac{1}{c\sqrt{\pi}} e^{-t^2/c^2} \tag{3-C-14}$$

2. $$\delta(t) = \lim_{c \to 0} \frac{\sin(t/c)}{\pi t} \tag{3-C-15}$$

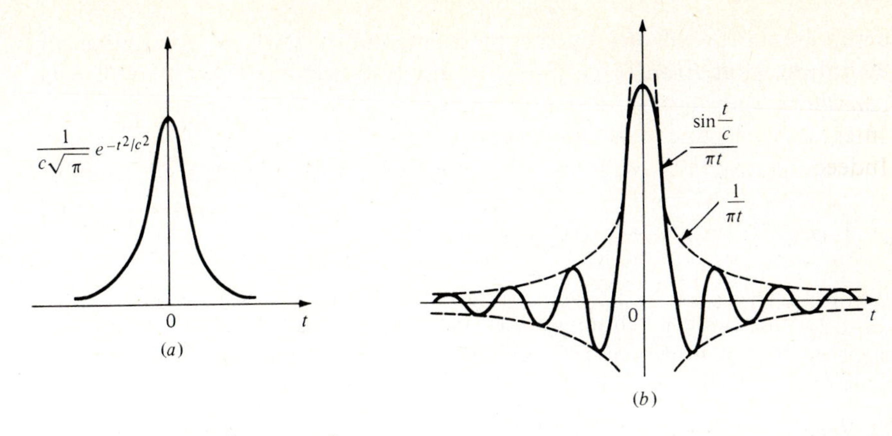

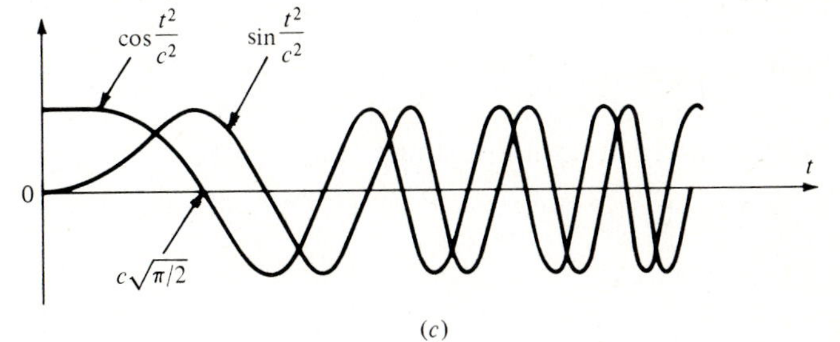

Figure 3-24

3. $$\delta(t) = \lim_{c \to 0} \frac{1}{c\sqrt{j\pi}} e^{jt^2/c^2} \tag{3-C-16}$$

In the first case, (3-C-13) is true because the function

$$f_c(t) = \frac{1}{c\sqrt{\pi}} e^{-t^2/c^2} \tag{3-C-17}$$

tends to zero with $c \to 0$ for $|t| > \epsilon$ (Fig. 3-24a).

In the second case, the function

$$f_c(t) = \frac{1}{\pi t} \sin\left(\frac{t}{c}\right) \tag{3-C-18}$$

oscillates rapidly for large c (Fig. 3-24*b*), and (3-C-13) follows from the Riemann lemma (3-B-4) with $f(t) = \varphi(t)/\pi t$ and $\omega = 1/c$ (elaborate).

In the last case,

$$f_c(t) = \frac{1}{c\sqrt{j\pi}} e^{jt^2/c^2} \tag{3-C-19}$$

and (3-C-13) is again a consequence of the oscillatory nature of $f_c(t)$ (Fig. 3-24*c*).

The representation of $\delta(t)$ by the preceding limits leads to a number of useful identities. Thus, from Eqs. (3-C-14) and (3-C-16) it follows that, if c is sufficiently small, then

$$\int_{-\infty}^{\infty} \varphi(t)e^{-(t-t_0)^2/c^2}\, dt \simeq c\sqrt{\pi}\,\varphi(t_0)$$
$$\int_{-\infty}^{\infty} \varphi(t)e^{j(t-t_0)^2/c^2}\, dt \simeq c\sqrt{j\pi}\,\varphi(t_0) \qquad \text{(3-C-20)}$$

In Sec. 4-1 we discuss refinements of these approximations. We note here only that, for their validity, $\varphi(t)$ should be sufficiently smooth. If $\varphi(t)$ is discontinuous at a point $t = t_1$, then (3-C-20) does not hold for $|t_1 - t_0| < c$.

With $A = 1/c$, (3-C-15) yields the identity

$$\delta(t) = \lim_{A\to\infty} \frac{\sin At}{\pi t} = \lim_{A\to\infty} \frac{1}{2\pi}\int_{-A}^{A} e^{j\omega t}\, d\omega = \frac{1}{2\pi}\int_{-\infty}^{\infty} e^{j\omega t}\, d\omega \qquad \text{(3-C-21)}$$

This result leads to the Fourier inversion formula [see (1-53)].

CHAPTER

FOUR

CONTINUOUS SYSTEMS

Assuming that the reader is familiar with the elementary properties of Laplace transforms and linear differential equations, we present, selectively, various topics from the theory of continuous (analog) systems.[1]

In Sec. 4-1, we develop the moment expansion of the convolution integral and use it in the study of differentiators, integrators, and spectrum analyzers.

In Sec. 4-2, we discuss the properties of low-pass filters, bandpass filters, and modulated signals. As preparation, we introduce the Hilbert transform for continuous signals, reserving its various extensions to Sec. 7-4.

In Sec. 4-3, we cover finite-order systems, stressing the analog version of the concepts introduced in Sec. 2-3 for discrete systems.

In Appendix 4-A, we prove the Cauchy-Schwarz inequality. We show that this important result, properly modified, leads to the determination of the maximum response of linear systems under various constraints.[2] *Other applications are given in subsequent chapters.*

---

[1] M. Van Valkenberg, "Network Analysis," Prentice-Hall, Inc., Englewood Cliffs, N.J., 1964.

[2] A. Papoulis, Maximum Response with Input Energy Constraints and Matched Filter Principle, *IEEE Trans. Circuit Theory*, vol. CT-17, no. 2, pp. 175–182, March, 1970.

4-1 MOMENT EXPANSION AND SPECTRUM ANALYZERS

In Sec. 1-2 we showed that the response $g(t)$ of a linear system to an arbitrary input $f(t)$ is given by

$$g(t) = \int_{-\infty}^{\infty} f(\tau)h(t - \tau)\, d\tau = \int_{-\infty}^{\infty} f(t - \tau)h(\tau)\, d\tau \tag{4-1}$$

where $h(t)$ is the impulse response of the system. This follows formally from the identity

$$f(t) = \int_{-\infty}^{\infty} f(\tau)\, \delta(t - \tau)\, d\tau \tag{4-2}$$

and the fact that $L[\delta(t - \tau)] = h(t - \tau)$.

If $f(t) = e^{j\omega t}$, then (4-1) yields

$$L[e^{j\omega t}] = H(\omega)e^{j\omega t} \tag{4-3}$$

where $H(\omega)$ is the system function.

We give another proof of these results: The response of the system to $e^{j\omega t}$ is a function

$$y(t, \omega) = L[e^{j\omega t}]$$

depending on t and on the parameter ω. Clearly (time-invariance)

$$y(t + t_0, \omega) = L[e^{j\omega(t + t_0)}]$$

for any t_0. But (linearity)

$$L[e^{j\omega(t + t_0)}] = e^{j\omega t_0}L[e^{j\omega t}] = e^{j\omega t_0}y(t, \omega)$$

Hence,

$$y(t + t_0, \omega) = e^{j\omega t_0}y(t, \omega)$$

With $t = 0$, the above yields

$$y(t_0, \omega) = e^{j\omega t_0}y(0, \omega)$$

Since this is true for any t_0, we conclude, with $H(\omega) = y(0, \omega)$, that $y(t, \omega) = H(\omega)e^{j\omega t}$. Thus, the response to an exponential is an exponential.

To derive (4-1), we express the signal $f(t)$ in terms of its Fourier transform $F(\omega)$:

$$f(t) = \frac{1}{2\pi}\int_{-\infty}^{\infty} F(\omega)e^{j\omega t}\, d\omega$$

Since the response to $e^{j\omega t}$ equals $H(\omega)e^{j\omega t}$, it follows from the linearity of the system that the response to $f(t)$ is given by

$$g(t) = \frac{1}{2\pi}\int_{-\infty}^{\infty} F(\omega)H(\omega)e^{j\omega t}\, d\omega \tag{4-4}$$

With $G(\omega)$ the transform of $g(t)$, we conclude from (4-4) that

$$G(\omega) = F(\omega)H(\omega) \tag{4-5}$$

and Eq. (4-1) results from the convolution theorem (3-28).

Moment Expansion

If $h(t)$ takes significant values for $|t| < \epsilon$ only, and $f(t)$ is sufficiently smooth so that $f(t - \tau) \simeq f(t)$ for $|\tau| < \epsilon$, then

$$g(t) = \int_{-\infty}^{\infty} f(t - \tau)h(\tau)\, d\tau \simeq f(t) \int_{-\infty}^{\infty} h(\tau)\, d\tau = H(0)f(t) \tag{4-6}$$

Similarly, if $f(t)$ is of short duration relative to $h(t)$, then

$$g(t) \simeq F(0)h(t) \tag{4-7}$$

Example 4-1 In Fig. 4-1 we show the response of a system with input $f(t)$ a triangular pulse of area $F(0) = AT$, and $h(t)$ an exponential of area $H(0) = B$. For $T = 10^2$, $h(t)$ is of short duration relative to $f(t)$; hence, $g(t) \simeq Bf(t)$. For $T = 10^{-2}$, $f(t)$ is of short duration relative to $h(t)$; hence, $g(t) \simeq ATh(t)$.

The approximation (4-6) is the first term in the expansion of $g(t)$ in terms of the derivatives of $f(t)$ and the moments [see (3-43)]

$$m_n = \int_{-\infty}^{\infty} t^n h(t)\, dt = j^n H^{(n)}(0) \tag{4-8}$$

of $h(t)$. We maintain that

$$g(t) = m_0 f(t) - m_1 f'(t) + \frac{m_2}{2} f''(t) + \cdots + \frac{(-1)^{n-1}}{(n-1)!} m_{n-1} f^{(n-1)}(t) + E_n \tag{4-9}$$

where E_n is an error term to be determined.

Proof 1 We expand the function $f(t - \tau)$ into a Taylor series about the point $\tau = 0$ [see Eq. (3-A-7)]:

$$f(t - \tau) = f(t) - \tau f'(t) + \frac{\tau^2}{2} f''(t) + \cdots + \frac{(-\tau)^{n-1}}{(n-1)!} f^{(n-1)}(t) + e_n \tag{4-10}$$

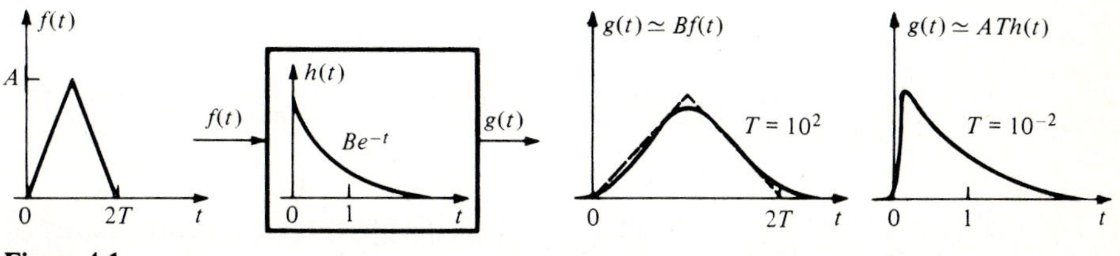

Figure 4-1

Inserting into the first integral in (4-6) and integrating termwise, we obtain (4-9).

Proof 2 We expand the function $H(\omega)$ into a Taylor series [see (3-43)]

$$H(\omega) = m_0 - j\omega m_1 + \frac{(j\omega)^2}{2} m_2 + \cdots + \frac{(-1)^n}{n!} (j\omega)^n m_n + \cdots \tag{4-11}$$

and insert into (4-5):

$$G(\omega) = m_0 F(\omega) - j\omega m_1 F(\omega) + \cdots + \frac{(-1)^n}{n!} (j\omega)^n m_n F(\omega) + \cdots$$

Taking the inverse transforms of both sides, we obtain (4-9) because $(j\omega)^n F(\omega)$ is the transform of $f^{(n)}(t)$.

Truncation error From Eq. (3-A-7), it follows that the term e_n in (4-10) is given by

$$e_n = \frac{(-\tau)^n}{n!} f^{(n)}(t - \tau_1) \qquad \text{where} \qquad 0 \le \tau_1 \le \tau$$

Inserting into (4-6), we conclude that the error E_n in (4-9) is given by

$$E_n = \frac{1}{n!} \int_{-\infty}^{\infty} (-\tau)^n f^{(n)}(t - \tau_1) h(\tau)\, d\tau \tag{4-12}$$

Since τ_1 depends on τ, the factor $f^{(n)}(t - \tau_1)$ cannot be taken outside the integral. However, if $f^{(n)}(t)$ is continuous and $t^n h(t) \ge 0$, then [see (3-A-1)]

$$E_n = \frac{1}{n!} f^{(n)}(t - \tau_0) \int_{-\infty}^{\infty} (-\tau)^n h(\tau)\, d\tau = \frac{(-1)^n m_n}{n!} f^{(n)}(t - \tau_0) \tag{4-13}$$

where τ_0 is some constant in the interval of integration.

Example 4-2 If $h(t)$ is a rectangular pulse of unit area,

$$h(t) = \begin{cases} \dfrac{1}{2a} & |t| \le a \\ 0 & |t| > a \end{cases}$$

then $m_0 = 1$, $m_1 = 0$, and $m_2 = a^2/3$. With $n = 2$, it follows from (4-9) and (4-13) that

$$g(t) = \frac{1}{2a} \int_{-a}^{a} f(t - \tau)\, d\tau = f(t) + \frac{a^2}{6} f''(t - \tau_0) \qquad |\tau_0| \le a \tag{4-14}$$

The moment expansion (4-9) can be used to derive the asymptotic series of certain integrals. This is illustrated in the next example.

Example 4-3

(*a*) If $h(t)$ is a normal curve of unit area,

$$h(t) = \frac{1}{c\sqrt{\pi}} e^{-t^2/c^2}$$

then
$$m_n = \frac{1}{c\sqrt{\pi}} \int_{-\infty}^{\infty} t^n e^{-t^2/c^2}\, dt = \begin{cases} \dfrac{1 \cdot 3 \cdots (n-1)c^n}{\sqrt{2^n}} & n \text{ even} \\ 0 & n \text{ odd} \end{cases}$$

This follows by repeated differentiation of the integral [see Eq. (8-2)]

$$\int_{-\infty}^{\infty} e^{-\alpha t^2}\, dt = \sqrt{\frac{\pi}{\alpha}} \tag{4-15}$$

Inserting into (4-9), we obtain

$$\int_{-\infty}^{\infty} f(t-\tau) \frac{e^{-\tau^2/c^2}}{c\sqrt{\pi}}\, d\tau = f(t) + \frac{c^2}{4} f''(t) + \cdots \tag{4-16}$$

(*b*) The preceding holds also if c is replaced with $c\sqrt{j}$; hence,

$$\int_{-\infty}^{\infty} f(t-\tau) \frac{e^{j\tau^2/c^2}}{c\sqrt{j\pi}}\, d\tau = f(t) + \frac{jc^2}{4} f''(t) + \cdots \tag{4-17}$$

From the above it follows that the delta function is a limit of a normal curve and of a linear FM signal

$$\frac{1}{c\sqrt{\pi}} e^{-t^2/c^2} \underset{c\to 0}{\rightarrow} \delta(t) \qquad \frac{1}{c\sqrt{j\pi}} e^{jt^2/c^2} \underset{c\to 0}{\rightarrow} \delta(t) \tag{4-18}$$

[see (1-20)] because the right sides of (4-16) and (4-17) tend to $f(t)$ as $c \to 0$.

Notes:

1. From the identity $F(\omega)H(\omega) = F(\omega)e^{-j\omega t_0}H(\omega)e^{j\omega t_0}$, it follows that $f(t) * h(t) = f(t - t_0) * h(t + t_0)$. Hence, in the expansion (4-9), we can replace $f(t)$ with $f(t - t_0)$ and the moments m_n of $h(t)$ with the moments

$$\mu_n = \int_{-\infty}^{\infty} t^n h(t + t_0)\, dt = \int_{-\infty}^{\infty} (t - t_0)^n h(t)\, dt$$

of $h(t + t_0)$. If $t_0 = m_1/m_0 = \eta$ is the "center of gravity" of $h(t)$, then $\mu_1 = 0$ and

$$\mu_2 = \int_{-\infty}^{\infty} (t-\eta)^2 h(t)\, dt = m_2 - \frac{m_1{}^2}{m_0} \tag{4-19}$$

is the "central moment of inertia" of $h(t)$. This choice of t_0 is desirable because the second term of the resulting moment expansion of $g(t)$ is zero.

2. We shall relate the parameters m_0, η, and μ_2 of a real signal $h(t)$ to the amplitude $A(\omega)$ and phase $\varphi(\omega)$ of its transform

$$H(\omega) = A(\omega)e^{j\varphi(\omega)}$$

Since $A(-\omega) = A(\omega)$ and $\varphi(-\omega) = -\varphi(\omega)$, we conclude that $A'(0) = 0$ and $\varphi(0) = 0$. Hence [see (4-11)],

$$A(\omega)e^{j\varphi(\omega)} = \left[A(0) + \frac{A''(0)}{2}\omega^2 + \cdots\right]e^{j\varphi'(0)\omega + \cdots}$$

$$= m_0 - j\omega m_1 - \frac{\omega^2}{2}m_2 + \cdots$$

Expanding the exponential and equating coefficients of equal powers of ω, we obtain

$$A(0) = m_0 \qquad A(0)\varphi'(0) = -m_1 \qquad A''(0) - A(0)[\varphi'(0)]^2 = -m_2$$

Hence,

$$\varphi'(0) = -\frac{m_1}{m_0} = -\eta \qquad A''(0) = \frac{m_1^2}{m_0} - m_2 = -\mu_2 \tag{4-20}$$

In the above, we have assumed that $m_0 \neq 0$.

Systems as Measuring Devices

Analog systems are often used to measure various functionals of a given signal $f(t)$. We shall examine conditions of approximating the following ideal systems with real, causal devices.

Proportional system:	$g(t) = Af(t)$	$H(\omega) = A$
Differentiator:	$g(t) = Af'(t)$	$H(\omega) = Aj\omega$
Integrator:	$g(t) = A\int_{-\infty}^{t} f(\tau)\,d\tau$	$h(t) = AU(t)$

Proportional systems If $h(t)$ is of short duration relative to $f(t)$ or, equivalently, if $F(\omega)$ is of short duration relative to $H(\omega)$ (Fig. 4-2) and the area $m_0 = H(0) = A(0)$

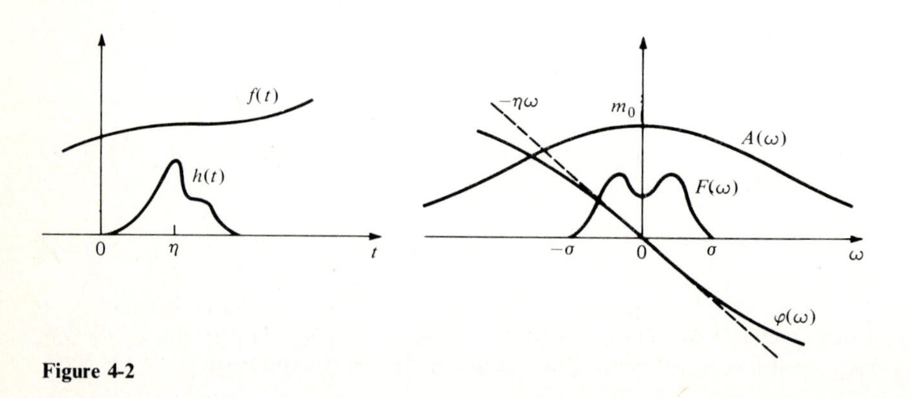

Figure 4-2

of $h(t)$ is not zero, then $H(\omega)$ can be approximated by $H(0)$ in the band $(-\sigma, \sigma)$ of the input. Hence,

$$G(\omega) \simeq H(0)F(\omega) \qquad g(t) \simeq m_0 f(t) \qquad m_0 \neq 0$$

as in (4-6). The approximation is improved if $A(\omega)$ is replaced with a constant, and $\varphi(\omega)$ with a straight line [see Eq. (4-20)]:

$$G(\omega) \simeq m_0 e^{-j\eta\omega} F(\omega) \qquad g(t) \simeq m_0 f(t-\eta) \tag{4-21}$$

Thus, under the above conditions, the output of the system is proportional to the delayed input. To determine the first-order term of the resulting error, we approximate $A(\omega)$ with a parabola in the band $(-\sigma, \sigma)$ of the input:

$$G(\omega) \simeq \left(m_0 - \frac{\mu_2}{2}\omega^2\right) e^{-j\eta\omega} F(\omega) \qquad g(t) \simeq m_0 f(t-\eta) + \frac{\mu_2}{2} f''(t-\eta) \tag{4-22}$$

The above shows that, for the validity of (4-21), the condition $\mu_2 f''(t) \ll m_0 f(t)$ must hold.

Example 4-4 If

$$h(t) = e^{-\alpha t}U(t) \qquad H(\omega) = \frac{1}{\alpha + j\omega}$$

then

$$m_n = \int_0^\infty t^n e^{-\alpha t}\, dt = \frac{n!}{\alpha^{n+1}}$$

Hence,

$$m_0 = \frac{1}{\alpha} \qquad \eta = \frac{m_1}{m_0} = \frac{1}{\alpha} \qquad \mu_2 = \frac{2}{\alpha^3} - \frac{1}{\alpha^3} = \frac{1}{\alpha^3}$$

and (4-22) yields

$$g(t) \simeq \frac{1}{\alpha} f\left(t - \frac{1}{\alpha}\right) + \frac{1}{2\alpha^3} f''\left(t - \frac{1}{\alpha}\right)$$

Step response The output $g(t)$ is nearly proportional to $f(t)$ only if $f(t)$ is continuous. To examine the behavior of $g(t)$ in the vicinity of discontinuity points of $f(t)$, we shall consider the special case $f(t) = U(t)$. This is sufficient because an arbitrary signal can be written as a sum of a discontinuous function and suitable steps.

We denote by $a(t)$ the *step response* of the system:

$$a(t) = L[U(t)] = \int_0^t h(\tau)\, d\tau \tag{4-23}$$

Assuming that $h(t)$ does not contain any singularities, we observe that $a(0) = 0$ and $a(\infty) = m_0$.

If $\varphi(\omega) = -\eta\omega$, then $h(t)$ has axis symmetry, $a(t)$ has point symmetry at $t = \eta$ as in Fig. 4-3, and $a(\eta) = m_0/2$. The constant η is known as the *delay time* of the filter, and the constant $t_r = m_0/h(\eta)$ *the rise time.*

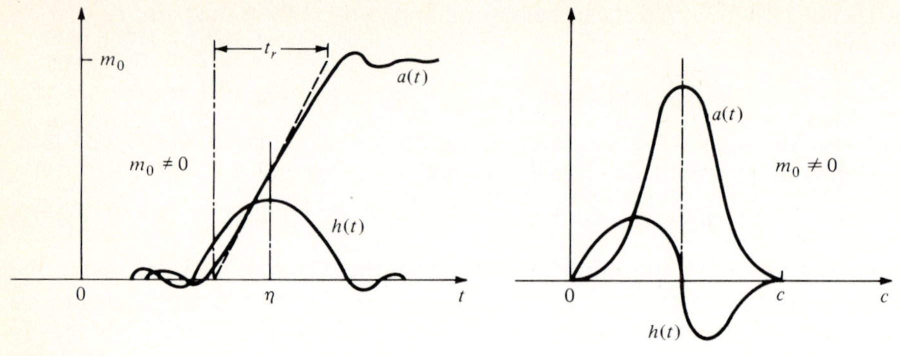

Figure 4-3

The response $g(t)$ of a system to an arbitrary causal input $f(t)$ can be expressed in terms of $a(t)$. Indeed, since $a'(t) = h(t)$, we conclude that

$$\int_0^t f(\tau)h(t-\tau)\,d\tau = -\int_0^t f(\tau)a'(t-\tau)\,d\tau = -f(\tau)a(t-\tau)\Big|_0^t + \int_0^t f'(\tau)a(t-\tau)\,d\tau$$

Hence,

$$g(t) = f(0)a(t) + \int_0^t f'(\tau)a(t-\tau)\,d\tau \tag{4-24}$$

The above follows also when $f(t)$ is expressed as sum of steps (Fig. 4-4):

$$f(t) \simeq f(0)U(t) + \sum_i f'(\tau_i)\,\Delta\tau\,U(t-\tau_i)$$

$$g(t) \simeq f(0)a(t) + \sum_i f'(\tau_i)\,\Delta\tau\,a(t-\tau_i)$$

Differentiators If $h(t)$ is of short duration relative to $f'(t)$ and its area $m_0 = H(0)$ is zero, then [see (4-9)]

$$g(t) \simeq -m_1 f'(t)$$

Using (4-24), we shall show that the properties of differentiators can be deduced from the properties of proportional systems.

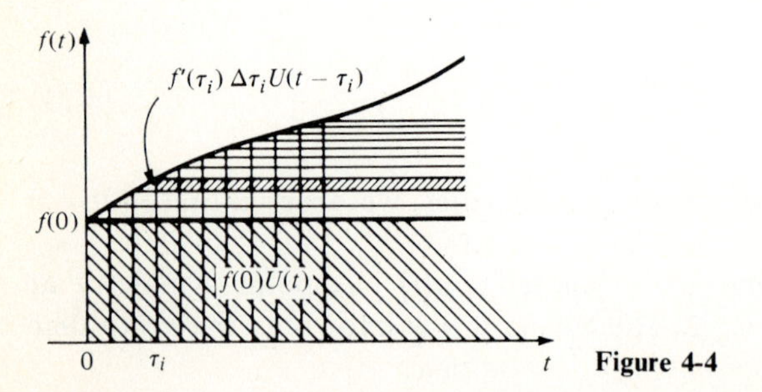

Figure 4-4

We shall assume that $h(t) = 0$ for $t > c$ (Fig. 4-3). Since $m_0 = 0$, it follows from Eq. (4-23) that $a(t) = 0$ for $t > c$ and (4-24) yields

$$g(t) = \int_0^t f'(\tau)a(t-\tau)\,d\tau \qquad t > c$$

The above shows that for $t > c$, $g(t)$ equals the output of a system with input $f'(t)$ and impulse response $a(t)$. Hence, it satisfies (4-22) provided that $f(t)$ is replaced with $f'(t)$, and the moments m_n of $h(t)$ with the moments $\overline{m}_n$ of $a(t)$. Using the fact that $a'(t) = h(t)$ and integrating by parts, we find

$$\overline{m}_n = \int_{-\infty}^{\infty} t^n a(t)\,dt = -\frac{1}{n+1}\int_{-\infty}^{\infty} t^{n+1}h(t)\,dt = -\frac{m_{n+1}}{n+1} \tag{4-25}$$

and (4-22) yields

$$g(t) \simeq \overline{m}_0\, f'(t-\overline{\eta}) + \frac{\overline{\mu}_2}{2} f'''(t-\overline{\eta}) \tag{4-26}$$

where $\overline{m}_0 = -m_1 \qquad \overline{\eta} = \dfrac{\overline{m}_1}{\overline{m}_0} = \dfrac{m_2}{2m_1} \qquad \overline{\mu}_2 = \overline{m}_2 - \dfrac{\overline{m}_1{}^2}{\overline{m}_0} = -\dfrac{m_3}{3} + \dfrac{m_2{}^2}{4m_1}$

Example 4-5 If $g(t) = (1/T)[f(t) - f(t-T)]$, then

$$H(\omega) = \frac{1 - e^{-j\omega T}}{T} = j\omega - \frac{(j\omega)^2}{2!}T + \frac{(j\omega)^3}{3!}T^2 + \cdots$$

hence, $m_1 = -1$, $m_2 = -T$, and $m_3 = -T^2$. Inserting into (4-9), we obtain

$$g(t) \simeq f'(t) - \frac{T}{2}f''(t) + \frac{T^2}{6}f'''(t)$$

As we see from Eq. (4-26), $\overline{m}_0 = 1$, $\overline{m}_1 = T/2$, and $\overline{\mu}_2 = T^2/12$, and the central-moment expansion (4-26) yields

$$g(t) \simeq f'\left(t - \frac{T}{2}\right) + \frac{T^2}{24} f'''\left(t - \frac{T}{2}\right)$$

Integrators Any system with $h(0) \neq 0$ can be used to measure the integral of a signal applied at $t = 0$, provided that the integration interval is sufficiently small. Indeed, if b is such that $h(t) \simeq h(0)$ for $t < b$, then

$$g(t) = \int_0^t f(\tau)h(t-\tau)\,d\tau \simeq h(0)\int_0^t f(\tau)\,d\tau \qquad \text{for} \qquad t < b \tag{4-27}$$

because $t - \tau < t < b$.

If $f(t) = 0$ for $t > b$, and $h(t)$ is approximately constant in any interval of length b (Fig. 4-5), then $h(t-\tau) \simeq h(t)$ for $\tau < b$. Hence,

$$g(t) = \int_0^b f(\tau)h(t-\tau)\,d\tau \simeq h(t)\int_0^b f(\tau)\,d\tau = F(0)h(t) \qquad t > b \tag{4-28}$$

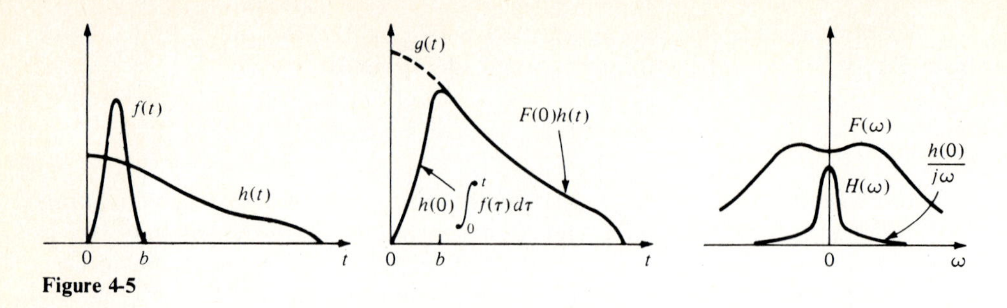

Figure 4-5

The coefficient $F(0)$ of the resulting response thus equals the total area of $f(t)$. This result can be improved as in (4-22).

> **Note:** The approximations (4-27) and (4-28) can also be deduced from frequency considerations. Indeed, since $h(t)$ is slowly varying, $H(\omega)$ is a low-pass filter (Fig. 4-5) whose band is sufficiently small so that $F(\omega)H(\omega) \simeq F(0)H(\omega)$, and (4-28) results. The justification of (4-27) involves a more subtle argument. It stems from the fact that the behavior of a signal near the origin is based on the behavior of its transform for large ω. Furthermore, if $h(t) = 0$ for $t < 0$ and $h(0) \neq 0$, then, for large ω, $H(\omega) \simeq h(0)/j\omega$ (asymptotic theorems). Thus, to determine $g(t)$ for small t, we can use the approximation $G(\omega) \simeq h(0)F(\omega)/j\omega$ whose inverse yields (4-27).

Spectrum Analyzers

A spectrum analyzer is an analog system designed to measure the Fourier transform $F(\omega)$ of a physical signal $f(t)$. Since $F(\omega)$ is the integral of the product $f(t)e^{-j\omega t}$, an analyzer is in fact an integrator. We shall discuss three schemes for determining $F(\omega)$ (see also Sec. 5-6):

1. Modulation We multiply $f(t)$ with the output $s(t) = e^{-j\omega_0 t}$ of a signal generator and use the resulting product

$$f_s(t) = f(t)e^{-j\omega_0 t}$$

as the input to a low-pass filter $H(\omega)$ (Fig. 4-6). Since $F_s(\omega) = F(\omega + \omega_0)$, we conclude that, if the band $(-\omega_c, \omega_c)$ of the filter is sufficiently small, then

$$G(\omega) = F(\omega + \omega_0)H(\omega) \simeq F(\omega_0)H(\omega) \tag{4-29}$$

Hence, $g(t) \simeq F(\omega_0)h(t)$ as in (4-28). For this to hold, the frequency ω_0 of the modulator must remain constant during the entire duration of $f(t)$.

To determine $F(\omega)$ for every ω, we must assign to ω_0 all values in the band of interest. This can be done if the frequency of the signal generator is changed and the signal $f(t)$ is stored and used repeatedly.

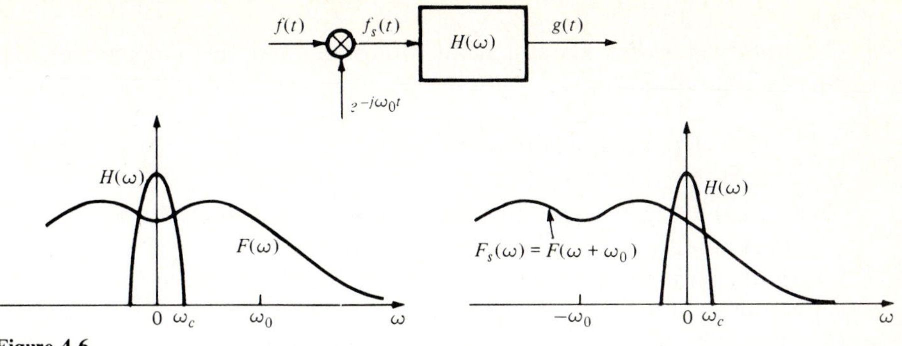

Figure 4-6

Real-time spectrum analyzer We shall show that if the output $s(t)$ of the signal generator and the impulse response $h(t)$ of the filter are linear FM signals whose instantaneous frequencies vary at opposite rates:

$$s(t) = e^{j(\omega_0 t - \alpha t^2)} \qquad h(t) = e^{j(\omega_0 t + \alpha t^2)} \tag{4-30}$$

then

$$g(t) = F(2\alpha t)h(t) \tag{4-31}$$

Proof The input to the system $h(t)$ is the product $f(t)s(t)$. Hence,

$$\begin{aligned} g(t) &= \int_{-\infty}^{\infty} f(\tau)e^{j(\omega_0 \tau - \alpha\tau^2)}e^{j[\omega_0(t-\tau)+\alpha(t-\tau)^2]}\,d\tau \\ &= e^{j(\omega_0 t + \alpha t^2)}\int_{-\infty}^{\infty} f(\tau)e^{-j2\alpha t\tau}\,d\tau \end{aligned}$$

and (4-31) results because the last integral equals $F(2\alpha t)$.

Thus, the envelope of the response equals the transform of the input. Hence, $F(\omega)$ can be determined by a simple measurement as a time signal. However, the required filter $h(t)$ is difficult to realize physically.

Example 4-6 If $f(t)$ is a pulse as in Fig. 4-7, then $F(\omega) = 2 \sin T\omega/\omega$. Hence, the envelope of the output equals

$$F(2\alpha t) = \frac{\sin 2\alpha T t}{\alpha t}$$

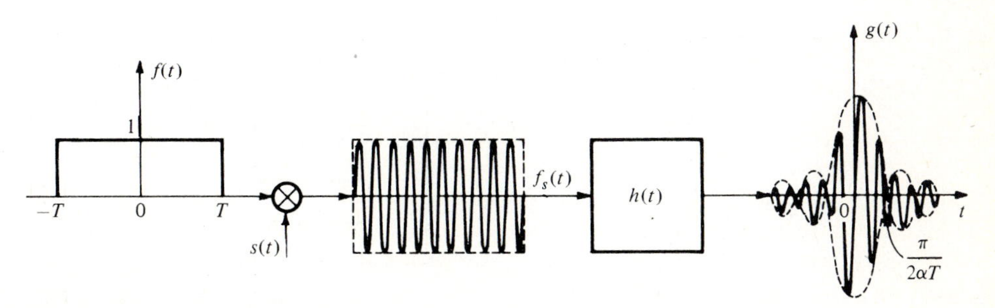

Figure 4-7

Notes:

1. *Quadratic phase filter.* The system function of the analyzer of Fig. 4-7 is given by [see Eq. (8-6)]

$$H(\omega) = \int_{-\infty}^{\infty} e^{j(\omega_0 t + \alpha t^2)} e^{-j\omega t}\, dt = \sqrt{\frac{j\pi}{\alpha}}\, e^{-j(\omega-\omega_0)^2/4\alpha} \tag{4-32}$$

Thus, $A(\omega) = \sqrt{\pi/\alpha}$ = constant, and $\varphi(\omega) = -(\omega - \omega_0)^2/4\alpha$; that is, the system is an all-pass filter with quadratic phase.

2. *Causal approximation.* Since $h(t)$ is not causal, a real-time spectrum analyzer cannot be realized physically. However, it can be approximated arbitrarily closely if sufficient delay is introduced. The envelope of the resulting output is then approximately equal to $F[2\alpha(t - t_0)]$.

3. *Pulse compression.* If α is sufficiently large, then the envelope $F(2\alpha t)$ of $g(t)$ can be made arbitrarily narrow. Thus, the analyzer of Fig. 4-7 can be used to "compress" the input signal $f(t)$. The compression ratio is proportional to the slope 2α of the instantaneous frequency $\omega_0 - 2\alpha t$ of the FM generator $s(t)$. If $f(t)$ is a pulse of duration $2T$ as in Example 4-6, and as a measure of the duration of $F(2\alpha t)$ we use the size $\pi/\alpha T$ of its main lobe, then the compression ratio equals $2\alpha T^2/\pi$ [see also Eq. (8-138)].

2. Bandpass filtering The signal $f(t)$ is the input to a bank of n bandpass filters connected in parallel as in Fig. 4-8. If the band $\Delta\omega_i$ of the ith filter is sufficiently small and is centered at ω_i, then its output $g_i(t)$ is given by

$$g_i(t) \simeq F(\omega_i)h_i(t) \qquad \text{because} \qquad G_i(\omega) = F(\omega)H_i(\omega) \simeq F(\omega_i)H_i(\omega) \tag{4-33}$$

From the above, it follows that the energy

$$E_i = \int_{-\infty}^{\infty} |g_i(t)|^2\, dt \simeq |F(\omega_i)|^2 \int_{-\infty}^{\infty} |h_i(t)|^2\, dt \tag{4-34}$$

of the output of the ith filter is proportional to $|F(\omega_i)|^2$. Hence, $|F(\omega_i)|$ can be determined by measuring E_i. The precise value of E_i is the weighted average

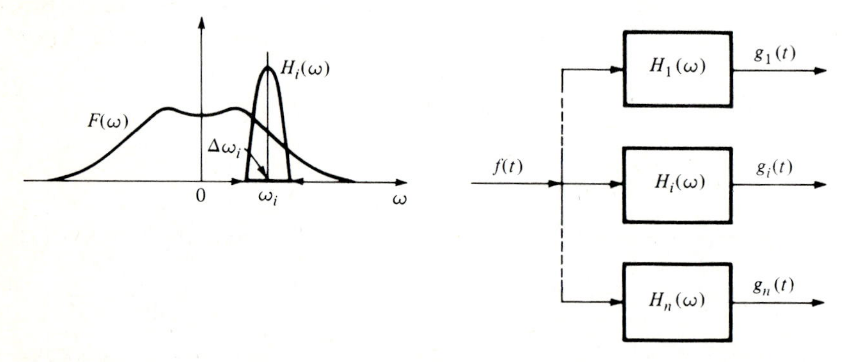

Figure 4-8

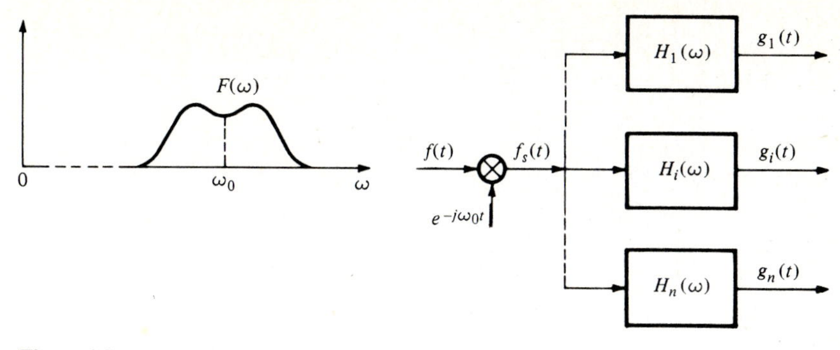

Figure 4-9

of $F(\omega)$ in the band of the ith filter. If the filter is ideal, i.e., if $H_i(\omega) = A$ is constant in the band (ω_a, ω_b) and zero otherwise, then

$$E_i = \frac{1}{2\pi}\int_{-\infty}^{\infty} |F(\omega)H(\omega)|^2\, d\omega = \frac{A^2}{2\pi}\int_{\omega_a}^{\omega_b} |F(\omega)|^2\, d\omega \tag{4-35}$$

3. Heterodyne In the analyzer of Fig. 4-8, the center frequencies ω_i of the filters $H_i(\omega)$ must cover the entire band of interest of the input. For high-frequency signals, it is more convenient to combine methods 1 and 2 as in Fig. 4-9. The signal $f(t)$ is multiplied by $e^{-j\omega_0 t}$, and the resulting product $f_s(t)$ is used as the input to m bandpass filters. The output $g_i(t)$ of the ith filter is given by $g_i(t) \simeq F(\omega_i + \omega_0)h_i(t)$ because

$$G_i(\omega) = F_s(\omega)H_i(\omega) = F(\omega + \omega_0)H_i(\omega) \simeq F(\omega_i + \omega_0)H_i(\omega)$$

Line spectra We wish to determine the amplitude a and the frequency ω_s of a harmonic signal

$$f(t) = ae^{j\omega_s t}$$

If $f(t)$ is the input t[illegible] analyzer of Fig. 4-8, then $g_i(t) = aH_i(\omega_s)e^{j\omega_s t}$. Hence, if the bands of the [illegible] do not overlap, and if they cover the entire range of possible values of ω_s, then one and only one response will be different from zero. Suppose that $H_i(\omega_s) \neq 0$. It then follows that ω_s is in the band (ω_a, ω_b) of the ith filter, and its amplitude can be determined in terms of $g_i(t)$.

If the adjacent filters overlap as in Fig. 4-10, then the ratio

$$\frac{g_{i+1}(t)}{g_i(t)} = \frac{H_{i+1}(\omega_s)}{H_i(\omega_s)}$$

of the resulting responses can be used to determine more accurately the value of ω_s.

Signals consisting of several harmonic components can be analyzed similarly.

The transient system function In the preceding, we have used the fact that the response of a filter to $e^{j\omega t}$ equals $H(\omega)e^{j\omega t}$. This is true only if $f(t) = e^{j\omega t}$ for all t.

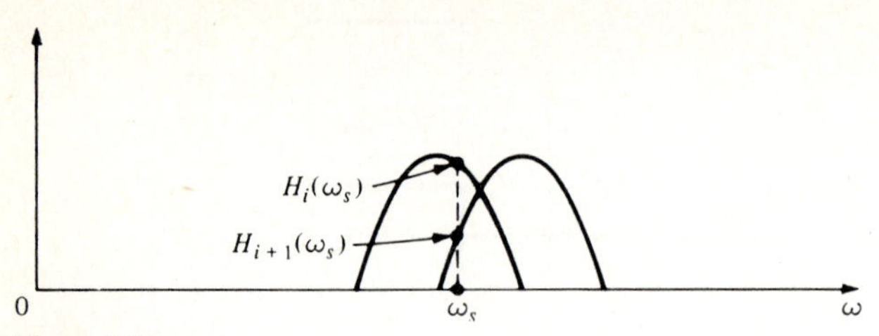

Figure 4-10

In practice, the input is applied at $t = 0$ (the time origin is arbitrary). Hence, to determine the selectivity properties of the filter, we must consider the response to a truncated exponential:

$$L[e^{j\omega t}U(t)] = \int_{-\infty}^{t} h(\tau)e^{j\omega(t-\tau)}\, d\tau = e^{j\omega t} \int_{-\infty}^{t} h(\tau)e^{-j\omega\tau}\, d\tau$$

Introducing the function

$$H_t(\omega) = \int_{-\infty}^{t} h(\tau)e^{-j\omega\tau}\, d\tau \tag{4-36}$$

we conclude that

$$L[e^{j\omega t}U(t)] = H_t(\omega)e^{j\omega t} \tag{4-37}$$

Thus, the response to an exponential applied at $t = 0$ is an exponential multiplied by $H_t(\omega)$. The factor $H_t(\omega)$ we shall call the *transient system function* of the given system. This function depends not only on ω, but also on t. At a given t, the selectivity properties of the filter are determined by the ω dependence of $|H_t(\omega)|$. Clearly, $H_t(\omega) \to H(\omega)$ for $t \to \infty$; hence, if t is sufficiently large, then $H_t(\omega)$ can be replaced with $H(\omega)$. The need for using $H_t(\omega)$ arises when the time of observation is limited.

Example 4-7 We shall determine the transient system function of the system

$$h(t) = e^{-\alpha t}U(t) \leftrightarrow H(\omega) = \frac{1}{\alpha + j\omega}$$

Since $h(t) = 0$ for $t < 0$, it follows from Eq. (4-36) that

$$H_t(\omega) = \int_{0}^{t} e^{-\alpha\tau}e^{-j\omega\tau}\, d\tau = \frac{1 - e^{-(\alpha + j\omega)t}}{\alpha + j\omega} \tag{4-38}$$

In Fig. 4-11, we plot the amplitude

$$|H_t(\omega)| = \sqrt{\frac{1 + e^{-2\alpha t} - 2e^{-\alpha t}\cos \omega t}{\alpha^2 + \omega^2}} \tag{4-39}$$

as a function of ω for $t = 0.4/\alpha$, $0.8/\alpha$, $1.6/\alpha$, and ∞. We note that the width of $|H_t(\omega)|$ decreases with increasing t; for sufficiently large t (of the order of $5/\alpha$), it approaches the width of $|H(\omega)|$.

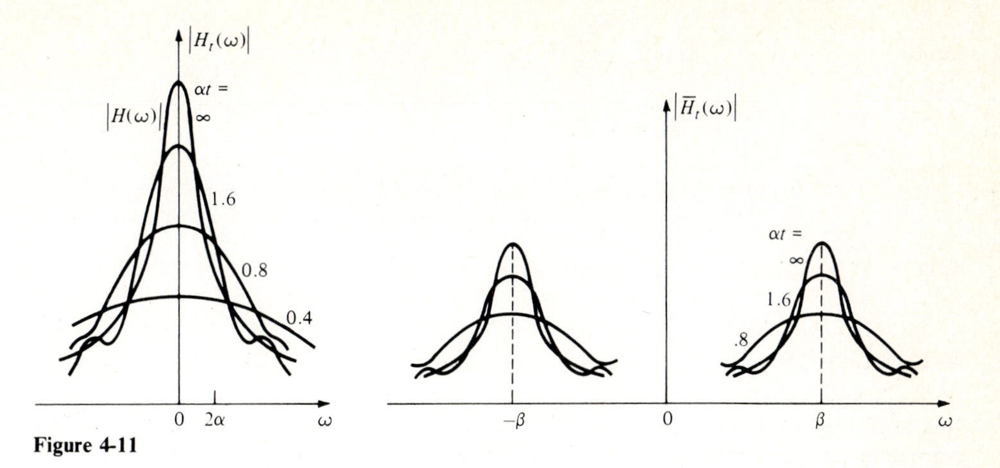

Figure 4-11

Notes:

1. If the bandwidth of a filter, measured in some realistic way, is $\Delta\omega_i$, then the duration T_i of its impulse response must be of the order of $1/\Delta\omega_i$ at least. This is evident in the preceding example and can be established in general (see Sec. 8-2). Therefore, in the selection of the filters $H_i(\omega)$ of an analyzer, a compromise must be sought between the time and bandwidth requirements: The analyzer localizes the frequency ω_s of a harmonic signal in the band (ω_a, ω_b) of the filter, and the energy of its response is proportional to the average of $|F(\omega)|^2$ in this band [see (4-35)]. Hence, the bandwidth $\omega_b - \omega_a$ must be small. The time t of observation must be larger than T_i. Otherwise, the selectivity of the filter is specified not by $H_i(\omega)$ but by the broader curve $H_t(\omega)$, and the output energy is no longer given by (4-35).
2. In the preceding treatment of analyzers, we considered (for simplicity) complex signals and systems. As we show in the next section, real signals and systems lead to similar results.
3. If the transient system function of a system $H(\omega)$ equals $H_t(\omega)$, then the corresponding function of the system $H_1(\omega) = H(\omega - \beta)$ equals $H_t(\omega - \beta)$. This follows when the impulse response $h_1(t) = h(t)e^{j\beta t}$ is inserted into Eq. (4-36).

Example 4-8 We wish to find the transient system function $\overline{H}_t(\omega)$ of the bandpass filter

$$\overline{H}(\omega) = \frac{1}{(j\omega + \alpha)^2 + \beta^2}$$

With $H(\omega) = 1/(\alpha + j\beta)$ as in Example 4-7, we have

$$\overline{H}(\omega) = \frac{1/2j\beta}{\alpha + j(\omega - \beta)} - \frac{1/2j\beta}{\alpha + j(\omega + \beta)} = \frac{1}{2j\beta}[H(\omega - \beta) - H(\omega + \beta)]$$

Hence, $$\bar{H}_t(\omega) = \frac{1}{2j\beta}[H_t(\omega - \beta) - H_t(\omega + \beta)]$$

where $H_t(\omega)$ is the function in (4-38).

If $\alpha \ll \beta$, then

$$|\bar{H}_t(\omega)| \simeq \frac{1}{2\beta}\sqrt{\frac{1 + e^{-2\alpha t} - 2e^{-\alpha t}\cos(\omega - \beta)t}{\alpha^2 + (\omega - \beta)^2}} \qquad \omega > 0$$

(Fig. 4-11) because $H_t(\omega + \beta)$ is negligible for $\omega > 0$.

4-2 FILTERS

We shall say that an analog system is a *filter* if its frequency response is zero outside a certain region B of the frequency axis. If B is the interval $(-\omega_c, \omega_c)$, then the filter is low-pass.

The response of a filter can be determined either from the convolution integral (4-1) or from the inversion formula (4-4). However, in special cases, other methods are available. Suppose, for example, that the filter is low-pass with linear phase:

$$H(\omega) = A(\omega)e^{-jt_0\omega} \tag{4-40}$$

Expanding its amplitude $A(\omega)$ into a Fourier series in the interval $(-\omega_c, \omega_c)$, we obtain

$$A(\omega) = \sum_{n=-\infty}^{\infty} a_n e^{jnT\omega} \qquad |\omega| \le \omega_c = \frac{\pi}{T} \tag{4-41}$$

and since $A(\omega) = 0$ for $|\omega| > \omega_c$ and $G(\omega) = F(\omega)H(\omega)$, we conclude with (4-40) that

$$G(\omega) = F(\omega)p_{\omega_c}(\omega) \sum_{n=-\infty}^{\infty} a_n e^{j(nT - t_0)\omega} \tag{4-42}$$

The inverse of the product $F(\omega)p_{\omega_c}(\omega)$ equals

$$f_c(t) = f(t) * \frac{\sin \omega_c t}{\pi t}$$

Hence (Fig. 4-12),

$$g(t) = \sum_{n=-\infty}^{\infty} a_n f_c(t + nT - t_0) \tag{4-43}$$

Causality It is known [see the Paley-Wiener condition, Eq. (7-15)] that if $h(t) = 0$ for $t < 0$, then $H(\omega)$ cannot be zero for every ω in an interval. Thus, a filter, as defined earlier, is not a causal system. However, as we show next, it can be approximated by a causal filter, provided that sufficient delay is introduced.

We wish to approximate the ideal low-pass filter $H(\omega)$ of Fig. 4-13 with a

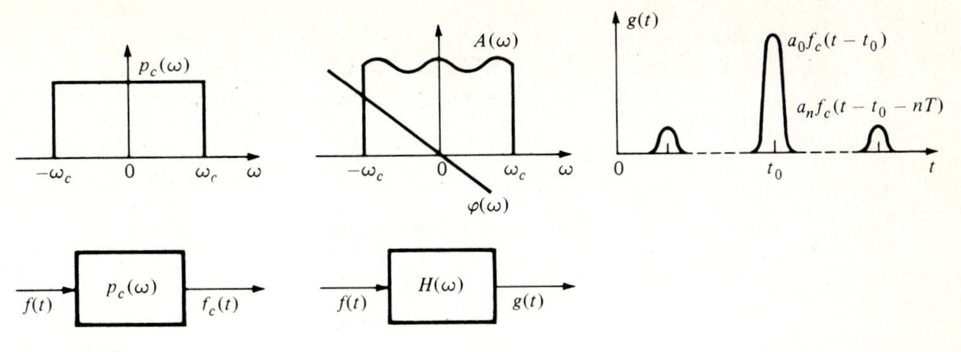

Figure 4-12

causal filter. With $h(t) = \sin \omega_c t/\pi t$ the impulse response of $H(\omega)$, and $q_a(t)$ a triangle, we form the functions

$$h_1(t) = h(t)q_a(t) \qquad h_2(t) = h_1(t - a)$$

and their transforms

$$H_1(\omega) = H(\omega) * \frac{2\sin^2(a\omega/2)}{\pi a \omega^2} = \int_{\omega - \omega_c}^{\omega + \omega_c} \frac{2\sin^2(ay/2)}{\pi a y^2}\,dy \qquad H_2(\omega) = H_1(\omega)e^{-ja\omega}$$

Clearly, $h_2(t) = 0$ for $t < 0$ because $h_1(t) = 0$ for $|t| > a$. Hence, $H_2(\omega)$ is a causal filter. Furthermore, if $a \gg \pi/\omega_c$, then $h_1(t) \simeq h(t)$. Hence, $|H_2(\omega)| \simeq H(\omega)$. If $a \to \infty$, then $|H_2(\omega)| \to H(\omega)$ but the phase lag $a\omega$ tends to infinity.

Bandpass Filters

Given a real, low-pass filter

$$h_c(t) \leftrightarrow H_c(\omega) = A_c(\omega)e^{j\varphi_c(\omega)} \tag{4-44}$$

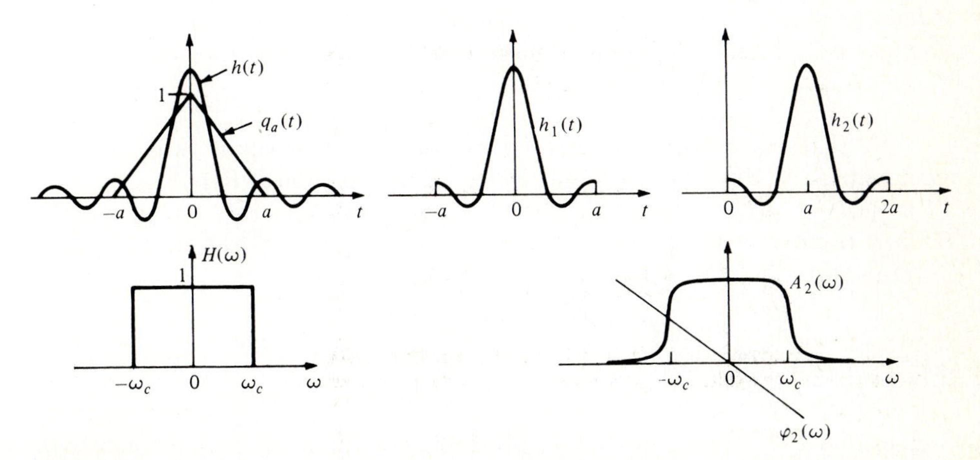

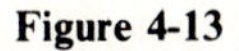
Figure 4-13

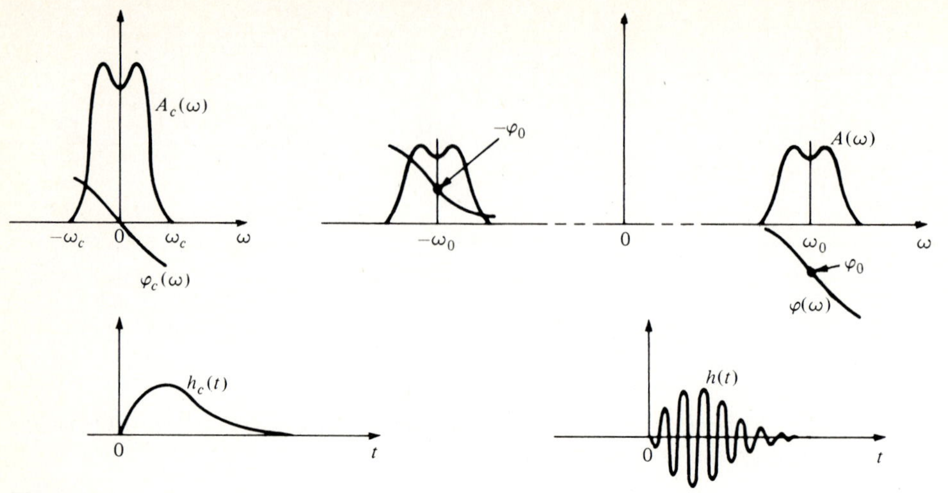

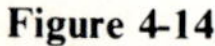
Figure 4-14

with cut-off frequency ω_c as in Fig. 4-14 and two constants φ_0 and $\omega_0 > \omega_c$, we form the filter

$$h(t) = h_c(t) \cos(\omega_0 t + \varphi_0) \qquad H(\omega) = \tfrac{1}{2}H_c(\omega - \omega_0)e^{j\varphi_0} + \tfrac{1}{2}H_c(\omega + \omega_0)e^{-j\varphi_0} \tag{4-45}$$

Since $H_c(\omega) = 0$ for $|\omega| > \omega_c$, it follows that $H(\omega) = 0$ outside the interval $(\omega_0 - \omega_c, \omega_0 + \omega_c)$ and its image; hence, $H(\omega)$ is a bandpass filter with center frequency ω_0. The filter so formed is symmetrical. That is, in the positive band $(\omega_0 - \omega_c, \omega_0 + \omega_c)$, the amplitude $A(\omega)$ has mirror symmetry, and the phase $\varphi(\omega) - \varphi_0$ point symmetry, about the center frequency ω_0. This follows from the fact that $H_c(-\omega) = H^*(\omega)$ and the assumption that $\omega_0 > \omega_c$ (elaborate). It is easy to show that the converse is also true: The impulse response $h(t)$ of a bandpass, symmetrical filter is a modulated signal, as in Eq. (4-45), with carrier frequency ω_0 and envelope the impulse response $h_c(t)$ of the low-pass filter

$$H_c(\omega) = Z_h(\omega + \omega_0)e^{-j\varphi_0} \qquad \text{where} \qquad Z_h(\omega) = 2H(\omega)U(\omega) \tag{4-46}$$

We shall show that a similar result holds for nonsymmetrical filters. As preparation, we discuss first the properties of the function $Z_h(\omega)$.

Hilbert transforms Given a real function $f(t)$ with Fourier transform $F(\omega)$, we form the function (Fig. 4-15)

$$Z_f(\omega) = 2F(\omega)U(\omega) \tag{4-47}$$

and its inverse transform

$$z_f(t) = \frac{1}{\pi}\int_0^\infty F(\omega)e^{j\omega t}\, d\omega \tag{4-48}$$

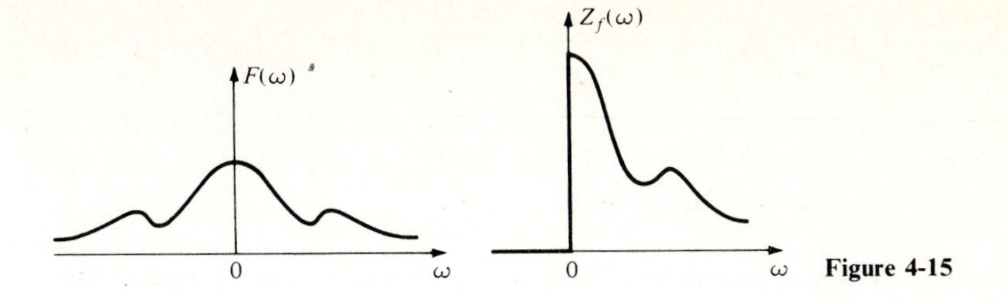

Figure 4-15

The function $z_f(t)$ is complex with real part $f(t)$ because

$$f(t) = \frac{1}{\pi} \text{Re} \int_0^{\infty} F(\omega)e^{j\omega t}\, d\omega$$

as it is easy to see from (3-8). The imaginary part of $z_f(t)$, denoted by $\hat{f}(t)$, is known as the *Hilbert transform* of $f(t)$ (see also Sec. 7-4):

$$z_f(t) = f(t) + j\hat{f}(t) \tag{4-49}$$

The function $z_f(t)$ we shall call the *analytic part* of $f(t)$.

Since $2U(\omega) = 1 + \text{sgn}\, \omega$, we conclude that

$$f(t) + j\hat{f}(t) \leftrightarrow Z_f(\omega) = F(\omega)(1 + \text{sgn}\, \omega) \qquad \hat{f}(t) \leftrightarrow -j\, \text{sgn}\, \omega F(\omega) \tag{4-50}$$

Therefore,

$$\hat{f}(t) = f(t) * \frac{1}{\pi t} = \frac{1}{\pi} \int_{-\infty}^{\infty} \frac{f(\tau)}{t - \tau}\, d\tau \tag{4-51}$$

because the inverse of $-j\, \text{sgn}\, \omega$ equals $1/\pi t$ [see (3-12)].

Quadrature filter From (4-51) it follows that $\hat{f}(t)$ is the output of the system of Fig. 4-16 with

$$h(t) = 1/\pi t \qquad H(\omega) = -j\, \text{sgn}\, \omega$$

and input $f(t)$. This system is known as a *quadrature filter* because its response to $\cos \omega_0 t$ equals $\sin \omega_0 t$. Indeed, if $f(t) = \cos \omega_0 t$, then $F(\omega) = \pi\, \delta(\omega - \omega_0) + \pi\, \delta(\omega + \omega_0)$; hence, $Z_f(\omega) = 2\pi\, \delta(\omega - \omega_0)$, $z_f(t) = e^{j\omega_0 t}$, and $\hat{f}(t) = \sin \omega_0 t$.

Note: If $z(t) = f_1(t) + jf_2(t)$ is a complex function with transform $Z(\omega)$, then the transforms of $f_1(t)$ and $f_2(t)$ are given by

$$F_1(\omega) = \frac{Z(\omega) + Z^*(-\omega)}{2} \qquad F_2(\omega) = \frac{Z(\omega) - Z^*(-\omega)}{2j} \tag{4-52}$$

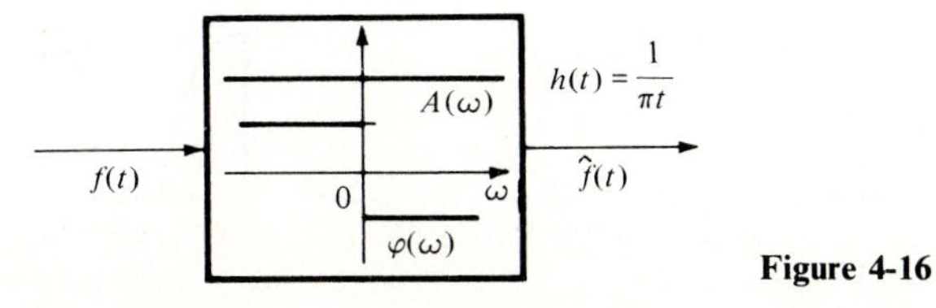

Figure 4-16

Indeed, $Z(\omega) = F_1(\omega) + jF_2(\omega)$, $F_1(-\omega) = F_1^*(\omega)$, and $F_2(-\omega) = F_2^*(\omega)$ (elaborate).

Suppose that $Z(\omega) = 0$ for $\omega < 0$. It then follows that $Z(\omega) = 2F_1(\omega)U(\omega)$ because, then, $Z^*(-\omega) = 0$ for $\omega > 0$. Hence, $z(t)$ is the analytic part of $f_1(t)$, and $f_2(t)$ is the Hilbert transform of $f_1(t)$.

Modulated signals and single sideband If $y(t)$ is a real signal whose transform $Y(\omega)$ is such that

$$Y(\omega) = 0 \qquad \text{for} \qquad |\omega| > \omega_0 \tag{4-53}$$

and if

$$f(t) = y(t) \cos \omega_0 t$$

then

$$\hat{f}(t) = y(t) \sin \omega_0 t \tag{4-54}$$

Proof Clearly,

$$F(\omega) = \tfrac{1}{2}Y(\omega - \omega_0) + \tfrac{1}{2}Y(\omega + \omega_0)$$

But [see Eq. (4-53)], $Y(\omega - \omega_0) = 0$ for $\omega < 0$, and $Y(\omega + \omega_0) = 0$ for $\omega > 0$. Hence (Fig. 4-17),

$$Z_f(\omega) = Y(\omega - \omega_0) \qquad z_f(t) = y(t)e^{j\omega_0 t} \tag{4-55}$$

and (4-54) results.

The function $Z_f(\omega)$ is symmetric about ω_0 because $Y(-\omega) = Y^*(\omega)$; therefore, it is uniquely determined in terms of its upper half. This leads to the representation of $f(t)$ in terms of the *single-sideband* signal $s(t)$ determined as follows:

With $Z_y(\omega) = 2Y(\omega)U(\omega)$ the analytic part of $Y(\omega)$, we form the signal

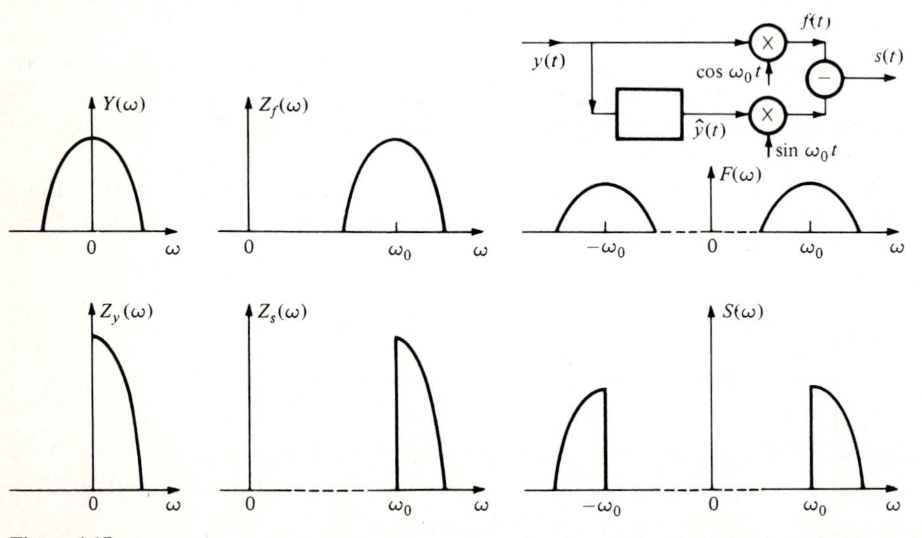

Figure 4-17

$Z_s(\omega) = Z_y(\omega - \omega_0)$ obtained by shifting $Z_y(\omega)$ to the right as in Fig. 4-17. The inverse

$$z_s(t) = z_y(t)e^{j\omega_0 t} = [y(t) + j\hat{y}(t)]e^{j\omega_0 t} \tag{4-56}$$

of $Z_s(\omega)$ is the analytic part of the signal (see note above)

$$s(t) = \text{Re } z_s(t) = y(t) \cos \omega_0 t - \hat{y}(t) \sin \omega_0 t \tag{4-57}$$

In polar form,

$$s(t) = a(t) \cos [\omega_0 t + \theta(t)] \qquad a(t) = \sqrt{y^2(t) + \hat{y}^2(t)} \qquad \tan \theta(t) = \frac{\hat{y}(t)}{y(t)} \tag{4-58}$$

Thus, an amplitude-modulated signal $f(t)$ of band B is represented by a signal $s(t)$ of band $B/2$. However, as we see from Eqs. (4-58), $s(t)$ is amplitude- and phase-modulated.

We note that the amplitude $y(t)$ of $f(t)$ can be determined from $s(t)$ by forming the product

$$2s(t) \cos \omega_0 t = y(t) + y(t) \cos 2\omega_0 t - \hat{y}(t) \sin 2\omega_0 t \tag{4-59}$$

and passing it through a low-pass filter (elaborate).

Hilbert transforms in system analysis Consider a real system with input a real signal $f(t)$ and output $g(t) = f(t) * h(t)$. With

$$Z_f(\omega) = 2F(\omega)U(\omega) \qquad Z_h(\omega) = 2H(\omega)U(\omega) \qquad Z_g(\omega) = 2G(\omega)U(\omega)$$

it follows from $G(\omega) = F(\omega)H(\omega)$ that

$$Z_g(\omega) = F(\omega)Z_h(\omega) = Z_f(\omega)H(\omega) = \tfrac{1}{2}Z_f(\omega)Z_h(\omega) \tag{4-60}$$

Denoting by $z_f(t)$, $z_h(t)$, and $z_g(t)$ the analytic parts of $f(t)$, $h(t)$, and $g(t)$, respectively, we conclude that

$$z_g(t) = f(t) * z_h(t) = z_f(t) * h(t) = \tfrac{1}{2}z_f(t) * z_h(t) \tag{4-61}$$

We can, thus, determine $g(t)$ by performing one of the above convolutions because $g(t) = \text{Re } z_g(t)$.

Example 4-9 If the input to a real system is a modulated signal $f(t)$ with band-limited envelope $y(t)$ as in (4-53),

$$f(t) = y(t) \cos \omega_0 t$$

then $z_f(t) = y(t)e^{j\omega_0 t}$ and (4-61) yields

$$g(t) = \text{Re } [y(t)e^{j\omega_0 t} * h(t)] = \tfrac{1}{2} \text{Re } [y(t)e^{j\omega_0 t} * z_h(t)] \tag{4-62}$$

Low-pass–bandpass transformation Given a bandpass system $H(\omega)$ with amplitude $A(\omega)$ and phase $\varphi(\omega)$ as in Fig. 4-18, we form the *equivalent low-pass filter*

$$H_c(\omega) = Z_h(\omega + \omega_0)e^{-j\varphi_0} \qquad \varphi_0 = \varphi(\omega_0) \tag{4-63}$$

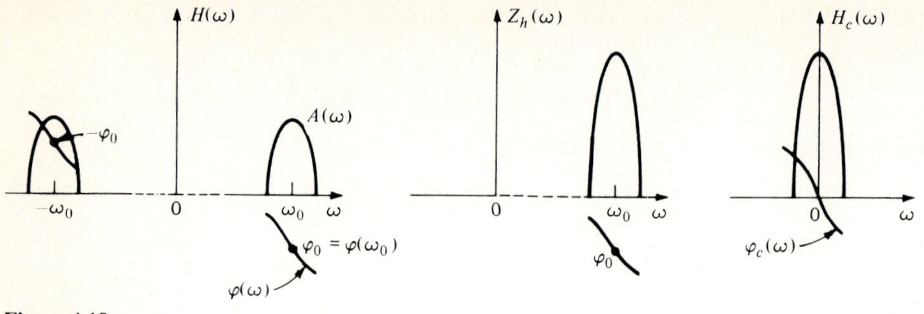

Figure 4-18

where $Z_h(\omega) = 2H(\omega)U(\omega)$ is the analytic part of $H(\omega)$. We shall say that the filter $H(\omega)$ is *symmetrical* if $H_c(\omega)$ has a real inverse $h_c(t)$, that is, if $H_c(-\omega) = H_c^*(\omega)$.

Theorem 1 The impulse response $h(t)$ of a symmetrical bandpass filter is an amplitude-modulated signal with envelope the impulse response $h_c(t)$ of the equivalent low-pass filter:

$$h(t) = h_c(t) \cos(\omega_0 t + \varphi_0) \tag{4-64}$$

Proof From Eq. (4-63) it follows that

$$Z_h(\omega) = H_c(\omega - \omega_0)e^{j\varphi_0}$$

Hence,

$$z_h(t) = h_c(t)e^{j(\omega_0 t + \varphi_0)} \tag{4-65}$$

and (4-64) results because $h(t) = \text{Re}\, z_h(t)$, and $h_c(t)$ is real by assumption.

Theorem 2 If the input to the above filter is a modulated signal with band-limited envelope $y(t)$ as in (4-53), then the response $g(t)$ is a modulated signal with envelope $\frac{1}{2}g_c(t)$, where (Fig. 4-19)

$$g_c(t) = y(t) * h_c(t)$$

is the response of the equivalent low-pass filter $H_c(\omega)$ to $y(t)$:

$$f(t) = y(t) \cos \omega_0 t \qquad g(t) = \tfrac{1}{2}g_c(t) \cos(\omega_0 t + \varphi_0) \tag{4-66}$$

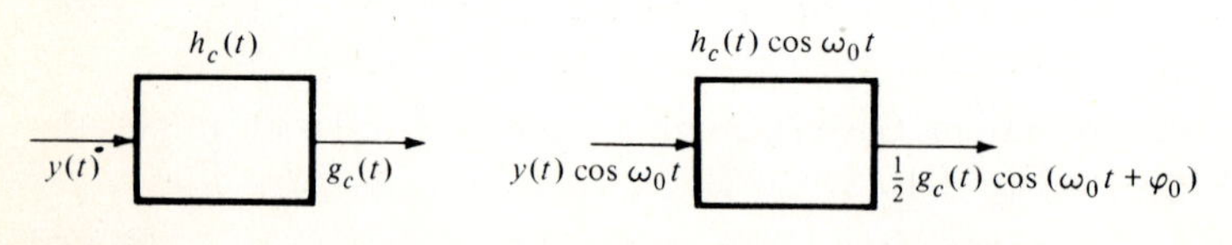

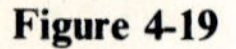

Figure 4-19

Proof Since $Z_f(\omega) = Y(\omega - \omega_0)$ [see (4-55)], we conclude from Eqs. (4-60) and (4-65) that

$$Z_g(\omega) = \tfrac{1}{2}Y(\omega - \omega_0)H_c(\omega - \omega_0)e^{j\varphi_0} = \tfrac{1}{2}G_c(\omega - \omega_0)e^{j\varphi_0}$$

because $G_c(\omega) = Y(\omega)H_c(\omega)$. Hence,

$$g(t) = \text{Re } z_g(t) = \tfrac{1}{2} \text{ Re } [g_c(t)e^{j(\omega_0 t + \varphi_0)}] \tag{4-67}$$

and (4-66) results.

Asymmetrical filters We define the equivalent low-pass filter $H_c(\omega)$ of an arbitrary bandpass filter $H(\omega)$ by (4-63), where ω_0 is the center of the band of $H(\omega)$. The impulse response $h_c(t)$ of this filter is now complex:

$$h_c(t) = p(t) + jq(t) \tag{4-68}$$

The real part $p(t)$ and imaginary part $q(t)$ of $h_c(t)$ we shall call its *in-phase* component and *quadrature* component, respectively.

Theorem 3 The impulse response of the asymmetrical filter $H(\omega)$ is an amplitude- and phase-modulated signal

$$h(t) = p(t) \cos (\omega_0 t + \varphi_0) - q(t) \sin (\omega_0 t + \varphi_0) \tag{4-69}$$

Proof Inserting (4-68) into (4-65), we obtain

$$h(t) = \text{Re } [p(t) + jq(t)]e^{j(\omega_0 t + \varphi_0)}$$

and (4-69) follows.

Theorem 4 The response $g(t)$ of the asymmetrical filter $H(\omega)$ to the modulated signal $f(t)$ of Theorem 2 is given by

$$g(t) = \tfrac{1}{2}g_p(t) \cos (\omega_0 t + \varphi_0) - \tfrac{1}{2}g_q(t) \sin (\omega_0 t + \varphi_0) \tag{4-70}$$

where
$$g_p(t) = y(t) * p(t) \qquad g_q(t) = y(t) * q(t) \tag{4-71}$$

Proof The response of the low-pass filter $H_c(\omega)$ to $y(t)$ is given by

$$g_c(t) = y(t) * [p(t) + jq(t)] = g_p(t) + jg_q(t) \tag{4-72}$$

Inserting into Eq. (4-67), we obtain Eq. (4-70).

Group and phase delay We have shown in (4-21) that if the input to a system is a slowly varying signal $f(t)$, then the response equals $H(0)f(t - \eta)$, where $\eta = -\varphi'(0)$ is the slope of $-\varphi(\omega)$ at the origin. We shall derive a related result for the response of a system to a modulated input. For this purpose, we introduce the *group delay* t_g and *phase delay* t_p

$$t_g = -\varphi'(\omega_0) \qquad t_p = -\varphi(\omega_0)/\omega_0 \tag{4-73}$$

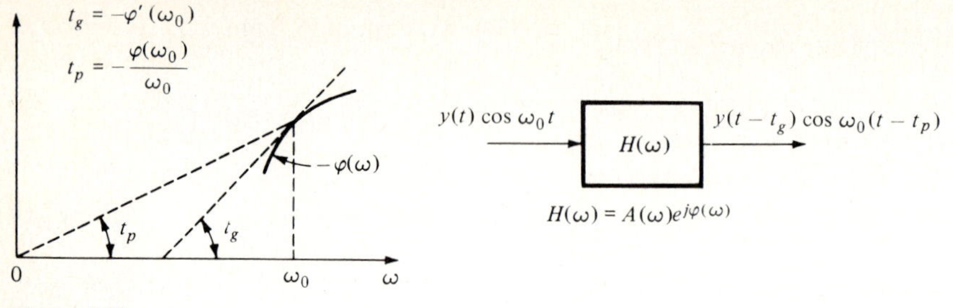

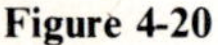

Figure 4-20

of the filter $H(\omega)$ at the carrier frequency ω_0 of the input (Fig. 4-20). We maintain that, if the envelope $y(t)$ of the input $f(t) = y(t) \cos \omega_0 t$ varies sufficiently slowly, then, the response $g(t)$ of the system is given by

$$g(t) \simeq A(\omega_0)y(t - t_g) \cos \omega_0(t - t_p) \tag{4-74}$$

Thus, t_g is the delay of the envelope of $f(t)$, and t_p the delay of the carrier.

Proof From (4-63) it follows that the phase $\varphi_c(\omega)$ of the equivalent low-pass filter $H_c(\omega)$ is given by $\varphi_c(\omega) = \varphi(\omega + \omega_0) - \varphi(\omega_0)$. Hence, $\varphi_c'(0) = \varphi'(\omega_0) = -t_g$. Furthermore, $H_c(0) = Z_h(\omega_0)e^{-j\varphi_0} = 2A(\omega_0)$. Applying (4-21) to the response $g_c(t) = y(t) * h_c(t)$, we obtain

$$g_c(t) \simeq 2A(\omega_0)y(t - t_g)$$

Since this is real, the component $g_q(t)$ in (4-72) is negligible. Therefore [see Eq. (4-70)],

$$g(t) \simeq A(\omega_0)y(t - t_g) \cos (\omega_0 t + \varphi_0)$$

and (4-74) results because $\varphi_0 = \varphi(\omega_0)$.

4-3 FINITE-ORDER SYSTEMS

Unlike our approach to discrete systems and difference equations, we do not intend here to develop systematically the theory of lumped-parameter systems, differential equations, and Laplace transforms; we shall assume that the reader is familiar with the elements of this theory. We plan merely to draw certain parallels with the concepts developed in Sec. 2-3, stressing topics that are not covered in beginning courses.

The Laplace Transform[1]

Deviating from our previous practice, we shall use in this section the notation $F(s)$ for the Laplace transform of the function $f(t)$:

$$F(s) = \int_{-\infty}^{\infty} e^{-st} f(t)\, dt \qquad s = \gamma + j\omega \tag{4-75}$$

In Chaps. 6 and 7 we discuss the analytic properties of $F(s)$. We mention here only that the region of convergence of the integral in Eq. (4-75) is a vertical strip R, as in Fig. 4-21*a*, with left boundary the line $\operatorname{Re} s = \gamma_1$ and right boundary the line $\operatorname{Re} s = \gamma_2$. The function $F(s)$ is analytic in the region R.

If the function $f(t)$ is zero for $t < 0$, then R is the half-plane $\operatorname{Re} s > \gamma_1$ (Fig. 4-21*b*). If $f(t)$ is zero outside a finite interval (time-limited), then (see Sec. 7-1) R is the entire s plane.

We illustrate with three examples:

$$f(t) = e^{-\alpha|t|} \qquad F(s) = \int_{-\infty}^{\infty} e^{-st} e^{-\alpha|t|}\, dt = \frac{2\alpha}{\alpha^2 - s^2} \qquad R\colon -\alpha < \operatorname{Re} s < \alpha$$

$$f(t) = e^{-\alpha t} U(t) \qquad F(s) = \int_{0}^{\infty} e^{-st} e^{-\alpha t}\, dt = \frac{1}{s + \alpha} \qquad R\colon \operatorname{Re} s > -\alpha$$

$$f(t) = p_a(t) \qquad F(s) = \int_{-a}^{a} e^{-st}\, dt = \frac{e^{-as} - e^{as}}{-s} \qquad R\colon \text{all } s$$

If the region R includes the $j\omega$ axis in its *interior*, i.e., if $\gamma_1 < 0 < \gamma_2$, then $F(j\omega)$ exists and equals the Fourier transform of $f(t)$. In this case, various properties of $F(s)$ can be deduced from the corresponding properties of Fourier transforms by changing ω to s/j. For example, from the derivative theorem (3-25), it follows that the Laplace transform of $f^{(n)}(t)$ equals $s^n F(s)$. From the pair $te^{-\alpha t}U(t) \leftrightarrow 1/(\alpha + j\omega)^2$, it follows that the Laplace transform of $te^{-\alpha t}U(t)$ equals $1/(\alpha + s)^2$.

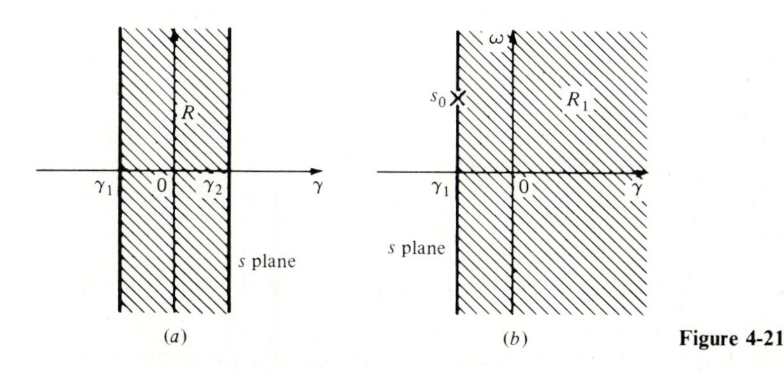

Figure 4-21

[1] G. Doetsch, "Guide to the Applications of Laplace Transforms" (translated). D. Van Nostrand Company, Inc., Princeton, N.J., 1963.

We should point out that this correspondence might not hold if the $j\omega$ axis is on the boundary of R. Indeed, the Laplace transform of $U(t)$ equals $1/s$ for $\operatorname{Re} s > 0$; the Fourier transform of $U(t)$ equals $\pi\,\delta(\omega) + 1/j\omega$.

It can be readily seen (Sec. 7-1) that the Fourier inversion formula (3-2), properly modified, leads to an integral along a vertical line in R, expressing $f(t)$ in terms of $F(s)$. This result will not, however, be used here. For the determination of the inverse transforms of rational functions, it suffices to use partial-fraction expansions.

As is the case for z transforms, to find $f(t)$ we need to know not only $F(s)$ but also the region R of convergence of (4-75); otherwise, $f(t)$ is not unique (see Prob. 60). However, if $f(t)$ is a causal function, this problem does not arise because then R is the half-plane $\operatorname{Re} s > \gamma_1$.

The system function as Laplace transform Consider a linear system with impulse response $h(t)$. If the input equals e^{st}, then the resulting response is given by

$$e^{st} * h(t) = \int_{-\infty}^{\infty} e^{s(t-\tau)} h(\tau)\, d\tau = e^{st} \int_{-\infty}^{\infty} e^{-s\tau} h(\tau)\, d\tau = H(s)e^{st}$$

where $H(s)$ is the Laplace transform of $h(t)$. Thus,

$$L[e^{st}] = H(s)e^{st} \tag{4-76}$$

That is, the exponential e^{st} is an eigenfunction of linear systems, and the corresponding eigenvalue equals $H(s)$. This is true for every s in the region R of existence of $H(s)$. If this region includes the $j\omega$ axis, then $H(j\omega)$ exists and is the system function as defined in Sec. 1-2.

Finite-Order Systems

We now assume that the systems under consideration are causal with rational system function

$$H(s) = \frac{N(s)}{D(s)} = \frac{d_0 s^r + d_1 s^{r-1} + \cdots + d_r}{s^m + c_1 s^{m-1} + \cdots + c_m} \tag{4-77}$$

From the assumed causality it follows that $H(s)$ exists in the half-plane $\operatorname{Re} s > \gamma_1$ and all its poles are to the left of the line $\operatorname{Re} s = \gamma_1$.

It is easy to see that the output $g(t)$ of the system so defined satisfies the differential equation

$$g^{(m)}(t) + c_1 g^{(m-1)}(t) + \cdots + c_m g(t) = d_0 f^{(r)}(t) + d_1 f^{(r-1)}(t) + \cdots + d_r f(t) \tag{4-78}$$

The impulse response $h(t)$ of the system is obtained by expanding $H(s)$ into partial fractions and using the pair [see (3-26)]

$$t^n e^{s_i t} U(t) \leftrightarrow \frac{n!}{(s - s_i)^{n+1}}$$

If $H(s)$ is a proper fraction and all its poles s_i are simple, then

$$H(s) = \sum_{i=1}^{m} \frac{A_i}{s - s_i} \qquad h(t) = \sum_{i=1}^{m} A_i e^{s_i t} U(t) \tag{4-79}$$

Example 4-10

$$H(s) = \frac{1}{(s + \alpha)^2 + \beta^2} = \frac{1/2j\beta}{s + \alpha - j\beta} - \frac{1/2j\beta}{s + \alpha + j\beta}$$

$$h(t) = \frac{1}{2j\beta}\left[e^{(-\alpha + j\beta)t} - e^{(-\alpha - j\beta)t}\right]U(t) = \frac{1}{\beta} e^{-\alpha t} \sin \beta t U(t)$$

Example 4-11

$$H(s) = \frac{1}{s(s + 1)^2} = \frac{1}{s} - \frac{1}{(s + 1)^2} - \frac{1}{s + 1}$$

$$h(t) = (1 - te^{-t} - e^{-t})U(t)$$

We shall say that a system is *stable* if all its poles s_i have negative real part. In this case, $h(t) \to 0$ as $t \to \infty$. We shall say that a system is *minimum-phase* if all its poles s_i and zeros z_i have negative real part.

Frequency response If a system is stable, then the region R contains the $j\omega$ axis. Hence, $H(s)$ exists for $s = j\omega$ and is a rational function of ω:

$$H(j\omega) = \frac{d_0(j\omega)^r + \cdots + d_r}{(j\omega)^m + \cdots + c_m} = A(\omega)e^{j\varphi(\omega)} \tag{4-80}$$

Thus,

$$L[e^{j\omega t}] = H(j\omega)e^{j\omega t} = \sum_{i=1}^{m} \frac{A_i}{j\omega - s_i} e^{j\omega t}$$

We shall next determine the response to a truncated exponential [see also Eq. (4-37)]. Since

$$\int_0^t e^{j\omega(t-\tau)} e^{s_i \tau} \, d\tau = \frac{1}{j\omega - s_i} (e^{j\omega t} - e^{s_i t})$$

it follows from (4-1) and (4-79) that

$$L[e^{j\omega t}U(t)] = \int_0^t e^{\,j\omega(t-\tau)} h(\tau) \, d\tau = \sum_{i=1}^{m} \frac{A_i}{j\omega - s_i} (e^{j\omega t} - e^{s_i t})$$

$$= H(j\omega)e^{j\omega t} - \sum_{i=1}^{m} \frac{A_i}{j\omega - s_i} e^{s_i t}$$

If the system is stable, then the last sum (transients) tends to zero as $t \to \infty$.

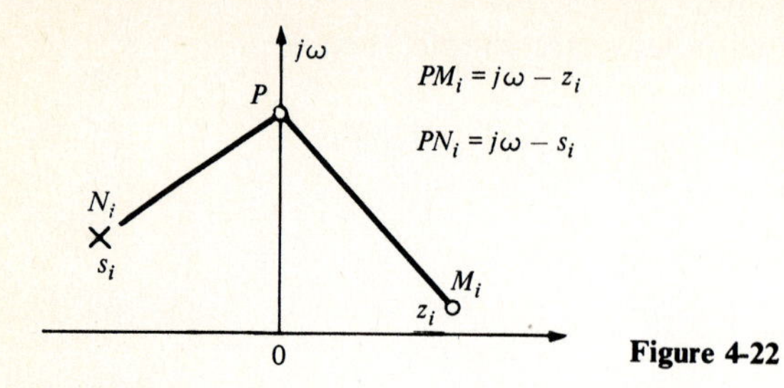

Figure 4-22

The determination of the amplitude $A(\omega)$ and phase $\varphi(\omega)$ of $H(j\omega)$ is often facilitated if its numerator and denominator are factored:

$$H(j\omega) = d_0 \frac{(PM_1)\cdots(PM_r)}{(PN_1)\cdots(PN_m)} \tag{4-81}$$

In the above, $PM_i = j\omega - z_i$ and $PN_i = j\omega - s_i$ are the vectors from the zeros z_i and poles s_i, respectively, of $H(s)$ to the point $j\omega$ of the imaginary axis (see Fig. 4-22). This factorization is particularly useful for narrow-band filters. In this case, only the zeros and poles that are near the $j\omega$ axis in the vicinity of the band of the filter are significant; all other factors can be assumed to be constant.[1] The following example is an illustration.

Example 4-12 The function

$$H(s) = \frac{s + \gamma}{(s + \alpha)^2 + \beta^2}$$

has one real zero $z_1 = -\gamma$ and two complex poles, $s_1 = -\alpha + j\beta$ and $s_2 = -\alpha - j\beta$ (Fig. 4-23). If $\alpha \ll \beta$, then $H(j\omega)$ takes significant values only if $j\omega$ is close to $j\beta$ (resonance). In this vicinity, the vector $j\omega - z_1$ is approximately equal to $j\beta + \gamma$, and the vector $j\omega - s_2$ is approximately equal to $\alpha + 2j\beta \simeq 2j\beta$. The form of $H(j\omega)$ depends, therefore, only on the term $j\omega - s_1$:

$$H(j\omega) \simeq \frac{j\beta + \gamma}{2j\beta(j\omega + \alpha - j\beta)} \qquad A(\omega) \simeq \frac{\sqrt{\beta^2 + \gamma^2}}{2\beta\sqrt{\alpha^2 + (\omega - \beta)^2}}$$

We shall now assume that the system is real, i.e., that the coefficients c_i and d_i of $H(s)$ are real, and examine the properties of $A(\omega)$.

Theorem The energy spectrum $A^2(\omega) = |H(j\omega)|^2$ of a real system of finite order is a nonnegative rational function of ω^2:

$$A^2(\omega) = \frac{x(\omega^2)}{y(\omega^2)} \tag{4-82}$$

[1] E. J. Angelo, Jr., and A. Papoulis, "Pole-Zero Patterns," McGraw-Hill Book Company, New York, 1964.

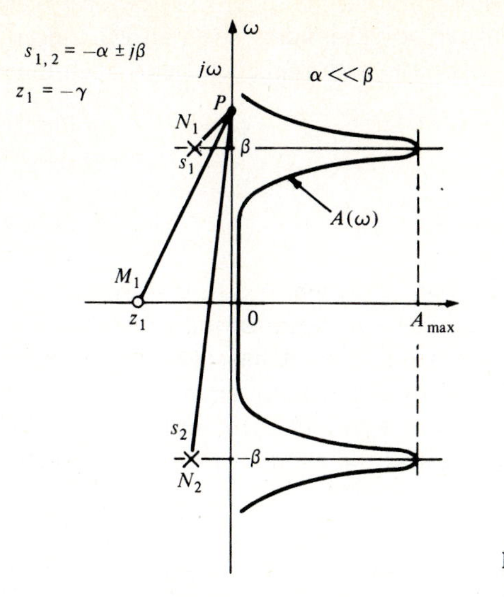

Figure 4-23

Proof If p_i is a zero or pole of $H(s)$, then $-p_i$ is a zero or pole of $H(-s)$. Hence, the product $H(s)H(-s)$ contains the factor

$$(s - p_i)(-s - p_i) = -s^2 + p_i^2$$

Since this is true for all zeros and poles, we conclude that $H(s)H(-s)$ is a rational function of s^2:

$$H(s)H(-s) = \frac{N(s)\,N(-s)}{D(s)\,D(-s)} = V(s^2) \tag{4-83}$$

Furthermore, $H^*(j\omega) = H(-j\omega)$ because the system is real. Hence,

$$A^2(\omega) = H(j\omega)H^*(j\omega) = H(s)H(-s)\Big|_{s=j\omega} = V(-\omega^2) \tag{4-84}$$

and Eq. (4-82) follows.

Example 4-13

$$H(s) = \frac{s+1}{2s^2 + 2s + 1} \qquad V(s^2) = \frac{1 - s^2}{1 + 4s^4} \qquad A^2(\omega) = \frac{1 + \omega^2}{1 + 4\omega^4}$$

The factorization problem A problem of some importance in signal analysis is the determination of a causal function $h(t)$ such that the amplitude $A(\omega)$ of its Laplace transform $H(s)$ is a specified nonnegative function. As we show in Sec. 7-2, this problem has a solution only if $A(\omega)$ satisfies certain conditions (Paley-Wiener). In the following, we show that, for rational functions, a solution always exists. In fact, if we impose the condition that $H(s)$ be minimum phase, then the solution is unique.

We first phrase and solve the problem for polynomials: Given an even, nonnegative polynomial $y(\omega^2)$, find a polynomial $D(s)$ with real coefficients such that

$$|D(j\omega)|^2 = y(\omega^2) \tag{4-85}$$

or, equivalently, such that

$$D(s)D(-s) = y(-s^2) \tag{4-86}$$

Solution If s_i is a root of the given function $y(-s^2)$, then $-s_i$ is, obviously, also a root; if s_i is a complex root, then s_i^* is also a root because the coefficients of $y(-s^2)$ are real; if $s_i = j\omega_i$ is an imaginary root, then it has even multiplicity because $y(\omega^2) \geq 0$. The polynomial $D(s)$ is formed by taking all roots of $y(-s^2)$ with negative real part and the imaginary roots with half their multiplicity. The remaining roots form the polynomial $D(-s)$, and the product $D(s)D(-s)$ equals $y(-s^2)$.

Example 4-14

$$y(\omega^2) = 100 + 29\omega^2 + \omega^4$$

$$y(-s^2) = 100 - 29s^2 + s^4 = (s+2)(s-2)(s+5)(s-5)$$

Hence, $\quad D(s) = (s+2)(s+5) \qquad D(-s) = (-s+2)(-s+5)$

Note: The complex roots of $y(-s^2)$ result from biquadratic factors of the form $s^4 + Vs^2 + W$. The corresponding factor in $D(s)$ is $s^2 + vs + w$, where $w = \sqrt{W}$ and $v = \sqrt{2w - V}$, as is easily seen. This simplifies the determination of $D(s)$.

Example 4-15

$$y(\omega^2) = 1 + \omega^4 \qquad y(-s^2) = 1 + s^4 \qquad V = 0 \qquad W = 1$$

Hence, $w = 1$, $v = \sqrt{2}$, and $D(s) = s^2 + \sqrt{2}s + 1$.

We return to the factorization problem for rational functions: Given an even, nonnegative rational function $x(\omega^2)/y(\omega^2)$, find a minimum-phase system $H_m(s)$ such that

$$|H_m(j\omega)|^2 = \frac{x(\omega^2)}{y(\omega^2)} \tag{4-87}$$

Solution We determine the polynomials $N(s)$ and $D(s)$ such that $|N(j\omega)|^2 = x(\omega^2)$ and $|D(j\omega)|^2 = y(\omega^2)$, as in Eq. (4-86). Clearly, the ratio $H_m(s) = N(s)/D(s)$ satisfies (4-87). It is minimum phase because the roots of $N(s)$ and $D(s)$ have negative real parts by construction.

Example 4-16 Find $H_m(s)$ such that

$$|H_m(j\omega)|^2 = \frac{\omega^2 + 4}{\omega^4 + 10\omega^2 + 9}$$

Solution

$$H_m(s)H_m(-s) = \frac{-s^2+4}{s^4-10s^2+9} = \frac{s+2}{(s+1)(s+3)}\,\frac{-s+2}{(-s+1)(-s+3)}$$

Hence,
$$H_m(s) = \frac{s+2}{(s+1)(s+3)}$$

Example 4-17 Find $H_m(s)$ such that

$$|H_m(j\omega)|^2 = \frac{1}{1+\omega^6}$$

Solution

$$H_m(s)H_m(-s) = \frac{1}{1-s^6} = \frac{1}{(1-s^2)(1+s^2+s^4)}$$

The left-hand plane roots of $1 + s^2 + s^4$ form the quadratic $1 + s + s^2$ (see preceding note); hence,

$$H_m(s) = \frac{1}{(1+s)(1+s+s^2)} = \frac{1}{1+2s+2s^2+s^3}$$

All-pass systems An all-pass system $H_0(s)$ is, by definition, a causal, stable system with unit amplitude:

$$|H_0(j\omega)| = 1$$

We shall show that such a system has the following properties:
Property 1: Its zeros and poles are symmetric with respect to the $j\omega$ axis; i.e., if its poles are s_i, then its zeros are $-s_i^*$:

$$H_0(s) = \frac{(s+s_1^*)\cdots(s+s_m^*)}{(s-s_1)\cdots(s-s_m)} \qquad \text{Re}\, s_i < 0 \tag{4-88}$$

Proof If $H_0(s)$ is given by (4-88), then $|H_0(j\omega)| = 1$ because $|j\omega + s_i^*| = |j\omega - s_i|$ (Fig. 4-24). Conversely, suppose that $H_0(s) = N(s)/D(s)$ is all-pass. With $|N(j\omega)|^2 = x(\omega^2)$ and $|D(j\omega)|^2 = y(\omega^2)$, we must have $x(\omega^2) = y(\omega^2)$. And since the roots of $D(s)$ are the left-hand plane roots of $y(-s^2)$, the roots of $N(s)$ must be the right-hand plane roots of $x(-s^2) = y(-s^2)$; otherwise, there would be cancellation.

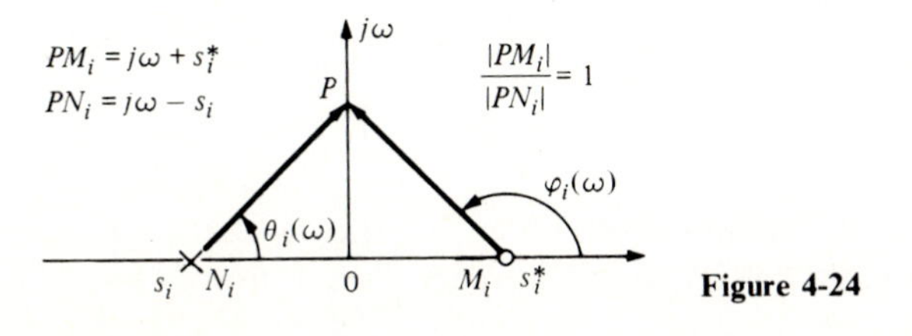

Figure 4-24

Property 2: The phase angle $\varphi(\omega)$ of an all-pass system is monotone decreasing from $2m\pi$ to zero as ω increases from $-\infty$ to ∞.

Proof The angle φ_i of $j\omega + s_i^*$ decreases from $3\pi/2$ to $\pi/2$ because $-s_i^*$ is on the right-hand plane; the angle θ_i of $j\omega - s_i$ increases from $-\pi/2$ to $\pi/2$ because s_i is on the left-hand plane. Hence, the angle of the ratio $(j\omega + s_i^*)/(j\omega - s_i)$ decreases from 2π to zero. This is true for every $i = 1, \ldots, m$; therefore, the angle of $H_0(j\omega)$ decreases from $2m\pi$ to zero.

Corollary From the monotonicity of $\varphi(\omega)$, it follows that the group delay $-\varphi'(\omega)$ of an all-pass system [see (4-73)] is positive for every ω.
Property 3: Suppose that the response of an all-pass system to an arbitrary input $y(t)$ is $g(t)$. We maintain that

$$\int_{-\infty}^{\infty} |y(t)|^2\,dt = \int_{-\infty}^{\infty} |g(t)|^2\,dt \tag{4-89}$$

and for any t_0

$$\int_{-\infty}^{t_0} |y(t)|^2\,dt \geq \int_{-\infty}^{t_0} |g(t)|^2\,dt \tag{4-90}$$

Proof Equation (4-89) follows from the energy theorem (3-37) because

$$|G(j\omega)| = |Y(j\omega)||H_0(j\omega)| = |Y(j\omega)|$$

To prove (4-90), we form the function

$$y_1(t) = \begin{cases} y(t) & t \leq t_0 \\ 0 & t > t_0 \end{cases} \tag{4-91}$$

obtained by truncating $y(t)$ for $t > t_0$ (Fig. 4-25). Denoting by $g_1(t)$ the response of the system $H_0(s)$ to the input $y_1(t)$, we conclude that (causality) for every $t \leq t_0$,

$$g_1(t) = \int_{-\infty}^{t} y_1(\tau)h_0(t-\tau)\,d\tau = \int_{-\infty}^{t} y(\tau)h_0(t-\tau)\,d\tau = g(t) \tag{4-92}$$

where $h_0(t)$ is the impulse response of the system. Applying (4-89) to the input $y_1(t)$ and output $g_1(t)$, we obtain

$$\int_{-\infty}^{t_0} |y_1(t)|^2\,dt = \int_{-\infty}^{\infty} |g_1(t)|^2\,dt = \int_{-\infty}^{t_0} |g_1(t)|^2\,dt + \int_{t_0}^{\infty} |g_1(t)|^2\,dt$$

and (4-90) results because $y(t) = y_1(t)$ and $g(t) = g_1(t)$ for $t \leq t_0$.

Minimum-phase systems A minimum-phase system $H_m(s)$ has the following properties.
Property 1: If $H(s)$ is an arbitrary system with poles on the left-hand plane (stable) but with zeros anywhere, and if its amplitude equals the amplitude of

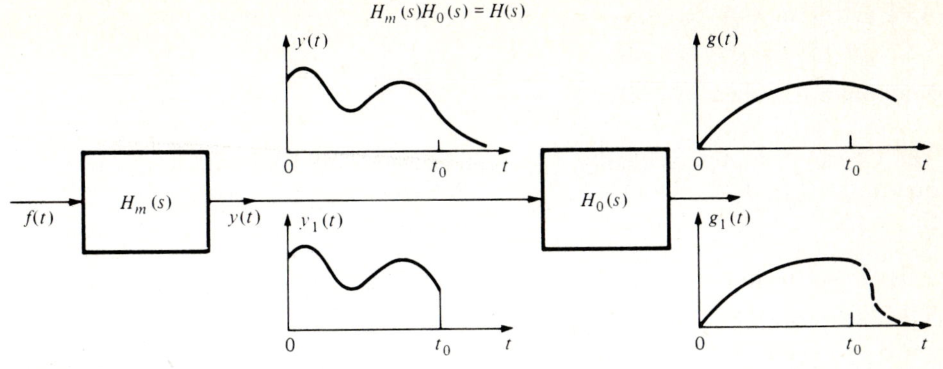

Figure 4-25

$H_m(s)$, then its group delay $-\varphi'(\omega)$ is larger than the group delay $-\varphi_m'(\omega)$ of $H_m(s)$:

$$\text{If} \quad |H(j\omega)| = |H_m(j\omega)| \quad \text{then} \quad -\varphi'(\omega) > -\varphi_m'(\omega) \tag{4-93}$$

Proof With

$$H_0(s) = \frac{H(s)}{H_m(s)} \tag{4-94}$$

it follows from Eq. (4-93) that $|H_0(j\omega)| = 1$. Furthermore, $H_0(s)$ is a stable system because its poles are either poles of $H(s)$ or zeros of $H_m(s)$. Therefore, $H_0(s)$ is all-pass; hence (property 2), its phase $\varphi_0(\omega)$ is monotone decreasing. Thus $\varphi_0'(\omega) < 0$, yielding (4-93) because $\varphi(\omega) = \varphi_0(\omega) + \varphi_m(\omega)$.

Property 2: If the same input $f(t)$ is applied to an arbitrary system $H(s)$ and to a minimum-phase system $H_m(s)$ with the same amplitude as in (4-93), then for any t_0,

$$\int_{-\infty}^{t_0} |y(t)|^2 \, dt \geq \int_{-\infty}^{t_0} |g(t)|^2 \, dt \tag{4-95}$$

where $y(t)$ and $g(t)$ are the responses of $H(s)$ and $H_m(s)$, respectively.

Proof Since $H(s) = H_m(s)H_0(s)$ as in (4-94), we conclude that the system $H(s)$ is equivalent to a minimum-phase system $H_m(s)$ in cascade with an all-pass system $H_0(s)$ (Fig. 4-25). Thus, the output $y(t)$ of $H_m(s)$ is the input to $H_0(s)$, and the resulting output is $g(t)$. Hence, (4-95) is a consequence of (4-90).

APPENDIX 4-A

Maximum Response of Linear Systems

The Cauchy-Schwarz inequality We shall use the following simple property of quadratics: If $I(y) = A - 2yB + y^2C$ and $I(y) \geq 0$ for every real y, then $B^2 - AC \leq 0$. If $B^2 - AC = 0$, then $I(y)$ has a real double root; that is, $I(k) = 0$ for some $y = k$.

Schwarz' inequality For any $z(x)$ and $w(x)$,

$$\left| \int_a^b z(x)w(x)\, dx \right|^2 \leq \int_a^b |z(x)|^2\, dx \int_a^b |w(x)|^2\, dx \tag{4-A-1}$$

Furthermore, (4-A-1) holds with the equality sign if and only if $z(x)$ is proportional to $w^*(x)$:

$$z(x) = kw^*(x) \tag{4-A-2}$$

(almost) everywhere in the interval of integration.

Proof We shall prove Eq. (4-A-1) for real functions; the reasoning can readily be extended to the complex case (see Prob. 61). We form the quadratic

$$I(y) = \int_a^b [z(x) - yw(x)]^2\, dx = \int_a^b z^2(x)\, dx - 2y \int_a^b z(x)w(x)\, dx + y^2 \int_a^b w^2(x)\, dx$$

Clearly, $I(y) \geq 0$ for every real y; hence, the coefficients of $I(y)$ must satisfy (4-A-1). To prove (4-A-2), we observe that, if (4-A-1) is an equality, then the quadratic $I(y)$ must have a real root; i.e., for some $y = k$,

$$I(k) = \int_a^b [z(x) - kw(x)]^2\, dx = 0$$

This is true only if the integrand $z(x) - kw(x)$ is identically zero, and the proof is complete.

Example 4-18 We shall show that if a function $f(x)$ is square-integrable in a finite interval (a, b), then it is also absolutely integrable in this interval. Indeed, from (4-A-1) it follows, with $z(x) = |f(x)|$ and $w(x) = 1$, that

$$\left| \int_a^b |f(x)|\, dx \right|^2 \leq \int_a^b |f(x)|^2\, dx \int_a^b dx = (b - a) \int_a^b |f(x)|^2\, dx \tag{4-A-3}$$

and since the right side is finite, the statement is proved.

Cauchy inequality For any z_n and w_n,

$$\left| \sum_n z_n w_n \right|^2 \leq \sum_n |z_n|^2 \sum_n |w_n|^2 \tag{4-A-4}$$

Furthermore, Eq. (4-A-4) holds with the equality sign if and only if

$$z_n = kw_n^* \tag{4-A-5}$$

To prove the above, we form the quadratic

$$I(y) = \sum_n |z_n - yw_n|^2$$

and reason as in the proof of (4-A-1).

The preceding inequalities lead to simple solutions of a variety of problems in system theory; we give next several illustrations. All the results have obvious digital versions.

Maximum system response We wish to determine the input $f(t)$ to a given system $h(t)$ such that its response

$$g(t_0) = \int_{-\infty}^{\infty} f(t)h(t_0 - t)\,dt = \frac{1}{2\pi}\int_{-\infty}^{\infty} F(\omega)H(\omega)e^{j\omega t_0}\,d\omega \tag{4-A-6}$$

is maximum at $t = t_0$. We shall assume that $f(t)$ satisfies various integral constraints. In the first three cases, the constraints involve the energy spectrum $|F(\omega)|^2$ of $f(t)$; the last case deals with an amplitude constraint.

The matched-filter principle If the energy

$$E = \int_{-\infty}^{\infty} |f(t)|^2\,dt = \frac{1}{2\pi}\int_{-\infty}^{\infty} |F(\omega)|^2\,d\omega$$

of $f(t)$ is specified, then

$$|g(t_0)|^2 \le \frac{E}{2\pi}\int_{-\infty}^{\infty} |H(\omega)|^2\,d\omega \tag{4-A-7}$$

The right side is the maximum of $g(t_0)$ and is reached only if

$$f(t) = kh^*(t_0 - t) \tag{4-A-8}$$

Proof From (4-A-1) it follows that

$$\left|\int_{-\infty}^{\infty} F(\omega)H(\omega)e^{j\omega t_0}\,d\omega\right|^2 \le \int_{-\infty}^{\infty} |F(\omega)|^2\,d\omega \int_{-\infty}^{\infty} |H(\omega)e^{j\omega t_0}|^2\,d\omega$$

yielding (4-A-7). Equality holds only if

$$F(\omega) = kH^*(\omega)e^{-j\omega t_0} \tag{4-A-9}$$

and Eq. (4-A-8) results because $h^*(-t) \leftrightarrow H^*(\omega)$.

Generalization We now assume that

$$E_i = \frac{1}{2\pi}\int_{-\infty}^{\infty} R(\omega)|F(\omega)|^2\,d\omega \qquad R(\omega) \ge 0 \tag{4-A-10}$$

where E_i is a given number, and $R(\omega)$ is a known function. This case arises if $f(t)$ is a current source and $R(\omega)$ is the real part of the input impedance of the system. The constant E_i is then the energy delivered to the system.

We maintain that

$$|g(t_0)|^2 \leq \frac{E_i}{2\pi} \int_{-\infty}^{\infty} \frac{|H(\omega)|^2}{R(\omega)} d\omega \tag{4-A-11}$$

The maximum is reached only if

$$F(\omega) = k \frac{H^*(\omega)}{R(\omega)} e^{-j\omega t_0} \tag{4-A-12}$$

Proof We multiply and divide the integrand of the last integral in Eq. (4-A-6) by $\sqrt{R(\omega)}$ and apply (4-A-1). The result,

$$\left| \int_{-\infty}^{\infty} \sqrt{R(\omega)}\, F(\omega) \frac{H(\omega)}{\sqrt{R(\omega)}} e^{j\omega t_0} d\omega \right|^2 \leq \int_{-\infty}^{\infty} R(\omega)|F(\omega)|^2 d\omega \int_{-\infty}^{\infty} \frac{|H(\omega)|^2}{R(\omega)} d\omega$$

yields (4-A-11). Equality holds only if

$$\sqrt{R(\omega)}\, F(\omega) = k \frac{H^*(\omega)}{\sqrt{R(\omega)}} e^{-j\omega t_0}$$

and (4-A-12) follows.

Multiple constraints The preceding approach, properly modified, can be used to solve the problem of maximizing $g(t_0)$ subject to several constraints of the form (4-A-10). We give only an illustration.

Szökefalvi-Nagy's inequality If the energy of a signal $f(t)$ equals E and the energy of its derivative equals E_1:

$$E_1 = \int_{-\infty}^{\infty} |f'(t)|^2 dt = \frac{1}{2\pi} \int_{-\infty}^{\infty} \omega^2 |F(\omega)|^2 d\omega \tag{4-A-13}$$

then

$$|f(t)| \leq \sqrt[4]{EE_1} \tag{4-A-14}$$

Equality holds for $t = t_0$ if and only if

$$f(t) = \sqrt[4]{EE_1}\, e^{-\alpha|t - t_0|} \qquad \alpha = \sqrt{\frac{E_1}{E}} \tag{4-A-15}$$

Proof From Eq. (4-A-1) it follows that, for any $\alpha > 0$ and any t_0,

$$\left| \int_{-\infty}^{\infty} \sqrt{\alpha^2 + \omega^2}\, F(\omega) \frac{e^{j\omega t_0}}{\sqrt{\alpha^2 + \omega^2}} d\omega \right|^2$$

$$\leq \int_{-\infty}^{\infty} (\alpha^2 + \omega^2)|F(\omega)|^2 d\omega \int_{-\infty}^{\infty} \frac{d\omega}{\alpha^2 + \omega^2} = 2\pi(\alpha^2 E + E_1) \frac{\pi}{\alpha}$$

The first integral above equals $2\pi f(t_0)$; hence,

$$|f(t_0)|^2 \leq \frac{\alpha E}{2} + \frac{E_1}{2\alpha} \tag{4-A-16}$$

Equality holds only if

$$\sqrt{\alpha^2 + \omega^2}\, F(\omega) = \frac{ke^{-j\omega t_0}}{\sqrt{\alpha^2 + \omega^2}}$$

that is, if

$$f(t) = \frac{k}{2\alpha} e^{-\alpha|t - t_0|} \tag{4-A-17}$$

Clearly, $E = k^2/4\alpha^3$ and $E_1 = k^2/4\alpha$. Hence, $\alpha = \sqrt{E_1/E}$ and $k = 2\alpha\sqrt[4]{EE_1}$, and Eq. (4-A-15) results.

We note that, since (4-A-16) holds for any $\alpha > 0$, the right side must be minimum for $\alpha = \sqrt{E_1/E}$ (elaborate).

Amplitude constraints We wish to find a function $f(t)$ of energy E such that the response $g(t_0)$ of the system $h(t)$ is maximum subject to the constraints

$$\int_{-\infty}^{\infty} f(t)\varphi_i(t)\, dt = x_i \qquad i = 1, \ldots, n \tag{4-A-18}$$

where the functions $\varphi_i(t)$ and the numbers x_i are given. With

$$y_i(t) = \varphi_i(t) - \frac{x_i}{F(0)}$$

it follows from (4-A-18) that

$$\int_{-\infty}^{\infty} f(t)y_i(t)\, dt = 0 \qquad i = 1, \ldots, n \tag{4-A-19}$$

because $F(0)$ equals the area of $f(t)$. We can thus replace (4-A-18) with the homogeneous constraints (4-A-19).

From Eqs. (4-A-6) and (4-A-19), it follows that

$$g(t_0) = \int_{-\infty}^{\infty} f(t)[h(t_0 - t) + \alpha_1 y_1(t) + \cdots + \alpha_n y_n(t)]\, dt \tag{4-A-20}$$

for any $\alpha_1, \ldots, \alpha_n$. Applying (4-A-1) to the above integral, we obtain the inequality

$$|g(t_0)|^2 \leq E \int_{-\infty}^{\infty} |h(t_0 - t) + \alpha_1 y_1(t) + \cdots + \alpha_n y_n(t)|^2\, dt \tag{4-A-21}$$

Equality holds if

$$f(t) = k[h^*(t_0 - t) + \alpha_1^* y_1^*(t) + \cdots + \alpha_n^* y_n^*(t)] \tag{4-A-22}$$

Substituting into the conjugate of (4-A-19), we conclude that the constants α_i must satisfy the system

$$\int_{-\infty}^{\infty} [h(t_0 - t) + \alpha_1 y_1(t) + \cdots + \alpha_n y_n(t)] y_i^*(t)\, dt = 0 \qquad i = 1, \ldots, n \tag{4-A-23}$$

We note that, if $\varphi_i(t) = h(t - t_i)$, then Eq. (4-A-19) yields $g(t_i) = x_i$; that is, the n constraints (4-A-18) mean that $g(t)$ should reach specified levels x_i at the times t_i.

CHAPTER

FIVE

DIGITAL PROCESSING OF ANALOG SIGNALS

In this the last chapter of Part One, we attempt to unify a number of concepts from the theory of discrete systems, continuous systems, and Fourier transforms, stressing numerical considerations and approximation techniques.[1, 2]

In Sec. 5-1, we develop the sampling theorem and comment on its role in bandlimited approximations. This important theorem leads to the justification of the assumption, implicit in digital processing of analog signals, that various functions can be represented in terms of their values at a discrete set of points. Other forms of the theorem are given in Sec. 6-2.

In Sec. 5-2, we present briefly the theory of orthogonal expansions, elaborating on the exponential set $e^{-s_i t}U(t)$ *for continuous signals and the set* $z_i^{-n}U[n]$ *for discrete signals, and we relate the optimum solution (orthogonality principle) to the interpolation problem in the complex plane.*

In Sec. 5-3, we introduce the approximation problem in the simulation of an analog system by a realizable discrete system and discuss various optimality criteria.

In Sec. 5-4, we consider nonrecursive simulators. We show that mean-square optimum solutions lead to truncated Fourier series, and frequency-domain interpolators to discrete Fourier series.

In Sec. 5-5, we discuss the time-sampling technique and the frequency transformation method in the approximate simulation of narrow-band filters of finite order.

In Sec. 5-6, we introduce a new method of filter design based on recursive, real-time spectral analysis.

---

[1] R. Rabiner and C. N. Rader (eds.), "Digital Signal Processing," Institute of Electrical and Electronics Engineers, Inc., New York, 1972.

[2] Digital Signal Processing Committee (eds.), "Selected Papers in Digital Signal Processing," Institute of Electrical and Electronics Engineers, Inc., New York, 1976.

5-1 SAMPLING AND INTERPOLATION

We wish to represent a function $f(t)$ in terms of its values $f(nT)$ at a sequence of equidistant points. For this purpose, we form the sum

$$f_k(t) = \sum_{n=-\infty}^{\infty} Tf(nT)k(t - nT) \tag{5-1}$$

where $k(t)$ is a given function. If

$$Tk(0) = 1 \qquad \text{and} \qquad k(nT) = 0 \qquad \text{for } n \neq 0$$

then

$$f_k(nT) = f(nT) \tag{5-2}$$

that is, $f_k(t)$ interpolates $f(t)$.

For the determination of the properties of $f_k(t)$, it is convenient to introduce the function

$$f_*(t) = \sum_{n=-\infty}^{\infty} Tf(nT)\,\delta(t - nT) \tag{5-3}$$

consisting of impulses of area $Tf(nT)$ as in Fig. 5-1. From Eq. (3-52) and the Poisson sum formula (3-87), it follows that the transform $F_*(\omega)$ of $f_*(t)$ is a periodic function given by

$$F_*(\omega) = \sum_{n=-\infty}^{\infty} Tf(nT)e^{-jnT\omega} = \sum_{n=-\infty}^{\infty} F(\omega + 2n\sigma) \qquad \sigma = \frac{\pi}{T} \tag{5-4}$$

Clearly,

$$f_k(t) = f_*(t) * k(t) \tag{5-5}$$

because $\delta(t - nT) * k(t) = k(t - nT)$. Hence, $f_k(t)$ is the output of a system with input $f_*(t)$ and impulse response $k(t)$. With $K(\omega)$ the transform of $k(t)$, it follows that

$$F_k(\omega) = K(\omega)F_*(\omega) = K(\omega) \sum_{n=-\infty}^{\infty} F(\omega + 2n\sigma) \tag{5-6}$$

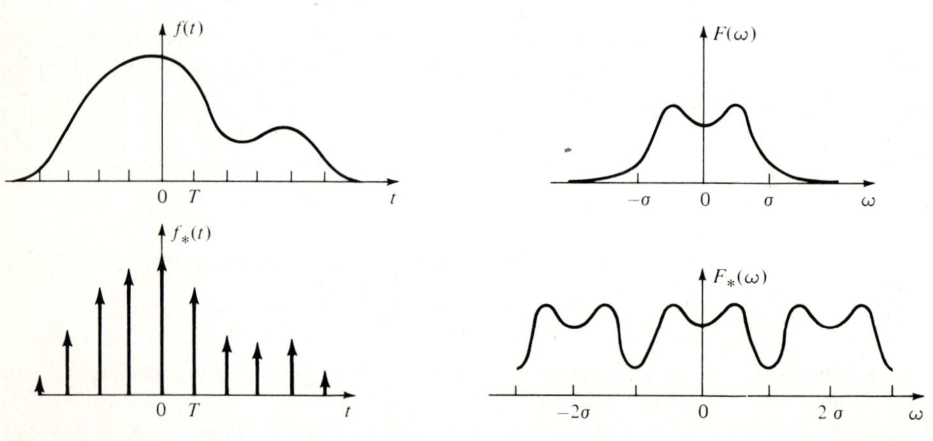

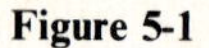

Figure 5-1

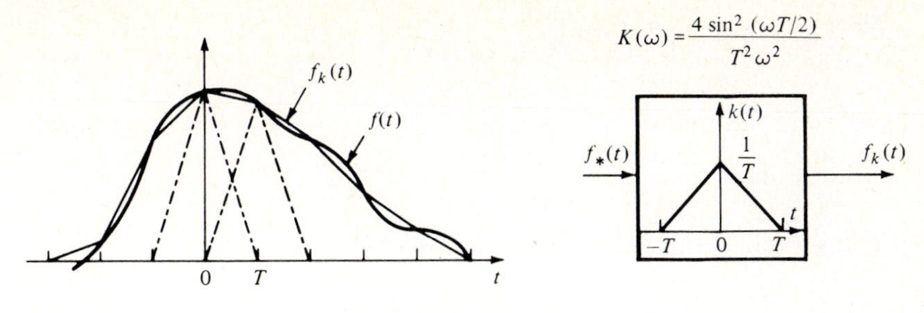

Figure 5-2

Example 5-1 A special case of (5-1) is the approximation of $f(t)$ with an inscribed polygon $f_k(t)$ as in Fig. 5-2. In this case, $k(t)$ is a triangle; hence [see Eq. (3-31)],

$$F_k(\omega) = \frac{4 \sin^2 (\omega T/2)}{T^2 \omega^2} \sum_{n=-\infty}^{\infty} F(\omega + 2n\sigma)$$

The Sampling Theorem

We shall say that a function $f(t)$ of finite energy is *bandlimited* if

$$F(\omega) = 0 \qquad \text{for } |\omega| > \sigma \tag{5-7}$$

and we shall use the abbreviation σ-BL, or simply BL, for this assumption. If $f(t)$ is σ-BL, then $F_*(\omega) = F(\omega)$ for $|\omega| < \sigma$ (Fig. 5-3). Hence,

$$F(\omega) = F_*(\omega) p_\sigma(\omega) \qquad \text{for all } \omega \tag{5-8}$$

Thus, the output $f_k(t)$ of the filter

$$k(t) = \frac{\sin \sigma t}{\pi t} \leftrightarrow p_\sigma(\omega) = K(\omega)$$

with input $f_*(t)$ equals $f(t)$. Therefore,

$$f(t) = \sum_{n=-\infty}^{\infty} f(nT) \frac{\sin \sigma(t - nT)}{\sigma(t - nT)} \qquad T = \frac{\pi}{\sigma} \tag{5-9}$$

The above fundamental result, known as the *sampling theorem*, expresses a function $f(t)$ in terms of its sample values $f(nT)$ at a sequence of equidistant

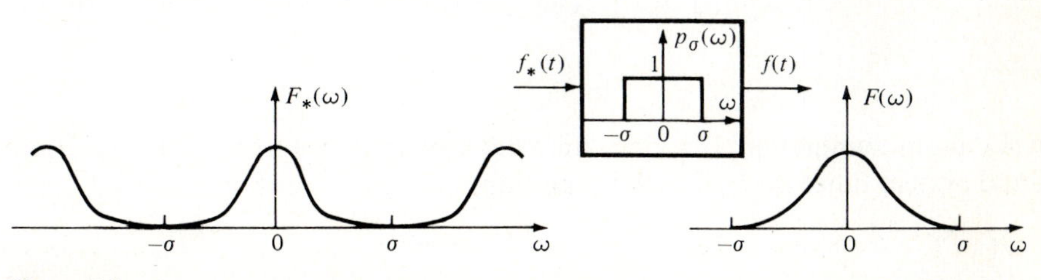

Figure 5-3

points $t = nT$. The sampling frequency

$$\frac{1}{T} = \frac{\sigma}{\pi}$$

is known as the *Nyquist rate*.

Note: The sampling expansion (5-9) can also be written in the form

$$f(t+\tau) = \sum_{n=-\infty}^{\infty} f(\tau + nT)\frac{\sin \sigma(t-nT)}{\sigma(t-nT)} \tag{5-10}$$

where τ is an arbitrary constant. Indeed, $f(t+\tau)$ is also a BL function, because its transform equals $F(\omega)e^{j\tau\omega}$. Hence, (5-10) follows when (5-9) is applied to the function $f(t+\tau)$.

Example 5-2 The following useful identity is a consequence of (5-10):

$$\sum_{n=-\infty}^{\infty} \frac{\sin a(\tau+nT)}{\tau+nT}\frac{\sin \sigma(t-nT)}{\sigma(t-nT)} = \frac{\sin a(t+\tau)}{t+\tau} \qquad a \le \sigma \le \frac{\pi}{T} \tag{5-11}$$

Proof If $f(t) = \sin at/t$, then $F(\omega) = \pi p_a(\omega) = 0$ for $|\omega| > \sigma$. Inserting into (5-10), we obtain (5-11).

Corollary Since (5-11) is true for any t and τ, we conclude, with $t = -\tau$, $a = \sigma$, and $\sigma t = \alpha$, that for any α,

$$\sum_{n=-\infty}^{\infty} \frac{\sin^2 \alpha}{(\alpha - n\pi)^2} = 1 \tag{5-12}$$

Truncation error[1] Suppose that the BL function $f(t)$ is approximated by the series

$$f_N(t) = \sum_{n=-N}^{N} f(nT)\frac{\sin \sigma(t-nT)}{\sigma(t-nT)}$$

The resulting mean-square error e_N is given by [see Eq. (5-56)]

$$e_N = \int_{-\infty}^{\infty} |f(t) - f_N(t)|^2\, dt = T \sum_{|n|>N} |f(nT)|^2 \tag{5-13}$$

The signal $f(t) - f_N(t)$ is bandlimited with energy e_N; hence [see Eq. (6-53)]

$$|f(t) - f_N(t)| \le \sqrt{\frac{\sigma e_N}{\pi}} \tag{5-14}$$

for every t.

Convolution as summation Consider an arbitrary system with impulse response $h_a(t)$ and system function $H_a(\omega)$. We maintain that, if the input $f(t)$ is σ-BL, then

[1] A. Papoulis, Error Analysis in Sampling Theory, *IEEE Proc.*, vol. 54, pp. 947–955, July, 1966.

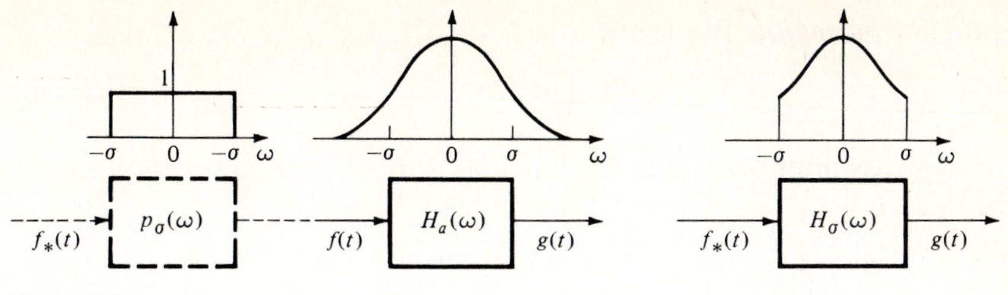

Figure 5-4

the response $g(t)$ can be written as a sum:

$$g(t) = \int_{-\infty}^{\infty} f(\tau)h_a(t - \tau)\, d\tau = \sum_{n=-\infty}^{\infty} Tf(nT)h_\sigma(t - nT) \tag{5-15}$$

where
$$h_\sigma(t) = \frac{1}{2\pi}\int_{-\sigma}^{\sigma} H_a(\omega)e^{j\omega t}\, d\omega \tag{5-16}$$

is the inverse transform of the function

$$H_\sigma(\omega) = p_\sigma(\omega)H_a(\omega) \tag{5-17}$$

resulting from truncating $H_a(\omega)$ as in Fig. 5-4.

Proof From (5-8) it follows that

$$G(\omega) = F(\omega)H_a(\omega) = F_*(\omega)p_\sigma(\omega)H_a(\omega) = F_*(\omega)H_\sigma(\omega)$$

Hence,
$$g(t) = f_*(t) * h_\sigma(t) = \sum_{n=-\infty}^{\infty} Tf(nT)\,\delta(t - nT) * h_\sigma(t)$$

and (5-15) follows.

Note: Since the response to $f(t + \tau)$ equals $g(t + \tau)$ and the function $f(t + \tau)$ is BL, we conclude, replacing $f(t)$ in (5-15) with $f(t + \tau)$, that

$$g(t + \tau) = \sum_{n=-\infty}^{\infty} Tf(\tau + nT)h_\sigma(t - nT) \tag{5-18}$$

Example 5-3 Suppose that $H_a(\omega)$ is an ideal differentiator

$$g(t) = f'(t) \qquad H_a(\omega) = j\omega$$

It then follows that

$$h_\sigma(t) = \frac{1}{2\pi}\int_{-\sigma}^{\sigma} j\omega e^{j\omega t}\, d\omega = \frac{\sigma t \cos \sigma t - \sin \sigma t}{\pi t^2}$$

From the above and Eq. (5-18), we conclude, with $t = T/2$, that

$$f'\left(\tau + \frac{T}{2}\right) = -\sigma \sum_{n=-\infty}^{\infty} \frac{(-1)^n}{(n\pi - \pi/2)^2} f(\tau + nT) \tag{5-19}$$

S. N. Bernstein's inequality If a function $f(t)$ is σ-BL and $|f(t)| \le M$, then

$$|f'(t)| \le \sigma M \tag{5-20}$$

Proof From the assumption, it follows that $|f(\tau + nT)| \le M$ for every τ and n; hence [see (5-19)],

$$\left| f'\left(\tau + \frac{T}{2}\right)\right| \le \sigma M \sum_{n=-\infty}^{\infty} \frac{1}{(n\pi - \pi/2)^2} = \sigma M$$

and (5-20) results. The last equality follows from (5-12) with $\alpha = \pi/2$.

Bandlimited interpolation Given an arbitrary function $f(t)$ and a constant T, we form the sum

$$f_i(t) = \sum_{n=-\infty}^{\infty} f(nT) \frac{\sin \sigma(t - nT)}{\sigma(t - nT)} \qquad \sigma = \frac{\pi}{T} \tag{5-21}$$

This sum is a special case of Eq. (5-1) obtained with $k(t) = \sin \sigma t/\pi t$, and its transform equals

$$F_i(\omega) = F_*(\omega)p_\sigma(\omega) \tag{5-22}$$

Thus, $f_i(t)$ is a BL function obtained by passing the signal $f_*(t)$ through an ideal low-pass filter (Fig. 5-5).

If $f(t)$ is approximated by $f_i(t)$, then the error $f(t) - f_i(t)$ is zero for $t = nT$ because $f_i(nT) = f(nT)$. It can be shown (see Prob. 62) that for any t,

$$|f(t) - f_i(t)| \le \frac{1}{\pi} \int_{|\omega| > \sigma} |F(\omega)| \, d\omega \tag{5-23}$$

Bandlimited mean-square approximation We wish to approximate the function $f(t)$ with a BL function $y(t)$ that minimizes the mean-square error

$$e = \int_{-\infty}^{\infty} |f(t) - y(t)|^2 \, dt = \frac{1}{2\pi} \int_{-\infty}^{\infty} |F(\omega) - Y(\omega)|^2 \, d\omega$$

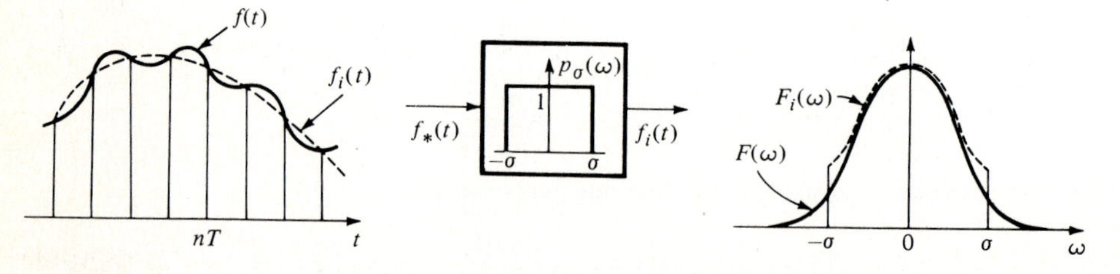

Figure 5-5

Since $Y(\omega) = 0$ for $|\omega| > \sigma$ by assumption,

$$e = \frac{1}{2\pi}\int_{|\omega|>\sigma} |F(\omega)|^2\, d\omega + \frac{1}{2\pi}\int_{|\omega|<\sigma} |F(\omega) - Y(\omega)|^2\, d\omega$$

Hence, e is minimum if

$$Y(\omega) = F_\sigma(\omega) = F(\omega)p_\sigma(\omega)$$

i.e., if

$$y(t) = f_\sigma(t) = \frac{1}{2\pi}\int_{-\sigma}^{\sigma} F(\omega)e^{j\omega t}\, d\omega \tag{5-24}$$

Thus, the mean-square optimum BL approximation of $f(t)$ is the output of an ideal low-pass filter with input $f(t)$. The following gives a bound for the resulting error: For any t,

$$|f(t) - f_\sigma(t)| \le \frac{1}{2\pi}\int_{|\omega|>\sigma} |F(\omega)|\, d\omega \tag{5-25}$$

Proof

$$|f(t) - f_\sigma(t)| = \frac{1}{2\pi}\left|\int_{-\infty}^{\infty} [F(\omega) - F_\sigma(\omega)]e^{j\omega t}\, d\omega\right| = \frac{1}{2\pi}\left|\int_{|\omega|>\sigma} F(\omega)e^{j\omega t}\, d\omega\right|$$

and Eq. (5-25) results.

Note: We have approximated the arbitrary function $f(t)$ with the functions $f_i(t)$ and $f_\sigma(t)$. Since both functions are BL, they are determined in terms of their samples $f_i(nT)$ and $f_\sigma(nT)$. The function $f_i(t)$ equals $f(t)$ at the sample points $t = nT$; this is not the case for $f_\sigma(t)$. The function $f_\sigma(t)$ yields the smallest mean-square error. The bound in (5-23) equals twice the bound in (5-25); however, the maximum of the error $|f(t) - f_i(t)|$ is not necessary larger than the maximum of the error $|f(t) - f_\sigma(t)|$.

If $f(t)$ is not BL, then a change in the sampling origin yields the sum

$$\sum_{n=-\infty}^{\infty} f(nT + \epsilon)\frac{\sin \sigma(t - nT)}{\sigma(t - nT)}$$

which is not equal to $f_i(t + \epsilon)$. Thus, a sampling jitter results in a different waveform. This is not the case for the approximation of $f(t)$ by the samples of $f_\sigma(t)$ because

$$\sum_{n=-\infty}^{\infty} f_\sigma(nT + \epsilon)\frac{\sin \sigma(t - nT)}{\sigma(t - nT)} = f_\sigma(t + \epsilon)$$

This shape-preserving property of $f_\sigma(t)$ is often desirable in the digital transmission of analog signals (television scanning, for example).

5-2 MEAN-SQUARE APPROXIMATIONS

We present briefly the elements of mean-square approximations, stressing exponential bases and the time-frequency equivalence. In this section, the Laplace transform of a signal $f(t)$ will be denoted by $F(s)$.

Inner products The inner product $\langle x, y \rangle$ of two functions $x(t)$ and $y(t)$ is, by definition,

$$\langle x, y \rangle = \int_{-\infty}^{\infty} x(t) y^*(t)\, dt \tag{5-26}$$

With $X(j\omega)$ and $Y(j\omega)$ the Fourier transforms of $x(t)$ and $y(t)$, respectively, we have (Parseval's formula)

$$\int_{-\infty}^{\infty} x(t) y^*(t)\, dt = \frac{1}{2\pi} \int_{-\infty}^{\infty} X(j\omega) Y^*(j\omega)\, d\omega \tag{5-27}$$

Defining inner products in the frequency domain by the right side of Eq. (5-27), we conclude that

$$\langle x, y \rangle = \langle X, Y \rangle \tag{5-28}$$

Note: In the definition of the inner product, the integration limits might be finite. This, however, can be considered as a special case of (5-26) by assuming that the integrands are zero outside the interval of integration.

Orthogonal functions We shall say that two functions $x(t)$ and $y(t)$ are orthogonal if their inner product is zero:

$$\langle x, y \rangle = 0 \tag{5-29}$$

From (5-28) it follows that the transforms $X(j\omega)$ and $Y(j\omega)$ of two orthogonal functions are orthogonal.

The Projection Problem

We wish to approximate an arbitrary function $f(t)$ with a linear combination

$$\hat{f}(t) = \sum_{k=1}^{n} a_k y_k(t) \tag{5-30}$$

of n linearly independent signals $y_k(t)$. The constants a_k are to be so chosen as to minimize the resulting mean-square error.

$$e = \int_{-\infty}^{\infty} \left| f(t) - \sum_{k=1}^{n} a_k y_k(t) \right|^2 dt = \langle f - \hat{f}, f - \hat{f} \rangle \tag{5-31}$$

Orthogonality principle The optimum coefficients a_i are such that the error $f(t) - \hat{f}(t)$ is orthogonal to the signals $y_i(t)$:

$$\langle f - \hat{f}, y_i \rangle = 0 \qquad i = 1, \ldots, n \tag{5-32}$$

Proof We assume for simplicity that all signals are real. Clearly, the mean-square error e is a function of the coefficients a_i and is minimum if

$$\frac{\partial e}{\partial a_i} = -2 \int_{-\infty}^{\infty} \left[f(t) - \sum_{k=1}^{n} a_k y_k(t) \right] y_i(t)\, dt = 0$$

i.e., if (5-32) holds.

Condition (5-32) is equivalent to the system

$$\langle y_1, y_i \rangle a_1 + \cdots + \langle y_n, y_i \rangle a_n = \langle f, y_i \rangle \qquad i = 1, \ldots, n \tag{5-33}$$

whose solution yields the optimum coefficients a_i.

Orthogonal expansions The system (5-33) is simplified if the functions $y_i(t)$ form an orthogonal set, i.e., if

$$\langle y_i, y_k \rangle = \begin{cases} E_i & i = k \\ 0 & i \neq k \end{cases} \tag{5-34}$$

In this case [see Eq. (5-33)],

$$a_i = \frac{1}{E_i} \langle f, y_i \rangle = \frac{1}{E_i} \int_{-\infty}^{\infty} f(t) y_i^*(t)\, dt \tag{5-35}$$

If the set $\{y_i(t)\}$ is not orthogonal, then, to simplify the solution of (5-33), we form a set of orthogonal functions $\varphi_i(t)$ that are linearly dependent on $y_i(t)$, and we approximate $f(t)$ with a linear combination of $\varphi_i(t)$. Since $\varphi_i(t)$ and $y_i(t)$ are linearly dependent, the two solutions are identical. The set $\{\varphi_i(t)\}$ is constructed by induction: With $\varphi_1(t) = y_1(t)$ and

$$\varphi_i(t) = A_1{}^i \varphi_1(t) + \cdots + A_k{}^i \varphi_k(t) + \cdots + A_{i-1}{}^i \varphi_{i-1}(t) + y_i(t) \tag{5-36}$$

it suffices to choose the coefficients $A_k{}^i$ such that

$$\langle \varphi_i, \varphi_k \rangle = 0 \qquad k = 1, \ldots, i - 1 \tag{5-37}$$

Inserting (5-36) into (5-37) and using the orthogonality of the functions $\varphi_k(t)$ for $k < i$ (induction assumption), we obtain

$$A_k{}^i = -\frac{\langle y_i, \varphi_k \rangle}{\langle \varphi_k, \varphi_k \rangle} \qquad k = 1, \ldots, i - 1 \tag{5-38}$$

The determination of the set $\{\varphi_i(t)\}$ is of computational value only if we need to find a_i for more than one function $f(t)$.

Notes:

1. The optimum mean-square approximation $\hat{f}(t)$ of $f(t)$ is called the *projection* of $f(t)$ on the space S_y of the functions $y_i(t)$ because, as we see from Eq. (5-32),

$$\langle f - \hat{f}, \hat{f} \rangle = 0 \qquad \text{and} \qquad \langle f, \hat{f} \rangle = \langle \hat{f}, \hat{f} \rangle \tag{5-39}$$

2. From the above and (5-31), it follows that

$$e = \langle f - \hat{f}, f - \hat{f} \rangle = \langle f - \hat{f}, f \rangle = \langle f, f \rangle - \langle f, \hat{f} \rangle = \langle f, f \rangle - \langle \hat{f}, \hat{f} \rangle \tag{5-40}$$

3. A set of functions $\{y_i(t)\}$ is called *orthonormal* if it is orthogonal and

$$E_i = \langle y_i, y_i \rangle = 1$$

4. If $\{y_i(t)\}$ is an orthonormal set and $\hat{f}(t)$ is a sum as in (5-30), then

$$\langle \hat{f}, \hat{f} \rangle = \sum_{k=1}^{n} |a_k|^2 \tag{5-41}$$

   If the sum $\hat{f}(t)$ is the projection of $f(t)$ on the space S_y, then the mean-square error e is given by [see (5-40)]

$$e = \int_{-\infty}^{\infty} |f(t)|^2 \, dt - \sum_{k=1}^{n} |a_k|^2 \tag{5-42}$$

5. An important problem in the theory of approximations is the investigation of the *completeness* of the set $\{y_i(t)\}$ in a space S, that is, the determination of the conditions under which a function $f(t)$ of a certain class S can be written as a linear combination of the functions $y_i(t)$. This problem has been investigated extensively for various special cases. The Fourier series is of particular interest.

Example 5-4 Given $f(t)$, we wish to find the $2N + 1$ coefficients a_k of the trigonometric polynomial

$$f_N(t) = \sum_{k=-N}^{N} a_k \, e^{jk\omega_0 t} \tag{5-43}$$

such that the integral

$$e_N = \int_{-T/2}^{T/2} \left| f(t) - \sum_{k=-N}^{N} a_k \, e^{jk\omega_0 t} \right|^2 dt \qquad T = \frac{2\pi}{\omega_0} \tag{5-44}$$

is minimum. This is a mean-square approximation problem in the interval $(-T/2, T/2)$ where $y_k(t) = e^{jk\omega_0 t}$. It is easy to see that

$$\int_{-T/2}^{T/2} e^{jk\omega_0 t} e^{-jm\omega_0 t} \, dt = \begin{cases} T & k \neq m \\ 0 & k = m \end{cases} \tag{5-45}$$

i.e., the set $\{y_i(t)\}$ is orthogonal. Hence, as we see from Eq. (5-35), the optimum coefficients a_k are given by

$$a_k = \frac{1}{T}\langle f, e^{jk\omega_0 t}\rangle = \frac{1}{T}\int_{-T/2}^{T/2} f(t)e^{-jk\omega_0 t}\,dt \tag{5-46}$$

i.e., they are the coefficients of the Fourier-series expansion of the function $f(t)$ in the interval $(-T/2, T/2)$.

Since [see Eq. (3-77)]

$$\int_{-T/2}^{T/2} |f(t)|^2\,dt = T\sum_{k=-\infty}^{\infty} |a_k|^2 \tag{5-47}$$

it follows from (5-42) that the mean-square error in the approximation of $f(t)$ by $\hat{f}(t)$ is

$$e_N = T\sum_{k=-\infty}^{\infty} |a_k|^2 - T\sum_{k=-N}^{N} |a_k|^2 = T\sum_{|k|>N} |a_k|^2 \tag{5-48}$$

As we know from the theory of Fourier series, $f_N(t) \to f(t)$ with $N \to \infty$; that is, the infinite set of exponentials $\{e^{jk\omega_0 t}\}$ is complete in the interval $(-T/2, T/2)$. This is not the case for every $f(t)$, but it holds for a large class of functions. This class includes all signals of bounded variation with finitely many discontinuity points.

Time-frequency equivalence The inner-product-preserving property (5-28) of Fourier transforms leads to the following conclusion: The minimum mean-square approximation of the transform $F(j\omega)$ of $f(t)$ by a linear combination of the transforms $Y_i(j\omega)$ of $y_i(t)$ is the transform

$$\hat{F}(j\omega) = \sum_{k=1}^{n} a_k Y_k(j\omega) \tag{5-49}$$

of $\hat{f}(t)$.

Proof From (5-32) and (5-28) it follows that

$$\langle F - \hat{F}, Y_i\rangle = 0 \qquad i = 1, \ldots, n \tag{5-50}$$

i.e., the coefficients a_i satisfy the orthogonality principle in the frequency domain. Hence, they are optimum.

As we show in the next example, this result establishes the equivalence between the sampling expansion (5-9) and the Fourier-series expansion.

We note that, if the set $\{y_i(t)\}$ is orthogonal, then the set $\{Y_i(j\omega)\}$ is also orthogonal. If $\{y_i(t)\}$ is complete in a space S_t, then $\{Y_i(j\omega)\}$ is complete in the space S_ω consisting of the transforms of the functions in S_t.

Example 5-5 Consider the set

$$y_k(t) = \frac{\sin \sigma(t - kT)}{\pi(t - kT)} \leftrightarrow p_\sigma(\omega)e^{-jkT\omega} = Y_k(j\omega) \qquad \sigma T = \pi \tag{5-51}$$

Since

$$\langle Y_k, Y_m \rangle = \frac{1}{2\pi}\int_{-\infty}^{\infty} Y_k(j\omega)Y_m^*(j\omega)\,d\omega = \frac{1}{2\pi}\int_{-\sigma}^{\sigma} e^{-jkT\omega}e^{jmT\omega}\,d\omega = \begin{cases} \dfrac{1}{T} & k = m \\ 0 & k \neq m \end{cases}$$

it follows that

$$\langle y_k, y_m \rangle = \int_{-\infty}^{\infty} \frac{\sin \sigma(t-kT)}{\pi(t-kT)} \frac{\sin \sigma(t-mT)}{\pi(t-mT)}\,dt = \begin{cases} \dfrac{1}{T} & k = m \\ 0 & k \neq m \end{cases} \tag{5-52}$$

Hence, the set $\{y_k(t)\}$ is orthogonal.

We shall approximate a σ-BL function $f(t)$ with the sum

$$f_N(t) = \sum_{k=-N}^{N} a_k \frac{\sin \sigma(t-kT)}{\pi(t-kT)} \tag{5-53}$$

From Eq. (5-35) it follows that, if

$$a_k = T\int_{-\infty}^{\infty} f(t)\frac{\sin \sigma(t-kT)}{\pi(t-kT)}\,dt = \frac{T}{2\pi}\int_{-\sigma}^{\sigma} F(j\omega)e^{jkT\omega}\,d\omega = Tf(kT) \tag{5-54}$$

then the mean-square error is minimum. The last equality in (5-54) is a consequence of the inversion formula

$$f(t) = \frac{1}{2\pi}\int_{-\sigma}^{\sigma} F(j\omega)e^{j\omega t}\,d\omega$$

As we know from the theory of Fourier series, the infinite set of exponentials $\{e^{jkT\omega}\}$ is complete in the interval $(-\sigma, \sigma)$. Hence, it is also complete in the interval $(-\infty, \infty)$ for the set of all functions $F(\omega)$ such that $F(\omega) = 0$ for $|\omega| > \sigma$. From the above, it follows that the set $\{\sin \sigma(t-kT)/\pi(t-kT)\}$ is complete in the space of all σ-BL functions. Hence, the sum $f_N(t)$ in (5-53) tends to $f(t)$ as $N \to \infty$. The result is the sampling expansion (5-9).

We note that (elaborate)

$$\int_{-\infty}^{\infty} |f(t)|^2\,dt = \frac{1}{2\pi}\int_{-\sigma}^{\sigma} |F(j\omega)|^2\,d\omega = \sum_{k=-\infty}^{\infty} T|f(kT)|^2 \tag{5-55}$$

Hence, in the approximation of $f(t)$ by $f_N(t)$, the mean-square error is [see (5-40)]

$$e = \int_{-\infty}^{\infty} |f(t)|^2\,dt - T\sum_{k=-N}^{N} |f(kT)|^2 = T\sum_{|k|>N} |f(kT)|^2 \tag{5-56}$$

The exponential set In the application of orthogonal expansions to linear systems, the set

$$y_i(t) = e^{-s_i t}U(t) \tag{5-57}$$

is of particular interest. It is used in approximating a given signal $f(t)$ with the impulse response of a finite-order system. As we show next, the properties of the corresponding orthogonal set $\{\varphi_i(t)\}$ and the sum $\hat{f}(t)$ are best described in terms of their Laplace transforms $\Phi_i(s)$ and $\hat{F}(s)$. We shall assume that all signals are zero for $t < 0$.

The results are based on the fact that the inner product of the functions $f(t)$ and $y_i(t)$ is given by

$$\langle f, y_i \rangle = \int_0^\infty f(t) y_i^*(t)\, dt = \int_0^\infty f(t) e^{-s_i^* t}\, dt = F(s_i^*) \tag{5-58}$$

Theorem 1 If

$$\hat{f}(t) = \sum_{k=1}^{n} a_k e^{-s_k t} U(t) \tag{5-59}$$

and the constants a_k are such that the integral

$$\int_0^\infty |f(t) - \hat{f}(t)|^2\, dt$$

is minimum, then

$$\hat{F}(s_i^*) = F(s_i^*) \qquad i = 1, 2, \ldots, n \tag{5-60}$$

Proof The above follows from the orthogonality principle (5-32):

$$\int_0^\infty [f(t) - \hat{f}(t)] e^{-s_i^* t}\, dt = 0 = F(s_i^*) - \hat{F}(s_i^*)$$

Interpolation Since the transform of $e^{-s_i t}U(t)$ equals $1/(s + s_i)$, it follows from Eq. (5-59) that

$$\hat{F}(s) = \frac{N(s)}{(s + s_1)(s + s_2) \cdots (s + s_n)} \tag{5-61}$$

where $N(s)$ is a polynomial of degree $n - 1$ whose coefficients can be determined from the n equations of (5-60). Thus, the mean-square approximation problem in the time domain is equivalent to an interpolation problem in the s domain.

Theorem 2 The transforms $\Phi_i(s)$ of the orthogonal set $\varphi_i(t)$ are given by

$$\Phi_i(s) = \frac{(s - s_1^*) \cdots (s - s_{i-1}^*)}{(s + s_1)(s + s_2) \cdots (s + s_i)} \qquad i = 1, 2, \ldots, n \tag{5-62}$$

Proof Clearly, the inverse $\varphi_i(t)$ of the above fraction is a linear combination of exponentials [see (4-84)]. As we see from (5-58) and (5-62),

$$\langle \varphi_i, e^{-s_k t} \rangle = \int_0^\infty \varphi_i(t) e^{-s_k^* t}\, dt = \Phi_i(s_k^*) = 0 \qquad \text{for } k < i$$

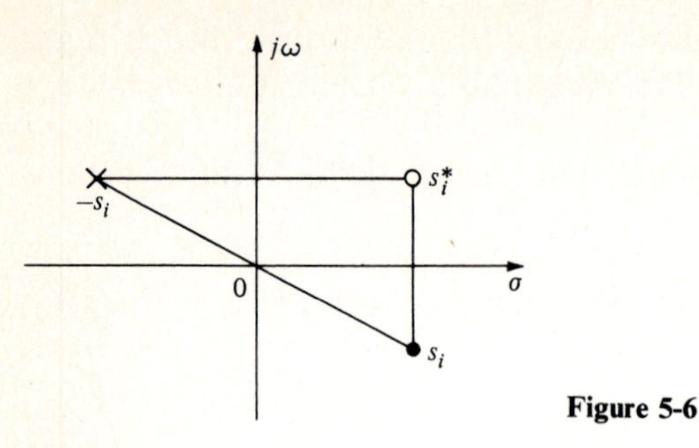

Figure 5-6

Hence, $$\langle \varphi_i, \varphi_m \rangle = \int_0^\infty \varphi_i(t)\varphi_m^*(t)\,dt = 0 \qquad \text{for } m \neq i \tag{5-63}$$

and the proof is complete.

Thus, $\Phi_i(s)$ has $i - 1$ zeros $s_1^*, \ldots, s_{i-1}^*$. These zeros are the mirror images of the poles $-s_1, \ldots, -s_{i-1}$ with respect to the $j\omega$ axis (Fig. 5-6).

Discrete Signals

We now give briefly the discrete version of the preceding results. The *inner product* of two sequences $x[n]$ and $y[n]$ is, by definition,

$$\langle x, y \rangle = \sum_{n=-\infty}^{\infty} x[n]y^*[n] \tag{5-64}$$

With $X(z)$ and $Y(z)$ the z transforms of $x[n]$ and $y[n]$, respectively, we have [see Eq. (2-18)]

$$\sum_{n=-\infty}^{\infty} x[n]y^*[n] = \frac{1}{2\sigma}\int_{-\sigma}^{\sigma} X(e^{j\omega T})Y^*(e^{j\omega T})\,d\omega \tag{5-65}$$

Defining the inner product of X and Y by

$$\langle X, Y \rangle = \frac{1}{2\sigma}\int_{-\sigma}^{\sigma} X(e^{j\omega T})Y^*(e^{j\omega T})\,d\omega \tag{5-66}$$

we conclude that

$$\langle x, y \rangle = \langle X, Y \rangle \tag{5-67}$$

We say that the sequences $x[n]$ and $y[n]$ are *orthogonal* if

$$\langle x, y \rangle = 0 \tag{5-68}$$

We wish to approximate an arbitrary sequence $f[n]$ with a linear combination of N linearly independent sequences $y_k[n]$

$$\hat{f}[n] = \sum_{k=1}^{N} a_k y_k[n] \tag{5-69}$$

so as to minimize the mean-square error

$$e = \sum_{n=-\infty}^{\infty} |f[n] - \hat{f}[n]|^2 \tag{5-70}$$

Orthogonality principle The optimum coefficients a_k are such that the error $f[n] - \hat{f}[n]$ is orthogonal to the sequences $y_i[n]$:

$$<f - \hat{f}, y_i> = 0 \qquad i = 1, \ldots, N \tag{5-71}$$

Proof For real signals,

$$\frac{\partial e}{\partial a_i} = -2 \sum_{n=-\infty}^{\infty} \{f[n] - \hat{f}[n]\} y_i[n]$$

yielding (5-71). Condition (5-71) leads to the system (5-33) for the determination of the N coefficients a_i.

If the sequences $y_i[n]$ are orthogonal as in (5-34), then

$$a_i = \frac{1}{E_i} \sum_{n=-\infty}^{\infty} f[n] y_i^*[n] \tag{5-72}$$

Geometric progressions If $f[n]$ is a causal sequence and

$$y_i[n] = z_i^{-n} U[n]$$

then

$$\langle f, y_i \rangle = \sum_{n=0}^{\infty} f[n](z_i^*)^{-n} = F(z_i^*) \tag{5-73}$$

Theorem 3 If

$$\hat{f}[n] = \sum_{k=1}^{N} a_k z_k^{-n} U[n]$$

and the constants a_k minimize the mean-square error e [see (5-70)], then

$$\hat{F}(z_i^*) = F(z_i^*) \qquad i = 1, 2, \ldots, N \tag{5-74}$$

where $\hat{F}(z)$ and $F(z)$ are the z transforms of $\hat{f}[n]$ and $f[n]$, respectively.

Proof From (5-73) it follows that [see Eq. (5-71)]

$$\sum_{n=0}^{\infty} (f[n] - \hat{f}[n])(z_i^*)^{-n} = 0 = F(z_i^*) - \hat{F}(z_i^*)$$

Theorem 4 If

$$\Phi_i(z) = \frac{z(z - z_1^*) \cdots (z - z_{i-1}^*)}{(z - z_1^{-1})(z - z_2^{-1}) \cdots (z - z_i^{-1})} \qquad i = 1, 2, \ldots, N \tag{5-75}$$

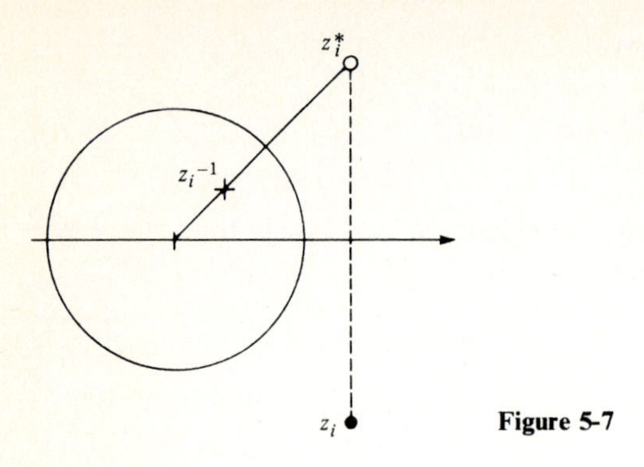

Figure 5-7

and $\varphi_i[n]$ is the inverse z transform of $\Phi_i(z)$, then

$$\sum_{n=0}^{\infty} \varphi_i[n]\varphi_m^*[n] = 0 \qquad \text{for} \qquad m \neq i \tag{5-76}$$

Proof It suffices to assume that $m < i$. Clearly [see (5-75)],

$$\sum_{n=0}^{\infty} \varphi_i[n](z_k^*)^{-n} = \Phi_i(z_k^*) = 0 \qquad 1 \leq k < i$$

This yields Eq. (5-76) because the sequence $\varphi_m[n]$ is a linear combination of the sequences $z_k{}^{-n}U[n]$, where $1 \leq k \leq m$ (partial-fraction expansion).

Note: The function $\varphi_i(z)$ has $i - 1$ zeros $z_1^*, \ldots, z_{i-1}^*$ symmetrical to its first $i - 1$ poles $z_1{}^{-1}, \ldots, z_{i-1}{}^{-1}$ with respect to the unit circle, as in Fig. 5-7.

5-3 DIGITAL SIMULATION OF ANALOG SYSTEMS

We have shown that, if the input to an analog system $H_a(\omega)$ is a σ-BL function $f(t)$, then the resulting output $g(t)$ can be written as a sum [see (5-15)]

$$g(t) = \sum_{k=-\infty}^{\infty} Tf(kT)h_\sigma(t - kT) \qquad T = \frac{\pi}{\sigma} \tag{5-77}$$

where $h_\sigma(t)$ is the impulse response of the truncated system $H_\sigma(\omega)$ (Fig. 5-8). Since $H_\sigma(\omega) = H_a(\omega)p_\sigma(\omega)$, it follows from the convolution theorem that

$$h_\sigma(t) = h_a(t) * \frac{\sin \sigma t}{\pi t} \qquad h_a(t) \leftrightarrow H_a(\omega) \tag{5-78}$$

With $t = nT$, (5-77) yields

$$g(nT) = \sum_{k=-\infty}^{\infty} Tf(kT)h_\sigma(nT - kT) \tag{5-79}$$

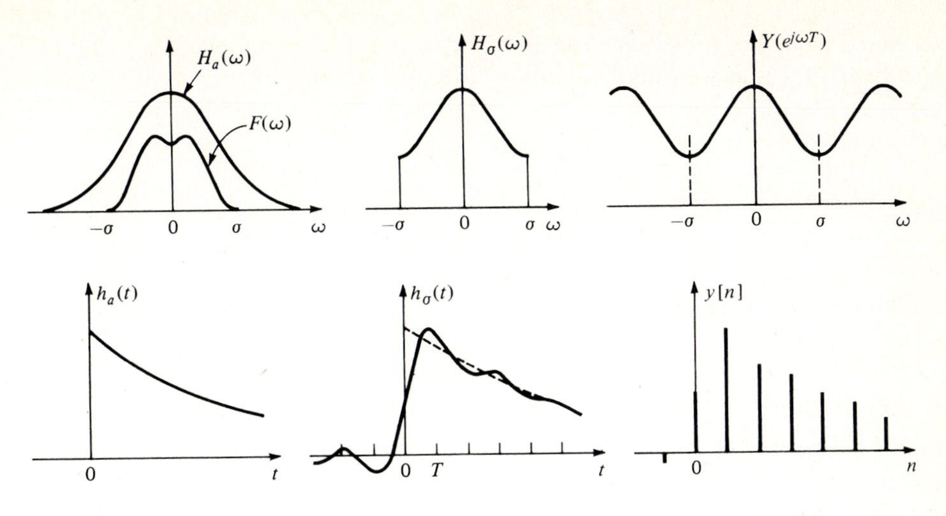

Figure 5-8

Introducing the sequences

$$f[n] = f(nT) \qquad y[n] = Th_\sigma(nT) \qquad g[n] = g(nT) \tag{5-80}$$

obtained by sampling the signals $f(t)$, $Th_\sigma(t)$, and $g(t)$, respectively, we conclude from Eq. (5-79) that

$$g[n] = \sum_{k=-\infty}^{\infty} f[k]y[n-k] \tag{5-81}$$

Thus, $g[n]$ is the output of a discrete system with input $f[n]$ and delta response $y[n]$.

The system function $Y(z)$ of the discrete system so formed is given by

$$Y(z) = \sum_{n=-\infty}^{\infty} y[n]z^{-n} = \sum_{n=-\infty}^{\infty} Th_\sigma(nT)z^{-n} \tag{5-82}$$

Hence [see (3-87)],

$$Y(e^{j\omega T}) = \sum_{n=-\infty}^{\infty} Th_\sigma(nT)e^{-jnT\omega} = \sum_{n=-\infty}^{\infty} H_\sigma(\omega + 2n\sigma) \tag{5-83}$$

Therefore, $Y(e^{j\omega T})$ is periodic (Fig. 5-8), and

$$Y(e^{j\omega T}) = H_\sigma(\omega) = H_a(\omega) \qquad \text{for } |\omega| < \sigma \tag{5-84}$$

The above leads to the following method for simulating an analog system (see also Sec. 1-3).

We are given the system function $H_a(\omega)$, usually as a set of specifications, and the input $f(t)$, often in real time. We wish to find the output $g(t)$.

We form a discrete system, either in terms of its delta response $y[n]$ determined by sampling the function $Th_\sigma(t)$, or in terms of its system function $Y(z)$

determined from Eq. (5-84). We use as input to the discrete system the samples of the given $f(t)$, and we obtain as output the samples $g[n]$ of the unknown $g(t)$. Clearly, $g(t)$ is BL because $G(\omega) = F(\omega)H(\omega)$ and $F(\omega) = 0$ for $|\omega| > \sigma$ by assumption; hence,

$$g(t) = \sum_{n=-\infty}^{\infty} g[n] \frac{\sin \sigma(t - nT)}{\sigma(t - nT)} \tag{5-85}$$

The simulation is thus complete.

The Approximation Problem

A discrete system is computer-realizable if its system function $Y(z)$ is rational in z and its delta response $y[n]$ is causal (as we shall note presently, the causality requirement can be relaxed). If $Y(z)$ is obtained by simulating an analog system, then it is not, in general, rational; in fact, its inverse $y[n] = Th_\sigma(nT)$ is not even causal. The computer simulation can, thus, only be approximate.

The approximation problem is the determination of a rational function

$$\hat{Y}(z) \equiv H(z) = \frac{b_0 + b_1 z^{-1} + \cdots + b_m z^{-m}}{1 + a_1 z^{-1} + \cdots + a_m z^{-m}} = \frac{N(z)}{D(z)} \tag{5-86}$$

of specified order m, with causal inverse [see Eq. (2-38)]

$$\hat{y}[n] \equiv h[n] = A_0 \delta[n] + \sum_{k=1}^{m} A_k z_k{}^n U[n] \tag{5-87}$$

such that the discrete-time error $|y[n] - h[n]|$ or the frequency-domain error $|Y(e^{j\omega T}) - H(e^{j\omega T})|$ is small in some sense. The unknowns are the $2m + 1$ coefficients a_k and b_k or, equivalently, the $m + 1$ coefficients A_k and the m roots z_k of $D(z)$.

The following two error criteria are used often:

Tchebycheff The digital filter $H(z)$ is optimum in the Tchebycheff sense if the discrete-time error

$$\max_n |y[n] - h[n]| \tag{5-88a}$$

or the frequency-domain error

$$\max_\omega |Y(e^{j\omega T}) - H(e^{j\omega T})| \tag{5-88b}$$

is minimum.

Mean square The digital filter $H(z)$ is optimum in the mean-square sense if the mean-square error

$$e = \sum_{n=-\infty}^{\infty} |y[n] - h[n]|^2 = \frac{1}{2\sigma}\int_{-\sigma}^{\sigma} |Y(e^{j\omega T}) - H(e^{j\omega T})|^2 \, d\omega \tag{5-89}$$

is minimum.

Notes:

1. Optimum closed-form solutions do not, in general, exist for either criterion. Optima can, however, be determined with iteration methods.
2. If it is assumed that the denominator $D(z)$ of $H(z)$ is known, then the mean-square optimum numerator $N(z)$ can be found with the techniques of Sec. 5-2. This is done either by forming the orthonormal set $\Phi_i(z)$ corresponding to the zeros of the given $D(z)$ [see (5-75)] or from the interpolation condition

$$H\left(\frac{1}{z_i^*}\right) = Y\left(\frac{1}{z_i^*}\right) \qquad i = 1, \ldots, m \qquad H(\infty) = Y(\infty) = b_0 \tag{5-90}$$

resulting readily from (5-74) if $\hat{F}(z)$ is replaced with $H(z) - A_0$, and z_i with $1/z_i$ (elaborate).
3. We have used as our criterion the minimization of the delta-response error $y[n] - h[n]$. This is desirable if the filter is designed for a whole class of inputs. For specific inputs, the error $g[n] - \hat{g}[n]$ must be minimized, where

$$g[n] = \frac{1}{2\pi}\int_{-\sigma}^{\sigma} F(\omega)Y(e^{j\omega T})e^{jnT\omega} \, d\omega \qquad \hat{g}[n] = \frac{1}{2\pi}\int_{-\sigma}^{\sigma} F(\omega)H(e^{j\omega T})e^{jnT\omega} \, d\omega \tag{5-91}$$

are the responses of the discrete simulator $Y(z)$ and the approximating filter $H(z)$, respectively, to the input $f[n] = f(nT)$ (see Prob. 22). The mean-square output error is

$$\sum_{n=-\infty}^{\infty} |g[n] - \hat{g}[n]|^2 = \frac{1}{2\pi T}\int_{-\sigma}^{\sigma} |F(\omega)|^2 |Y(e^{j\omega T}) - H(e^{j\omega T})|^2 \, d\omega \tag{5-92}$$

and its minimum yields the optimum $H(z)$ for the input $f(t)$.

Shifting As we show next, a reduction of the error is possible if the impulse response $h_a(t)$ of the system to be simulated is suitably shifted.

Causality The discrete system $y[n]$ is not, in general, causal even if it is a simulator of a causal system $h_a(t)$. If, for example, $h_a(t) = e^{-\alpha t}U(t)$, then

$$h_\sigma(t) = \int_{-\infty}^{t} e^{-\alpha(t-\tau)} \frac{\sin \sigma\tau}{\pi\tau} \, d\tau$$

This function is obviously not zero for $t < 0$ (Fig. 5-8). In the approximation of $y[n]$ with a causal filter $h[n]$, the mean-square error consists, therefore, of two sums:

$$e = \sum_{n=-\infty}^{\infty} |y[n] - h[n]|^2 = \sum_{n=-\infty}^{-1} |y[n]|^2 + \sum_{n=0}^{\infty} |y[n] - h[n]|^2 \tag{5-93}$$

The sum from $-\infty$ to -1 is a term that must be added to the mean-square error of the approximation of the causal part $y[n]U[n]$ of $y[n]$ by $h[n]$. This term can be reduced if $h_a(t)$ is replaced with $h_a(t + t_1)$, where t_1 is a suitable constant—i.e., if the digital filter $H(z)$ simulates not $H_a(\omega)$ but the analog system $H_a(\omega)e^{j\omega t_1}$. With this shift, the output $g[n]$ of $H(z)$ equals the samples $g(nT + t_1)$ of the delayed output $g(t + t_1)$ of the original system. Such correction is not possible if the simulation is to be performed in real time.

The result of shifting is equivalent to the assumption that the simulator $H(z)$ of $H_a(\omega)$ is an improper rational function of z. In the following, we shall so interpret the digital filter; i.e., we shall replace the assumption of causality with the assumption that $h[n] = 0$ for $n < -N < 0$.

Continuity The approximation error e depends on the behavior of $y[n]$ as $|n| \to \infty$. For example, if $H(z)$ is a polynomial, i.e., if $h[n] = 0$ for $|n| > N$ (nonrecursive filter), then the sum [see Eq. (5-93)]

$$\sum_{|n|>N} |y[n]|^2$$

is part of the error e. In the recursive case with preshifting, the sum is from $-\infty$ to $-N - 1$.

We must, therefore, attempt to reduce $y[n]$ for large n. This is done by sampling not $h_\sigma(t)$, but the shifted signal $h_\sigma(t + t_0)$:

$$y[n] = Th_\sigma(nT + t_0) \tag{5-94}$$

where t_0 is determined as follows:

As we see from (5-84), the samples $y[n] = Th_\sigma(nT)$ are the coefficients of the Fourier-series expansion of the function $H_a(\omega)$ in the interval $(-\sigma, \sigma)$:

$$H_a(\omega) = \sum_{n=-\infty}^{\infty} y[n]e^{-jnT\omega} \qquad |\omega| < \sigma \tag{5-95}$$

In Sec. 6-1, we prove that the coefficients of the Fourier-series expansion of a periodic function (of bounded variation) tend to zero at least as fast as $1/n$ as $|n| \to \infty$; if the function is also continuous, then they tend to zero as $1/n^2$. The function

$$H_a(\omega) = A(\omega)e^{j\varphi(\omega)}$$

is, in many cases of simulation, continuous; hence, the sum in Eq. (5-95) satisfies the continuity conditions for $|\omega| < \sigma$. At the endpoints $\omega = \pm\sigma$, its amplitude is continuous because $A(-\sigma) = A(\sigma)$. However, its phase is discontinuous if

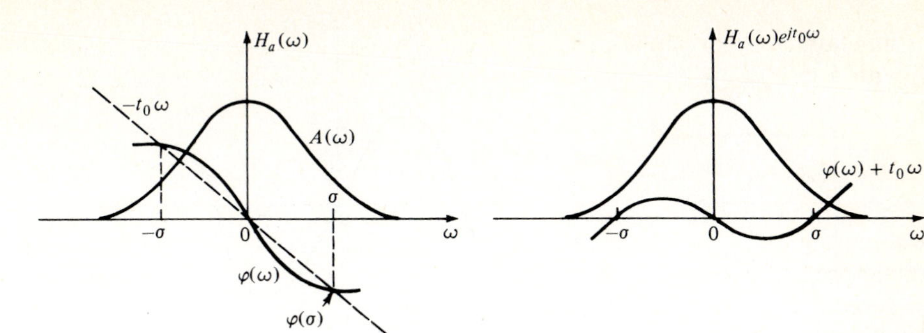

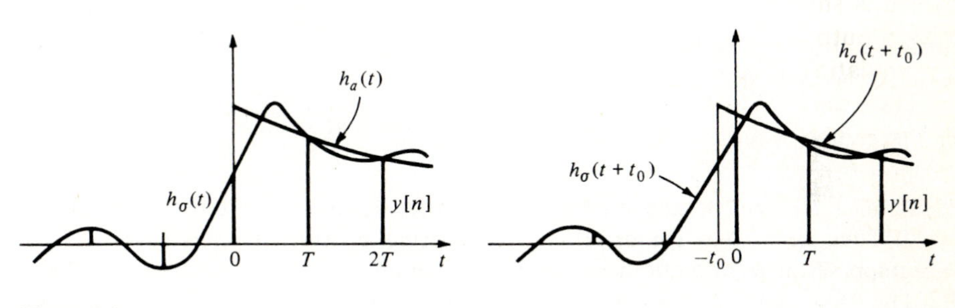

Figure 5-9

$\varphi(-\sigma) = -\varphi(\sigma) \neq \varphi(\sigma)$, that is, if $\varphi(\sigma) \neq 0$. To establish endpoint continuity, we form the function[1] (Fig. 5-9)

$$H_a(\omega)e^{jt_0\omega} \qquad t_0 = -\frac{\varphi(\sigma)}{\sigma} \tag{5-96}$$

whose phase $\varphi(\omega) + t_0\omega$ equals zero at $\omega = \sigma$. If, therefore,

$$y[n] = h_\sigma(nT + t_0) \tag{5-97}$$

then $y[n]$ tends to zero as $1/n^2$ for $|n| \to \infty$, because then the corresponding sum in (5-95) is continuous for all ω.

Note: The optimum constant t_0 equals the *phase delay* $-\varphi(\omega)/\omega$ of the system $H_a(\omega)$ at $\omega = \sigma$ [see Eq. (4-73)].

Example 5-6

$$H_a(\omega) = \frac{1}{\alpha + j\omega} \qquad A(\omega) = \frac{1}{\sqrt{\alpha^2 + \omega^2}} \qquad \varphi(\omega) = -\tan^{-1}\frac{\omega}{\alpha} \qquad t_0 = \frac{1}{\sigma}\tan^{-1}\frac{\sigma}{\alpha}$$

(Fig. 5-9). If $\sigma \gg \alpha$, then $t_0 \simeq \pi/2\sigma = T/2$.

[1] D. Youla, private communication.

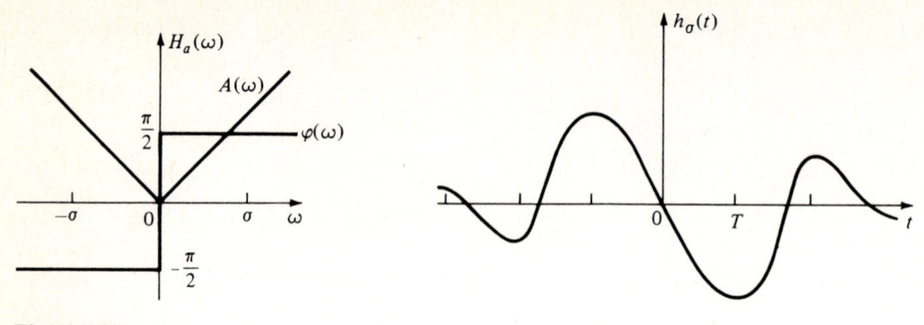

Figure 5-10

Example 5-7 If the analog system to be simulated is a differentiator, then $H_a(\omega) = j\omega$ and $h_a(t) = \delta'(t)$; hence (Fig. 5-10),

$$h_\sigma(t) = \delta'(t) * \frac{\sin \sigma t}{\pi t} = \frac{d}{dt}\frac{\sin \sigma t}{\pi t} = \frac{\sigma t \cos \sigma t - \sin \sigma t}{\pi t^2}$$

as in Example 1-21. From the above, it follows that $y[0] = 0$ and $y[n] = Th_\sigma(nT) = (-1)^n/nT$ for $n \neq 0$. Thus, $y[n] \to 0$ as $1/n$ for $|n| \to \infty$ as expected, because $\varphi(\sigma) = \pi/2 \neq 0$.

To improve the asymptotic behavior of $y[n]$, we introduce the delay $t_0 = -\pi/2\sigma = -T/2$. The delta response of the modified simulator is now given by

$$y[n] = Th_\sigma\left(nT - \frac{T}{2}\right) = \frac{4(-1)^n}{\pi T(2n-1)^2} \tag{5-98}$$

and it goes to zero as $1/n^2$ for $|n| \to \infty$.

We note that the output $g[n]$ of the modified simulator equals the sample values of $g(t + t_0) = f'(t - T/2)$. This leads to the identity

$$f'\left(nT - \frac{T}{2}\right) = \sum_{k=-\infty}^{\infty} \frac{4(-1)^k}{\pi T(2k-1)^2} f(nT - kT) \tag{5-99}$$

In the next two sections we discuss several approximation methods stressing computational simplicity.

5-4 NONRECURSIVE FILTERS

The system function

$$H(z) = \sum_{n=-N}^{N} b_n z^{-n} \tag{5-100}$$

defines a nonrecursive filter whose output $g[n]$ can be expressed directly in terms of the input $f[n]$:

$$g[n] = \sum_{k=-N}^{N} b_k f[n-k] \tag{5-101}$$

This filter is not causal, but causality is simply restored with the factor z^{-N}.

The output of the resulting causal filter $z^{-N}H(z)$ is then $g[n-N]$. Clearly,

$$H(e^{j\omega T}) = \sum_{n=-N}^{N} b_n e^{-jnT\omega} \tag{5-102}$$

is a trigonometric polynomial, and our problem is to find its $2N+1$ coefficients b_n so as to approximate a given function $H_a(\omega)$ in the band $(-\sigma, \sigma)$ of the input.

Mean-Square Optimum and Fourier Series

We wish to determine the coefficients b_n such that the mean-square approximation error

$$e_N = \frac{1}{2\sigma}\int_{-\sigma}^{\sigma} \left| H_a(\omega) - \sum_{n=-N}^{N} b_n e^{-jnT\omega} \right|^2 d\omega \tag{5-103}$$

is minimum. As we know (Example 5-4), e_N is minimum if the numbers b_n equal the Fourier-series coefficients of $H_a(\omega)$:

$$b_n = \frac{1}{2\sigma}\int_{-\sigma}^{\sigma} H_a(\omega)e^{jnT\omega}\, d\omega = Th_\sigma(nT) = y[n] \qquad |n| \le N \tag{5-104}$$

Thus, the digital filter $H(z)$ is obtained by truncating $y[n]$ for $|n| > N$ [see Eq. (5-82)]. The resulting mean-square error e_N is given by

$$e_N = \sum_{|n|>N} |y[n]|^2 \tag{5-105}$$

Notes:

1. Reasoning similarly, we can show that if

$$H(z) = \sum_{n=N_1}^{N_2} b_n z^{-n}$$

then the mean-square error e is again minimum if $b_n = y[n]$, and it is given by

$$e(N_1, N_2) = \sum_{n<N_1} + \sum_{n>N_2} |y[n]|^2 \tag{5-106}$$

For the same number of terms $N_2 - N_1 = 2N + 1$ as in (5-100), the error $e(N_1, N_2)$ can be reduced with a proper choice of N_1. The same result is obtained if the symmetrical filter (5-100) is used to approximate the shifted system $H_a(\omega)e^{jt_1\omega}$, where $t_1 = (N + N_1)T$. The value of t_1 minimizing $e(N_1, N_2)$ can only be found numerically. A reasonable choice is the center of gravity of $h_a(t)$,

$$t_1 = \eta = -\varphi'(0) \tag{5-107}$$

[see (4-20)]. This is also the center of gravity of $h_\sigma(t)$ because $H_\sigma(\omega) = H_a(\omega)$ for $|\omega| < \sigma$ and, hence, $\varphi'_\sigma(0) = \varphi'(0)$.

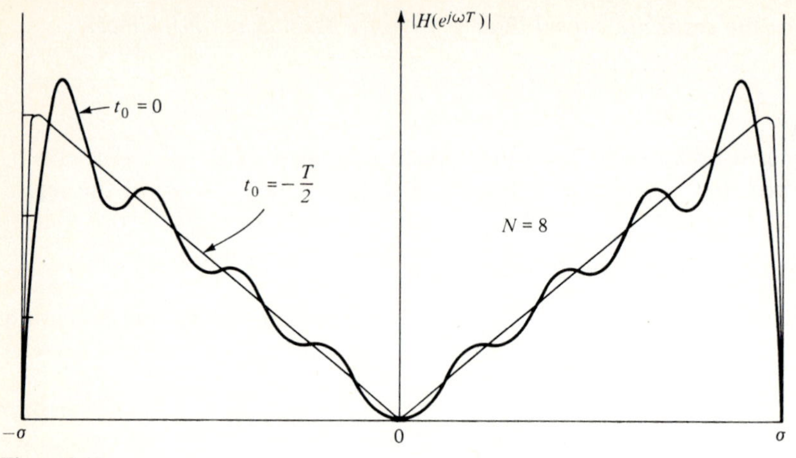

Figure 5-11

2. An additional error reduction is possible if a delay t_0 is introduced to ensure continuity at $\omega = \pm\sigma$:

$$H_a(\sigma) = H_a(-\sigma) \tag{5-108}$$

[see (5-96)]. This results in a reduction of $y[n]$ for large n, and it eliminates the Gibbs' phenomenon.

Example 5-8 If $H_a(\omega) = j\omega$ as in Example 5-7, then [see Eq. (1-75)]

$$H(z) = \sum_{\substack{n=-N \\ n \neq 0}}^{N} \frac{(-1)^n}{nT} z^{-n}$$

With the delay $t_0 = -T/2$, we obtain [see Eq. (5-98)]

$$H(z) = \sum_{n=-N}^{N} \frac{4(-1)^n}{\pi T(2n-1)^2} z^{-n}$$

In Fig. 5-11, we plot the function $H(e^{j\omega T})$ for $t_0 = 0$ and $t_0 = -T/2$. As we see from the figure, the introduction of a simple delay greatly improves the approximation.

Interpolation and Discrete Fourier Series

We shall now determine the $2N + 1$ coefficients b_n of $H(z)$ such that $H(e^{j\omega T})$ equals $H_a(\omega)$ at $2N + 1$ equidistant points:

$$H_a(n\omega_0) = H(e^{jn\omega_0 T}) \qquad |n| \le N \tag{5-109}$$

(Fig. 5-12), where

$$\omega_0 = \frac{2\sigma}{N_0} \qquad N_0 = 2N + 1 \tag{5-110}$$

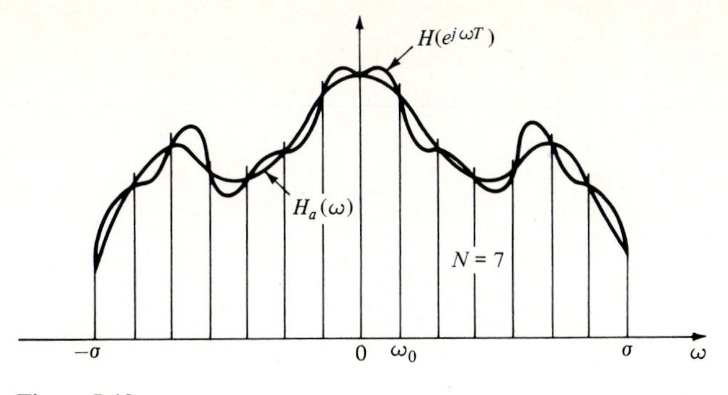

Figure 5-12

Inserting (5-100) into (5-109), we obtain the system

$$H_a(n\omega_0) = \sum_{m=-N}^{N} b_m w^{-mn} = H(w^n) \qquad w = e^{j2\pi/N_0} \tag{5-111}$$

because $n\omega_0 T = 2\pi n/N_0$. The above shows that the numbers $H_a(n\omega_0)$ and b_m form a DFS pair [see (3-104)]:

$$H_a(n\omega_0) \underset{N_0}{\leftrightarrow} N_0 b_m \tag{5-112}$$

Hence,

$$b_m = \frac{1}{N_0} \sum_{n=-N}^{N} H_a(n\omega_0) w^{mn} \tag{5-113}$$

Thus, if a digital filter $H(z)$ is such that its frequency response $H(e^{j\omega T})$ interpolates the frequency response $H_a(\omega)$ of an analog system as in Eq. (5-109), then its delta response $h[n] = b_n$, and the sample values $H_a(n\omega_0)$ of $H_a(\omega)$ form a DFS pair.

Note: If $H(z)$ is replaced with the causal equivalent

$$z^{-N}H(z) = \sum_{n=0}^{N_0-1} c_n z^{-n} \qquad c_n = b_{n-N} \tag{5-114}$$

then the summation in (5-111) is from 0 to $N_0 - 1$ as in (3-104).

Interpolating filters The filter $H(z)$ is obtained by computing the DFS b_m from (5-113). We shall show that the resulting sum (5-100) can be expressed directly in terms of the sample values $H_a(n\omega_0)$ of $H_a(\omega)$. For this purpose, we form the nonrecursive filter

$$K_r(z) = \frac{1}{2N+1} \sum_{n=-N}^{N} (w^r z^{-1})^n = \frac{(w^r z^{-1})^{N+1} - (w^r z^{-1})^{-N}}{(2N+1)(w^r z^{-1} - 1)} \tag{5-115}$$

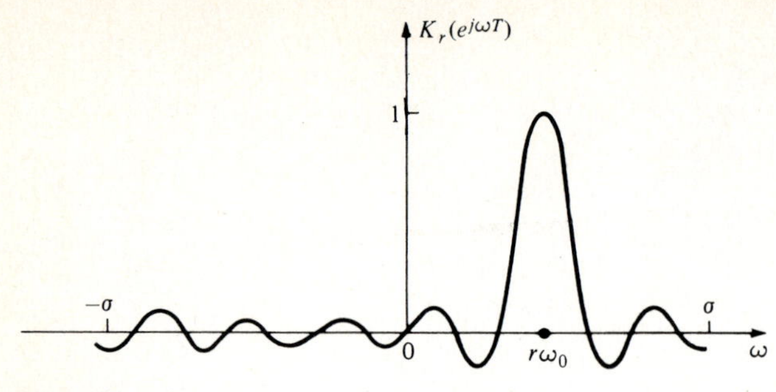

Figure 5-13

It is easy to see that

$$K_r(e^{j\omega T}) = \frac{\sin T(\omega - r\omega_0)(N + \frac{1}{2})}{(2N + 1)\sin T(\omega - r\omega_0)/2} \tag{5-116}$$

because $w^r = e^{jT\omega_0 r}$. Hence,

$$K_r(e^{jn\omega_0 T}) = \begin{cases} 1 & n = r \\ 0 & n \neq r \end{cases} \tag{5-117}$$

as in Fig. 5-13.

From the above, it follows that the interpolating filter is given by the sum

$$H(z) = \sum_{r=-N}^{N} H_a(r\omega_0)K_r(z) \tag{5-118}$$

(Fig. 5-14) because $H(e^{jn\omega_0 T}) = H_a(n\omega_0)$, as we can see for (5-117). This filter is further developed in Sec. 5-6.

Note: A consequence of Eq. (5-118) is the *sampling expansion*

$$H(e^{j\omega T}) = \sum_{r=-N}^{N} H(e^{jr\omega_0 T})K_r(e^{j\omega T}) \tag{5-119}$$

expressing a trigonometric polynomial of the form (5-102) in terms of its sample values at $2N + 1$ equidistant points.

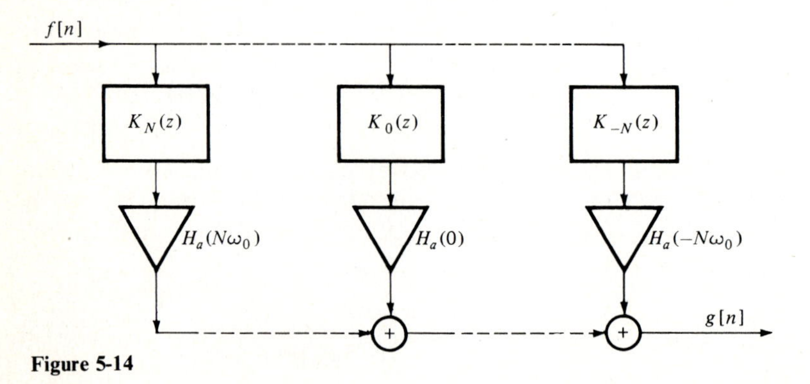

Figure 5-14

5-5 FILTERS

We have so far discussed the problem of approximating the analog system $H_a(\omega)$ with a digital filter $H(z)$,

$$H_a(\omega) \simeq H(e^{j\omega T}) = \frac{b_0 + b_1 e^{-j\omega T} + \cdots + b_m e^{-jm\omega T}}{1 + a_1 e^{-j\omega T} + \cdots + a_m e^{-jm\omega T}} \tag{5-120}$$

in the entire interval $(-\sigma, \sigma)$. The constant $\sigma = \pi/T$ is determined in terms of the sampling interval T, and it is assumed that the class of inputs is σ-BL. For a sufficiently small error, the required order m is, in general, large. The demands on the accuracy of the approximation can be relaxed if we *oversample* i.e., if we choose T such that the ratio $\sigma = \pi/T$ is larger than the band of the input $f(t)$. Assuming that

$$F(\omega) = 0 \qquad \text{for} \qquad |\omega| > \omega_c \qquad \text{and} \qquad \omega_c < \sigma$$

we conclude from Eq. (5-91) that the output $\hat{g}[n]$ of $H(z)$ and the samples $g[n] = g(nT)$ of the output $g(t)$ of $H_a(\omega)$ are given by

$$\begin{aligned} \hat{g}[n] &= \frac{1}{2\pi}\int_{-\omega_c}^{\omega_c} F(\omega)H(e^{j\omega T})e^{jnT\omega}\, d\omega \\ g[n] &= \frac{1}{2\pi}\int_{-\omega_c}^{\omega_c} F(\omega)H_a(\omega)e^{jnT\omega}\, d\omega \end{aligned} \tag{5-121}$$

respectively. Hence, to satisfy the simulation condition $\hat{g}[n] \simeq g[n]$, it suffices to choose $H(z)$ such that

$$H_a(\omega) \simeq H(e^{j\omega T}) \qquad \text{for } |\omega| \le \omega_c \tag{5-122}$$

Example 5-9 We shall examine the possibility of using the digital filter

$$H(z) = \frac{1}{T}(1 - z^{-1}) \tag{5-123}$$

to approximate the differentiator $H_a(\omega) = j\omega$. Clearly,

$$H(e^{j\omega T}) = \frac{2j}{T}\sin\frac{\omega T}{2}\, e^{-j\omega T/2} \simeq j\omega e^{-j\omega T/2}$$

for $|\omega T| \ll \pi$, that is, for $|\omega| \ll \sigma$ (Fig. 5-15). Therefore, the filter $H(z)$ in (5-123) is an

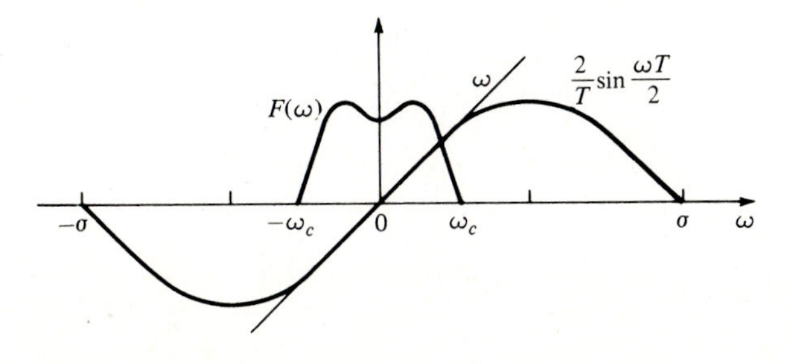

Figure 5-15

approximate simulator of the delayed differentiator $j\omega e^{-j\omega T/2}$ if the input $f(t)$ is w_c-BL and $\omega_c \ll \sigma$. Since the output of the differentiator equals $f'(t - T/2)$, we conclude that, under the above assumption,

$$f'\left(nT - \frac{T}{2}\right) \simeq \frac{f[n] - f[n-1]}{T} = \frac{f(nT) - f(nT - T)}{T}$$

We shall show next that the approximation problem is simplified if the analog system to be simulated is of finite order, i.e., if its system function is rational as in Eq. (4-77),

$$H_I(s) = \frac{d_0 s^r + d_1 s^{r-1} + \cdots + d_r}{s^m + c_1 s^{m-1} + \cdots + c_m} = \frac{N_a(s)}{D_a(s)} \qquad r < m \tag{5-124}$$

and its frequency response

$$H_I(j\omega) = H_a(\omega)$$

is negligible for $|\omega| > \sigma$. This assumption is useful in the following applications:

1. Numerical solution of differential equations [see (4-86)]
2. Digital simulation of physical systems consisting of a finite number of lumped parameters
3. Digital simulation of arbitrary systems having known rational approximations as in (5-124)

The last application is common in filter problems. The function $H_I(s)$ is then determined from available solutions of the analog approximation problem.

Time Sampling

As we know, the digital simulator $y[n]$ of an analog system $H_I(s)$ is obtained by sampling the impulse response $h_\sigma(t)$ of the truncated system $H_\sigma(\omega) = H_I(j\omega)p_\sigma(\omega)$. If

$$H_I(j\omega) \simeq 0 \qquad \text{for } |\omega| > \sigma \tag{5-125}$$

then
$$h_a(t) = \frac{1}{2\pi}\int_{-\infty}^{\infty} H_a(\omega)e^{j\omega t}\,d\omega \simeq \frac{1}{2\pi}\int_{-\sigma}^{\sigma} H_a(\omega)e^{j\omega t}\,d\omega = h_\sigma(t) \tag{5-126}$$

Hence,
$$h[n] = Th_a(nT) \simeq Th_\sigma(nT) = y[n] \tag{5-127}$$

As we see from Eq. (5-126), the resulting time-domain error $h[n] - y[n]$ is such that

$$|h[n] - y[n]| \le \frac{1}{2\pi}\int_{|\omega|>\sigma} |H_a(\omega)|\,d\omega \tag{5-128}$$

In the frequency domain,

$$H(e^{j\omega T}) = \sum_{n=-\infty}^{\infty} Th_a(nT)e^{-jnT\omega} = \sum_{n=-\infty}^{\infty} H_a(\omega + 2n\sigma) \tag{5-129}$$

This sum does not equal $H_a(\omega)$ for $|\omega| < \sigma$; the difference $H(e^{j\omega T}) - H_a(\omega)$ is the aliasing error.

Sampling $h_a(t)$ instead of $h_\sigma(t)$ is of use in the simulation of analog systems of finite order, as in (5-124), because, as we shall presently see, the resulting simulator $H(z)$ is a digital filter of the same order.

Indeed, suppose first that

$$H_I(s) = \frac{1}{s - s_i} \qquad s_i = -\alpha_i + j\beta_i \tag{5-130}$$

Clearly, $h_a(t) = e^{s_i t}U(t)$; hence,

$$h[n] = Te^{s_i nT}U[n] \tag{5-131}$$

Thus, the simulator of a first-order system with pole s_i is a first-order digital system

$$H(z) = \sum_{n=0}^{\infty} Te^{s_i nT}z^{-n} = \frac{Tz}{z - e^{s_i T}} \tag{5-132}$$

with pole $z_i = e^{s_i T}$. This is approximate, and it holds if $\alpha_i \ll \sigma - |\beta_i|$ because then

$$H_I(j\omega) = \frac{1}{\alpha_i + j(\omega - \beta_i)} \simeq 0 \qquad \text{for } |\omega| > \sigma$$

The general case can be reduced to the above by expanding $H_I(s)$ into partial fractions:

$$H_I(s) = \sum_{i=1}^{m} \frac{A_i}{s - s_i} = \frac{N_a(s)}{D_a(s)} \qquad h_a(t) = \sum_{i=1}^{m} A_i e^{s_i t}U(t) \tag{5-133}$$

$$H(z) = \sum_{i=1}^{m} \frac{A_i Tz}{z - e^{s_i T}} = \frac{N(z)}{D(z)} \qquad h[n] = \sum_{i=1}^{m} A_i Te^{s_i nT}U[n] \tag{5-134}$$

Thus, the simulator of a finite-order system $H_I(s)$, obtained by sampling $h_a(t)$, is a digital filter $H(z)$ of the same order.

Since

$$z_i = e^{s_i T} \qquad \text{where} \qquad D(z_i) = 0 \qquad \text{and} \qquad D_a(s_i) = 0 \tag{5-135}$$

it follows that, if $\text{Re}\, s_i < 0$, then $|z_i| < 1$. Hence, a stable analog system yields a stable digital simulator.

Example 5-10 In Fig. 5-16 we show a second-order system

$$H_I(s) = \frac{1}{(s + \alpha)^2 + \beta^2} = \frac{1/2j\beta}{s - s_1} - \frac{1/2j\beta}{s - s_2} \qquad s_{12} = -\alpha \pm j\beta$$

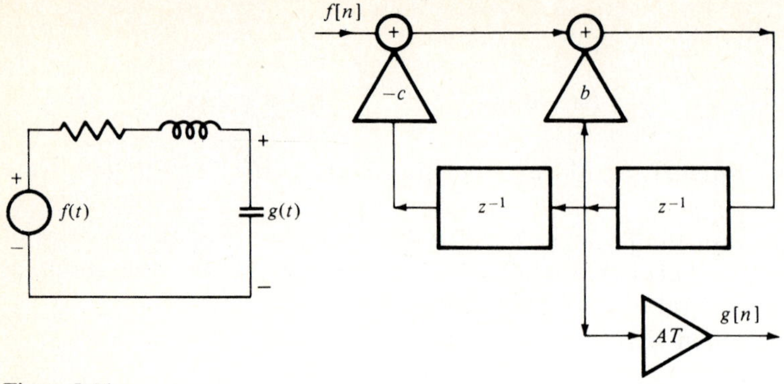

Figure 5-16

and the corresponding digital simulator

$$H(z) = \frac{ATz}{z^2 - bz + c} \qquad A = e^{-\alpha T} \frac{\sin \beta T}{\beta} \qquad b = 2e^{-\alpha T} \cos \beta T \qquad c = e^{-2\alpha T}$$

The approximation condition (5-125) holds if $\alpha \ll \sigma - \beta$ (elaborate).

The following special cases are of interest:

(*a*) *Resonant circuit*. If $\alpha \ll \beta$, then the system is a narrow-band bandpass filter with center frequency $\omega_0 = \beta$ (Fig. 5-17).
(*b*) *Butterworth filter*. If $\alpha = \beta = \omega_c/\sqrt{2} + j\omega_c/\sqrt{2}$, then

$$H_I(s) = \frac{1}{s^2 + \sqrt{2}\omega_c s + \omega_c^2} \qquad |H_I(j\omega)| = \frac{1}{\sqrt{\omega^4 + \omega_c^4}}$$

i.e., the system is low-pass filter with cut-off frequency ω_c (Fig. 5-18).

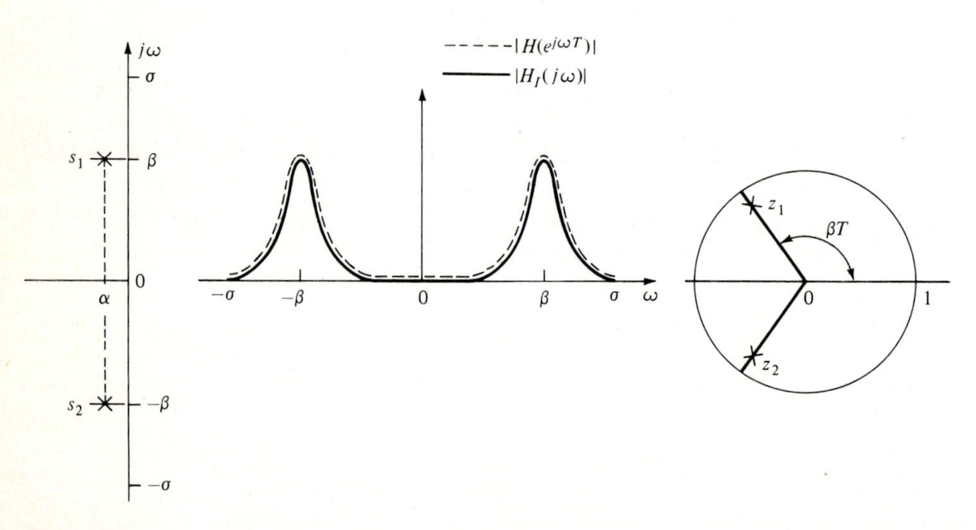

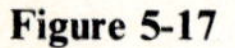
Figure 5-17

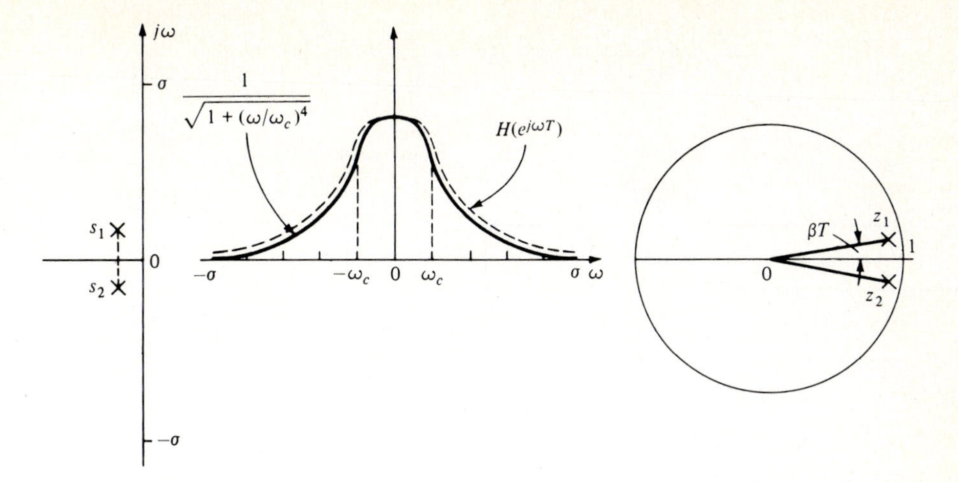

Figure 5-18

Notes:

1. We assumed in Eq. (5-133) that all poles s_i of $H_I(s)$ are simple. Multiple roots lead to similar results. Suppose, for example, that s_i is a double root:

$$H_I(s) = \frac{1}{(s - s_i)^2} \qquad h_a(t) = te^{s_i t}U(t)$$

The corresponding simulator is [see (2-4)]

$$H(z) = \sum_{n=0}^{\infty} T^2 n e^{s_i n T} z^{-n} = T^2 e^{s_i T} \frac{z}{(z - e^{s_i T})^2}$$

2. As we know (page 75), if $h_a(t)$ is discontinuous at $t = nT$, then for the validity of (5-129) we must assume that $h_a(nT) = h_a(nT^+)/2 + h_a(nT^-)/2$. Therefore, in order for the simulator of $H_I(s)$ to satisfy (5-129), a constant must be added to the function $H(z)$ in (5-134). Indeed, $h_a(t)$ is continuous for $t \neq 0$ and $h_a(0^-) = 0$; hence, $h[0] = Th_a(0)$ must be changed from $Th_a(0^+)$ to $Th_a(0^+)/2$, where [see Eqs. (5-133) and (5-124)]

$$h_a(0^+) = \sum_{i=1}^{m} A_i = \lim_{s \to \infty} sH_I(s) = \begin{cases} d_0 & \text{if } r = m - 1 \\ 0 & \text{if } r < m - 1 \end{cases}$$

Hence, for (5-129) to hold, $H(z)$ must be given by

$$H(z) = \sum_{i=1}^{m} \frac{A_i Tz}{z - e^{s_i T}} - \frac{T}{2} d_0 \tag{5-136}$$

No such correction is needed if $h_a(0) = 0$, that is, if the degree of $N(s)$ is less than $m - 1$ as in Example 5-10.

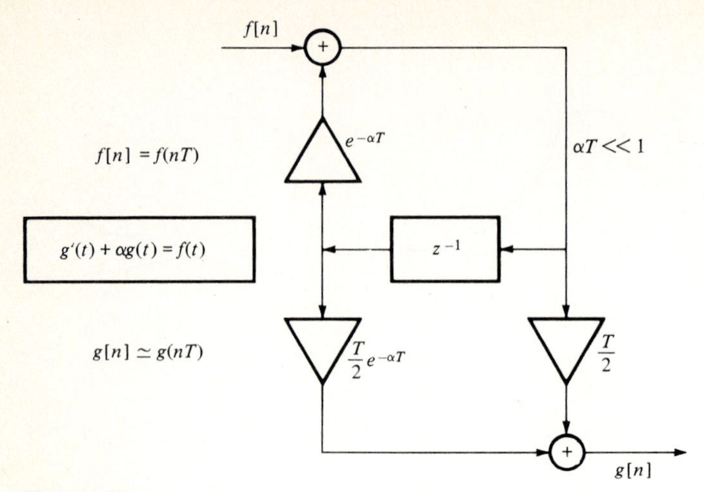

Figure 5-19

Example 5-11 We wish to solve numerically the differential equation

$$g'(t) + \alpha g(t) = f(t)$$

where $f(t)$ is a σ-BL function. This equation is equivalent to the system

$$H_I(s) = \frac{1}{s + \alpha} \qquad \text{with} \qquad s_1 = -\alpha \qquad h_a(t) = e^{-\alpha t}U(t) \qquad h_a(0^+) = 1$$

Hence, its digital simulator is

$$H(z) = \frac{Tz}{z - e^{-\alpha T}} - \frac{T}{2}$$

as in Fig. 5-19.

Frequency Transformations

If $H_I(s)$ is an analog system of order m as in (5-124), then

$$H(z) = H_I\left(c\,\frac{z - 1}{z + 1}\right) \qquad c > 0 \tag{5-137}$$

is a digital filter of the same order, and

$$H(e^{j\omega T}) = H_I\left(jc \tan \frac{\omega T}{2}\right) \tag{5-138}$$

because

$$\frac{e^{j\omega T} - 1}{e^{j\omega T} + 1} = j \tan \frac{\omega T}{2} \tag{5-139}$$

The above leads to a method for generating digital filters with useful properties. We illustrate with a special case.

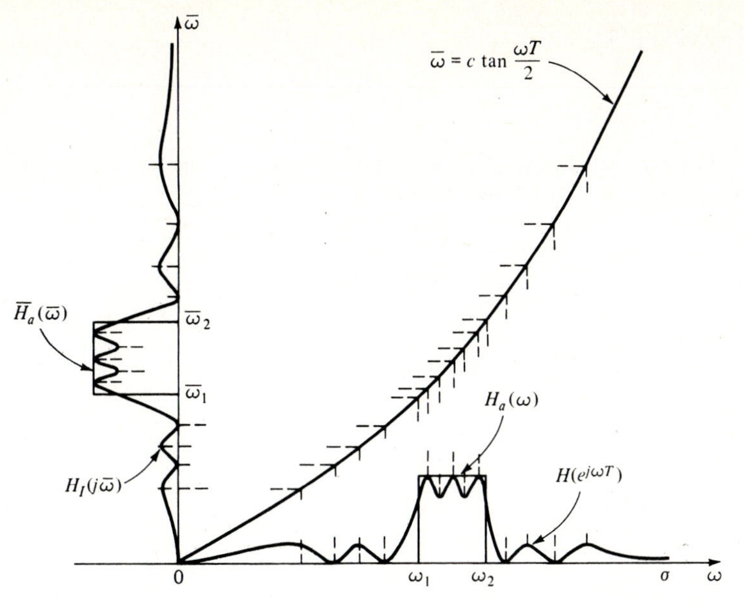

Figure 5-20

Suppose that we wish to simulate a bandpass filter $H_a(\omega)$ with band (ω_1, ω_2), as in Fig. 5-20. We determine, first, an analog system $H_I(s)$ of finite order whose frequency response $H_I(j\overline{\omega})$ approximates another bandpass filter $\overline{H}_a(\overline{\omega})$ with band $(\overline{\omega}_1, \overline{\omega}_2)$. The digital filter $H(z)$ obtained from $H_I(s)$ with the transformation

$$s = c\,\frac{z-1}{z+1} \tag{5-140}$$

as in Eq. (5-137), has a frequency response

$$H(e^{j\omega T}) = H_I(j\overline{\omega}) \qquad \overline{\omega} = c \tan \frac{\omega T}{2} \tag{5-141}$$

Hence, it is obtained from $H_I(j\overline{\omega})$ by a nonlinear frequency transformation as in Fig. 5-20.

If the cut-off frequencies $\overline{\omega}_1$ and $\overline{\omega}_2$ of the analog system $H_I(s)$ are such that

$$\overline{\omega}_1 = c \tan \frac{\omega_1 T}{2} \qquad \overline{\omega}_2 = c \tan \frac{\omega_2 T}{2} \tag{5-142}$$

then, as we see from the figure, $H(e^{j\omega T})$ approximates the desired response $H_a(\omega)$. If the scaling factor c is such that $c = 2/T$, then

$$\overline{\omega} = \frac{2}{T} \tan \frac{\omega T}{2} \simeq \omega \qquad \text{for } |\omega T| \ll 1 \tag{5-143}$$

In this case, if $H_I(s)$ is a low-pass filter, then $H(z)$ is a low-pass filter with the same cut-off frequency.

Example 5-12 The second-order system

$$H_I(s) = \frac{1}{s^2 + \sqrt{2}\,s + 1} \qquad |H_I(j\varpi)| = \frac{1}{\sqrt{1 + \varpi^4}}$$

is a low-pass filter (Butterworth) with cut-off frequency $\varpi_c = 1$. The transformation $s = 2(z - 1)/T(z + 1)$ yields the digital low-pass filter

$$H(z) = \frac{T^2(z+1)^2}{4(z-1)^2 + 2\sqrt{2}\,T(z^2 - 1) + T^2(z+1)^2}$$

$$|H(e^{j\omega T})| = \frac{1}{\sqrt{1 + [(2/T)\tan(\omega T/2)]^4}}$$

with cut-off frequency $\omega_c \simeq 1$.

Notes:

1. The transformation

$$s = c\,\frac{z - 1}{z + 1}$$

maps the region $\operatorname{Re} s \leq 0$ of the s plane into the region $|z| \leq 1$ of the z plane. Hence, if s_i is a pole of $H_I(s)$ and $\operatorname{Re} s < 0$, then

$$z_i = \frac{1 + s_i/c}{1 - s_i/c}$$

is a pole of $H(z)$, and $|z_i| < 1$; that is, a stable analog system $H_I(s)$ yields a stable digital simulator $H(z)$.

2. Frequency transformations are used mainly in the design of narrow-band filters.
3. If $H_I(j\varpi)$ is equiripple (Fig. 5-20), then $H(e^{j\omega T})$ is also equiripple. However, symmetry is not preserved.
4. The digital filter $H(z)$ is determined either from $H_I(s)$ through the complex transformation $s = c(z - 1)/(z + 1)$, or from $H_I(j\varpi)$ through the real transformation $\varpi = c \tan(\omega T/2)$. The two approaches are equivalent. The following generalizations are possible.

Complex-frequency transformation If $r(z)$ is a rational function of z, and $H_I(s)$ is an analog system of finite order, then

$$H(z) = H_I[r(z)] \tag{5-144}$$

is, obviously, a realizable filter. This is a generalization of Eq. (5-137).

Real-frequency transformation From the theorem on page 49 [see (2-46)], it follows that, if $r_1(x)$ is a rational function of x such that $r_1(x) \geq 0$ for $|x| \leq 1$ and

$$\varpi = \sqrt{r_1(\cos \omega T)} \tag{5-145}$$

then we can find a digital filter $H(z)$ such that

$$|H(e^{j\omega T})| = |H_I(j\bar{\omega})| \tag{5-146}$$

Since

$$\cos \omega T = \frac{1 - \tan^2 (\omega T/2)}{1 + \tan^2 (\omega T/2)}$$

(5-145) is equivalent to

$$\bar{\omega} = \sqrt{r_2\left(\tan^2 \frac{\omega T}{2}\right)} \tag{5-147}$$

where $r_2(x)$ is a nonnegative rational function of x. Clearly, (5-141) is a special case.

The function $H(z)$ can be determined from its amplitude $|H(e^{j\omega T})|$ with the techniques of Sec. 2-3. However, if (5-147) is of the form

$$\bar{\omega} = \left(\tan \frac{\omega T}{2}\right) r_3\left(\tan^2 \frac{\omega T}{2}\right) \tag{5-148}$$

and $r_3(x)$ is rational, then

$$H(z) = H_I(s) \qquad s = \frac{z-1}{z+1} r_3\left[-\left(\frac{z-1}{z+1}\right)^2\right] \tag{5-149}$$

as it is easy to show (elaborate).

5-6 RECURSIVE FREQUENCY-DOMAIN FILTERING

Suppose that $f(t)$ is a sophisticated signal of long duration, for example, a voice record. In principle, it can be uniquely determined in terms of its transform $F(\omega)$; however, any numerical processing directly involving $F(\omega)$ is unrealistic because $F(\omega)$ is too complex to be recovered with sufficient accuracy from its samples. To avoid this problem, we propose to describe the spectral properties of $f(t)$ in terms of the transform $F(t,\omega)$ of a moving segment of $f(t)$ [see Eq. (5-150) below]. We thus trade off a complex function $F(\omega)$ of one variable with a simple function $F(t,\omega)$ of two variables.

We show that the samples $F(nT_1, m\omega_0)$ of $F(t, \omega)$ satisfy a first-order recursion with input the difference $f(nT_1) - f(nT_1 - T)$ of the values of $f(t)$ at the endpoints of the segment under consideration. This leads to a simple method for recursive filtering in the frequency domain. The resulting system is the recursive equivalent of the interpolator[1] $H(z)$ introduced in Eq. (5-118). However, in this section its components $K_r(z)$ are not interpreted as interpolators (see Fig. 5-13), but as

[1] L. R. Rabiner and R. W. Schafer, Recursive and Nonrecursive Realizations of Digital Filters Designed by Frequency Sampling Techniques, *IEEE Trans. on Audio and Electroacoustics*, vol. AU-19, no. 3, pp. 200–207, September, 1971.

spectrum analyzers yielding in real time the discrete Fourier series of the samples $f(nT_1)$ of a moving segment of the input. This interpretation is of particular interest in the simulation of systems driven by sophisticated signals.

The Running Fourier Transform

We define the *running Fourier transform* of a signal $f(t)$ by the integral

$$F(t, \omega) = \int_{-c}^{c} f(t + \tau)e^{-j\omega\tau}\,d\tau = e^{j\omega t}\int_{t-c}^{t+c} f(\alpha)e^{-j\omega\alpha}\,d\alpha \tag{5-150}$$

where c is a given constant. For a fixed t, $F(t, \omega)$ is the Fourier transform in the variable τ of the segment $f(t + \tau)$ of $f(t)$ shown in Fig. 5-21.

Theorem The running transform satisfies a first-order differential equation in the variable t:

$$\frac{\partial F(t, \omega)}{\partial t} - j\omega F(t, \omega) = e^{-jc\omega}f(t + c) - e^{jc\omega}f(t - c) \tag{5-151}$$

Proof Differentiating (5-150) with respect to t, we obtain

$$\frac{\partial F(t, \omega)}{\partial t} = j\omega e^{j\omega t}\int_{t-c}^{t+c} f(\alpha)e^{-j\omega\alpha}\,d\alpha + e^{-jc\omega}f(t + c) - e^{jc\omega}f(t - c)$$

and (5-151) results.

Inversion formula We maintain that

$$f(t) = \frac{1}{2c}\sum_{m=-\infty}^{\infty} F(t, m\omega_0) \qquad \omega_0 = \frac{\pi}{c} \tag{5-152}$$

Proof We expand the function $f(t + \tau)$ into a Fourier series in the variable τ. The coefficients of the expansion in the interval $(-c, c)$ are given by [see Eq. (1-150)]

$$\frac{1}{2c}\int_{-c}^{c} f(t + \tau)e^{-jm\omega_0\tau}\,d\tau = \frac{1}{2c}F(t, m\omega_0)$$

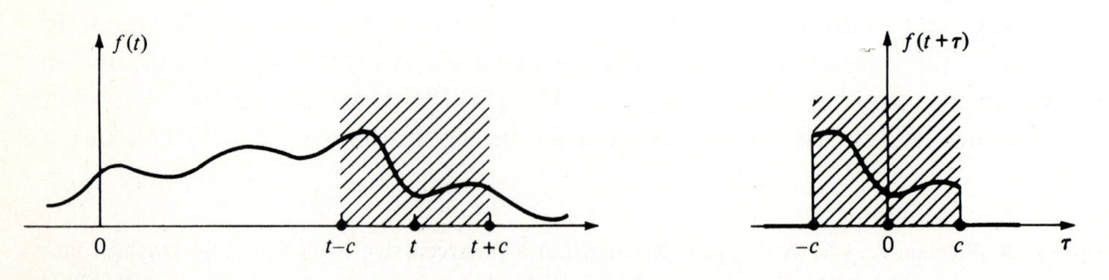

Figure 5-21

Hence,

$$f(t+\tau) = \frac{1}{2c} \sum_{m=-\infty}^{\infty} F(t, m\omega_0) e^{jm\omega_0 \tau} \qquad |\tau| < c \tag{5-153}$$

and (5-152) results with $\tau = 0$.

Thus, $f(t)$ is uniquely determined in terms of the samples $F(t, m\omega_0)$ of $F(t, \omega)$. From (5-151) it follows with $\omega = m\omega_0 = m\pi/c$ that these samples satisfy the equation

$$\frac{dF(t, m\omega_0)}{dt} - jm\omega_0 F(t, m\omega_0) = (-1)^m [f(t+c) - f(t-c)] \tag{5-154}$$

Convolution theorem Consider a linear system with time-limited impulse response:

$$h(t) = 0 \qquad \text{for } |t| > c$$

and system function $H_a(\omega)$. We shall show that its response

$$g(t) = \int_{-c}^{c} f(t-\alpha) h(\alpha)\, d\alpha \tag{5-155}$$

to an arbitrary input $f(t)$ with running transform $F(t, \omega)$ is given by

$$g(t) = \frac{1}{2c} \sum_{m=-\infty}^{\infty} F(t, m\omega_0) H_a(m\omega_0) \tag{5-156}$$

Proof Inserting (5-153) into (5-155), we obtain (5-156) because

$$\int_{-c}^{c} h(\alpha) e^{-jm\omega_0 \alpha}\, d\alpha = H_a(m\omega_0) \tag{5-157}$$

Filtering The preceding theorem permits us to evaluate the response of a time-limited system to an input $f(t)$ of long duration in terms of the running transform of $f(t)$. The result is given by the sum in (5-156) involving only the samples of $F(t, \omega)$. Thus, to realize $H_a(\omega)$ with this method, it suffices to solve a first-order equation for each m.

Recursive DFS

Given a sequence $f[n]$ and an integer N, we form the *running z transform*

$$\Phi(n, z) = \sum_{k=0}^{N-1} f[n-k] z^{-k} \tag{5-158}$$

of $f[n]$. For a fixed n, $\Phi(n, z)$ is the z transform in the variable k of the segment $f[n-k]$ of $f[n]$ shown in Fig. 5-22.

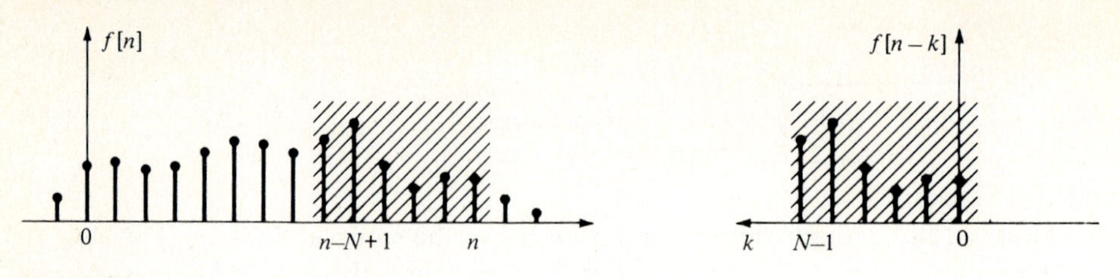

Figure 5-22

Theorem The function $\Phi(n, z)$ satisfies the first-order recursion equation

$$\Phi(n, z) - z^{-1}\Phi(n-1, z) = f[n] - z^{-N}f[n-N] \qquad (5\text{-}159)$$

Proof From (5-158) it follows with $k + 1 = r$ that

$$\Phi(n-1, z) = z\sum_{r=1}^{N} f[n-r]z^{-r}$$

and (5-159) results.

Inversion formula We maintain that

$$f[n] = \frac{1}{N}\sum_{m=0}^{N-1} \Phi(n, w^{-m}) \qquad w = e^{j2\pi/N} \qquad (5\text{-}160)$$

Proof Clearly,

$$\Phi(n, w^{-m}) = \sum_{k=0}^{N-1} f[n-k]w^{km} \qquad (5\text{-}161)$$

This equation shows that for a fixed n, the DFS of the N numbers $f[n-k]$ equals $\Phi(n, w^{-m})$. From the DFS inversion formula (3-104), it follows therefore that

$$f[n-k] = \frac{1}{N}\sum_{m=0}^{N-1} \Phi(n, w^{-m})w^{-km} \qquad k = 0, 1, \ldots, N-1 \qquad (5\text{-}162)$$

and (5-160) results with $k = 0$.

The *running DFS* $\Phi(n, w^{-m})$ of $f[n]$ satisfies the recursion equation

$$\Phi(n, w^{-m}) - w^m\Phi(n-1, w^{-m}) = f[n] - f[n-N] \qquad (5\text{-}163)$$

This follows from Eq. (5-159) with $z = w^{-m}$ because $w^N = 1$.

The DFS analyzer The recursion equation (5-163) defines a discrete system with input $f[n]$, output $\Phi(n, w^{-m})$, and system function

$$S(z, m) = \frac{1 - z^{-N}}{1 - w^m z^{-1}} \qquad (5\text{-}164)$$

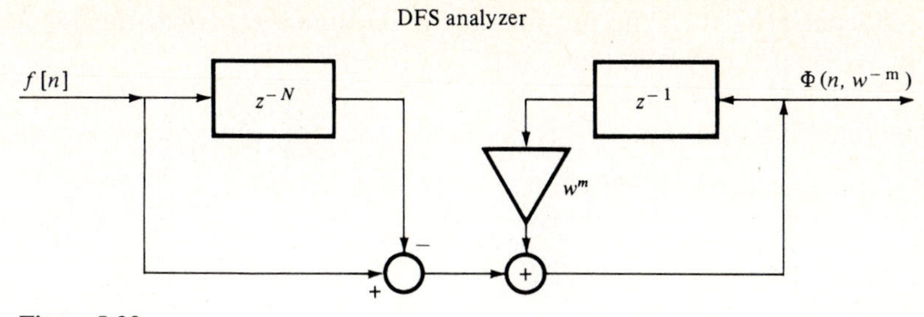

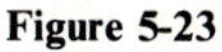
Figure 5-23

This system is shown in Fig. 5-23. It consists of one shift register with output $f[n-N]$, one delay element, and one multiplier. The shift register can be omitted if we have direct access not only to $f[n]$, but also to $f[n-N]$.

It is easy to see that $S(w^k, r) = N\delta[k-r]$. From this it follows that $S(z, r)$ is essentially equivalent to the system $K_r(z)$ defined in (5-115).

Connecting N such systems in parallel, we obtain the *real-time spectrum analyzer* shown in Fig. 5-24 (on the left of the line $F - F$). If the input to the

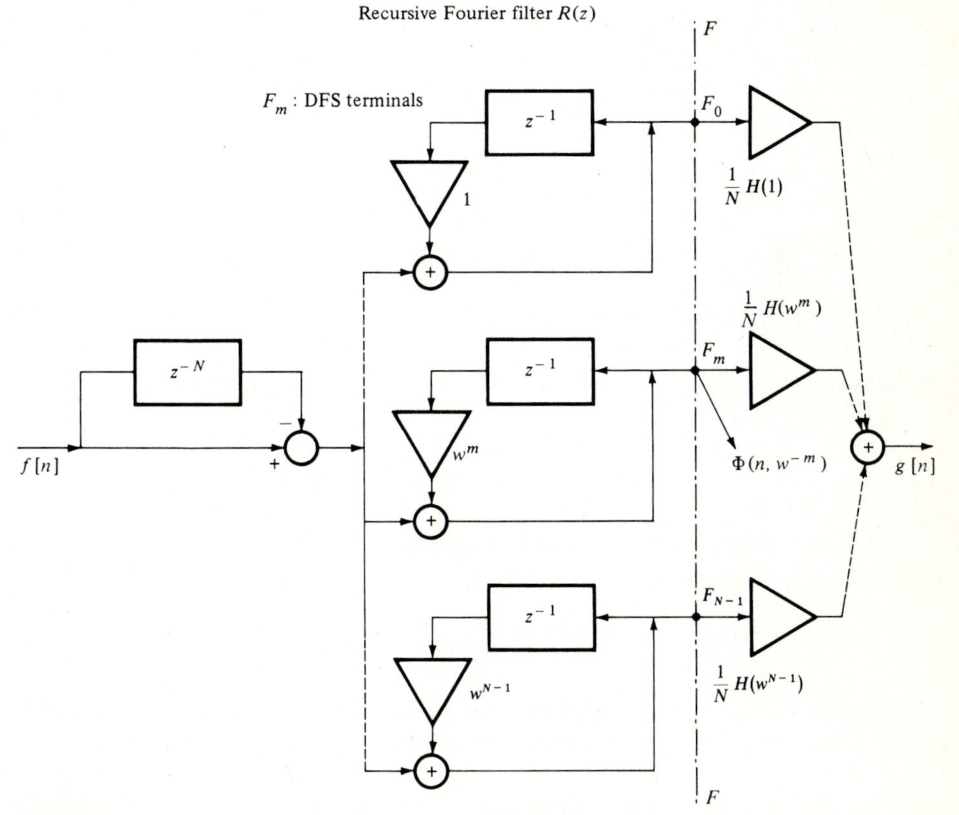

Figure 5-24

analyzer is a sequence $f[n]$, then the outputs at the terminals F_m equal the DFS coefficients $\Phi(n, w^{-m})$ of the N numbers $f[n], \ldots, f[n-N+1]$.

Convolution theorem Suppose that $f[n]$ is the input to a nonrecursive system of order N. With

$$H(z) = \sum_{n=0}^{N-1} h[n]z^{-n} \tag{5-165}$$

(the system function) and

$$g[n] = \sum_{k=0}^{N-1} f[n-k]h[k] \tag{5-166}$$

(the resulting output), we maintain that

$$g[n] = \frac{1}{N}\sum_{m=0}^{N-1} \Phi(n, w^{-m})H(w^m) \tag{5-167}$$

Proof Inserting $f[n-k]$ as given by (5-162) into (5-166) and changing the order of summation, we obtain (5-167).

The recursive Fourier filter We connect the terminals F_m of the analyzer of Fig. 5-24 with N multipliers (on the right of the F-F line) and an adder to form the system

$$R(z) = \frac{1}{N}\sum_{m=0}^{N-1} H(w^m)S(z, m) \tag{5-168}$$

Clearly, $R(z)$ is equivalent to the system $H(z)$ in Eq. (5-165) because its response to an arbitrary input $f[n]$ is the right side of (5-167). Thus we have a recursive realization of the nonrecursive filter $H(z)$.

Simulation Consider an arbitrary signal $f(t)$ with running transform $F(t, \omega)$, as in Eq. (5-150). With

$$f[n] = f(nT_1 + c) \qquad T_1 = \frac{T}{N} \qquad T = 2c$$

its samples and $\Phi(n, z)$ the running z transform of $f[n]$, it follows from (3-130) that

$$\Phi(n, w^{-m}) = \frac{N}{T} F(nT_1, m\omega_0) \qquad \omega_0 = \frac{2\pi}{T} \tag{5-169}$$

within aliasing errors. If, therefore, the system function $H_a(\omega)$ in (5-157) is negligible for $|\omega| \geq N\omega_0/2$ and the gains $H(w^m)$ of the system $R(z)$ are such that

$$H(w^m) = H_a(m\omega_0) \qquad |m| < \frac{N}{2}$$

then, [see Eqs. (5-167) and (5-156)]

$$g[n] = \frac{1}{T}\sum_{m=0}^{N-1} F(nT_1, m\omega_0)H(w^m) = g(nT_1) \tag{5-170}$$

Hence, the recursive Fourier filter $R(z)$ is a simulator of the time-limited analog system $H_a(\omega)$.

We note that the number of components of $R(z)$ need not equal N. Depending on the required resolution, we might use in (5-170) a multiple of ω_0 as the fundamental frequency. Furthermore, if $H_a(m\omega_0)$ is zero for some values of m then the corresponding DFS components $S(z, m)$ can be omitted. The recursive system $R(z)$ is, therefore, particularly simple if $H_a(\omega)$ is a narrow-band filter.

In general, $R(z)$ contains complex factors. Such factors can be avoided if the conjugate analyzers $S(z, \pm m)$ and the corresponding gains $H_a(\pm jm\omega_0)$ are combined. With

$$H_a(\pm jm\omega_0) = A_m e^{\pm j\varphi_m}$$

and $$\frac{1}{N}H_a(m\omega_0)S(z, m) + \frac{1}{N}H_a(-m\omega_0)S(z, -m) = (1 - z^{-N})D(z, m)$$

it follows readily that

$$ND(z, m) = \frac{A_m e^{j\varphi_m}}{1 - w^m z^{-1}} + \frac{A_m e^{-j\varphi_m}}{1 - w^{-m}z^{-1}}$$

$$= \frac{2A_m \cos\varphi_m - 2z^{-1}\cos(\varphi_m - 2\pi m/N)}{1 - 2z^{-1}\cos(2\pi m/N) + z^{-2}}$$

Hence, $D(z, m)$ is a real second-order system and

$$R(z) = (1 - z^{-N})\left[\frac{1}{N}H_a(0) + \sum_{m=1}^{N/2} D(z, m)\right]$$

The poles $w^{\pm m}$ of $D(z, m)$ are on the unit circle. For stability considerations, it is desirable to multiply them by a factor $r < 1$. The resulting error is small if r is close to 1.

PART TWO

SELECTED TOPICS

Part Two covers a number of rather sophisticated topics of theoretical and applied interest, selected not only for their utility but also because they contribute to the mastery of the material of Part One. It is written on a more advanced level and assumes some familiarity with the theory of analytic functions.

CHAPTER

SIX

BANDLIMITED FUNCTIONS

In this chapter, we discuss various topics related to bandlimited and time-limited functions,[1] *stressing areas of general interest.*

In Sec. 6-1, we examine their analytic and asymptotic properties, and we apply the results to Fourier series and trigonometric polynomials.

In Sec. 6-2, we present a generalized version of the sampling theorem introduced in Sec. 5-1.

In Sec. 6-3, we solve a variety of problems involving extrema of system responses and bounds of bandlimited functions.[2] *The solutions are based on two theorems proved in Appendix 6-A.*

In Sec. 6-4, we introduce the digital and analog prolate spheroidal functions and develop their main properties.

---

[1] G. C. Temes et al., The Optimization of Bandlimited Systems, *IEEE Proc.*, vol. 61, no. 2, pp. 196–234, February, 1973.

[2] A. Papoulis, Limits on Bandlimited Signals, *IEEE Proc.*, vol. 55, no. 10, pp. 1677–1686, October, 1967.

6-1 PROPERTIES OF BANDLIMITED FUNCTIONS

We shall say that a function $f(t)$ is *bandlimited* if its Fourier transform is zero outside a finite interval

$$F(\omega) = 0 \qquad \text{for } |\omega| > \sigma \tag{6-1}$$

and its energy E is finite

$$E = \int_{-\infty}^{\infty} |f(t)|^2 \, dt = \frac{1}{2\pi} \int_{-\sigma}^{\sigma} |F(\omega)|^2 \, d\omega < \infty \tag{6-2}$$

We shall say that a function $f(t)$ is *time-limited* if

$$f(t) = 0 \qquad \text{for} \qquad |t| > \tau \qquad \text{and} \qquad E < \infty \tag{6-3}$$

The abbreviations σ-BL and τ-TL, or simply BL and TL, will mean that the signals under consideration are bandlimited or time-limited, as in Fig. 6-1. Such signals are of interest for the following reasons:

1. In the applications of Fourier transforms, various functions are either BL or TL or they yield BL or TL responses owing to transducer limitations: imaging with finite size lenses, interferometers with limited maximum delay, antenna patterns, transmission through BL channels.
2. In the numerical processing of analog signals, the computations are carried out in terms of the values of $f(t)$ at a discrete set of points t_i, and the results are correct only if $f(t)$ is smooth in some sense. In mathematics, the smoothness requirement is usually expressed in terms of bounds on derivatives or on polynomial approximations. For physical signals, bandlimitedness is a more realistic assumption reflecting, if only approximately, natural properties of responses of dynamic systems.[1] Furthermore, it leads to very convenient

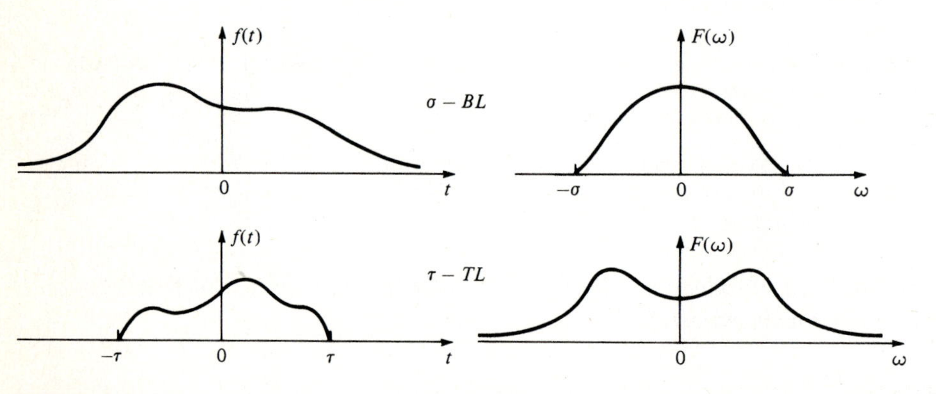

Figure 6-1

[1] D. Slepian, On Bandwidth, *IEEE Proc.*, vol. 64, no. 3, pp. 292–300, March, 1976.

relationships in digital signal processing. For example, it permits the unique determination of a signal $f(t)$ from its sample values as in the sampling expansion (5-9); it yields simple bounds of various functionals of $f(t)$ (Sec. 6-3); and it reduces integrals into sums as in the following illustrations:

If $f(t)$ is σ-BL, then [see Eq. (5-8)]

$$\int_{-\infty}^{\infty} f(t)e^{-j\omega t}\,dt = F(\omega) = Tp_\sigma(\omega) \sum_{n=-\infty}^{\infty} f(nT)e^{-jnT\omega} \tag{6-4}$$

Hence,

$$\int_{-\infty}^{\infty} f(t)\,dt = F(0) = T \sum_{n=-\infty}^{\infty} f(nT) \tag{6-5}$$

As we see from (6-4), the coefficients of the Fourier-series expansion of $F(\omega)$ in the interval $(-\sigma, \sigma)$ equal $Tf(nT)$. Applying Parseval's formula for Fourier integrals and Fourier series, we thus obtain

$$\int_{-\infty}^{\infty} |f(t)|^2\,dt = \frac{1}{2\pi}\int_{-\sigma}^{\sigma} |F(\omega)|^2\,d\omega = T \sum_{n=-\infty}^{\infty} |f(nT)|^2 \tag{6-6}$$

Suppose, finally, that the function $h(t)$ is also σ-BL. We then conclude from (5-15) that

$$\int_{-\infty}^{\infty} f(t-\tau)h(\tau)\,d\tau = T \sum_{n=-\infty}^{\infty} f(nT)h(t-nT) \tag{6-7}$$

Note: From the symmetry of Fourier transforms, it follows that all theorems about BL functions have corresponding TL versions. For example, the TL version of the theorem "A BL signal $f(t)$ is an entire function of the complex variable t" is the theorem "The transform $F(\omega)$ of a TL function $f(t)$ is an entire function of the complex variable ω." We shall discuss, therefore, only one version of each theorem.

Analyticity

Suppose that the function $f(t)$ is σ-BL. It then follows from (6-1) and the inversion formula that

$$f(t) = \frac{1}{2\pi}\int_{-\sigma}^{\sigma} F(\omega)e^{j\omega t}\,d\omega \tag{6-8}$$

Thus, $f(t)$ is expressed as an integral of a square-integrable function with finite limits. The following theorems are consequences.

Theorem 1 The transform $F(\omega)$ of a BL function is absolutely integrable:

$$\int_{-\sigma}^{\sigma} |F(\omega)|\,d\omega < \infty \tag{6-9}$$

Proof Schwarz' inequality (4-A-1) yields

$$\left| \int_{-\sigma}^{\sigma} |F(\omega)| \, d\omega \right|^2 \le 2\sigma \int_{-\sigma}^{\sigma} |F(\omega)|^2 \, d\omega = 4\pi\sigma E$$

and (6-9) results because $E < \infty$.

Theorem 2 For any t, real or complex,

$$|f(t)| \le \sqrt{\frac{\sigma E}{\pi}} \, e^{\sigma|t|} \tag{6-10}$$

Proof If $|\omega| \le \sigma$, then $|e^{-j\omega t}| \le e^{\sigma|t|}$. Hence,

$$|f(t)| \le \frac{1}{2\pi} \int_{-\sigma}^{\sigma} |F(\omega)e^{-j\omega t}| \, d\omega \le \frac{e^{\sigma|t|}}{2\pi} \int_{-\sigma}^{\sigma} |F(\omega)| \, d\omega \le \sqrt{\frac{\sigma E}{\pi}} \, e^{\sigma|t|}$$

Theorem 3 A BL function $f(t)$ is analytic in the entire t plane, and its derivative is obtained by differentiating (6-8) under the integral sign:

$$f'(t) = \frac{1}{2\pi} \int_{-\sigma}^{\sigma} j\omega F(\omega) e^{j\omega t} \, d\omega \tag{6-11}$$

Proof It suffices to show that the difference

$$D = \frac{f(t+\epsilon) - f(t)}{\epsilon} - \frac{1}{2\pi} \int_{-\sigma}^{\sigma} j\omega F(\omega) e^{j\omega t} \, d\omega = \frac{1}{2\pi} \int_{-\sigma}^{\sigma} \left(\frac{e^{j\omega\epsilon} - 1}{\epsilon} - j\omega \right) F(\omega) e^{j\omega t} \, d\omega$$

tends to zero as $\epsilon \to 0$. For any z, real or complex,

$$\left| \frac{e^z - 1 - z}{z^2} \right| = \left| \frac{1}{2!} + \frac{z}{3!} + \cdots \right| < 1 + |z| + \frac{|z|^2}{2!} + \cdots = e^{|z|}$$

Hence,
$$\left| \frac{e^{j\omega\epsilon} - 1 - j\omega\epsilon}{\epsilon} \right| \le \omega^2 |\epsilon| e^{|\omega\epsilon|}$$

Furthermore, if $|\omega| < \sigma$, then $|e^{j\omega t}| \le e^{\sigma|t|}$ and $e^{|\omega\epsilon|} \le e^{\sigma|\epsilon|}$. Therefore,

$$|D| \le \sigma^2 |\epsilon| \sqrt{\frac{\sigma E}{\pi}} \, e^{\sigma(|t| + |\epsilon|)} \underset{\epsilon \to 0}{\longrightarrow} 0$$

Functions of exponential type A function $f(t)$ is of *exponential type* if there exist two constants A and σ such that

$$|f(t)| < Ae^{\sigma|t|} \tag{6-12}$$

for every t. From the last two theorems, it follows that a σ-BL function is analytic in the entire t plane (entire function) and of exponential type.

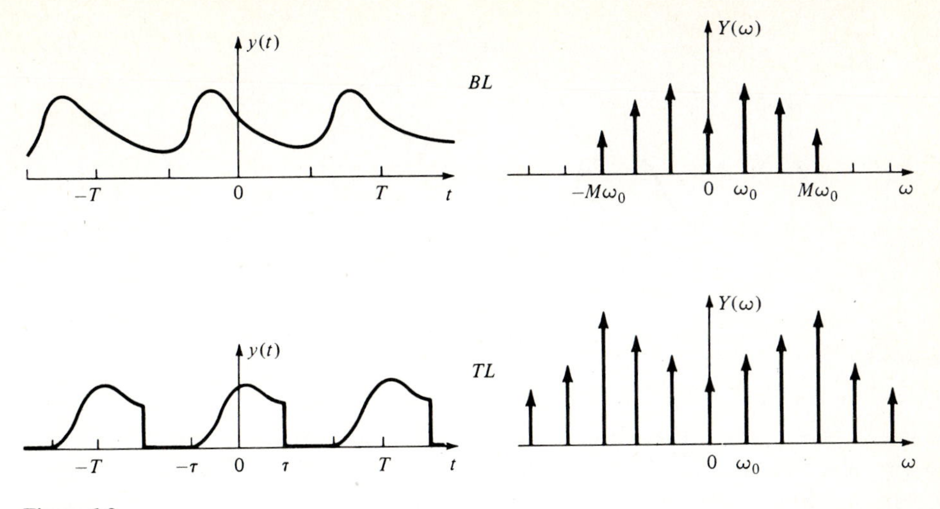

Figure 6-2

The following important theorem is the converse of Theorems 2 and 3. It establishes the equivalence between σ-BL functions and entire functions of exponential type.

Theorem 4 (Paley-Wiener) If the function $f(t)$ has finite energy, is analytic in the entire t plane, and is of exponential type as in Eq. (6-12), then it is σ-BL; that is, its Fourier transform $F(\omega)$ exists and $F(\omega) = 0$ for $|\omega| > \sigma$. The proof is given in Appendix 7-A.

Periodic functions We shall say that a function $y(t)$ is *periodic BL* if it is a trigonometric polynomial (Fig. 6-2*a*)

$$y(t) = \sum_{n=-M}^{M} a_n e^{jn\omega_0 t} \tag{6-13}$$

and its energy E in the interval $(0, T)$ is finite.

We shall say that a periodic function $y(t)$ is *periodic TL* if $E < \infty$ and (Fig. 6-2*b*)

$$y(t) = 0 \qquad \text{for } \tau < |t| < \frac{T}{2} \tag{6-14}$$

Theorem 5 A periodic TL function cannot be a trigonometric polynomial.

Proof Suppose that $y(t)$ is a trigonometric polynomial as in (6-13). We wish to show that it cannot be identically zero in any interval. For this purpose, we form the function

$$G(z) = \sum_{n=-M}^{M} a_n z^n \tag{6-15}$$

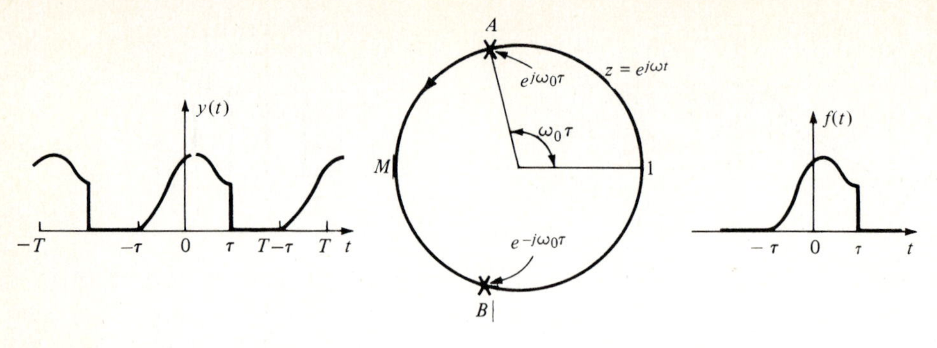

Figure 6-3

Clearly, $y(t) = G(e^{j\omega_o t})$; therefore, if $y(t_i) = 0$, then $G(z_i) = 0$ for $z_i = e^{j\omega_o t_i}$. From the above it follows that, if $y(t) = 0$ in the interval $(\tau, T - \tau)$, then $G(z)$ must be zero for every z on the arc AMB of the unit circle (Fig. 6-3); that is, $G(z)$ must have infinitely many roots. But this is impossible because $z^M G(z)$ is a polynomial. Hence, $y(t)$ cannot be TL.

Theorem 6 A function $f(t)$ cannot be bandlimited and time-limited.

Proof 1 As we have shown in Theorem 3, if $f(t)$ is BL, then it is analytic in the entire t plane (entire function). It is known that an entire function vanishing in an interval must be identically zero; hence, $f(t) \equiv 0$.

Proof 2 The preceding proof is based on function-theoretical considerations. The following proof is a consequence of Theorem 5; it uses only properties of polynomials:

Suppose that the function $f(t)$ is τ-TL as in Eq. (6-3). We select a constant T larger than 2τ and form the function

$$y(t) = \sum_{n=-\infty}^{\infty} f(t + nT)$$

Clearly, $y(t)$ is a periodic TL function (Fig. 6-3), and its Fourier-series coefficients a_n equal $F(n\omega_o)/T$ [see (3-69)]. If, therefore, $F(\omega) = 0$ for $|\omega| > \sigma$, then $a_n = 0$ for $|n| > \sigma/\omega_o$. Hence, $y(t)$ is a trigonometric polynomial. This is, however, impossible (Theorem 5).

Boundary Values and Asymptotic Properties

We have shown in Appendix 3-B that the behavior of $F(\omega)$ for large ω depends on the continuity properties of $f(t)$ and its derivatives:

If $f(t)$ and all its derivatives of order up to n exist and are bounded, then

$$F(\omega) = O\left(\frac{1}{\omega^{n+1}}\right) \qquad \omega \to \infty \tag{6-16}$$

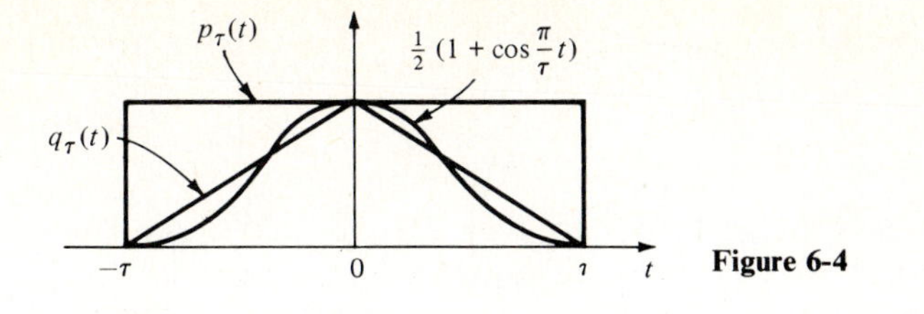

Figure 6-4

We shall show that, for TL functions, the asymptotic properties of $F(\omega)$ depend on the behavior of $f(t)$ near the endpoints $t = \pm\tau$ of the interval $(-\tau, \tau)$. At these points, the derivatives are defined as interior limits:

$$f^{(k)}(-\tau) = \lim f^{(k)}(-\tau + \epsilon) \qquad f^{(k)}(\tau) = \lim f^{(k)}(\tau - \epsilon) \qquad \epsilon \to 0 \qquad \epsilon > 0$$

Since $f(t) = 0$ for $|t| > \tau$, if $f^{(k)}(\pm\tau) \neq 0$, then $f^{(k+1)}(t)$ contains impulses at $t = \pm\tau$. Hence, for a rapid asymptotic attenuation of $F(\omega)$, $f(t)$ must have a high order of tangency with the t axis. This is illustrated with the three examples of Fig. 6-4:

$$p_\tau(t) \leftrightarrow \frac{2\sin\tau\omega}{\omega} \qquad q_\tau(t) \leftrightarrow \frac{4\sin^2(\tau\omega/2)}{\tau\omega^2} \qquad \frac{1}{2}\left(1 + \cos\frac{\pi}{\tau}t\right)p_\tau(t) \leftrightarrow \frac{\pi^2\sin\tau\omega}{\omega(\pi^2 - \tau^2\omega^2)}$$

The pulse $p_\tau(t)$ is discontinuous at $t = \pm\tau$; hence, its transform $F_1(\omega)$ goes to zero only as $1/\omega$ for $\omega \to \infty$. The triangle $q_\tau(t)$ is continuous, but its derivative is discontinuous at $t = \pm\tau$; hence, $F_2(\omega) = O(1/\omega^2)$. Finally, the raised cosine and its derivative are continuous, but the second derivative is discontinuous at $t = \pm\tau$; hence, $F_3(\omega) = O(1/\omega^3)$.

Theorem 7 If a function $f(t)$ is TL and has bounded derivatives of order up to n for every $|t| < \tau$, then its Fourier transform

$$F(\omega) = \int_{-\tau}^{\tau} f(t)e^{-j\omega t}\,dt \tag{6-17}$$

has the following asymptotic expansion for $\omega \to \infty$:

$$F(\omega) = \frac{1}{j\omega}[f(-\tau)e^{j\tau\omega} - f(\tau)e^{-j\tau\omega}] + \cdots$$
$$+ \frac{1}{(j\omega)^n}[f^{(n-1)}(-\tau)e^{j\tau\omega} - f^{(n-1)}(\tau)e^{-j\tau\omega}] + O\left(\frac{1}{\omega^{n+1}}\right) \tag{6-18}$$

Proof The proof is based on the Riemann lemma, Eq. (3-B-4): If $z(t)$ is a function of bounded variation, then

$$Z(\omega) = \int_{-\tau}^{\tau} z(t)e^{-j\omega t}\,dt = O\left(\frac{1}{\omega}\right) \qquad \omega \to \infty \tag{6-19}$$

Integrating (6-17) by parts, we obtain

$$F(\omega) = \int_{-\tau}^{\tau} f(t) \frac{de^{-j\omega t}}{-j\omega} = \frac{f(t)e^{-j\omega t}}{-j\omega}\Bigg|_{-\tau}^{\tau} + \frac{1}{j\omega}\int_{-\tau}^{\tau} f'(t)e^{-j\omega t}\,dt \tag{6-20}$$

Since $f'(t)$ is bounded, the last integral goes to zero as $1/\omega$, as we see from (6-19) with $z(t) = f'(t)$. The last term in (6-20) is, therefore, of the order of $1/\omega^2$, and (6-18) follows for $n = 1$. Integrating the last integral in (6-20) by parts, we establish (6-18) for $n = 2$. Repeated integration by parts yields (6-18) for any n.

Theorem 8 If the TL function $f(t)$ has bounded derivatives of order up to n for every $|t| < \tau$, and at the endpoints $t = \pm\tau$,

$$f^{(k)}(-\tau) = 0 \qquad f^{(k)}(\tau) = 0 \qquad k = 0, \ldots, n-1 \tag{6-21}$$

Then

$$F(\omega) = O\left(\frac{1}{\omega^{n+1}}\right) \qquad \omega \to \infty \tag{6-22}$$

Proof Inserting (6-21) into (6-18), we obtain (6-22).

Theorem 9 If the TL function $f(t)$ has bounded derivatives of order up to n for every $|t| < \tau$, and

$$f^{(k)}(-\tau) = f^{(k)}(\tau) \qquad k = 0, \ldots, n-1 \tag{6-23}$$

then the sample values $F(m\pi/\tau)$ of $F(\omega)$ go to zero as $1/m^{n+1}$ as $m \to \infty$:

$$F\left(\frac{m\pi}{\tau}\right) = O\left(\frac{1}{m^{n+1}}\right) \qquad m \to \infty \tag{6-24}$$

Proof Clearly, if $\omega = m\pi/\tau$, then

$$e^{j\tau\omega} = e^{jm\pi} = (-1)^m \qquad e^{-j\tau\omega} = e^{-jm\pi} = (-1)^m$$

and (6-24) follows from (6-18).

Corollary If the function $f(t)$ of Theorem 9 is expanded into a Fourier series in the interval $(-\tau, \tau)$:

$$f(t) = \sum_{m=-\infty}^{\infty} b_m e^{jm\omega_0 t} \qquad |t| < \tau = \frac{\pi}{\omega_0} \tag{6-25}$$

then

$$b_m = O\left(\frac{1}{m^{n+1}}\right) \qquad m \to \infty \tag{6-26}$$

Proof The coefficients b_m of Eq. (6-25) are given by

$$b_m = \frac{1}{2\tau}\int_{-\tau}^{\tau} f(t)e^{-jm\omega_0 t}\,dt = \frac{1}{2\tau}F(m\omega_0) = \frac{1}{2\tau}F\left(\frac{m\pi}{\tau}\right)$$

[see (6-17) and (3-69)]. Hence, (6-26) follows from (6-24).

Fourier series We shall, finally, relate the asymptotic properties of the Fourier-series coefficients a_m of a periodic function

$$y(t) = \sum_{m=-\infty}^{\infty} a_m e^{jm\omega_0 t} \qquad \omega_0 = \frac{2\pi}{T} \tag{6-27}$$

to the continuity properties of $y(t)$ and its derivatives.

Theorem 10 If $y(t)$ has bounded derivatives of order up to n for every t, then

$$a_m = O\left(\frac{1}{m^{n+1}}\right) \qquad m \to \infty \tag{6-28}$$

Proof From the assumption of the theorem, it follows that the function $y(t)$ and its derivatives of order up to $n-1$ are continuous for every t. Since they are also periodic, they must satisfy the symmetry condition

$$y^{(k)}\left(\frac{T}{2}\right) = y^{(k)}\left(-\frac{T}{2}\right) \qquad k = 0, \ldots, n-1 \tag{6-29}$$

We form the τ-TL function

$$f(t) = y(t)p_\tau(t) \qquad \text{where } \tau = \frac{T}{2}$$

Clearly, $f(t) = y(t)$ for $|t| \le T/2$; hence, $f(t)$ satisfies Eq. (6-23) for $\tau = T/2$, and the coefficients of its Fourier-series expansion in the interval $(-T/2, T/2)$ equal a_m. Therefore, (6-28) follows from (6-26).

6-2 GENERALIZED SAMPLING

In Sec. 5-1 we showed that a σ-BL function $f(t)$ can be expressed in terms of its sample values $f(nT)$, where $1/T = \sigma/\pi$ is the Nyquist rate. This fundamental result can also be obtained by expanding the exponential $e^{j\omega t}$ into a Fourier series in the variable ω in the interval $(-\sigma, \sigma)$. The coefficients of the resulting expansion are functions of t,

$$e^{j\omega t} = \sum_{n=-\infty}^{\infty} a_n(t)e^{jnT\omega} \qquad |\omega| < \sigma \tag{6-30}$$

and are given by

$$a_n(t) = \frac{1}{2\sigma}\int_{-\sigma}^{\sigma} e^{j\omega t}e^{-jnT\omega}\,d\omega = \frac{\sin\sigma(t-nT)}{\sigma(t-nT)} \tag{6-31}$$

Inserting Eq. (6-30) into the Fourier inversion formula, we obtain

$$f(t) = \frac{1}{2\pi} \int_{-\sigma}^{\sigma} F(\omega) e^{j\omega t} \, d\omega = \frac{1}{2\pi} \int_{-\sigma}^{\sigma} F(\omega) \sum_{n=-\infty}^{\infty} a_n(t) e^{jnT\omega} \, d\omega$$

Hence,

$$f(t) = \sum_{n=-\infty}^{\infty} f(nT) \frac{\sin \sigma(t - nT)}{\sigma(t - nT)} \tag{6-32}$$

In this section, we generalize this important result. As preparation, we express the signal $f(t)$ in terms of the sample values $g(nT)$ of the output

$$g(t) = \frac{1}{2\pi} \int_{-\sigma}^{\sigma} F(\omega) H(\omega) e^{j\omega t} \, d\omega \tag{6-33}$$

of a system $H(\omega)$ driven by $f(t)$.

Theorem 1 We maintain that

$$f(t) = \sum_{n=-\infty}^{\infty} g(nT) y(t - nT) \tag{6-34}$$

where

$$y(t) = \frac{1}{2\sigma} \int_{-\sigma}^{\sigma} \frac{e^{j\omega t}}{H(\omega)} \, d\omega \tag{6-35}$$

Proof We expand the function $e^{j\omega t}/H(\omega)$ into a Fourier series in the interval $(-\sigma, \sigma)$:

$$\frac{e^{j\omega t}}{H(\omega)} = \sum_{n=-\infty}^{\infty} b_n(t) e^{jnT\omega} \qquad |\omega| < \sigma \tag{6-36}$$

The coefficients $b_n(t)$ are given by

$$b_n(t) = \frac{1}{2\sigma} \int_{-\sigma}^{\sigma} \frac{e^{j\omega t}}{H(\omega)} e^{-jnT\omega} \, d\omega = y(t - nT)$$

Hence,

$$e^{j\omega t} = H(\omega) \sum_{n=-\infty}^{\infty} y(t - nT) e^{jnT\omega}$$

Inserting into the Fourier inversion formula, we obtain

$$f(t) = \frac{1}{2\pi} \int_{-\sigma}^{\sigma} F(\omega) e^{j\omega t} \, d\omega = \frac{1}{2\pi} \int_{-\sigma}^{\sigma} F(\omega) H(\omega) \sum_{n=-\infty}^{\infty} y(t - nT) e^{jnT\omega} \, d\omega$$

and Eq. (6-34) results [see (6-33)].

Example 6-1 We wish to express the function $f(t)$ in terms of the samples $g(nT)$ of the integral

$$g(t) = \alpha \int_{0}^{\infty} f(t - \tau) e^{-\alpha\tau} \, d\tau$$

This case arises in the reconstruction of $f(t)$ from its samples when the proportional system for measuring $f(nT)$ is not ideal. Clearly, $g(t)$ is the output of the system

$$H(\omega) = \frac{\alpha}{\alpha + j\omega}$$

with input $f(t)$. Hence, $f(t)$ is given by (6-34), where

$$y(t) = \frac{1}{2\sigma}\int_{-\sigma}^{\sigma}\left(1 + \frac{j\omega}{\alpha}\right)e^{j\omega t}\,d\omega = \frac{\sigma t\cos\sigma t + (\alpha t - 1)\sin\sigma t}{\sigma\alpha t^2}$$

Generalized Sampling Expansion[1]

Suppose now that $f(t)$ is a common input to m systems $H_1(\omega), \ldots, H_m(\omega)$ as in Fig. 6-5. We shall show that it can be expressed in terms of the samples $g_k(nT)$ of the resulting outputs

$$g_k(t) = \frac{1}{2\pi}\int_{-\sigma}^{\sigma} F(\omega)H_k(\omega)e^{j\omega t}\,d\omega \tag{6-37}$$

sampled at the rate

$$\frac{1}{T} = \frac{\sigma}{m\pi} \tag{6-38}$$

For this purpose, we introduce the constant

$$c = \frac{2\sigma}{m} = \frac{2\pi}{T} \tag{6-39}$$

and the m functions $Y_1(\omega, t), \ldots, Y_m(\omega, t)$ determined by solving the system

$$H_1(\omega)Y_1(\omega, t) + \cdots + H_m(\omega)Y_m(\omega, t) = 1$$

$$H_1(\omega + c)Y_1(\omega, t) + \cdots + H_m(\omega + c)Y_m(\omega, t) = e^{jct}$$

$$\cdots\cdots\cdots\cdots\cdots\cdots\cdots\cdots\cdots\cdots\cdots\cdots$$

$$H_1[\omega + (m-1)c]Y_1(\omega, t) + \cdots + H_m[\omega + (m-1)c]Y_m(\omega, t) = e^{j(m-1)ct} \tag{6-40}$$

where t is arbitrary, and ω is in the interval $(-\sigma, -\sigma + c)$.

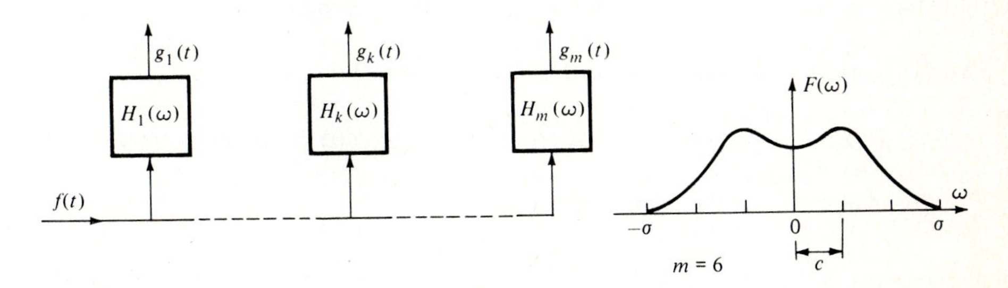

Figure 6-5

[1] A. Papoulis, Generalized Sampling Expansion, *IEEE Trans. on Circuits and Systems*, November, 1977.

Theorem 2 We maintain that

$$f(t) = \sum_{n=-\infty}^{\infty} [g_1(nT)y_1(t - nT) + \cdots + g_m(nT)y_m(t - nT)] \tag{6-41}$$

where
$$y_k(t) = \frac{1}{c}\int_{-\sigma}^{-\sigma+c} Y_k(\omega, t)e^{j\omega t}\, d\omega \qquad k = 1, \ldots, m \tag{6-42}$$

This is the generalized sampling expansion, to be proved presently. We illustrate, first, with two examples.

Example 6-2 The preceding theorem permits us to express a function $f(t)$ in terms of its samples and samples of its $m - 1$ derivatives taken at $1/m$ times the Nyquist rate. For this application, $H_k(\omega) = (j\omega)^{k-1}$ and $g_k(t) = f^{(k-1)}(t)$, where $k = 1, \ldots, m$.

We shall carry out the computations for $m = 2$. We now have

$$T = \frac{2\pi}{\sigma} \qquad c = \sigma \qquad g_1(t) = f(t) \qquad g_2(t) = f'(t) \qquad H_1(\omega) = 1 \qquad H_2(\omega) = j\omega$$

The system (6-40) takes the form

$$Y_1(\omega, t) + j\omega Y_2(\omega, t) = 1$$
$$Y_1(\omega, t) + j(\omega + \sigma)Y_2(\omega, t) = e^{j\sigma t}$$

Hence,
$$Y_1(\omega, t) = 1 - \frac{\omega}{\sigma}(e^{j\sigma t} - 1) \qquad Y_2(\omega, t) = \frac{1}{j\sigma}(e^{j\sigma t} - 1)$$

Inserting into Eq. (6-42), we obtain

$$y_1(t) = \frac{4\sin^2(\sigma t/2)}{\sigma^2 t^2} \qquad y_2(t) = \frac{4\sin^2(\sigma t/2)}{\sigma^2 t}$$

and (6-41) yields (elaborate)

$$f(t) = 4\sin^2\frac{\sigma t}{2}\sum_{n=-\infty}^{\infty}\left[\frac{f(nT)}{(\sigma t - 2n\pi)^2} + \frac{f'(nT)}{\sigma(\sigma t - 2n\pi)}\right] \tag{6-43}$$

Example 6-3 (Bunched samples) Suppose that $f(t)$ is sampled at $1/m$ times the Nyquist rate, but in each sampling interval not one but m samples are used (Fig. 6-6). We maintain that $f(t)$ is uniquely determined in terms of the resulting samples

$$f(nT + \alpha_k) \qquad |\alpha_k| < \frac{T}{2} \qquad k = 1, \ldots, m \tag{6-44}$$

Indeed, if $H_k(\omega) = e^{j\alpha_k\omega}$, then $g_k(t) = f(t + \alpha_k)$. Hence, $f(t)$ is given by (6-41), where $g_k(nT) = f(nT + \alpha_k)$.

We shall carry out the computations for

$$m = 2 \qquad \alpha_1 = -\alpha_2 = \alpha$$

Figure 6-6

In this case, the system (6-40) yields

$$e^{j\alpha\omega}Y_1(\omega, t) + e^{-j\alpha\omega}Y_2(\omega, t) = 1$$

$$e^{j\alpha(\omega+\sigma)}Y_1(\omega, t) + e^{-j\alpha(\omega+\sigma)}Y_2(\omega, t) = e^{j\sigma t}$$

We solve the above system for $Y_1(\omega, t)$ and $Y_2(\omega, t)$ and determine $y_1(t)$ and $y_2(t)$ from (6-42). Inserting the results into (6-41), we obtain (elaborate)

$$f(t) = \frac{\cos \sigma\alpha - \cos \sigma t}{\sigma \sin \sigma\alpha} \sum_{n=-\infty}^{\infty} \left[\frac{f(nT+\alpha)}{t - nT - \alpha} - \frac{f(nT-\alpha)}{t - nT + \alpha}\right] \tag{6-45}$$

Lemma 1 The functions $Y_k(\omega, t)$ are periodic in t with period $T = 2\pi/c$:

$$Y_k(\omega, t+T) = Y_k(\omega, t), \qquad k = 1, \ldots, m \tag{6-46}$$

Proof The coefficients $H_k(\omega + rc)$ of the system (6-40) are independent of t, and the right side consists of periodic functions of t because

$$e^{jrc(t+T)} = e^{jrct}$$

Hence, its solutions $Y_k(\omega, t)$ are periodic.

Lemma 2 For every ω in the interval $(-\sigma, -\sigma + c)$ in which $Y_k(\omega, t)$ is defined and for every t,

$$Y_k(\omega, t)e^{j\omega t} = \sum_{n=-\infty}^{\infty} y_k(t - nT)e^{jnT\omega} \tag{6-47}$$

Proof From Eqs. (6-42) and (6-46) it follows that

$$y_k(t - nT) = \frac{1}{c}\int_{-\sigma}^{-\sigma+c} Y_k(\omega, t - nT)e^{j\omega(t-nT)}\, d\omega$$

$$= \frac{1}{c}\int_{-\sigma}^{-\sigma+c} Y_k(\omega, t)e^{j\omega t}e^{-jnT\omega}\, d\omega$$

The last integral shows that $y_k(t - nT)$ is the nth coefficient of the Fourier-series expansion of the function $Y_k(\omega, t)e^{j\omega t}$ in the interval $(-\sigma, -\sigma + c)$, and (6-47) results.

Lemma 3 For every ω in the interval $(-\sigma, \sigma)$,

$$e^{j\omega t} = H_1(\omega) \sum_{n=-\infty}^{\infty} y_1(t - nT)e^{jnT\omega} + \cdots + H_m(\omega) \sum_{n=-\infty}^{\infty} y_m(t - nT)e^{jnT\omega} \tag{6-48}$$

Proof Multiplying the first equation of (6-40) by $e^{j\omega t}$ and using (6-47), we conclude that (6-48) holds for every ω in the interval $(-\sigma, -\sigma + c)$. Multiplying the second equation by $e^{j\omega t}$ and using (6-47) and the identity $e^{jnT(\omega+c)} = e^{jnT\omega}$, we obtain

$$e^{j(\omega+c)t} = H_1(\omega+c) \sum_{n=-\infty}^{\infty} y_1(t-nT)e^{jnT(\omega+c)}$$

$$+ \cdots + H_m(\omega+c) \sum_{n=-\infty}^{\infty} y_m(t-nT)e^{jnT(\omega+c)}$$

for every ω in the interval $(-\sigma, \ -\sigma+c)$. But as ω varies in the interval $(-\sigma, -\sigma+c)$, $\omega+c$ varies in the interval $(-\sigma+c, \ -\sigma+2c)$. Hence, Eq. (6-48) holds in that interval. Using the rth equation in (6-40), we similarly show that (6-48) holds for every ω in the interval $[-\sigma+(r-1)c, \ -\sigma+rc]$.

Proof of Theorem 2 Replacing the function $e^{j\omega t}$ with the right side of (6-48) in the inversion formula

$$f(t) = \frac{1}{2\pi}\int_{-\sigma}^{\sigma} F(\omega)e^{j\omega t}\,d\omega$$

and using (6-37), we obtain (6-41).

Note: The sampling expansion (6-41) holds only if the determinant of the system (6-40) is different from zero for every ω in the interval $(-\sigma, \ -\sigma+c)$. This is the case for the functions $(j\omega)^{k-1}$ of Example 6-2 (Vandermonde determinant) and for the functions $e^{j\alpha_k\omega}$ of Example 6-3 because $|\alpha_k| < T/2$.

6-3 BOUNDS AND EXTREME VALUES OF BANDLIMITED FUNCTIONS

If a function $f(t)$ is BL and its energy E is specified, then it cannot take arbitrarily large values. In this section, we determine the maximum of $f(t)$ and of various functionals of $f(t)$. The results are based on the following theorem:

If $g(t)$ is the response of a linear system to a σ-BL input $f(t)$ of energy E:

$$g(t) = \frac{1}{2\pi}\int_{-\sigma}^{\sigma} F(\omega)H(\omega)e^{j\omega t}\,d\omega \qquad E = \frac{1}{2\pi}\int_{-\sigma}^{\sigma} |F(\omega)|^2\,d\omega \tag{6-49}$$

then, for any t,

$$|g(t)| \le \left[\frac{E}{2\pi}\int_{-\sigma}^{\sigma} |H(\omega)|^2\,d\omega\right]^{1/2} \tag{6-50}$$

Equality holds for $t = t_0$ only if

$$F(\omega) = kH^*(\omega)e^{-j\omega t_0} \qquad \text{for } |\omega| < \sigma \tag{6-51}$$

where k is a constant whose value is determined from the energy requirement.

Proof From Schwarz' inequality (4-A-1), it follows that, for any t_0,

$$\left|\int_{-\sigma}^{\sigma} F(\omega)H(\omega)e^{j\omega t_0}\,d\omega\right|^2 \le \int_{-\sigma}^{\sigma} |F(\omega)|^2\,d\omega \int_{-\sigma}^{\sigma} |H(\omega)|^2\,d\omega \qquad (6\text{-}52)$$

and Eq. (6-50) results readily. Furthermore [see (4-A-2)], (6-52) is an equality only if $F(\omega)$ satisfies (6-51).

Special cases

1. If $H(\omega) = 1$, then $g(t) = f(t)$ and (6-50) yields

$$|f(t)| \le \sqrt{\frac{E\sigma}{\pi}} \qquad (6\text{-}53)$$

Equality is reached at $t = t_0$ only if

$$f(t) = \sqrt{\frac{E\pi}{\sigma}}\frac{\sin\sigma(t-t_0)}{\pi(t-t_0)} \qquad (6\text{-}54)$$

as can easily be seen Eq. (6-51) (elaborate).

2. If $H(\omega) = j\omega$, then $g(t) = f'(t)$,

$$\int_{-\sigma}^{\sigma} |H(\omega)|^2\,d\omega = \int_{-\sigma}^{\sigma} |\omega|^2\,d\omega = \frac{2\sigma^3}{3}$$

and (6-50) yields

$$|f'(t)| \le \sigma\sqrt{\frac{E\sigma}{3\pi}} \qquad (6\text{-}55)$$

Note: From (6-55) we conclude that

$$|f(t_1) - f(t_2)| \le \sigma\sqrt{\frac{E\sigma}{3\pi}}|t_1 - t_2|$$

because
$$f(t_1) - f(t_2) = \int_{t_2}^{t_1} f'(t)\,dt$$

Trigonometric polynomials The preceding results, properly modified, hold also for trigonometric polynomials and discrete systems. We shall discuss the parallel of (6-53):

If $y(t)$ is a trigonometric polynomial with $N = 2M + 1$ terms, as in (6-A-5), and its energy equals E, then

$$|y(t)| \le \sqrt{NET} \qquad (6\text{-}56)$$

This follows from the Cauchy inequality (4-A-4):

$$|y(t)|^2 = \left|\sum_{n=-M}^{M} y_n e^{jn\omega_0 t}\right|^2 \le \sum_{n=-M}^{M} |y_n|^2 \sum_{n=-M}^{M} |e^{jn\omega_0 t}|^2 = TE(2M+1)$$

Extreme Values

In many applications, it is desirable to find BL functions or trigonometric polynomials such that their weighted integrals have extreme values. We give below several illustrations. The derivations are based on the two theorems of Appendix 7-A.

Maximum energy concentration An important problem in signal analysis is the determination of a σ-BL function $y(t)$ such that the ratio

$$\alpha_y = \frac{1}{E_y} \int_{-\tau}^{\tau} |y(t)|^2 \, dt \tag{6-57}$$

of its energy in a specified interval $(-\tau, \tau)$ to its total energy E_y is maximum.

The above ratio is a special case of Eq. (6-A-14) if

$$v(t) = p_\tau(t) = \begin{cases} 1 & |t| < \tau \\ 0 & |t| > \tau \end{cases} \tag{6-58}$$

Inserting into (6-A-15), we conclude that α_y is maximum if $y(t)$ equals the eigenfunction $\bar{y}(t)$ of the integral equation

$$\int_{-\tau}^{\tau} y(x) \frac{\sin \sigma(t-x)}{\pi(t-x)} \, dx = \lambda y(t) \tag{6-59}$$

corresponding to the maximum eigenvalue $\lambda_{\max}$.

The solutions of (6-59) are known as *prolate spheroidal wave functions*. In the next section, we discuss their properties in some detail. We note here only that the eigenvalues of (6-59) are real and tend to zero:

$$1 > \lambda_{\max} = \lambda_0 > \lambda_1 > \cdots > \lambda_n \to 0 \qquad \text{as } n \to \infty$$

Hence, α_y has a maximum $\bar{\alpha} = \lambda_0$, but no minimum. The energy of a BL function in any finite interval can be arbitrarily small.

Minimum moment of inertia[1] We wish to find a σ-BL function $y(t)$ such that the normalized moment of inertia

$$\alpha_y = \frac{1}{E_y} \int_{-\infty}^{\infty} t^2 |y(t)|^2 \, dt \tag{6-60}$$

of $|y(t)|^2$ is minimum. As we show in Sec. 7-3, this problem is of interest in the theory of windows.

Again, Eq. (6-60) is a special case of (6-A-15) if $v(t) = t^2$. Therefore, the

[1] A. Papoulis, Apodization for Optimum Imaging of Smooth Objects, *J. Opt. Soc. Am.*, vol. 62, no. 12, pp. 1423–1429, December, 1972.

unknown signal is the eigenfunction $y(t)$ of the integral equation [see (6-A-14)]

$$\int_{-\infty}^{\infty} x^2 y(x) \frac{\sin \sigma(t-x)}{\pi(t-x)} dx = \lambda y(t) \tag{6-61}$$

corresponding to the minimum eigenvalue $\lambda_{\min}$.

To solve (6-61), we take transforms of both sides and use the pairs

$$t^2 y(t) \leftrightarrow -Y''(\omega) \qquad \frac{\sin \sigma t}{\pi t} \leftrightarrow p_\sigma(\omega)$$

This leads to the differential equation

$$Y''(\omega) + \lambda Y(\omega) = 0 \tag{6-62}$$

with the requirement that $Y(\omega) = 0$ for $|\omega| > \sigma$. This requirement yields the boundary conditions

$$Y(\sigma) = 0 \qquad Y(-\sigma) = 0 \tag{6-63}$$

because, if $Y(\omega)$ is discontinuous, then $\alpha_y = \infty$ (elaborate).

The solutions of (6-62) are of the form $A \cos \sqrt{\lambda}\,\omega + B \sin \sqrt{\lambda}\,\omega$, and they satisfy (6-63) only if $B = 0$ and $\lambda = n^2\pi^2/4\sigma^2$. Thus, the minimum $\underline{\alpha}$ of α_y is given by

$$\underline{\alpha} = \lambda_{\min} = \frac{\pi^2}{4\sigma^2} \tag{6-64}$$

and the corresponding $\underline{y}(t)$ is the inverse of the truncated cosine

$$Y(\omega) = \begin{cases} k \cos \dfrac{\pi\omega}{2\sigma} & |\omega| \le \sigma \\ 0 & |\omega| > \sigma \end{cases} \tag{6-65}$$

Thus [see (3-23)],

$$\underline{y}(t) = \frac{k}{2\sigma} \frac{\cos \sigma t}{\pi^2/4\sigma^2 - t^2} \tag{6-66}$$

as in Fig. 6-7.

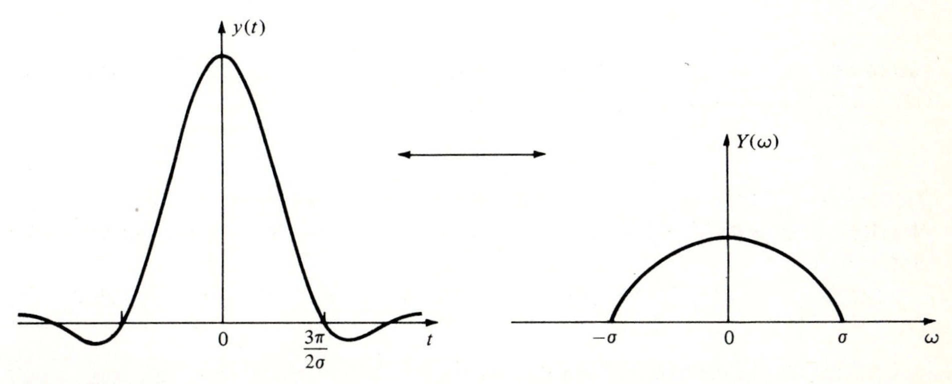

Figure 6-7

Positive signals[1] A related problem is the determination of a nonnegative function $w(t)$ whose transform $W(\omega)$ is zero for $|\omega| > 2\sigma$:

$$w(t) \geq 0 \qquad W(\omega) = 0 \qquad \text{for } |\omega| > 2\sigma$$

such that the normalized moment of inertia

$$\gamma_w = \frac{1}{W(0)} \int_{-\infty}^{\infty} t^2 w(t)\, dt \tag{6-67}$$

is minimum.

As we show in Appendix 6-A, the optimum $\underline{w}(t)$ is given by

$$\underline{w}(t) = |\underline{y}(t)|^2 = \frac{\cos^2 \sigma t}{4\sigma^2(\pi^2/4\sigma^2 - t^2)^2} \tag{6-68}$$

where we have chosen for k the value 1. The minimum of γ_w equals $\pi^2/4\sigma^2$ [see Eq. (6-64)].

Since $\underline{Y}(\omega)$ is a truncated cosine and

$$\underline{W}(\omega) = \frac{1}{2\pi} \underline{Y}(\omega) * \underline{Y}(-\omega)$$

it follows that (see Prob. 4)

$$\underline{W}(\omega) = \frac{\sigma}{2\pi^2} \left| \sin \frac{\pi}{2\sigma} \omega \right| + \frac{\sigma}{2\pi} \left(1 - \frac{|\omega|}{2\sigma} \right) \cos \frac{\pi}{2\sigma} \omega \tag{6-69}$$

for $|\omega| < 2\sigma$, and $\underline{W}(\omega) = 0$ for $|\omega| > 2\sigma$ (Fig. 6-8).

Note: From the preceding discussion, it follows readily that

$$\int_{-\infty}^{\infty} t^2 |y(t)|^2\, dt \geq \frac{\pi^2}{4\sigma^2} \int_{-\infty}^{\infty} |y(t)|^2\, dt \tag{6-70}$$

for every σ-BL function. Equality holds only if $y(t)$ is given by (6-66). Similarly,

$$\int_{-\infty}^{\infty} t^2 w(t)\, dt \geq \frac{\pi^2}{4\sigma^2} \int_{-\infty}^{\infty} w(t)\, dt \tag{6-71}$$

for every nonnegative 2σ-BL function. Equality holds only if $w(t)$ is given by (6-68).

Trigonometric polynomials with minimum moment of inertia[2] Consider the class of trigonometric polynomials of order M and energy E_y, as in (6-A-5), such

[1] A. Papoulis, Minimum Bias Windows for High Resolution Spectral Estimates, *IEEE Trans. Inf. Theory*, vol. IT-19, pp. 9–12, 1973.

[2] A. Papoulis, A New Class of Fourier Series Kernels, *IEEE Trans. Circuit Theory*, vol. CT-20, no. 2, pp. 101–107, March, 1973.

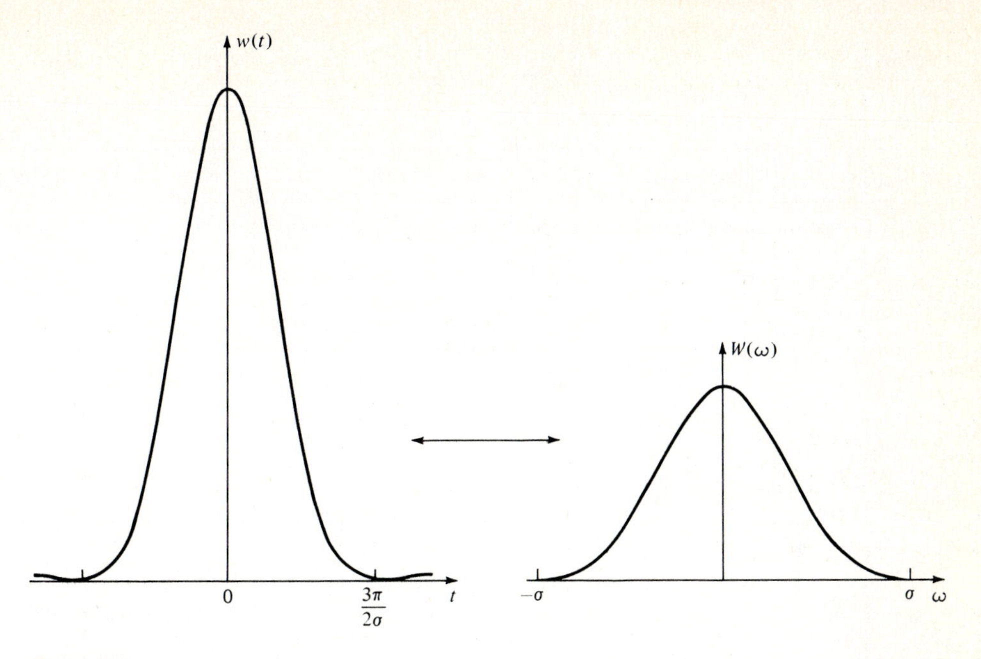

Figure 6-8

that the ratio

$$\alpha_y = \frac{1}{E_y}\int_{-T/2}^{T/2} t^2 |y(t)|^2 \, dt \tag{6-72}$$

is minimum. This is a special case of Eq. (6-A-4) with $v(t) = t^2$. The coefficients v_n of the Fourier-series expansion of the function t^2 in the interval $(-T/2, T/2)$ are given by

$$v_n = \frac{1}{T}\int_{-T/2}^{T/2} t^2 e^{-jn\omega_0 t} \, dt = \begin{cases} \dfrac{(-1)^n T^2}{2\pi^2 n^2} & n \neq 0 \\ \dfrac{T^2}{12} & n = 0 \end{cases}$$

Inserting into (6-A-7), we obtain the system

$$\frac{T^2}{12} + \frac{T^2}{2\pi^2}\sum_{\substack{k=-M \\ k\neq 0}}^{M} \frac{(-1)^{n-k}}{(n-k)^2} y_k = \lambda y_n \qquad n = -M, \ldots, M \tag{6-73}$$

As we show in Theorem 1 of Appendix 6-A, the minimum value of α_y equals the minimum eigenvalue $\lambda_{\min}$ of (6-73); the minimum is reached if the coefficients of $y(t)$ form the corresponding eigenvector $\{\underline{y}_k\}$. The system (6-73) has been solved numerically for M from 1 to 10, and the resulting values of $\lambda_{\min}$ and $\{\underline{y}_k\}$ are shown in Table 6-1.

Table 6-1

M	1	2	3	4	5	6	7	8	9	10
$10^3\lambda_{min}$	17.74	7.542	4.153	2.625	1.808	1.320	1.006	0.7924	0.6401	0.5278
$\underline{y}_0$	0.7376	0.5876	0.5084	0.4524	0.4126	0.3811	0.3563	0.3355	0.3182	0.3031
$\underline{y}_1$	0.4775	0.5062	0.4659	0.4298	0.3976	0.3715	0.3491	0.3304	0.3141	0.3000
$\underline{y}_2$		0.2666	0.3514	0.3616	0.3552	0.3421	0.3284	0.3147	0.3022	0.2906
$\underline{y}_3$			0.1738	0.2584	0.2865	0.2954	0.2943	0.2895	0.2826	0.2752
$\underline{y}_4$				0.1242	0.1991	0.2327	0.2490	0.2549	0.2560	0.2540
$\underline{y}_5$					0.0944	0.1591	0.1933	0.2128	0.2227	0.2276
$\underline{y}_6$						0.0748	0.1308	0.1635	0.1841	0.1963
$\underline{y}_7$							0.0611	0.1099	0.1405	0.1612
$\underline{y}_8$								0.0511	0.0940	0.1223
$\underline{y}_9$									0.0436	0.0815
$\underline{y}_{10}$										0.0376

[1] A. Papoulis, A New Class of Fourier Series Kernels, *IEEE Trans. Circuit Theory*, vol. CT-20, no. 2, pp. 101–107, March, 1973.

Positive signals A related problem is the determination of a nonnegative trigonometric polynomial of order $2M$, as in (6-A-11), such that its normalized moment of inertia

$$\gamma_w = \frac{1}{Tw_0}\int_{-T/2}^{T/2} t^2 w(t)\,dt \tag{6-74}$$

is minimum.

This is a special case (6-A-12) with $v(t) = t^2$. Hence, the optimum $\underline{w}(t)$ is given by

$$\underline{w}(t) = \left|\sum_{k=-M}^{M} \underline{y}_k\, e^{jk\omega_0 t}\right|^2 = \sum_{k=-2M}^{2M} \underline{w}_k\, e^{jk\omega_0 t} \tag{6-75}$$

where $\underline{y}_k$ is the eigenvector of (6-73), and the resulting $\underline{\gamma}_w$ equals the corresponding eigenvalue λ_{min}.

It is easy to see from Eq. (6-75) that

$$\underline{w}_n = \begin{cases} \displaystyle\sum_{k=-M}^{M-n} \underline{y}_{n+k}\,\underline{y}_k & 0 \le n \le M \\ \displaystyle\sum_{k=-M-n}^{M} \underline{y}_{n+k}\,\underline{y}_k & -M \le n \le 0 \end{cases} \tag{6-76}$$

Table 6-2

N	2	4	6	8	10	12	14	16	18	20
$\underline{w}_0$	1	1	1	1	1	1	1	1	1	1
$\underline{w}_1$	0.7044	0.8649	0.9233	0.9507	0.9657	0.9748	0.9807	0.9847	0.9876	0.9898
$\underline{w}_2$	0.2280	0.5696	0.7363	0.8238	0.8745	0.9063	0.9274	0.9422	0.9528	0.9608
$\underline{w}_3$		0.2699	0.5041	0.6514	0.7442	0.8053	0.8472	0.8771	0.8990	0.9157
$\underline{w}_4$		0.0711	0.2854	0.4652	0.5933	0.6833	0.7476	0.7946	0.8299	0.8570
$\underline{w}_5$			0.1221	0.2936	0.4397	0.5519	0.6363	0.7003	0.7494	0.7877
$\underline{w}_6$			0.0302	0.1566	0.2986	0.4217	0.5209	0.5994	0.6614	0.7108
$\underline{w}_7$				0.0642	0.1811	0.3019	0.4083	0.4971	0.5698	0.6291
$\underline{w}_8$				0.0154	0.0937	0.1993	0.3043	0.3980	0.4781	0.5456
$\underline{w}_9$					0.0376	0.1182	0.2133	0.3061	0.3898	0.4628
$\underline{w}_{10}$					0.0089	0.0601	0.1384	0.2244	0.3075	0.3831
$\underline{w}_{11}$						0.0238	0.0810	0.1552	0.2334	0.3086
$\underline{w}_{12}$						0.0056	0.0407	0.0996	0.1692	0.2408
$\underline{w}_{13}$							0.0160	0.0577	0.1159	0.1811
$\underline{w}_{14}$							0.0037	0.0288	0.0738	0.1303
$\underline{w}_{15}$								0.0112	0.0425	0.0886
$\underline{w}_{16}$								0.0026	0.0211	0.0561
$\underline{w}_{17}$									0.0082	0.0321
$\underline{w}_{18}$									0.0019	0.0159
$\underline{w}_{19}$										0.0062
$\underline{w}_{20}$										0.0014

[1] A. Papoulis, A New Class of Fourier Series Kernels, *IEEE Trans. Circuit Theory*, vol. CT-20, no. 2, pp. 101–107, March, 1973.

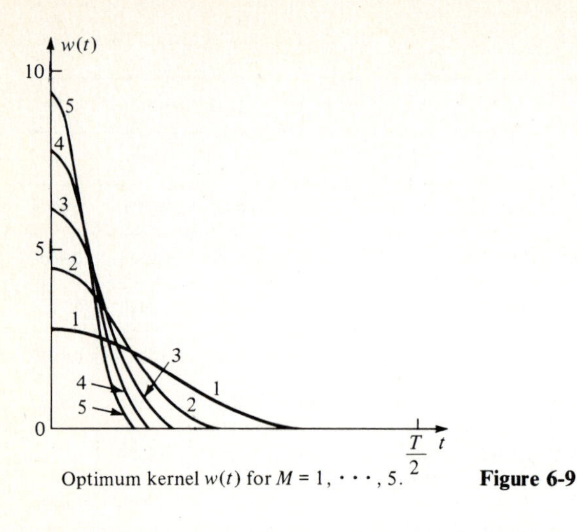

Optimum kernel $w(t)$ for $M = 1, \cdots, 5$. **Figure 6-9**

In Table 6-2 we show the values of $\{\underline{w}_n\}$ computed from (6-76) for M from 1 to 10, and Fig. 6-9 we plot the corresponding $\underline{w}(t)$ for M from 1 to 5.

Maximum response of systems with time-limited inputs of given energy[1] The input to a known linear system is a function $f(t)$ vanishing for $|t| > \tau$ and of energy E. We wish to find the form of $f(t)$ such that, for given values of τ and E, the energy E_g of the resulting response $g(t)$ is maximum.

With $H(\omega)$ the system function, we know that $G(\omega) = F(\omega)H(\omega)$. Hence,

$$E_g = \frac{1}{2\pi}\int_{-\infty}^{\infty} |H(\omega)|^2 |F(\omega)|^2 \, d\omega \tag{6-77}$$

Thus, our problem is to maximize the above integral or, since E is given, the ratio E_g/E. This problem is identical with the problem solved in Theorem 2 of Appendix 6-A if t is interchanged with ω, $v(t)$ with $|H(\omega)|^2$, and $y(t)$ with $F(\omega)$.

As we have shown, the optimum $F(\omega)$ satisfies (6-A-15), and its inverse $f(t)$ satisfies (6-A-31), properly modified. The function $V(\omega)$ in (6-A-31) is the transform of $v(t)$. In our case, $v(t)$ is replaced with $|H(\omega)|^2$; hence, we must replace $V(\omega)$ with the inverse transform $\rho(t)$ of $|H(\omega)|^2$:

$$\rho(t) = h(t) * h^*(-t) \leftrightarrow |H(\omega)|^2 \tag{6-78}$$

Replacing, finally, $Y(\omega)$ with $f(t)$, we conclude from (6-A-31) that the optimum $f(t)$ is the eigenfunction of the integral equation

$$\int_{-\tau}^{\tau} \rho(t - x) f(x) \, dx = \lambda f(t) \qquad |t| \le \tau \tag{6-79}$$

[1] J. H. H. Chalk, The Optimum Pulse-Shape for Pulse Communication, *Proc. Inst. Elec. Eng. London*, vol. 87, pp. 88–92, 1950.

corresponding to the maximum eigenvalue λ_{max}, and the resulting maximum of E_g equals $E\lambda_{max}$.

6-4 THE PROLATE SPHEROIDAL FUNCTIONS[1]

The prolate spheroidal functions are the solutions of the integral equation

$$\int_{-\tau}^{\tau} \varphi(x) \frac{\sin \sigma(t-x)}{\pi(t-x)}\, dx = \lambda\varphi(t) \tag{6-80}$$

We shall discuss their properties.

The first three properties follow from the general theory of integral equations and are stated without proof. All other properties will be proved.

Eigenvalues. Equation (6-80) has solutions only for certain values λ_n of λ. The numbers λ_n are real, positive, and such that

$$1 > \lambda_0 > \lambda_1 > \lambda_n > \cdots \underset{n\to\infty}{\to 0} \tag{6-81}$$

Eigenfunctions. To each λ_n there corresponds only one function $\varphi_n(t)$ within a constant factor. With a proper choice of this factor, the functions $\varphi_n(t)$ form a real orthonormal set in the interval $(-\infty, \infty)$:

$$\int_{-\infty}^{\infty} \varphi_k(t)\varphi_n(t)\, dt = \begin{cases} 1 & k = n \\ 0 & k \neq n \end{cases} \tag{6-82}$$

Furthermore,

$$\varphi_n(-t) = (-1)^n \varphi_n(t)$$

Completeness. An arbitrary σ-BL function $f(t)$ can be written as a sum

$$f(t) = \sum_{n=0}^{\infty} a_n \varphi_n(t) \qquad \text{for all } t \tag{6-83}$$

where

$$a_n = \int_{-\infty}^{\infty} f(t)\varphi_n(t)\, dt \tag{6-84}$$

Note: Equation (6-80) contains the parameters τ and σ. However, its solutions form essentially a one-parameter family of functions depending on the product

$$c = \tau\sigma \tag{6-85}$$

[1] D. Slepian, H. J. Landau, and H. O. Pollack, Prolate Spheroidal Wave Functions, Fourier Analysis and Uncertainty Principle I and II, *Bell Syst. Tech. J.*, vol. 40, no. 1, pp. 43–84, 1961.

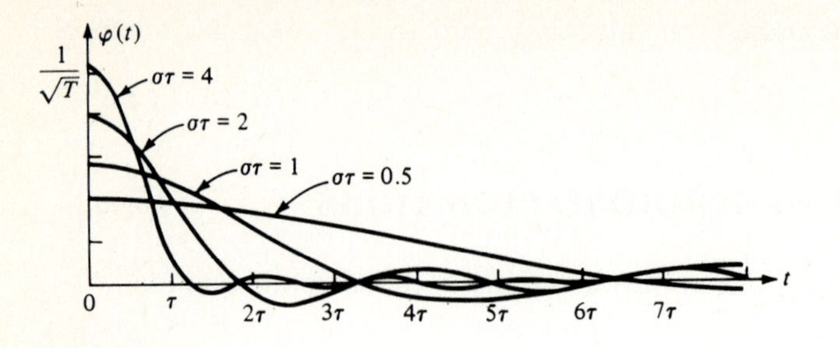

Figure 6-10

Indeed, substituting τx for x and τt for t in (6-80), we obtain the normalized equation

$$\int_{-1}^{1} \psi(x) \frac{\sin c(t-x)}{\pi(t-x)}\, dx = \lambda \psi(t) \qquad \psi(t) = \varphi(\tau t) \tag{6-86}$$

In Fig. 6-10 we show the eigenfunction $\varphi_0(t)$ of (6-86) corresponding to the maximum eigenvalue λ_0 for $c = 0.5$, 1, 2, and 4. In Fig. 6-11 we plot λ_0 as a function of c.

Fourier Transforms

The integral in (6-80) equals the convolution of the kernel $\sin \sigma t/\pi t$ with the segment

$$\varphi_\tau(t) = \varphi(t) p_\tau(t) \tag{6-87}$$

of $\varphi(t)$. Hence, (6-80) can be written in the form

$$\varphi_\tau(t) * \frac{\sin \sigma t}{\pi t} = \lambda \varphi(t) \tag{6-88}$$

Denoting by $\Phi(\omega)$ and $\Phi_\tau(\omega)$ the Fourier transforms of the functions $\varphi(t)$ and $\varphi_\tau(t)$, respectively,

$$\varphi(t) \leftrightarrow \Phi(\omega) \qquad \varphi_\tau(t) \leftrightarrow \Phi_\tau(\omega)$$

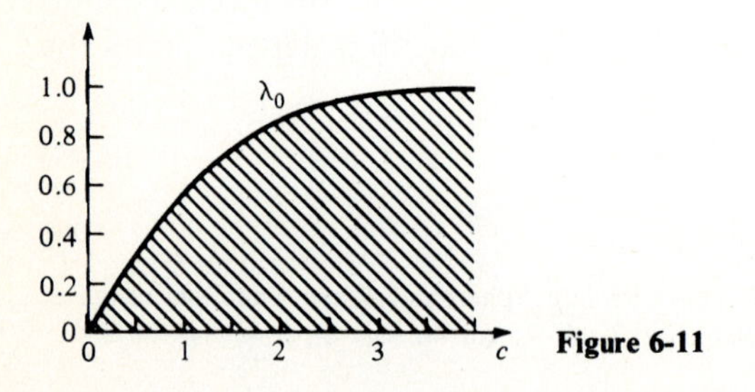

Figure 6-11

we conclude from (6-88) and the convolution theorem that

$$\Phi_\tau(\omega)p_\sigma(\omega) = \lambda\Phi(\omega) \tag{6-89}$$

The above shows that

$$\Phi(\omega) = 0 \qquad \text{for } |\omega| > \sigma \tag{6-90}$$

i.e., that the solutions $\varphi_n(t)$ of (6-80) are σ-BL. Furthermore, since $\Phi(\omega) = \Phi(\omega)p_\sigma(\omega)$, they also satisfy the equation

$$\int_{-\infty}^{\infty} \varphi(x) \frac{\sin \sigma(t-x)}{\pi(t-x)}\, dx = \varphi(t) \tag{6-91}$$

Double orthogonality As we noted, the functions $\varphi_n(t)$ are orthogonal in the interval $(-\infty, \infty)$. We maintain that they are also orthogonal in the interval $(-\tau, \tau)$:

$$\int_{-\tau}^{\tau} \varphi_k(t)\varphi_n(t)\, dt = \begin{cases} \lambda_n & k = n \\ 0 & k \neq n \end{cases} \tag{6-92}$$

Proof Clearly, $\varphi_n(t)$ satisfies Eq. (6-80) if $\lambda = \lambda_n$. Multiplying the resulting equation by $\varphi_k(t)$ and integrating from $-\infty$ to ∞, we conclude, after a change in the order of integration, that

$$\int_{-\tau}^{\tau} \varphi_n(x) \int_{-\infty}^{\infty} \varphi_k(t) \frac{\sin \sigma(t-x)}{\pi(t-x)}\, dt\, dx = \lambda_n \int_{-\infty}^{\infty} \varphi_n(t)\varphi_k(t)\, dt$$

and (6-92) follows from (6-82).

The next theorem shows that the functions $\varphi_n(t)$, suitably truncated and scaled, equal their transforms $\Phi_n(\omega)$ (Fig. 6-12).

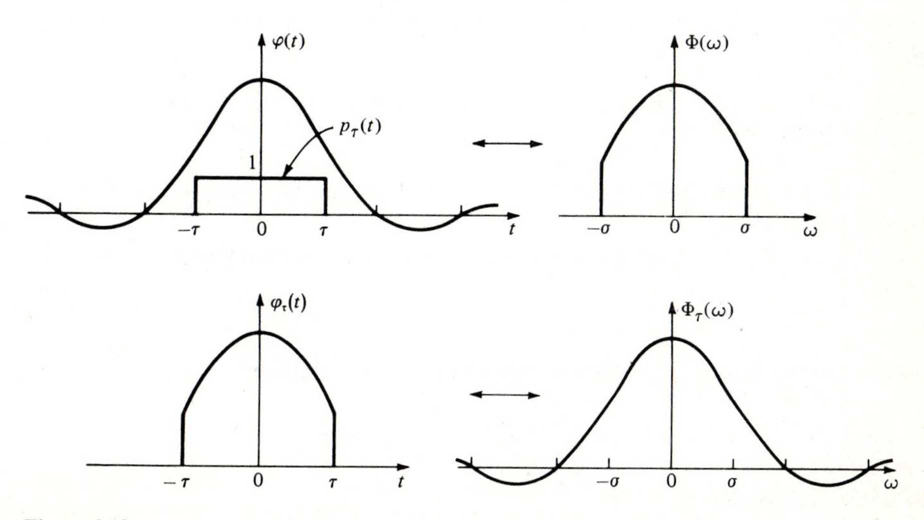

Figure 6-12

Theorem 1

$$\varphi_n(t) \leftrightarrow (-j)^n \sqrt{\frac{2\pi\tau}{\sigma\lambda_n}}\, \varphi_n\left(\frac{\tau}{\sigma}\,\omega\right) p_\sigma(\omega) \tag{6-93}$$

$$\varphi_n(t) p_\tau(t) \leftrightarrow (-j)^n \sqrt{\frac{2\pi\tau\lambda_n}{\sigma}}\, \varphi_n\left(\frac{\tau}{\sigma}\,\omega\right) \tag{6-94}$$

Proof Taking transforms of both sides of (6-87), we obtain

$$\Phi_\tau(\omega) = \int_{-\infty}^{\infty} \Phi(\eta) \frac{\sin \tau(\omega - \eta)}{\pi(\omega - \eta)}\, d\eta \tag{6-95}$$

Since $\Phi(\omega) = 0$ for $|\omega| > \sigma$ and $\Phi_\tau(\omega) = \lambda\Phi(\omega)$ for $|\omega| < \sigma$, Eq. (6-95) yields

$$\int_{-\sigma}^{\sigma} \Phi(\eta) \frac{\sin \tau(\omega - \eta)}{\pi(\omega - \eta)}\, d\eta = \lambda\Phi(\omega) \qquad |\omega| < \sigma \tag{6-96}$$

This equation is equivalent to (6-80) because, with $x = \tau\eta/\sigma$ and $t = \tau\omega/\sigma$, (6-80) takes the form

$$\int_{-\sigma}^{\sigma} \varphi\left(\frac{\tau}{\sigma}\,\eta\right) \frac{\sin \tau(\omega - \eta)}{\pi(\omega - \eta)}\, d\eta = \lambda\varphi\left(\frac{\tau}{\sigma}\,\omega\right) \tag{6-97}$$

And, since for each $\lambda = \lambda_n$ it has a unique solution within a constant factor, we conclude that

$$\Phi(\omega) = A\varphi\left(\frac{\tau\omega}{\sigma}\right) \qquad \text{for } |\omega| < \sigma$$

As we noted, the functions $\varphi_n(t)$ are real, and $\varphi_n(-t) = (-1)^n\varphi_n(t)$. Therefore, their transforms are real for even n and imaginary for odd n. Hence, with B a real number, $A = (-j)^n B$. To determine B, we use Eqs. (6-82) and (6-92) and Parseval's formula:

$$1 = \int_{-\infty}^{\infty} \varphi^2(t)\, dt = \frac{B^2}{2\pi} \int_{-\sigma}^{\sigma} \varphi^2\left(\frac{\tau}{\sigma}\,\omega\right) d\omega = \frac{B^2\sigma}{2\pi\tau} \int_{-\sigma}^{\sigma} \varphi^2(x)\, dx = \frac{B^2\sigma\lambda}{2\pi\tau}$$

Solving for B, we obtain (6-93). The sign of the radical can be $+$ or $-$, depending on how $\varphi_n(t)$ is defined.

Equation (6-94) is a consequence of (6-93) and the symmetry theorem (3-17) (elaborate).

Corollary The functions $\varphi_n(t)$ also satisfy the integral equation

$$\varphi_n(t) = (-j)^n \sqrt{\frac{\tau}{2\pi\sigma\lambda_n}} \int_{-\sigma}^{\sigma} \varphi_n\left(\frac{\tau}{\sigma}\,\omega\right) e^{j\omega t}\, d\omega \tag{6-98}$$

Proof It follows from (6-93) and the inversion formula (3-2).

Theorem 2 An arbitrary τ-TL function $y_\tau(t)$ can be written as a sum

$$y_\tau(t) = \sum_{n=0}^{\infty} b_n \varphi_n(t) p_\tau(t) \qquad \text{for all } t \tag{6-99}$$

where

$$b_n = \frac{1}{\lambda_n} \int_{-\tau}^{\tau} y_\tau(t) \varphi_n(t)\, dt \tag{6-100}$$

Proof The above is a consequence of Theorem 1 and the expansion (6-83). Indeed, taking transforms of both sides of (6-83), we conclude with (6-93) that

$$F(\omega) = \sqrt{\frac{2\pi\tau}{\sigma}} \sum_{n=0}^{\infty} (-j)^n \frac{a_n}{\sqrt{\lambda_n}} \varphi_n\left(\frac{\tau}{\sigma}\omega\right) p_\sigma(\omega) \tag{6-101}$$

This equation shows that an arbitrary function $F(\omega)$ that equals zero for $|\omega| > \sigma$ can be expanded into a series as in (6-101). Changing $\tau\omega/\sigma$ to t and $F(\omega)$ to $y_\tau(t)$, we obtain (6-99).

To determine the constants b_n, we multiply both sides of (6-99) by $\varphi_k(t)$ and integrate from $-\tau$ to τ. As we see from (6-92), the result yields (6-100).

Notes:

1. *Bandlimited segment.* As we noted, a BL function $f(t)$ can be written as a series of the form (6-83). Clearly, since the functions $\varphi_n(t)$ are orthonormal and the energy E of $f(t)$ is finite, we conclude from (6-83) that

$$E = \int_{-\infty}^{\infty} f^2(t)\, dt = \sum_{n=0}^{\infty} a_n^2 < \infty \tag{6-102}$$

From (6-99) and (6-92), it follows that the energy of the TL function $y_\tau(t)$ is given by

$$\int_{-\tau}^{\tau} y_\tau^2(t)\, dt = \sum_{n=0}^{\infty} \lambda_n b_n^2 < \infty \tag{6-103}$$

Suppose now that the coefficients b_n are such that

$$\sum_{n=0}^{\infty} b_n^2 < \infty \tag{6-104}$$

In this case, the sum in

$$y(t) = \sum_{n=0}^{\infty} b_n \varphi_n(t) \tag{6-105}$$

converges for every t, and the function $y(t)$ so formed is BL. Hence, if the coefficients b_n of the expansion (6-99) of $y_\tau(t)$ satisfy (6-104), then $y_\tau(t)$ is a segment of a BL function.

2. *Bandlimited approximation.* From the preceding note, it follows that we cannot, in general, find a BL function $y(t)$ agreeing with a given TL

function $y_\tau(t)$ in the interval $(-\tau, \tau)$, because the sum in Eq. (6-105) might diverge. However, the truncated sum

$$y_N(t) = \sum_{n=0}^{N} b_n \varphi_n(t) \tag{6-106}$$

is a BL function existing for all t. In the interval $(-\tau, \tau)$ it is close to $y_\tau(t)$ in the mean-square sense for large N, because

$$\int_{-\tau}^{\tau} [y_N(t) - y_\tau(t)]^2 \, dt = \sum_{n=N+1}^{\infty} \lambda_n b_n^2 \underset{N\to\infty}{\longrightarrow} 0 \tag{6-107}$$

If $y_\tau(t)$ is not a segment of a BL function, then $y_N(t)$ diverges for $|t| > \tau$ as $N \to \infty$, because

$$\int_{-\infty}^{\infty} y_N^2(t) \, dt = \sum_{n=0}^{N} b_n^2 \underset{N\to\infty}{\longrightarrow} \infty \tag{6-108}$$

Extremal Properties

Consider a function $f(t)$ with finite energy E. With τ and σ two specified numbers, we form the ratios

$$\alpha = \frac{1}{E} \int_{-\tau}^{\tau} |f(t)|^2 \, dt \qquad \beta = \frac{1}{2\pi E} \int_{-\sigma}^{\sigma} |F(\omega)|^2 \, d\omega \tag{6-109}$$

As $f(t)$ ranges over all functions, the numbers α and β take various values in the unit square of the $\alpha\beta$ plane. In the following, we determine the set of possible pairs (α, β). In particular, given one of these numbers, we determine the maximum of the other and the function $f(t)$ for which the maximum is reached (optimum). For simplicity, we assume that $E = 1$.

Case 1: Bandlimited signals If the function $f(t)$ is σ-BL, then

$$\beta = 1$$

As we have shown in Example 6-6, the maximum $\bar{\alpha}$ of α then equals the maximum eigenvalue λ_0 of (6-80). We shall rederive this result using the expansion (6-83) of $f(t)$:

$$f(t) = \sum_{n=0}^{\infty} a_n \varphi_n(t) \qquad \sum_{n=0}^{\infty} a_n^2 = E = 1$$

Clearly,

$$\int_{-\tau}^{\tau} f^2(t) \, dt = \sum_{n=0}^{\infty} \lambda_n a_n^2 \le \lambda_0 \sum_{n=0}^{\infty} a_n^2 = \lambda_0 \tag{6-110}$$

because $\lambda_0 \ge \lambda_n$ and the functions $\varphi_n(t)$ are orthogonal in the interval $(-\tau, \tau)$ [see Eq. (6-92)]. Hence,

$$\alpha = \int_{-\tau}^{\tau} f^2(t) \, dt \le \lambda_0 \tag{6-111}$$

If $\alpha = \lambda_0$, then $f(t)$ is unique and equals $\varphi_0(t)$:

$$f(t) = \varphi_0(t)$$

because then (6-110) is an equality; hence $a_n = 0$ for $n > 0$. If $\alpha < \lambda_0$, then (see Prob. 85) we can find a σ-BL function $f(t)$ whose energy concentration equals α. In this case, $f(t)$ is not unique.

Case 2: Time-limited signals If the function $f(t)$ is τ-TL, then

$$\alpha = 1$$

Reversing t and ω, we conclude that

$$\beta \le \lambda_0 \tag{6-112}$$

If $\beta = \lambda_0$, then

$$f(t) = \frac{1}{\sqrt{\lambda_0}} \varphi_0(t) p_\tau(t) \tag{6-113}$$

Case 3: Arbitrary signals Given a number $\alpha < 1$, we wish to find the maximum $\bar{\beta}$ of β and the corresponding $f(t)$.

If $\alpha \le \lambda_0$, then as we noted in case 1, we can find a σ-BL $f(t)$ with energy ratio α; hence, $\bar{\beta} = 1$. If $\alpha = \lambda_0$, then $f(t) = \varphi_0(t)$; if $\alpha < \lambda_0$, then $f(t)$ is not unique. It remains to consider the case $\alpha > \lambda_0$.

Theorem 3 The maximum $\bar{\beta}$ is given by

$$\bar{\beta} = \cos^2(\theta_0 - \theta_1) \tag{6-114}$$

where
$$\cos\theta_0 = \sqrt{\lambda_0} \qquad \cos\theta_1 = \sqrt{\alpha} \tag{6-115}$$

and the corresponding $f(t)$ is given by

$$f(t) = a\varphi_0(t) p_\tau(t) + b\varphi_0(t) \tag{6-116}$$

where
$$a = \sqrt{\frac{1-\alpha}{1-\lambda_0}} \qquad b = \sqrt{\frac{\alpha}{\lambda_0}} - a \tag{6-117}$$

Proof We introduce the functions

$$f_\tau(t) = f(t) p_\tau(t) \qquad \text{and} \qquad f_\sigma(t) \leftrightarrow F(\omega) p_\sigma(\omega) \tag{6-118}$$

obtained by time-limiting and bandlimiting $f(t)$. It can be shown (Prob. 86) that for β to be maximum, $f(t)$ must be a linear combination of $f_\tau(t)$ and $f_\sigma(t)$:

$$f(t) = Af_\tau(t) + Bf_\sigma(t) \tag{6-119}$$

Inserting into (6-118), we conclude that (elaborate)

$$\frac{1-B}{A} f_\sigma(t) = \frac{B}{1-A} \int_{-\tau}^{\tau} f_\sigma(x) \frac{\sin\sigma(t-x)}{\pi(t-x)} dx$$

Hence, $f_\sigma(t)$ is an eigenfunction of Eq. (6-80) corresponding to the eigenvalue

$$\lambda = \frac{(1-A)(1-B)}{AB}$$

That is, $f_\sigma(t)$ is proportional to $\varphi(t)$ and

$$f(t) = a\varphi(t) + b\varphi(t)p_\tau(t) \tag{6-120}$$

Equating the energy of $f(t)$ in the interval $(-\infty, \infty)$ to 1, and in the interval $(-\tau, \tau)$ to α, we obtain

$$1 = a^2 + \lambda b^2 + 2ab\lambda \qquad \alpha = (a+b)^2\lambda$$

With $\cos\theta = \sqrt{\lambda}$ and $\cos\theta_1 = \sqrt{\alpha}$, the above yields

$$a = \sqrt{\frac{1-\alpha}{1-\lambda}} = \frac{\sin\theta_1}{\sin\theta} \qquad b = \sqrt{\frac{\alpha}{\lambda}} - a = \frac{\cos\theta_1}{\cos\theta} - \frac{\sin\theta_1}{\sin\theta} \tag{6-121}$$

Furthermore [see Eq. (6-120) and Theorem 1],

$$\beta = \frac{1}{2\pi}\int_{-\sigma}^{\sigma} |F(\omega)|^2\, d\omega = (a + b\lambda)^2 \tag{6-122}$$

From (6-121) and (6-122) it follows that $\beta = \cos^2(\theta - \theta_1)$. This quantity is maximum if $\theta = \arccos\sqrt{\lambda}$ is minimum, i.e., if λ is maximum. Hence, $\lambda = \lambda_0$ and $\varphi(t) = \varphi_0(t)$, and the proof is complete.

Note: From Theorem 3 it follows that the set of possible pairs (α, β) is the region R_c of Fig. 6-13 bounded by coordinate lines and the curve

$$\bar{\beta}(\alpha) = \cos^2(\theta_0 - \theta_1)$$

as in (6-114). Since $\theta_0 = \arccos\sqrt{\lambda_0}$ and λ_0 is a function of c, the curve $\bar{\beta}(\alpha)$ depends on c.

Digital Prolate Functions[1,2]

For periodic functions, the maximum-energy-concentration problem takes the following form:

Given a number $\tau < T/2$, find a trigonometric polynomial

$$y(t) = \sum_{n=-M}^{M} y_n e^{jn\omega_0 t}$$

[1] D. W. Tufts and J. T. Francis, Designing Digital Low-pass Filters—Comparison of Some Methods and Criteria, *IEEE Trans. Audio Electroacoust.*, vol. AU-18, pp. 487–494, December, 1970.

[2] A. Papoulis and M. Bertran, Digital Filtering and Prolate Functions, *IEEE Trans. Circuit Theory* (special issue on nonlinear circuits), vol. CT-19, pp. 674–681, November, 1972.

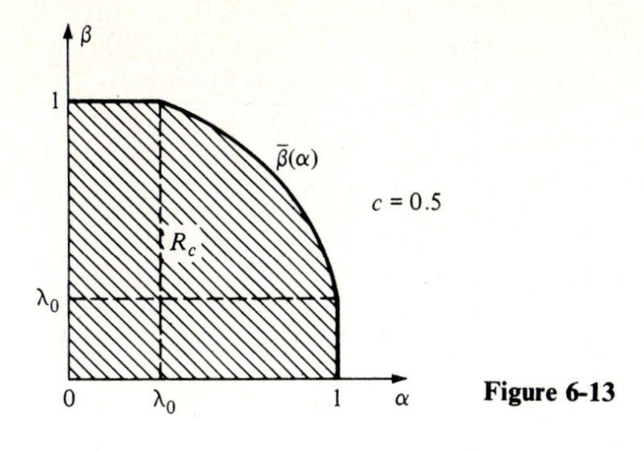

Figure 6-13

as in Eq. (6-A-5) such that the ratio

$$\alpha = \frac{1}{E_y} \int_{-\tau}^{\tau} |y(t)|^2 \, dt \tag{6-123}$$

is maximum or minimum.

The above ratio is a special case of the ratio α_y in (6-A-4) corresponding to

$$v(t) = p_\tau(t)$$

Since
$$v_n = \frac{1}{T} \int_{-T/2}^{T/2} p_\tau(t) e^{-jn\omega_0 t} \, dt = \frac{1}{T} \int_{-\tau}^{\tau} e^{-jn\omega_0 t} \, dt = \frac{\sin n\omega_0 \tau}{\pi n} \tag{6-124}$$

we conclude from (6-A-7) that the coefficients y_n of the optimum $y(t)$ must be such that

$$\sum_{k=-M}^{M} \frac{\sin \omega_0 \tau(n-k)}{\pi(n-k)} y_k = \lambda y_n \qquad |n| \leq M \tag{6-125}$$

This system is the discrete version of the integral equation (6-80).

As we know from the theory of linear equations, (6-125) has $2M + 1$ eigenvalues

$$1 > \lambda_0 > \cdots > \lambda_k > \cdots > \lambda_{2M} > 0 \tag{6-126}$$

that depend on M and τ. In Fig. 6-14, we plot the maximum $\lambda_{\max} = \lambda_0$ as a function of τ for $M = 1, 2, \ldots, 10$. The corresponding eigenvectors $\{\varphi_n{}^k\}$ form an orthonormal set

$$\sum_{n=-M}^{M} \varphi_n{}^k \varphi_n{}^i = \begin{cases} 1 & i = k \\ 0 & i \neq k \end{cases} \tag{6-127}$$

The trigonometric polynomials

$$\varphi^k(t) = \sum_{n=-M}^{M} \varphi_n{}^k e^{jn\omega_0 t} \qquad k = 0, \ldots, 2M \tag{6-128}$$

we call *digital prolate functions.* From Eq. (6-125) it follows that the energy ratio α of $\varphi^k(t)$ equals λ_k. Hence, if $y(t) = \varphi^0(t)$, then α is maximum and equals λ_0.

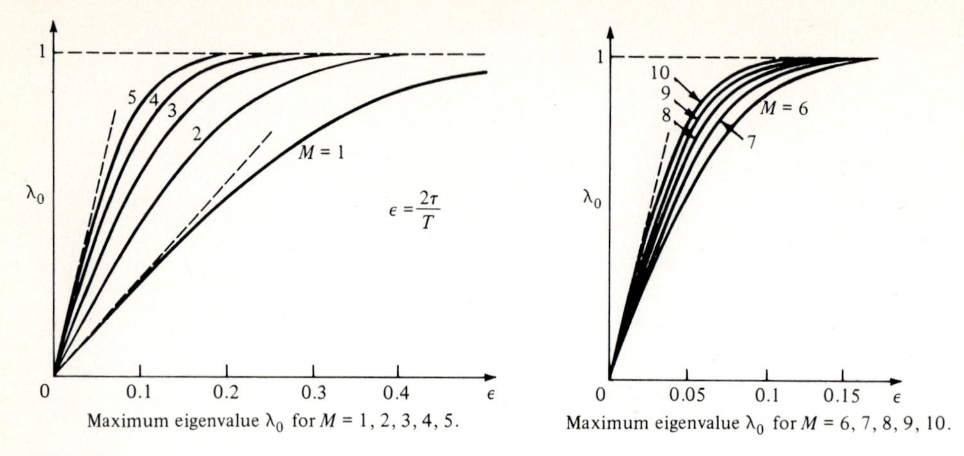

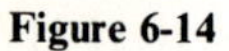

Figure 6-14

If $y(t) = \varphi^{2M}(t)$, then α is minimum and equals λ_{2M}. In Fig. 6-15, we plot $\varphi^0(t)$ for $\tau = 0.05T$ and $M = 2, 4, 6, 8, 10$.

We discuss next, briefly, the properties of the functions $\varphi^k(t)$.

An equivalent integral equation If $\{y_n\}$ is an eigenvector of (6-125) and

$$y(t) = \sum_{n=-M}^{M} y_n e^{jn\omega_0 t} \tag{6-129}$$

then $y(t)$ is an eigenfunction of the integral equation

$$\frac{1}{T}\int_{-\tau}^{\tau} y(x)\,\frac{\sin\,[(M+1/2)\omega_0(t-x)]}{\sin\,[\omega_0(t-x)/2]}\,dx = \lambda y(t) \tag{6-130}$$

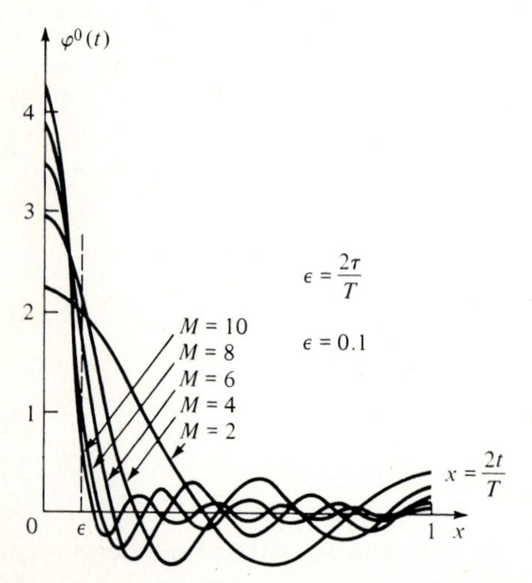

Optimum polynomial $\varphi^0(t)$ for $\epsilon = 0.1$, $M = 2, 4, 6, 8, 10$. **Figure 6-15**

Proof With

$$y(t)p_\tau(t) = \sum_{n=-\infty}^{\infty} b_n e^{jn\omega_0 t} \qquad \text{and} \qquad z(t) = \sum_{n=-M}^{M} b_n e^{jn\omega_0 t} \tag{6-131}$$

it follows from (6-124) and (6-125) that

$$b_n = \sum_{k=-M}^{M} v_{n-k} y_k = \begin{cases} \lambda y_n & |n| \le M \\ \displaystyle\sum_{k=-M}^{M} \frac{\sin \omega_0 \tau(n-k)}{\pi(n-k)} y_k & |n| > M \end{cases} \tag{6-132}$$

Hence,

$$z(t) = \lambda y(t) \tag{6-133}$$

and (6-130) results (elaborate).

Orthogonality From Eqs. (3-76) and (6-128) it follows that

$$\frac{1}{T}\int_{-T/2}^{T/2} \varphi^k(t)\varphi^i(t)\,dt = \sum_{n=-M}^{M} \varphi_n{}^k \varphi_n{}^i = \begin{cases} 1 & i = k \\ 0 & i \ne k \end{cases} \tag{6-134}$$

Expanding $\varphi^k(t)p_\tau(t)$ into a series and forming the functions $z^k(t)$ as in (6-131), we conclude that $z^k(t) = \lambda\varphi^k(t)$. Hence,

$$\frac{1}{T}\int_{-\tau}^{\tau} \varphi^k(t)\varphi^i(t)\,dt = \sum_{n=-M}^{M} \varphi_n{}^k \varphi_n{}^i \lambda_k = \begin{cases} \lambda_k & i = k \\ 0 & i \ne k \end{cases} \tag{6-135}$$

(elaborate). Thus, the trigonometric polynomials $\varphi^k(t)$ are orthogonal in the intervals $(-T/2, T/2)$ and $(-\tau, \tau)$.

APPENDIX 6-A

Extrema of Integrals of Trigonometric Polynomials and Bandlimited Functions

Theorems 1 and 2 of this appendix unify a number of seemingly unrelated problems.

Trigonometric Polynomials

Consider the inner product α of the vectors Ax and x:

$$\alpha = (Ax, x) = \sum_{i,j=1}^{m} a_{ij} x_i x_j^* \tag{6-A-1}$$

where A is a hermitian matrix. Clearly, α is real because $a_{ij} = a_{ji}^*$ and, as x ranges over all vectors of unit length, its values form a bounded set. It is well known from the theory of matrices that this set is between the largest and the smallest eigenvalue of A:

$$Ax = \lambda x \qquad \lambda_{\text{max}} \geq \alpha \geq \lambda_{\text{min}} \tag{6-A-2}$$

Furthermore, the extreme values $\bar{\alpha} = \lambda_{\text{max}}$ and $\underline{\alpha} = \lambda_{\text{min}}$ are reached if x equals the eigenvector of A corresponding to λ_{max} and λ_{min}, respectively, because if x is an eigenvector, then

$$\alpha = (Ax, x) = (\lambda x, x) = \lambda \tag{6-A-3}$$

Using the above, we shall determine the extreme values $\bar{\alpha}$ and $\underline{\alpha}$ of the ratio

$$\alpha_y = \frac{1}{E_y} \int_{-T/2}^{T/2} v(t) |y(t)|^2 \, dt \tag{6-A-4}$$

where $v(t)$ is a given real function, and $y(t)$ is a trigonometric polynomial with energy E_y:

$$y(t) = \sum_{n=-M}^{M} y_n e^{jn\omega_0 t} \qquad E_y = T \sum_{n=-M}^{M} |y_n|^2 \tag{6-A-5}$$

For this purpose, we expand the function $v(t)$ into a Fourier series in the interval $(-T/2, T/2)$:

$$v(t) = \sum_{n=-\infty}^{\infty} v_n e^{jn\omega_0 t} \tag{6-A-6}$$

Theorem 1 The maximum $\bar{\alpha}$ and minimum $\underline{\alpha}$ of α_y, as $y(t)$ ranges over all trigonometric polynomials of specified order M, equal the maximum and minimum eigenvalues, respectively, of the system

$$\sum_{k=-M}^{M} v_{n-k} y_k = \lambda y_n \qquad n = -M, -M+1, \ldots, M \tag{6-A-7}$$

The extreme values

$$\bar{\alpha} = \lambda_{\max} \qquad \text{and} \qquad \underline{\alpha} = \lambda_{\min} \tag{6-A-8}$$

are reached for $y(t) = \bar{y}(t)$ and $y(t) = \underline{y}(t)$, respectively, where $\bar{y}(t)$ and $\underline{y}(t)$ are two trigonometric polynomials whose Fourier-series coefficients equal the eigenvectors of Eq. (6-A-7) corresponding to the eigenvalues $\lambda_{\max}$ and $\lambda_{\min}$, respectively.

Proof We can assume that $E_y = T$ because the ratio α_y remains unchanged if $y(t)$ is multiplied by a constant. With this assumption, we have

$$\alpha_y = \sum_{i,\,j=-M}^{M} v_{j-i} y_i y_j^* \qquad \sum_{i=-M}^{M} |y_i|^2 = 1 \tag{6-A-9}$$

(see Prob. 73). The first sum is a quadratic form, as in (6-A-1), with $m = 2M + 1$ and with the indices suitably modified. Furthermore, $v_{i-j} = v_{j-i}^*$ because $v_{-n} = v_n^*$ since $v(t)$ is real. Therefore, the proof of the theorem follows from the stated property of matrices.

Note: From the convolution theorem (3-72), it follows that the system (6-A-7) is equivalent to the integral equation

$$\frac{1}{T}\int_{-T/2}^{T/2} v(x)y(x)\frac{\sin[(M + 1/2)\omega_0(t - x)]}{\sin[\omega_0(t - x)/2]}\,dx = \lambda y(t) \tag{6-A-10}$$

Positive signals Suppose, now, that $w(t)$ is a nonnegative trigonometric polynomial of order $2M$:

$$w(t) = \sum_{n=-2M}^{2M} w_n e^{jn\omega_0 t} \qquad w(t) \geq 0 \tag{6-A-11}$$

With $v(t)$ a given real function, we form the ratio

$$\gamma_w = \frac{1}{Tw_0}\int_{-T/2}^{T/2} v(t)w(t)\,dt \tag{6-A-12}$$

We shall determine its extrema as $w(t)$ ranges over all trigonometric polynomials of order $2M$.

The Fejér-Riesz Theorem It can be shown (see Sec. 7-2) that we can find a trigonometric polynomial $y(t)$ of order M, as in (6-A-5), such that

$$w(t) = |y(t)|^2 \tag{6-A-13}$$

Since

$$Tw_0 = \int_{-T/2}^{T/2} w(t)\,dt = \int_{-T/2}^{T/2} |y(t)|^2\,dt = E_y$$

it follows that $\gamma_w = \alpha_y$, where α_y is the ratio in (6-A-4). Hence, the extrema of γ_w equal the eigenvalues $\lambda_{\max}$ and $\lambda_{\min}$ of (6-A-7), and the corresponding polynomials are given by $\bar{w}(t) = |\bar{y}(t)|^2$ and $\underline{w}(t) = |\underline{y}(t)|^2$, respectively.

Bandlimited Functions

Given a real function $v(t)$, we form the ratio

$$\alpha_y = \frac{1}{E_y}\int_{-\infty}^{\infty} v(t)|y(t)|^2\,dt \tag{6-A-14}$$

where $y(t)$ is a σ-BL function with energy E_y. We shall determine the extreme values $\bar{\alpha}$ and $\underline{\alpha}$ of α_y as $y(t)$ ranges over all σ-BL functions.

Theorem 2 If the eigenvalues of the integral equation

$$\int_{-\infty}^{\infty} v(x)y(x)\frac{\sin\sigma(t-x)}{\pi(t-x)}\,dx = \lambda y(t) \tag{6-A-15}$$

have a maximum λ_{max} or a minimum λ_{min}, then

$$\bar{\alpha} = \lambda_{max} \qquad \underline{\alpha} = \lambda_{min} \tag{6-A-16}$$

and the extrema are reached if $y(t)$ equals the corresponding eigenfunctions $\bar{y}(t)$ and $\underline{y}(t)$.

Note: The existence of λ_{max} and λ_{min} depends on the form of $v(t)$ and must be established for each case.

Proof We shall assume that the given function $v(t)$ is positive and bounded. This assumption can be relaxed by adding a constant or by a limiting argument.

To any σ-BL function $y(t)$, we associate the function

$$z(t) = \int_{-\infty}^{\infty} v(x)y(x)\frac{\sin\sigma(t-x)}{\pi(t-x)}\,dx \tag{6-A-17}$$

obtained by convolving the product $v(t)y(t) = r(t)$ with the kernel $\sin\sigma t/\pi t$. With $r(t)\leftrightarrow R(\omega)$ and $z(t)\leftrightarrow Z(\omega)$, it follows that

$$Z(\omega) = R(\omega)p_\sigma(\omega) \tag{6-A-18}$$

Hence, $z(t)$ is σ-BL.

We shall show, first, that the ratio α_y of $y(t)$ cannot exceed the ratio

$$\alpha_z = \frac{1}{E_z}\int_{-\infty}^{\infty} v(t)|z(t)|^2\,dt \tag{6-A-19}$$

of the function $z(t)$ so formed:

$$\alpha_y \le \alpha_z \tag{6-A-20}$$

For this purpose, we form the identities [see Eqs. (6-A-14), (6-A-18), and (3-35)]

$$\alpha_y E_y = \int_{-\infty}^{\infty} r(t)y^*(t)\,dt = \frac{1}{2\pi}\int_{-\infty}^{\infty} R(\omega)Y^*(\omega)\,d\omega = \frac{1}{2\pi}\int_{-\sigma}^{\sigma} Z(\omega)Y^*(\omega)\,d\omega \tag{6-A-21}$$

$$E_z = \frac{1}{2\pi}\int_{-\sigma}^{\sigma} |Z(\omega)|^2 \, d\omega = \frac{1}{2\pi}\int_{-\sigma}^{\sigma} R(\omega)Z^*(\omega) \, d\omega = \int_{-\infty}^{\infty} v(t)y(t)z^*(t) \, dt \qquad \text{(6-A-22)}$$

Applying Schwarz' inequality (4-A-1) to the last integral in (6-A-21), we obtain

$$\left|\int_{-\sigma}^{\sigma} Z(\omega)Y^*(\omega) \, d\omega\right|^2 \le \int_{-\sigma}^{\sigma} |Z(\omega)|^2 \, d\omega \int_{-\sigma}^{\sigma} |Y(\omega)|^2 \, d\omega \qquad \text{(6-A-23)}$$

Hence,

$$(\alpha_y E_y)^2 \le E_z E_y \qquad \text{(6-A-24)}$$

Similarly, with $v(t) = \sqrt{v(t)}\sqrt{v(t)}$, the last integral in (6-A-22) yields

$$\left|\int_{-\infty}^{\infty} v(t)y(t)z^*(t) \, dt\right|^2 \le \int_{-\infty}^{\infty} v(t)|y(t)|^2 \, dt \int_{-\infty}^{\infty} v(t)|z(t)|^2 \, dt \qquad \text{(6-A-25)}$$

Therefore,

$$E_z^2 \le \alpha_y E_y \alpha_z E_z \qquad \text{(6-A-26)}$$

Combining (6-A-24) and (6-A-26), we obtain $\alpha_y^2 E_y \le E_z \le \alpha_y E_y \alpha_z$, and (6-A-20) results.

If $\alpha_y = \alpha_z$, then Eqs. (6-A-23) and (6-A-25) must be equalities. This is possible only if $z(t)$ is proportional to $y(t)$ [see (4-A-2)] and $E_z = \alpha_y^2 E_y$, that is, if

$$z(t) = \alpha_y y(t) \qquad \text{(6-A-27)}$$

From the above and (6-A-17), it follows that if (6-A-20) is an equality, then $y(t)$ is an eigenfunction of (6-A-15), and the corresponding eigenvalue λ equals α_y. Conversely, if $y(t)$ is an eigenfunction of (6-A-15), then $z(t) = \lambda y(t)$ and $\lambda = \alpha_y$ (elaborate).

Suppose, finally, that the ratio α_y has a maximum $\bar{\alpha}$ and it is reached for $y(t) = y_0(t)$. As we have shown, the ratio $\alpha_{\bar{y}}$ of the eigenfunction $\bar{y}(t)$ of (6-A-15) equals λ_{max}. Hence, $\lambda_{max} \le \bar{\alpha}$. To prove that $\bar{\alpha} = \lambda_{max}$, it suffices, therefore, to show that $y_0(t)$ is an eigenfunction of (6-A-15) or, equivalently, that with $z_0(t)$ defined as in (6-A-17), $\alpha_{z_0} = \bar{\alpha}$. This is obvious because $\alpha_{z_0} \ge \bar{\alpha}$ [see (6-A-20)], and $\alpha_{z_0} \le \bar{\alpha}$ because $\bar{\alpha}$ is maximum by assumption.

To prove the second part of the theorem, we assume that $v(t)$ is bounded by M, and we form the nonnegative function $s(t) = M - v(t)$ and the ratio

$$\beta_y = \frac{1}{E_y}\int_{-\infty}^{\infty} s(t)|y(t)|^2 \, dt = M - \alpha_y \qquad \text{(6-A-28)}$$

which is of the same form as (6-A-14) with $v(t)$ replaced by $s(t)$. From the first part of the theorem, it follows that $\beta_y = \bar{\beta}$ if $y(t)$ is an eigenfunction of the equation

$$\int_{-\infty}^{\infty} s(x)y(x)\frac{\sin \sigma(t-x)}{\pi(t-x)} \, dx = \mu y(t) \qquad \text{(6-A-29)}$$

corresponding to the maximum eigenvalue μ_{max}. With the substitution $s(t) =$

$M - v(t)$, Eq. (6-A-29) yields

$$\int_{-\infty}^{\infty} v(t)y(x)\,\frac{\sin\sigma(t-x)}{\pi(t-x)}\,dx = (M-\mu)y(t) \tag{6-A-30}$$

because, for σ-BL functions,

$$y(t) * \frac{\sin\sigma t}{\pi t} = y(t)$$

Clearly, (6-A-30) is equivalent to (6-A-15) if $\lambda = M - \mu$; hence, $\mu_{\max} = M - \lambda_{\min}$. Furthermore [see (6-A-28)], if $\beta_y = \bar{\beta} = \mu_{\max}$, then $\alpha_y = \underline{\alpha}$ is minimum, and $\underline{\alpha} = M - \bar{\beta} = M - \mu_{\max} = \lambda_{\min}$.

Note: Taking transforms of both sides of (6-A-15), we obtain $\lambda Y(\omega) = R(\omega)p_\sigma(\omega)$, where $R(\omega)$ is the transform of the product $v(t)y(t)$. Hence, $Y(\omega) = 0$ for $|\omega| > \sigma$; that is, all eigenfunctions of (6-A-15) are σ-BL. Furthermore, with $v(t) \leftrightarrow V(\omega)$, it follows from the convolution theorem (3-30) that

$$R(\omega) = \frac{1}{2\pi}\int_{-\sigma}^{\sigma} V(\omega-\eta)Y(\eta)\,d\eta$$

Therefore, the transforms $Y(\omega)$ of the eigenfunctions of (6-A-15) are eigenfunctions of the integral equation

$$\frac{1}{2\pi}\int_{-\sigma}^{\sigma} V(\omega-\eta)Y(\eta)\,d\eta = \lambda Y(\omega) \qquad \text{for} \qquad |\omega| < \sigma \tag{6-A-31}$$

Positive signals We shall, finally, determine the extreme values of the ratio

$$\gamma_w = \frac{1}{W(0)}\int_{-\infty}^{\infty} v(t)w(t)\,dt \tag{6-A-32}$$

where $v(t)$ is a given function, and $w(t)$ ranges over all 2σ-BL nonnegative functions.

The Akhiezer-Krein theorem It can be shown (see Sec. 7-2) that if

$$w(t) \leftrightarrow W(\omega) \qquad w(t) \geq 0 \qquad \text{and} \qquad W(\omega) = 0 \qquad \text{for } |\omega| > 2\sigma$$

then we can find a function $y(t)$ with transform $Y(\omega)$ such that

$$w(t) = |y(t)|^2 \qquad \text{and} \qquad Y(\omega) = 0 \qquad \text{for } |\omega| > \sigma \tag{6-A-33}$$

Since
$$W(0) = \int_{-\infty}^{\infty} w(t)\,dt = \int_{-\infty}^{\infty} |y(t)|^2\,dt = E_y$$

it follows that $\gamma_w = \alpha_y$, where α_y is the ratio in (6-A-14). Hence, the extrema of γ_w equal the eigenvalues $\lambda_{\max}$ and $\lambda_{\min}$ of (6-A-15). These extrema are reached for $w(t) = \bar{w}(t) = |\bar{y}(t)|^2$ and $w(t) = \underline{w}(t) = |\underline{y}(t)|^2$, respectively.

CHAPTER

SEVEN

FACTORIZATION, WINDOWS, HILBERT TRANSFORMS

In this chapter, we introduce several topics related to the analytic properties of Laplace transforms.[1,2]

In Sec. 7-1, we derive the complex inversion formula and use it to characterize causal or time-limited functions in terms of the asymptotic properties of their transforms.

In Sec. 7-2, we discuss various forms of the factorization problem for Fourier integrals and Fourier series and the Paley-Wiener condition. This problem was considered in Sec. 4-3 for rational transforms. The discussion includes the factorization problem for trigonometric polynomials (Fejér-Riesz theorem) and for time-limited functions (Akhiezer generalization).

In Sec. 7-3, we examine the truncation problem in the numerical evaluation of Fourier integrals. We present the method of windows and various extrapolation techniques, concluding with a brief explanation of the principle of maximum entropy.

In Sec. 7-4, we develop the two forms of Hilbert transforms (real and causal) for Fourier integrals, Fourier series, and discrete Fourier series.

In Appendix 7-A, we prove the Paley-Wiener theorem. This theorem establishes the equivalence between bandlimited functions and functions of exponential type. In Appendix 7-B, we discuss Jordan's lemma.

---

[1] G. Doetsch, "Laplace Transformation," Dover Publications, Inc., New York, 1943.

[2] D. V. Widder, "The Laplace Transform," Princeton University Press, Princeton, N.J., 1946.

7-1 ANALYTIC AND ASYMPTOTIC PROPERTIES OF LAPLACE TRANSFORMS

In the next two sections, we shall denote by $F(s)$ the Laplace transform of a function $f(t)$, and by $F(j\omega)$ its Fourier transform (see also Sec. 4-3):

$$F(s) = \int_{-\infty}^{\infty} f(t)e^{-st}\, dt \tag{7-1}$$

In this section, we examine the region of convergence and the asymptotic properties of $F(s)$ for the following classes of functions:

Time-limited: $f(t) = 0$ for $|t| > \tau$

Causal: $f(t) = 0$ for $t < 0$

Anticausal: $f(t) = 0$ for $t > 0$

This permits us to establish whether a function belongs to one of these classes merely by observing the properties of its transform.

Unless otherwise stated, it will be assumed that all functions are ordinary (no impulses) and their energy E is finite (they belong to L^2). As preparation, we develop the complex inversion formula for arbitrary functions.

The Complex Inversion Formula

With $s = \gamma + j\omega$, Eq. (7-1) yields

$$F(s) = F(\gamma + j\omega) = \int_{-\infty}^{\infty} f(t)e^{-\gamma t}e^{-j\omega t}\, dt$$

This shows that $F(s)$ is the Fourier transform of the function $f(t)e^{-\gamma t}$. Applying the Fourier inversion formula (3-2) to this function, we obtain

$$e^{-\gamma t}f(t) = \frac{1}{2\pi}\int_{-\infty}^{\infty} F(\gamma + j\omega)e^{j\omega t}\, d\omega$$

For a fixed γ, the variable $s = \gamma + j\omega$ varies along a vertical line L (Fig. 7-1) as ω varies from $-\infty$ to ∞. On this line, $ds = j\, d\omega$; hence,

$$f(t) = \frac{1}{2\pi j}\int_{L} F(s)e^{st}\, ds \tag{7-2}$$

Thus, $f(t)$ is given by a *line integral* along a vertical line L (Bromwich path) in the region of existence of $F(s)$.

The inversion formula as a contour integral Using Jordan's lemma (see Appendix 7-B), we shall express the line integral (7-2) as the limit of an integral along a closed curve.

Suppose that C is a closed curve $BAEB$ consisting of a vertical line segment BA and a circular arc Γ of radius r centered at the origin (Fig. 7-1).

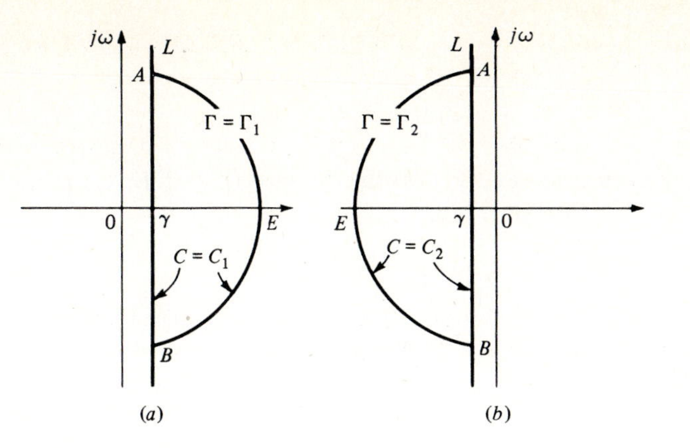

Figure 7-1

We maintain that, if $F(s) \to 0$ for every s on Γ as $r \to \infty$, then

$$f(t) = \lim_{r \to \infty} \frac{1}{2\pi j} \oint_C F(s) e^{st}\, ds \tag{7-3}$$

1. For every $t < 0$, if $\Gamma = \Gamma_1$ is on the right-hand plane, $\operatorname{Re} s \geq \gamma$ (Fig. 7-1*a*).
2. For every $t > 0$, if $\Gamma = \Gamma_2$ is on the left-hand plane, $\operatorname{Re} s \leq \gamma$ (Fig. 7-1*b*).

To prove (7-3), we write the contour integral in (7-3) as a sum of two line integrals:

$$\oint_C F(s) e^{st}\, ds = \int_{BA} F(s) e^{st}\, ds + \int_\Gamma F(s) e^{st}\, ds$$

As $r \to \infty$, the integral along BA tends to the integral along the Bromwich path L, and the integral along Γ tends to zero (Jordan's lemma). Hence, Eq. (7-3) follows from (7-2).

Note: If $F(s) \to 0$ as $s \to \infty$ in any direction, then (7-3) holds for any t, provided that $C = C_1$ if $t < 0$ and $C = C_2$ if $t > 0$.

Time-limited Functions and the Paley-Wiener Theorem

The Laplace transform of a TL function is given by the integral

$$F(s) = \int_{-\tau}^{\tau} f(t) e^{-st}\, dt \tag{7-4}$$

This integral is similar to (6-8) with t and s interchanged. Reasoning as in the proof of (6-11), we conclude that the nth derivative of $F(s)$ exists for any s and is given by

$$F^{(n)}(s) = \int_{-\tau}^{\tau} (-t)^n f(t) e^{-st}\, dt \tag{7-5}$$

Thus, $F(s)$ is an analytic function for every s (entire function), and $F^{(n)}(s)$ is the Laplace transform of $(-t)^n f(t)$. Furthermore,

$$|F(s)| \le \sqrt{2E\tau}\, e^{|s|\tau} \tag{7-6}$$

as we can see by applying Schwarz' inequality to Eq. (7-4). [See also (6-10).]

Thus, the transform of a TL function is an entire function of exponential type. It can be shown that the converse is also true: if $F(s)$ is an entire function of exponential type, then it is the transform of a TL function. The proof is given in Appendix 7-A.

Example 7-1 We shall apply the preceding results to the function

$$F(s) = \left(\frac{\sinh as}{s}\right)^3 = \frac{(e^{as} - e^{-as})^3}{(2s)^3}$$

Clearly, $F(s)$ is an entire function because the zero $s = 0$ of the denominator is canceled by the zero of the numerator. Furthermore, it satisfies (7-6) for some E and for $\tau = 3a$, as is easily seen. Hence, the function $F(j\omega) = (\sin a\omega/\omega)^3$ is the Fourier transform of a τ – TL function, where $\tau = 3a$.

Example 7-2 A more interesting application is the function

$$F(j\omega) = \frac{N(\omega) \sin \tau\omega}{\omega(\omega^2 - \pi^2/\tau^2) \cdots (\omega^2 - n^2\pi^2/\tau^2)}$$

where $N(\omega)$ is a polynomial of degree at most $2n$. As we can see, the function $F(s)$ is entire, and it satisfies (7-6) (elaborate). Hence, its inverse transform exists and it equals zero for $|t| > \tau$.

Causal Functions and Signal Front

The Laplace transform

$$F(s) = \int_{0}^{\infty} f(t) e^{-st}\, dt \tag{7-7}$$

(unilateral) of a causal function has the following properties:

If it exists for some $s = s_0$, then it exists for every s such that $\text{Re}\, s > \text{Re}\, s_0$; hence (Dedekind cut), the region $R = R_1$ of its existence is a half-plane $\text{Re}\, s > \gamma_1$. The value of γ_1 depends on the behavior of $f(t)$ as $t \to \infty$.

The function $F(s)$ is analytic in the region R_1, and its nth derivative is the transform of $(-t)^n f(t)$.

Furthermore, $F(s) \to 0$ as $s \to \infty$ along any ray in R_1, and $F(\gamma + j\omega) \to 0$ as $\omega \to \infty$ for every $\gamma > \gamma_1$ (see also Riemann's lemma).

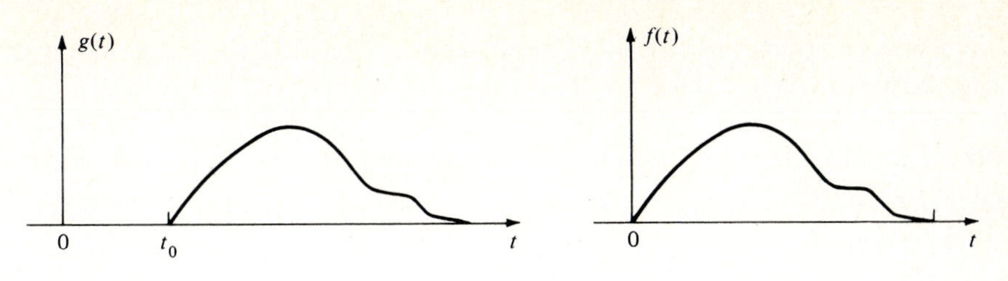

Figure 7-2

These properties are basic in the theory of Laplace transforms. Their proof is not too difficult but will be omitted.

Conversely, if the function $F(s)$ is analytic in the right-hand plane R_1 and tends to zero as $s \to \infty$ on the arc Γ_1 of Fig. 7-1*a*, then its inverse transform $f(t)$ is causal. This follows from the inversion formula (7-3) and Cauchy's theorem.[1]

Signal front The following is a useful application of the asymptotic properties of unilateral transforms:

Suppose that the transform $G(s)$ of a function $g(t)$ is analytic in R_1 and for some τ, positive or negative, $e^{\tau s}G(s) \to 0$ as $s \to \infty$ on Γ_1. We maintain that $g(t) = 0$ for $t < \tau$. To show this, we form the function $f(t) = g(t + \tau)$. Since $F(s) = e^{\tau s}G(s)$, it follows from our assumption that $F(s)$ is analytic in R_1 and tends to zero as $s \to \infty$ on Γ_1. Hence, $f(t) = 0$ for $t < 0$, from which it follows that $g(t) = 0$ for $t < \tau$.

If t_0 is the least upper bound of all numbers with this property (Fig. 7-2), then t_0 is the *front* of $g(t)$; that is, $g(t) = 0$ for $t < t_0$, and $g(t) \neq 0$ for some t arbitrarily close to t_0.

Example 7-3
(*a*) If $G(s) = (1/s)e^{-\sqrt{(a + bs)(c + ds)}}$, then $e^{\tau s}G(s) \to 0$ for every $\tau < t_0 = \sqrt{bd}$ as $s \to \infty$ on Γ_1. This is not true for any $\tau > t_0$; hence, $g(t)$ starts at $t = t_0 = \sqrt{bd}$.
(*b*) If $G(s) = (1/s)e^{-\sqrt{ads}}$, then $e^{\tau s}G(s) \to 0$ only if $\tau < 0$; hence, $g(t)$ starts at $t = 0$.

Anticausal Functions

Suppose, finally, that $f(t) = 0$ for $t > 0$. In this case,

$$F(s) = \int_{-\infty}^{0} f(t)e^{-st}\,dt \tag{7-8}$$

Changing t to $-t$ and s to $-s$, we conclude that $F(s)$ is analytic in the half-plane R_2 to the left of the line $\text{Re}\, s = \gamma_2$, where γ_2 is a constant depending on the behavior of $f(t)$ as $t \to -\infty$. Furthermore, $F(s) \to 0$ as $s \to \infty$ along any ray in R_2.

[1] If $G(s)$ is an analytic function in a region R containing the closed curve C and its interior, then $\oint_C G(s)\,ds = 0$.

Conversely, if $F(s)$ is analytic in R_2 and tends to zero as $s \to \infty$ on the arc Γ_2 of Fig. 7-1*b*, then $f(t) = 0$ for $t > 0$.

Note: The transform $F(s)$ of an arbitrary signal $f(t)$ can be written as the sum of two integrals:

$$F(s) = \int_0^{\infty} f(t)e^{-st}\,dt + \int_{-\infty}^{0} f(t)e^{-st}\,dt$$

As we have seen, the first integral exists for $\text{Re}\, s > \gamma_1$ (region R_1), and the second for $\text{Re}\, s < \gamma_2$ (region R_2). If $\gamma_1 < \gamma_2$, then the function $F(s)$ is analytic in the vertical strip $\gamma_1 < \text{Re}\, s < \gamma_2$. Furthermore, $F(\gamma + j\omega) \to 0$ as $|\omega| \to \infty$ for every γ between γ_1 and γ_2 (Riemann lemma). However, there is no certainty that $F(s) \to 0$ as $s \to \infty$ along any ray that is not parallel to the $j\omega$ axis.

z Transforms

The discrete version of the preceding results relates the analytic and asymptotic properties of the function

$$F(z) = \sum_{n=-\infty}^{\infty} f[n]z^{-n} \tag{7-9}$$

to the properties of the sequence $f[n]$. We start with a modified form of the inversion formula (2-12).

With $z = re^{j\theta}$, Eq. (7-9) yields

$$F(re^{j\theta}) = \sum_{n=-\infty}^{\infty} f[n]r^{-n}e^{-jn\theta}$$

This shows that, for each r, $F(z)$ is a periodic function of θ with period 2π, and its Fourier-series coefficients equal $f[n]r^{-n}$. Hence [see (3-69)],

$$f[n]r^{-n} = \frac{1}{2\pi}\int_{-\pi}^{\pi} F(re^{j\theta})e^{jn\theta}\,d\theta$$

For a fixed r, the variable $z = re^{j\theta}$ varies along a circle C of radius r centered at the origin as θ varies from $-\pi$ to π. Along this circle, $dz = jre^{j\theta}\,d\theta$. Hence, $f[n]$ is given by a contour integral

$$f[n] = \frac{1}{2\pi j}\int_C F(z)z^{n-1}\,dz \tag{7-10}$$

along any circle in the region of existence of $F(z)$.

If $f[n] = 0$ for $|n| > M$, then

$$F(z) = \sum_{n=-M}^{M} f[n]z^{-n} \tag{7-11}$$

Clearly, $F(z)$ is analytic for every z except $z = 0$ (open plane), and

$$|F(z)| \le \sqrt{(2M+1)E}\,|z|^{\pm M} \qquad E = \sum_{n=-M}^{M} |f[n]|^2 \tag{7-12}$$

The above follows when the Cauchy inequality is applied to the sum in (7-11).

Conversely, if $F(z)$ is analytic in the open plane, and $|F(z)| < A|z|^{\pm M}$ for any z, then $f[n] = 0$ for $|n| > M$. This follows from Eq. (7-10) when the radius r of C is made arbitrarily large or small (elaborate).

If $f[n] = 0$ for $n < 0$, then

$$F(z) = \sum_{n=0}^{\infty} f[n]z^{-n}$$

This function is analytic for $|z| > r_1$ and tends to $f[0]$ as $z \to \infty$.

Conversely, if $F(z)$ is analytic for $|z| > r_1$, and $z^m F(z) \to 0$ as $z \to \infty$, then $f[n] = 0$ for $n < m$. This follows from (7-10) when the radius r of C is made arbitrarily large.

7-2 THE FACTORIZATION PROBLEM

As we have seen, if a function $f(t)$ is causal and belongs to L^2, then its Laplace transform $F(s)$ is analytic for $\text{Re}\, s > 0$. We shall say that a function $F(s)$ is *minimum phase* if it is analytic and its inverse $1/F(s)$ is also analytic for $\text{Re}\, s > 0$. (See also Sec. 4-3.)

The problem The factorization problem can be phrased in the following three equivalent forms:

1. Given a real, nonnegative function $A(\omega)$ belonging to L^2, find a causal function $f(t)$ such that $|F(j\omega)| = A(\omega)$.
2. With $A(\omega)$ real, nonnegative, and belonging to L^2, find a function $F(s)$ analytic for $\text{Re}\, s > 0$ and such that $|F(j\omega)| = A(\omega)$.
3. Given a function $g(t)$ whose transform $G(s)$ is nonnegative on the $j\omega$ axis, find a causal function $f(t)$ such that

$$g(t) = f(t) * f^*(-t) \tag{7-13}$$

The term *factorization* has its origin in the third form of the problem, because Eq. (7-13) is equivalent to

$$G(j\omega) = |F(j\omega)|^2 \qquad G(s) = F(s)F_-(s) \tag{7-14}$$

where by $F_-(s)$ we mean the transform of $f^*(-t)$. It is easy to see that $F_-(s) = F^*(-s^*)$ and, if $f(t)$ is real, then $F_-(s) = F(-s)$.

We shall show that the problem has a solution if the function $A(\omega)$ satisfies the *Paley-Wiener* condition[1]

$$\int_{-\infty}^{\infty} \frac{|\ln A(\omega)|}{1+\omega^2}\, d\omega < \infty \tag{7-15}$$

Notes:

1. Extending the terminology of Sec. 4-3, we shall say that a function $F_0(s)$ is *all-pass* if it is analytic for Re $s > 0$ and $|F_0(j\omega)| = 1$. Clearly, if $F(s)$ is a solution of the factorization problem, and $F_0(s)$ is an arbitrary all-pass function, then the product $F(s)F_0(s)$ is also a solution. However, as we shall see, the problem has a unique minimum-phase solution.
2. It can be shown (Prob. 74) that the Paley-Wiener condition is also necessary; i.e., all functions $F(s)$ analytic for Re $s > 0$ and with finite energy on the $j\omega$ axis must satisfy (7-15) for $A(\omega) = |F(j\omega)|$.
3. Condition (7-15) shows that the spectrum of causal systems cannot be identically zero in any interval, no matter how small.

Discrete form of the problem The discrete form of the problem involves discrete sequences y_n and their z transforms $Y(z)$. As we know, if y_n is a causal sequence with finite energy, then $Y(z)$ is analytic for $|z| > 1$. We shall say that $Y(z)$ is *minimum phase* if its inverse $1/Y(z)$ is also analytic for $|z| > 1$ (see also Sec. 2-3).

As in the continuous case, the problem can be phrased in three equivalent forms:

1. Given a real, nonnegative function $a(\theta)$, find a causal sequence y_n such that the amplitude of its z transform $Y(z)$ on the unit circle equals $a(\theta)$:

$$|Y(e^{j\theta})| = \left|\sum_{n=0}^{\infty} y_n e^{-jn\theta}\right| = a(\theta) \tag{7-16}$$

2. Given $a(\theta)$ real and nonnegative, find a function $Y(z)$ analytic for $|z| > 1$ and such that $|Y(e^{j\theta})| = a(\theta)$.
3. Given a sequence w_n whose z transform $W(z)$ is nonnegative on the unit circle,

$$W(e^{j\theta}) = \sum_{n=-\infty}^{\infty} w_n e^{-jn\theta} = a^2(\theta) \geq 0 \tag{7-17}$$

find a causal sequence y_n with z transform $Y(z)$ such that

$$W(e^{j\theta}) = |Y(e^{j\theta})|^2 \qquad W(z) = Y(z)Y_-(z) \tag{7-18}$$

where by $Y_-(z)$ we mean the z transform of w^*_{-n}. It is easy to see that $Y_-(z) = Y^*(1/z^*)$ and, if y_n is real, then $Y_-(z) = Y(1/z)$.

[1] R. E. A. C. Paley and N. Wiener, Fourier Transforms in the Complex Domain, *Amer. Math. Soc. Coll.*, no. 19, 1934.

We shall show that the problem has a solution if $a(\theta)$ satisfies the discrete form of the Paley-Wiener condition,

$$\int_{-\pi}^{\pi} |\ln a(\theta)|\, d\theta < \infty \tag{7-19}$$

and that the solution is unique if $Y(z)$ is minimum phase.

Discrete-continuous equivalence The bilinear transformation

$$s = \frac{z-1}{z+1} \qquad j\omega = \frac{e^{j\theta}-1}{e^{j\theta}+1} = j\tan\frac{\theta}{2} \qquad d\theta = \frac{2\, d\omega}{1+\omega^2} \tag{7-20}$$

maps the region $|z| \geq 1$ of the z plane on the region $\text{Re}\, s \geq 0$ of the s plane.

Suppose that $W(z)$ is a function such that $W(e^{j\theta}) \geq 0$. With

$$G(s) = W\left(\frac{1+s}{1-s}\right) \tag{7-21}$$

Eqs. (7-20) yield $G(j\omega) = W(e^{j\theta}) \geq 0$. Hence,

$$\int_{-\pi}^{\pi} |\ln W(e^{j\theta})|\, d\theta = 2\int_{-\infty}^{\infty} \frac{|\ln G(j\omega)|}{1+\omega^2}\, d\omega \tag{7-22}$$

If, therefore, $W(z)$ satisfies the discrete factorization condition (7-19), then $G(s)$ satisfies the continuous factorization condition (7-15).

Suppose, finally, that $G(s) = F(s)F_{-}(s)$ as in (7-14). With

$$Y(z) = F\left(\frac{z-1}{z+1}\right) \tag{7-23}$$

it follows that $W(z) = G(s) = F(s)F_{-}(s) = Y(z)Y_{-}(z)$ as in (7-18).

We have thus shown that the discrete problem can be reduced to the continuous problem.

Note: In the factorization problem we determine the functions $F(j\omega)$ and $W(e^{j\theta})$ in terms of their amplitudes. The corresponding *real-part* problem is also of interest:

Given a real function $R(\omega)$ belonging to L^1, find a function $F(s)$ analytic for $\text{Re}\, s > 0$ and such that $\text{Re}\, F(j\omega) = R(\omega)$; given a function $r(\theta)$, find a function $Y(z)$ analytic for $|z| > 1$ and such that $\text{Re}\, Y(e^{j\theta}) = r(\theta)$.

This problem always has a unique solution. The continuous case is treated in Prob. 75; the discrete case will be solved presently (see also Hilbert transforms, Sec. 7-4).

The solution We are given a nonnegative function $A(\omega)$ satisfying the Paley-Wiener condition (7-15). We wish to find a function $F(s)$ analytic for $\text{Re}\, s > 0$ and such that

$$|F(j\omega)| = A(\omega) \tag{7-24}$$

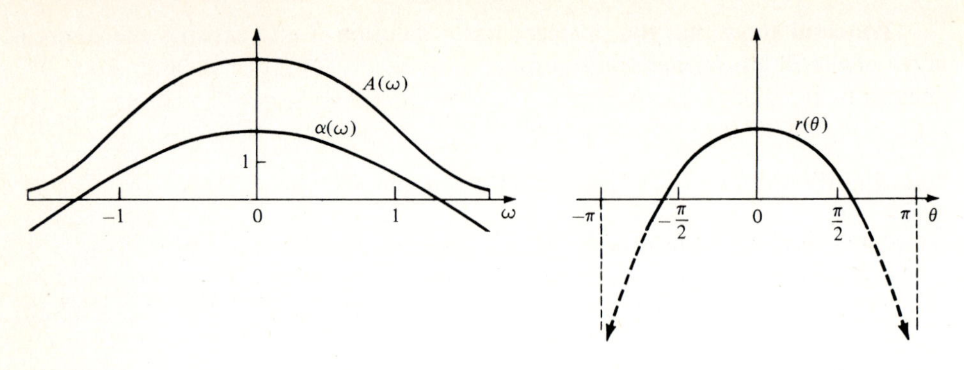

Figure 7-3

To solve this problem, we shall first reduce it to a discrete real-part problem. For this purpose, we form the functions (Fig. 7-3)

$$\alpha(\omega) = \ln A(\omega) \qquad r(\theta) = \alpha\left(\tan\frac{\theta}{2}\right) \tag{7-25}$$

Clearly, $r(\theta)$ is a real periodic function belonging to L^1, because [see Eq. (7-15)]

$$\int_{-\pi}^{\pi} |r(\theta)|\, d\theta = 2\int_{-\infty}^{\infty} \frac{|\alpha(\omega)|}{1+\omega^2}\, d\omega < \infty \tag{7-26}$$

Therefore, it can be expanded into a convergent Fourier series

$$r(\theta) = \sum_{n=-\infty}^{\infty} r_n e^{-jn\theta} \qquad r_n = \frac{1}{2\pi}\int_{-\pi}^{\pi} r(\theta)e^{jn\theta}\, d\theta \tag{7-27}$$

With

$$Y(z) = r_0 + 2\sum_{n=1}^{\infty} r_n z^{-n} \tag{7-28}$$

it follows that $Y(e^{j\theta})$ exists and

$$\text{Re } Y(e^{j\theta}) = r(\theta) \tag{7-29}$$

because $r_{-n} = r_n^*$ (elaborate). Furthermore, $Y(z)$ is analytic for $|z| > 1$ and can be expressed directly in terms of $r(\theta)$:

$$Y(z) = -r_0 + \frac{1}{\pi}\int_{-\pi}^{\pi} r(\theta)\frac{z}{z - e^{j\theta}}\, d\theta$$

This follows from inserting r_n, as given in (7-27), into (7-28) (see also Prob. 18).

From the analyticity of $Y(z)$, we conclude that the functions

$$c(s) = Y\left(\frac{1+s}{1-s}\right) \qquad F(s) = e^{c(s)} \tag{7-30}$$

are analytic for Re $s > 0$. Furthermore, $F(s)$ is minimum phase because its inverse $e^{-c(s)}$ is also analytic for Re $s > 0$.

Thus, starting from the given $A(\omega)$, we formed in succession the functions $\alpha(\omega)$ and $r(\theta)$, the sequence r_n, and the functions $Y(z)$, $c(s)$, and $F(s)$. We maintain that the function $F(s)$ solves our problem. Indeed, it is analytic for $\text{Re}\, s > 0$, and $|F(j\omega)| = A(\omega)$ because

$$c(j\omega) = Y(e^{j\theta}) \qquad \text{Re}\, c(j\omega) = r(\theta) = \alpha(\omega) \qquad |F(j\omega)| = e^{\text{Re}\, c(j\omega)} = e^{\alpha(\omega)}$$

Note: The following is an obvious consequence of the factorization problem: If $H(s)$ is the Laplace transform of a causal function, then we can associate with it a unique minimum-phase function $F(s)$ such that $|F(j\omega)| = |H(j\omega)|$.

The Factorization Problem for Time-limited Functions and Trigonometric Polynomials[1]

We have seen that, if $g(t) \leftrightarrow G(j\omega)$ and $G(j\omega) \geq 0$, then we can find a causal function $f(t)$ such that $g(t) = f(t) * f^*(-t)$. We shall next show that, if $g(t) = 0$ for $|t| > \tau$, the function $f(t)$ equals zero not only for $t < 0$ but also for $t > \tau$. We discuss first the discrete version of the problem.

The Fejér-Riesz theorem If

$$W(z) = \sum_{n=-m}^{m} w_n z^{-n} \qquad \text{and} \qquad W(e^{j\omega T}) \geq 0 \tag{7-31}$$

then we can find a function

$$Y(z) = \sum_{n=0}^{m} y_n z^{-n} \tag{7-32}$$

such that

$$W(e^{j\omega T}) = |Y(e^{j\omega T})|^2 \tag{7-33}$$

The function $Y(z)$ is unique if we impose the condition that all its roots be in the circle $|z| \leq 1$ (minimum phase).

Proof From the assumption that $W(e^{j\omega T})$ is real, we conclude that $w_{-n} = w_n^*$; hence,

$$W^*(z) = W\left(\frac{1}{z^*}\right) \tag{7-34}$$

as is easy to show (elaborate). Suppose that z_i is a root of $W(z)$. It then follows that

$$W(z_i) = 0 = W^*(z_i) = W\left(\frac{1}{z_i^*}\right) \tag{7-35}$$

Therefore, $1/z_i^*$ is also a root. If $z_i = e^{j\theta_i}$ is a root on the unit circle, then it must have even multiplicity because $W(e^{j\theta}) \geq 0$.

[1] N. I. Akhiezer, "Theory of Approximations," Frederick Ungar Publishing Co., New York, 1956.

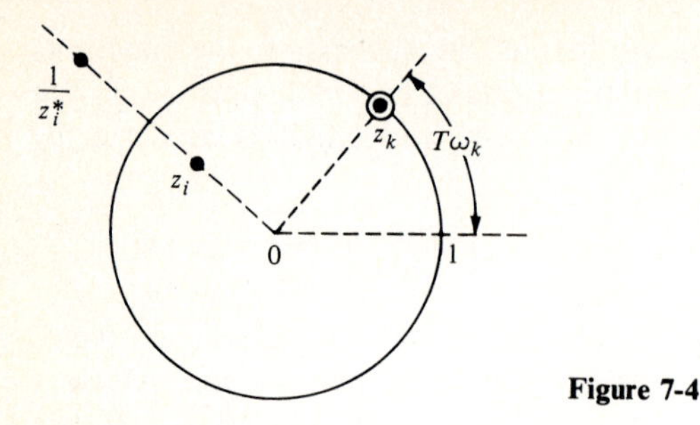

Figure 7-4

We denote by z_i all roots of $W(z)$ that are inside the unit circle or on its boundary. There are obviously m such roots if we count the roots on the boundary with half their multiplicity. The remaining m roots $1/z_i^*$ are outside the unit circle or on its boundary (Fig. 7-4). Factoring $W(z)$, we thus obtain

$$W(z) = Az^{-m} \prod_{i=1}^{m} (z - z_i) \prod_{i=1}^{m} (zz_i^* - 1) \tag{7-36}$$

where A is some constant.

We finally form the function

$$Y(z) = \sqrt{A}\, z^{-m} \prod_{i=1}^{m} (z - z_i) \tag{7-37}$$

This function is of the form (7-32), and it satisfies (7-33) because

$$|e^{j\omega T} - z_i| = |e^{j\omega T} z_i^* - 1|$$

The proof of the theorem is thus complete.

Note: To determine $Y(z)$ from Eq. (7-37), we must factor the polynomial $W(z)$. The following approach avoids factorization:

As we know (page 80), the values of $Y(z)$ at the points $z_k = e^{jT\omega_k}$, $T\omega_k = 2\pi k/(m + 1)$, and its coefficients y_n form a DFS pair. With

$$\alpha(\omega) + j\varphi(\omega) = \ln Y(e^{jT\omega})$$

the function $\alpha(\omega)$ is known because it equals $\frac{1}{2} \ln W(e^{jT\omega})$. To determine $Y(z)$, it suffices, therefore, to compute $\varphi(\omega_k)$. If $W(z)$ has no roots on the unit circle, then $Y(z)$ has no roots in the region $|z| \geq 1$. Hence, the function $\ln Y(z)$ is analytic in this region and can be expanded into a convergent power series

$$\ln Y(z) = \sum_{n=0}^{\infty} a_n z^{-n}$$

From the above, it follows that if the coefficients a_n are real, then

$$\alpha(\omega) = \sum_{n=0}^{\infty} a_n \cos nT\omega \qquad \varphi(\omega) = -\sum_{n=1}^{\infty} a_n \sin nT\omega$$

That is, $\varphi(\omega)$ is the Hilbert transform of $\alpha(\omega)$ [see Eq. (7-135)].

Thus, to determine the coefficients y_n of $Y(z)$, we first compute $\varphi(\omega)$ in terms of the known $\alpha(\omega)$, using two DFS of order N. The DFS of order $m+1$ of the resulting numbers $\alpha(\omega_k) + j\varphi(\omega_k)$ yields y_n. The order N must be sufficiently high to reduce aliasing errors.

The Akhiezer-Krein theorem[1] Given a time-limited function $g(t)$ with nonnegative Fourier transform

$$G(\omega) = \int_{-\tau}^{\tau} g(t)e^{-j\omega t}\,dt \qquad G(\omega) \geq 0 \tag{7-38}$$

(we return to the earlier notation for Fourier transforms), we can find a function

$$F(\omega) = \int_{0}^{\tau} f(t)e^{-j\omega t}\,dt \tag{7-39}$$

such that

$$G(\omega) = |F(\omega)|^2 \tag{7-40}$$

Proof We inscribe into the given function $g(t)$ a polygon $g_m(t)$, as in Fig. 7-5. Thus, $g_m(t)$ consists of $2m$ straight line segments and equals $g(t)$ for $t = nT$, where $T = \tau/m$. As we have shown in Example 5-1, the transform $G_m(\omega)$ of $g_m(t)$ is given by

$$G_m(\omega) = \frac{4\sin^2(\omega T/2)}{T^2\omega^2} \sum_{n=-\infty}^{\infty} G(\omega + 2n\sigma) \tag{7-41}$$

where $\sigma = \pi/T$.

From the Poisson sum formula (3-87), it follows that

$$\sum_{n=-\infty}^{\infty} G(\omega + 2n\sigma) = T \sum_{n=-m}^{m} g(nT)e^{-jnT\omega} \tag{7-42}$$

The summation is from $-m$ to m because $g(nT) = 0$ for $|n| > m$. The above sums are nonnegative because $G(\omega) \geq 0$ by assumption. Hence [see (7-33)], we can find $m+1$ numbers y_n such that

$$T \sum_{n=-m}^{m} g(nT)e^{-jnT\omega} = \left| \sum_{n=0}^{m} y_n e^{-jnT\omega} \right|^2$$

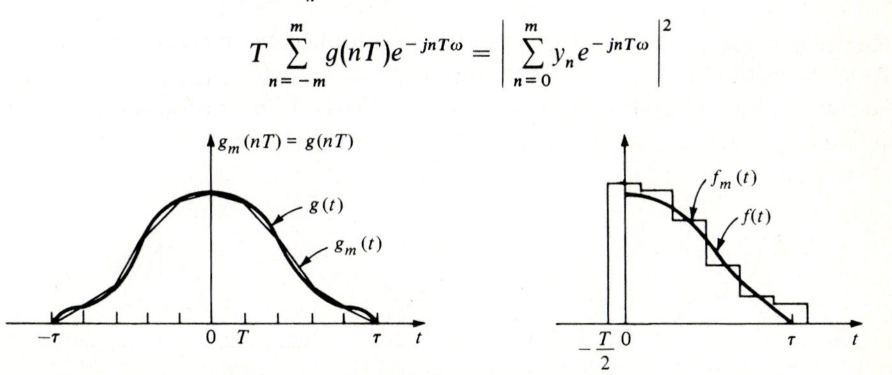

Figure 7-5

[1] A. Papoulis, The factorization problem for time-limited functions and trigonometric polynomials, *IEEE Trans. on Circuits and Systems*, Vol. 25, No. 1, Jan., 1978.

Inserting into (7-42) and (7-41), we obtain

$$G_m(\omega) = |F_m(\omega)|^2 \tag{7-43}$$

where

$$F_m(\omega) = \frac{2 \sin (\omega T/2)}{T\omega} \sum_{n=0}^{m} y_n e^{-jnT\omega} \tag{7-44}$$

Clearly, the inverse transform $f_m(t)$ of $F_m(\omega)$ is a staircase function as in Fig. 7-5, vanishing outside the interval $(-T/2, \tau + T/2)$ and such that $f_m(nT) = y_n/T$.

We can thus satisfy the requirements of the theorem arbitrarily closely by selecting a sufficiently large integer m, because then $T \ll \tau$ and $g_m(t) \simeq g(t)$.

We shall, finally, consider the behavior of the functions $g_m(t)$ and $f_m(t)$ as $m \to \infty$, that is, as $T \to 0$. The function $g_m(t)$ is continuous by construction. The function $g(t)$ is continuous because its transform $G(\omega)$ is nonnegative (Prob. 76). Furthermore, $g_m(nT) = g(nT)$; hence, $g_m(t) \to g(t)$ as $T \to 0$.

We know from the earlier discussion [see Eq. (7-14)] that a minimum-phase function $F(\omega)$ can be found such that $G(\omega) = |F(\omega)|^2$. It is not difficult to see, therefore, that $F_m(\omega) \to F(\omega)$ as $m \to \infty$. And since the interval $(-T/2, \tau + T/2)$ tends to the interval $(0, \tau)$, we conclude that the inverse $f(t)$ of $F(\omega)$ is zero not only for $t < 0$ but also for $t > \tau$, because it is the limit of the functions $f_m(t)$.

7-3 WINDOWS AND EXTRAPOLATION

A central problem in Fourier analysis is the determination of the Fourier transform $F(\omega)$ of a signal $f(t)$ in terms of the segment

$$f_\tau(t) = f(t)p_\tau(t) \tag{7-45}$$

of $f(t)$ (Fig. 7-6). Since $f(t)$ is not known for $|t| > \tau$, its transform $F(\omega)$ cannot be determined exactly; it can only be estimated. If the transform $F_\tau(\omega)$ of $f_\tau(t)$ is used to estimate $F(\omega)$, then the error $F_\tau(\omega) - F(\omega)$ results. In this section, we discuss two methods for reducing this error.

Windows The known segment $f_\tau(t)$ of $f(t)$ is multiplied by a τ-TL function $w(t)$, and the transform $F_w(\omega)$ of the resulting product $f_\tau(t)w(t) = f_w(t)$ is used as the estimate of $F(\omega)$. The signal $w(t)$ and its transform $W(\omega)$ are chosen so as to

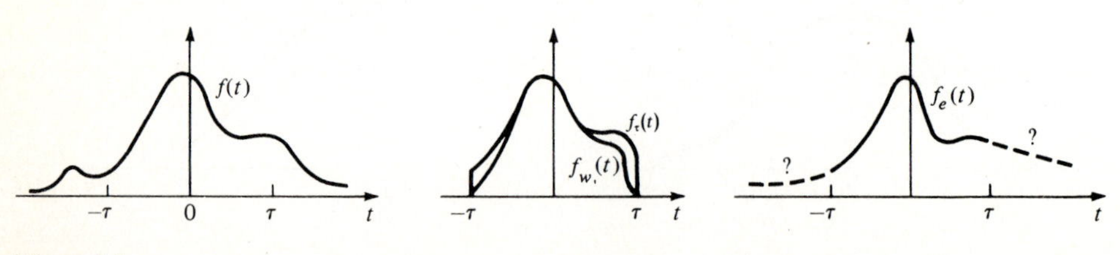

Figure 7-6

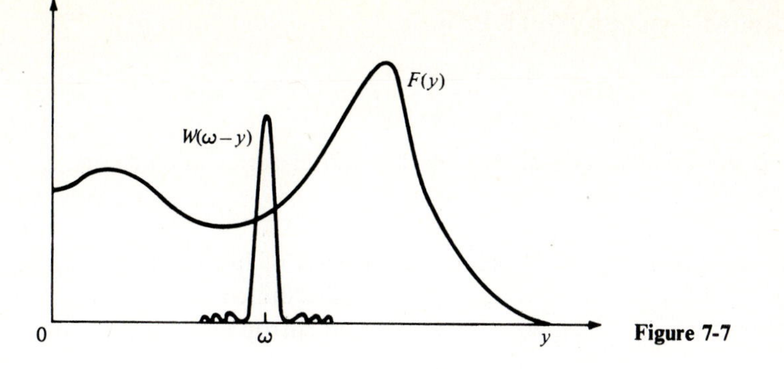

Figure 7-7

reduce in some sense the error $F_w(\omega) - F(\omega)$. Since (Fig. 7-7)

$$F_w(\omega) = \int_{-\tau}^{\tau} w(t)f(t)e^{-j\omega t}\,dt = \frac{1}{2\pi}\int_{-\infty}^{\infty} F(y)W(\omega - y)\,dy \tag{7-46}$$

the function $W(\omega)$ must be concentrated near the origin.

This method is computationally simple, involving only FFT's. However, it replaces the unknown part of $f(t)$ with zero.

Extrapolation The unknown $F(\omega)$ is estimated by the Fourier transform $F_e(\omega)$ of a function $f_e(t)$ that equals $f(t)$ for $|t| < \tau$ (Fig. 7-6) and that extrapolates the given segment of $f(t)$ for $|t| > \tau$. The extrapolation scheme is based on various assumptions about $F(\omega)$. If, for example, it is known that $f(t)$ is in the space of a family of functions $\{y_n(t)\}$, then the sum

$$g_e(t) = \sum_{n=1}^{N} a_n y_n(t) \tag{7-47}$$

can be used to extrapolate $f_\tau(t)$. The constants a_n are determined so that $g_e(t) \simeq f(t)$ for $|t| < \tau$.

This method is based on a priori assumptions and is, in general, computationally complex.

> **Note:** The truncation of $f(t)$ might be due to measurement restrictions, to computational economy, or to other constraints. However, if $f(t)$ contains noise, then it is intentionally truncated to reduce the effect of the noise (Blackman-Tukey). This case arises in spectral estimation.

Fourier series The same considerations hold in the determination of a periodic function

$$F(\omega) = \sum_{n=-\infty}^{\infty} f[n]e^{-jnT\omega} \tag{7-48}$$

if the coefficients $f[n]$ are known for $|n| \leq M$ only.

Windows The unknown function $F(\omega)$ is estimated by the sum [see Eq. (3-75)]

$$F_x(\omega) = \sum_{n=-M}^{M} x_n f[n] e^{-jnT\omega} = \frac{1}{2\sigma}\int_{-\sigma}^{\sigma} F(\omega - y)X(y)\,dy \qquad (7\text{-}49)$$

where $X(\omega)$ is a trigonometric polynomial with Fourier-series coefficients x_n, and $\sigma = \pi/T$.

Extrapolation It is assumed that $F(\omega)$ belongs to a class of functions involving $2M + 1$ parameters. The unknown parameters are determined in terms of the $2M + 1$ given coefficients $f[n]$. The method of maximum entropy is an illustration.

Windows

We shall say that the functions

$$w(t) \leftrightarrow W(\omega)$$

form a *window pair* if they are real, even, normalized:

$$w(0) = \frac{1}{2\pi}\int_{-\infty}^{\infty} W(\omega)\,d\omega = 1 \qquad (7\text{-}50)$$

time-limited:

$$w(0) = 0 \qquad \text{for } |t| > \tau \qquad (7\text{-}51)$$

and such that $W(\omega)$ is of "short duration." The last requirement is imprecise and subject to various interpretations depending on the applications.

In the selection of window pairs, the following quantities are of interest:

Energy

$$E = \frac{1}{2\pi}\int_{-\infty}^{\infty} W^2(\omega)\,d\omega = \int_{-\tau}^{\tau} w^2(t)\,dt \qquad (7\text{-}52)$$

Energy moment

$$M_2 = \frac{1}{2\pi}\int_{-\infty}^{\infty} \omega^2 W^2(\omega)\,d\omega = \int_{-T}^{T} [w'(t)]^2\,dt \qquad (7\text{-}53)$$

In the above, the last equality follows when Parseval's formula is applied to the pair $w'(t) \leftrightarrow j\omega W(\omega)$.

Amplitude moment

$$m_2 = \frac{1}{2\pi}\int_{-\infty}^{\infty} \omega^2 W(\omega)\,d\omega = -w''(0) \qquad (7\text{-}54)$$

The last equality follows because $-w''(t) \leftrightarrow \omega^2 W(\omega)$. We note that the moment m_2 gives a meaningful measure of the concentration of $W(\omega)$ near the origin only if

$$W(\omega) \geq 0 \qquad (7\text{-}55)$$

Asymptotic attenuation. The integer n such that

$$W(\omega) = O\left(\frac{1}{\omega^n}\right) \qquad \text{as } |\omega| \to \infty \tag{7-56}$$

As we have seen in Sec. 6-1, n is also the largest integer such that

$$w^{(k)}(\pm\tau) = 0 \qquad k \le n-2$$

Thus, to obtain a window $W(\omega)$ with high rate of attenuation for large ω, we must use a time signal $w(t)$ with high order of tangency at the endpoints of the interval $(-\tau, \tau)$. For example, if $w(t) = (1 - t^2/\tau^2)^n p_\tau(t)$, then

$$W(\omega) = O\left(\frac{1}{\omega^{n+1}}\right) \qquad \text{as } \omega \to \infty$$

We give next several examples of windows that are used extensively (Fig. 7-8). These windows have good overall properties but are not optimum in any specific sense.

Bartlett

$$q_\tau(t) \leftrightarrow \frac{4\sin^2(\tau\omega/2)}{\tau\omega^2} \tag{7-57}$$

$$E = \frac{2}{3}\tau \qquad M_2 = \frac{2}{\tau} \qquad m_2 = \infty \qquad n = 2 \qquad W(\omega) \underset{\omega\to\infty}{\simeq} \frac{4}{\tau\omega^2}$$

The time function $w(t)$ is a triangle, and the frequency function $W(\omega)$ equals the Fejér kernel.

Tukey

$$\frac{1}{2}\left(1 + \cos\frac{\pi}{\tau}t\right)p_\tau(t) \leftrightarrow \frac{\pi^2 \sin\tau\omega}{\omega(\pi^2 - \tau^2\omega^2)} \tag{7-58}$$

$$E = \frac{3}{4}\tau \qquad M_2 = \frac{\pi^2}{4\tau} \qquad m_2 = \frac{\pi^2}{2\tau^2} \qquad n = 3 \qquad W(\omega) \underset{\omega\to\infty}{\simeq} \frac{\pi^2}{\tau^2\omega^3}$$

The time function is a raised cosine, truncated for $|t| > \tau$ [see Eq. (3-24)].

Hamming

$$\left(0.54 + 0.46\cos\frac{\pi}{\tau}t\right)p_\tau(t) \leftrightarrow \frac{(1.08\pi^2 - 0.16\tau^2\omega^2)\sin\tau\omega}{\omega(\pi^2 - \tau^2\omega^2)} \tag{7-59}$$

$$E = 0.795\tau \qquad M_2 = \infty \qquad n = 1 \qquad W(\omega) \underset{\omega\to\infty}{\simeq} \frac{0.16}{\omega}$$

This window is obtained by adding a constant to a raised cosine and normalizing.

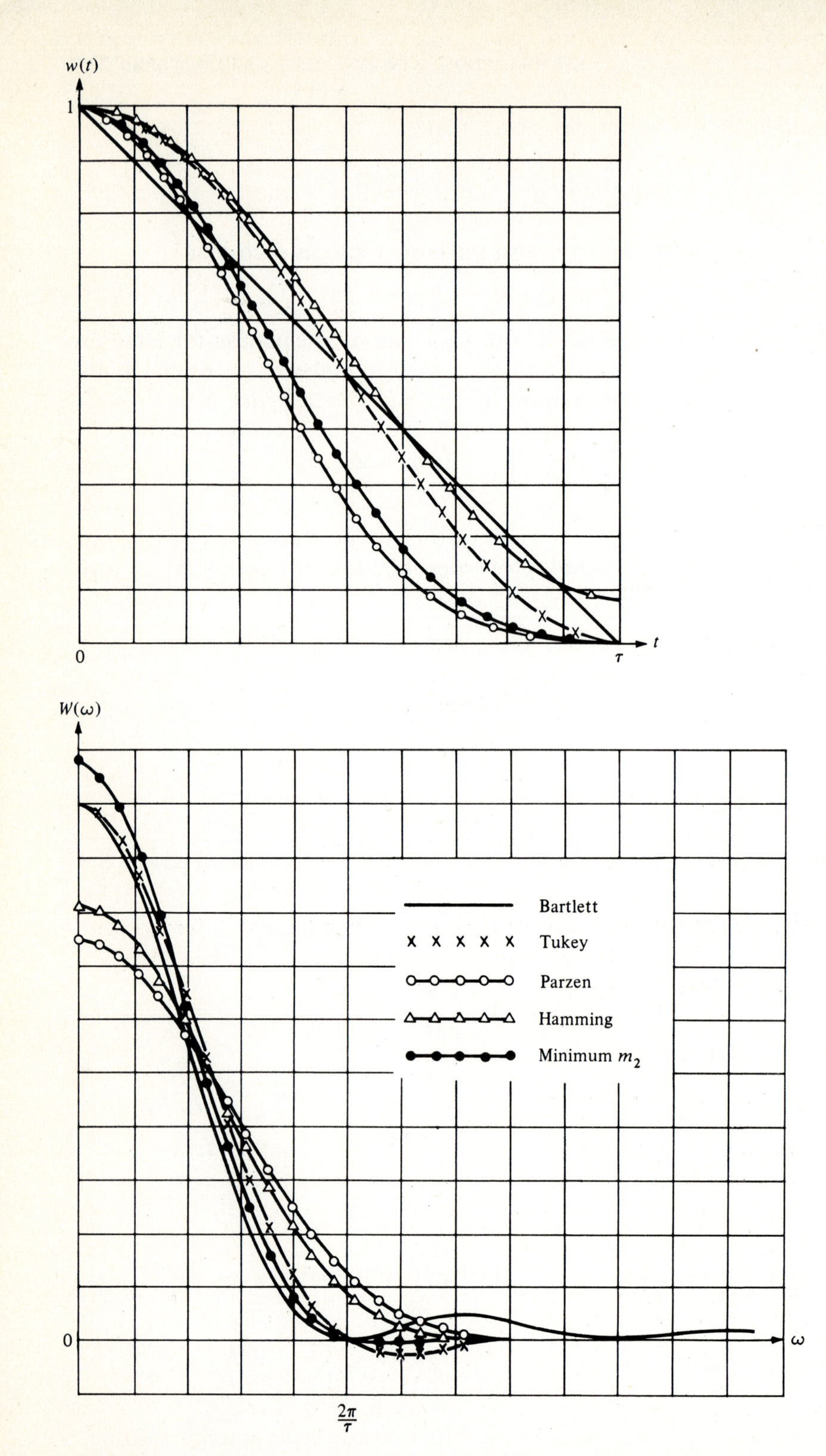

$w(t)$
1
0
τ
t
$W(\omega)$
0
ω
$\frac{2\pi}{\tau}$
Bartlett
Tukey
Parzen
Hamming
Minimum m_2

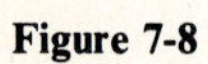

Figure 7-8

Parzen

$$\frac{3}{\tau} q_{\tau/2}(t) * q_{\tau/2}(t) \leftrightarrow \frac{3\tau}{4}\left(\frac{\sin \tau\omega/4}{\tau\omega/4}\right)^4 \tag{7-60}$$

$$E = 0.539\tau \qquad M_2 = \frac{3}{\tau} \qquad m_2 = \frac{12}{\tau^2} \qquad n = 4 \qquad W(\omega) \underset{\omega \to \infty}{\simeq} \frac{192}{\tau^3\omega^4}$$

The frequency function $W(\omega)$ equals the square of the Bartlett window normalized.

Optimum windows The following windows have extremal properties that are of interest in various applications:

Maximum energy concentration

$$k\varphi_0(t)p_\tau(t) \leftrightarrow k\sqrt{\frac{2\pi\tau\lambda_0}{\sigma}}\,\varphi_0\left(\frac{\tau}{\sigma}\,\omega\right) \tag{7-61}$$

In the above, $\varphi_0(t)$ is the solution of the integral equation (6-80) (prolate spheroidal function) corresponding to the maximum eigenvalue λ_0. As we have shown in Sec. 6-4, its transform has the largest concentration of energy in a given interval $(-\sigma, \sigma)$.

Specified zero crossings

$$w(t) = k(1 + 2\alpha \cos at)p_\tau(t) \qquad \alpha = \frac{-2 \sin a\tau}{2a\tau + \sin 2a\tau} \tag{7-62}$$

It can be show (see Prob. 70) that the ratio $W(0)/\sqrt{E}$ is maximum subject to the constraint $W(\pm a) = 0$. The Hamming window is a special case. This window is of interest in optics (apodization).

Minimum energy moment

$$\left(\cos \frac{\pi t}{2\tau}\right)p_\tau(t) \leftrightarrow 4\pi\tau \frac{\cos \tau\omega}{\pi^2 - 4\tau^2\omega^2} \tag{7-63}$$

This pair is such that the ratio M_2/E is minimum and equals $\pi^2/4\tau^2$ [see (6-70)].

Minimum amplitude moment

$$\left[\frac{1}{\pi}\left|\sin \frac{\pi}{\tau} t\right| + \left(1 - \frac{|t|}{\tau}\right) \cos \frac{\pi}{\tau} t\right]p_\tau(t) \leftrightarrow 4\tau\pi^2 \frac{1 + \cos \tau\omega}{(\pi^2 - \tau^2\omega^2)^2} \tag{7-64}$$

$$E \simeq 0.587\tau \qquad M_2 \simeq \frac{2.79}{\tau} \qquad m_2 = \frac{\pi^2}{\tau^2} \qquad n = 4 \qquad W(\omega) \underset{\omega \to \infty}{\simeq} \frac{8\pi^2}{\tau^3\omega^4}$$

As we have seen [see (6-71)], the second moment m_2 of $W(\omega)$ is minimum subject to the constraint $W(\omega) \geq 0$.

We show next that the above window minimizes the truncation error $F_w(\omega) - F(\omega)$ in high-resolution computations.

The moment expansion (4-9), applied to the convolution integral in Eq. (7-46), yields the following approximation for the truncation error [see (7-54)]:

$$F_w(\omega) - F(\omega) = \frac{1}{2\pi}\int_{-\infty}^{\infty} F(\omega - y)W(y)\,dy - F(\omega) \simeq \frac{F''(\omega)}{2} m_2 \qquad (7\text{-}65)$$

because the function $W(\omega)$ is even and of area 2π. It might appear from the above that the error is minimum if the second moment m_2 of $W(\omega)$ is minimum. However, this is not the case. Equation (7-65) is only valid as an approximation if $F(\omega)$ is sufficiently smooth and $m_2 \neq 0$. If $m_2 = 0$, then the next term of the expansion of $F_w(\omega) - F(\omega)$ becomes significant.

The minimization of m_2 leads to a minimum error only if we use positive windows. Indeed, if $W(\omega) \geq 0$, then [see (4-13)]

$$F_w(\omega) - F(\omega) = \frac{F''(\omega - y_0)}{4\pi}\int_{-\infty}^{\infty} y^2 W(y)\,dy \qquad (7\text{-}66)$$

where y_0 is some constant of the order of $1/\tau$ because the duration of $W(\omega)$ is of the order of $1/\tau$ (see Sec. 8-2). Therefore, for sufficiently large τ, we can use the approximation $F''(\omega - y_0) \simeq F''(\omega)$. Under these assumptions, the truncation error is minimum if m_2 is minimum, i.e., if $W(\omega)$ is the window in (7-64). In this case,

$$F_w(\omega) - F(\omega) \simeq \frac{\pi^2}{2\tau^2} F''(\omega) \qquad (7\text{-}67)$$

A general class of windows The preceding windows are special cases of the following class of functions:

$$W(\omega) = \frac{N(\omega)P(\omega)}{(\omega^2 - \omega_1{}^2)(\omega^2 - \omega_2{}^2)\cdots(\omega^2 - \omega_n{}^2)} \qquad |\omega_i| < \sigma \qquad (7\text{-}68)$$

$N(\omega)$ is a polynomial of degree less than $2n$. $P(\omega)$ is a trigonometric polynomial with period 2σ, vanishing at $\omega = \pm\omega_i$:

$$P(\pm\omega_i) = 0 \qquad i = 1, \ldots, n \qquad (7\text{-}69)$$

$N(\omega)$ and $P(\omega)$ are either both even

$$N(-\omega) = N(\omega) \qquad P(\omega) = \sum_{k=0}^{M} b_k \cos ka\omega \qquad a = \frac{\pi}{\sigma} \qquad (7\text{-}70)$$

or both odd

$$N(-\omega) = -N(\omega) \qquad P(\omega) = \sum_{k=1}^{M} c_k \sin ka\omega \qquad (7\text{-}71)$$

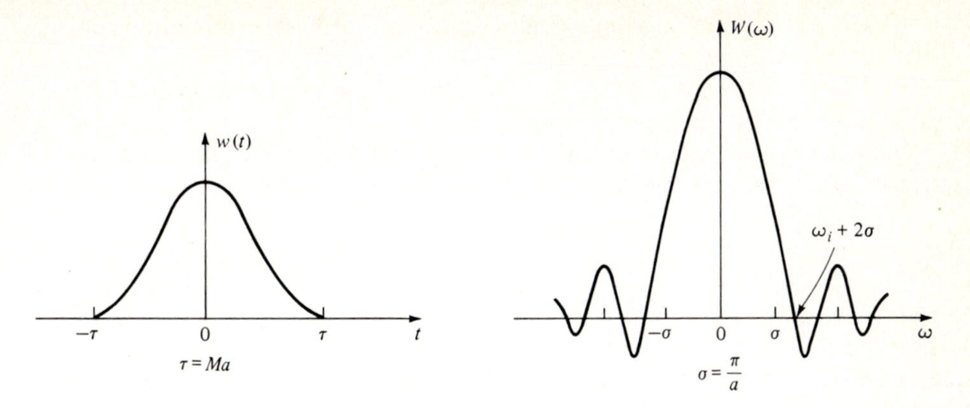

Figure 7-9

Clearly, $W(\omega)$ has no poles because the zeros $\pm\omega_i$ of its denominator are canceled by the zeros of the numerator $P(\omega)$ in the main lobe $(-\sigma, \sigma)$. Furthermore, we can find a constant A such that

$$|W(\omega)| < Ae^{Ma|\omega|} \qquad \text{for all } \omega \tag{7-72}$$

This follows readily from Eq. (7-68) (elaborate). Thus, $W(\omega)$ is an entire function of exponential type. Hence [see (7-A-6)],

$$w(t) = 0 \qquad \text{for } |t| > \tau = Ma \tag{7-73}$$

We note also that $W(\omega)$ has infinitely many zeros outside the interval $(-\sigma, \sigma)$ (see Fig. 7-9):

$$W(\omega_i + 2r\sigma) = 0 \qquad \text{for } r \neq 0 \tag{7-74}$$

This follows from Eq. (7-69) and the periodicity of $P(\omega)$.

The parabolic window We mention also the window

$$W(\omega) = \frac{3\pi}{2\sigma}\left[1 - \left(\frac{\omega}{\sigma}\right)^2\right]p_\sigma(\omega) \tag{7-75}$$

This window does not have a TL inverse as in (7-51); however, it has the following optimal property[1] (Prob. 78):

In the class of all positive signals $W(\omega) \geq 0$ of unit area and specified energy E, the truncated parabola (7-75) has the smallest second moment m_2. For this signal, $m_2 = 1/5\sigma$, where $\sigma = 6E/5$.

In Sec. 10-2 (data smoothing), we discuss an application.

[1] A. Papoulis, Estimation of the Average Density of a Nonuniform Poisson Process, *IEEE Trans. Commun.*, vol. COM-22, no. 2, pp. 162–167, February, 1974.

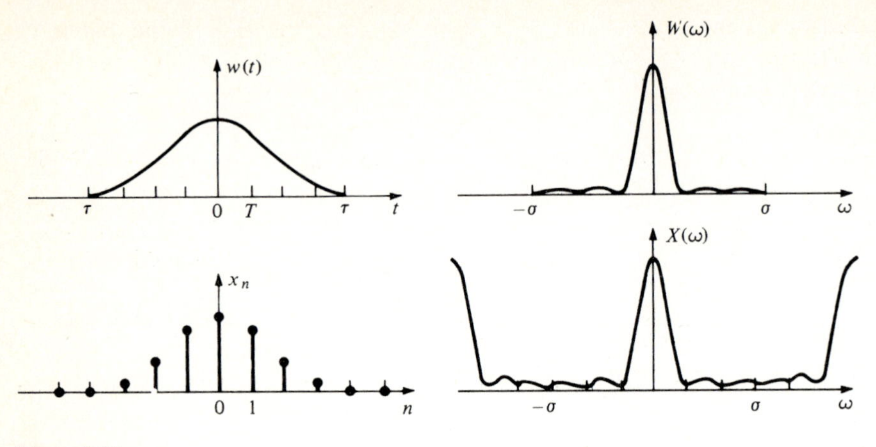

Figure 7-10

Digital windows A *digital window* is a real, even, normalized trigonometric polynomial

$$X(\omega) = \sum_{n=-M}^{M} x_n e^{-jnT\omega} \qquad x_0 = \frac{1}{2\sigma}\int_{-\sigma}^{\sigma} X(\omega)\, d\omega = 1 \qquad \sigma = \frac{\pi}{T} \tag{7-76}$$

of "short duration" relative to its period 2σ (Fig. 7-10). A digital window can be obtained by sampling an analog window. This is done as follows:

We select an analog window $w(t)$ with

$$\tau = MT$$

and form its samples

$$x_n = w(nT) \qquad |n| \leq M \tag{7-77}$$

Inserting into (7-76), we obtain [see (3-87)]

$$X(\omega) = \sum_{n=-M}^{M} w(nT)e^{-jnT\omega} = \frac{1}{T}\sum_{n=-\infty}^{\infty} W(\omega + 2n\sigma) \tag{7-78}$$

This approach yields a satisfactory solution if $M \gg 1$. Indeed, since the duration of $W(\omega)$ is of the order of π/τ (see Sec. 8-2) and

$$\frac{\pi}{\tau} = \frac{\pi}{MT} = \frac{\sigma}{M} \ll \sigma$$

we conclude that $W(\omega)$ is negligible for $|\omega| > \sigma$. Hence,

$$X(\omega) \simeq \frac{W(\omega)}{T} \qquad \text{for } |\omega| < \sigma$$

Digital windows can be designed directly by phrasing the various optimality criteria in terms of trigonometric polynomials. We give as an illustration the digital form of the minimum-amplitude-moment window (7-64):

We reason as in (7-66). With $X(\omega) \geq 0$, we conclude that if the periodic function $F(\omega)$ is approximated by the finite sum $F_x(\omega)$ as in (7-49), then the resulting truncation error is given by

$$F_x(\omega) - F(\omega) \simeq \frac{F''(\omega)}{4\sigma} \int_{-\sigma}^{\sigma} y^2 X(y)\, dy \tag{7-79}$$

To minimize the above, we must find a window $X(\omega)$ with minimum second moment. The solution is given in Eq. (6-75) with t and ω interchanged.

Extrapolation of Bandlimited Functions

We have shown in Sec. 6-1 that a BL function $f(t)$ is entire; hence, the Taylor series

$$f(t) = \sum_{n=-\infty}^{\infty} f^{(n)}(0) \frac{t^n}{n!} \tag{7-80}$$

converges for every t. Clearly, the derivatives $f^{(n)}(0)$ of $f(t)$ at $t = 0$ are determined in terms of any finite segment $f_\tau(t)$ of $f(t)$. Inserting their values into (7-80), we can, therefore, find $f(t)$ for any t. We have thus, in principle, a method for extrapolating $f_\tau(t)$ (analytic continuation).

The above approach is not useful for determining the transform $F(\omega)$ of $f(t)$ in terms of the finite segment $f_\tau(t)$. First, it is not possible to evaluate $f^{(n)}(0)$ precisely if noise is present. Furthermore, the truncation of (7-80) yields large errors for large t; since knowledge of $f(t)$ for all t is needed to evaluate $F(\omega)$, the errors are unacceptable.

In the following, we discuss two realistic methods for determining the transform $F(\omega)$ of a real BL function $f(t)$ in terms of $f_\tau(t)$.

Prolate functions As we have shown in Sec. 6-4, a BL function can be expanded into a series

$$f(t) = \sum_{k=0}^{\infty} a_k \varphi_k(t) \tag{7-81}$$

where $\varphi_k(t)$ are the prolate spheroidal functions and

$$a_k = \int_{-\infty}^{\infty} f(t)\varphi_k(t)\, dt = \frac{1}{\lambda_k} \int_{-\tau}^{\tau} f(t)\varphi_k(t)\, dt \tag{7-82}$$

Furthermore [see Eq. (6-93)], the Fourier transform $F(\omega)$ of $f(t)$ is given by

$$F(\omega) = \sqrt{\frac{2\pi\tau}{\sigma}} \sum_{k=0}^{\infty} (-j)^k \frac{a_k}{\sqrt{\lambda_k}} \varphi_k\left(\frac{\tau}{\sigma}\,\omega\right) p_\sigma(\omega) \tag{7-83}$$

Thus, to find $F(\omega)$ from the segment $f(t)p_\tau(t)$ of $f(t)$, we compute a_k from the last equality in (7-82) and insert the result into (7-83).

An iteration method [1,2] In this method, the required computations are Fourier transforms.

1. We find the Fourier transform $F_\tau(\omega)$ of the given segment $f_\tau(t)$.
2. *First iteration.* We form the function

$$F_1(\omega) = F_\tau(\omega)p_\sigma(\omega) \tag{7-84}$$

by truncating $F_\tau(\omega)$ for $|\omega| > \sigma$ (Fig. 7-11), compute its inverse transform $f_1(t)$,

$$f_1(t) \leftrightarrow F_1(\omega)$$

form the function

$$g_1(t) = f_1(t) + [f(t) - f_1(t)]p_\tau(t) = \begin{cases} f_\tau(t) & |t| \le \tau \\ f_1(t) & |t| > \tau \end{cases} \tag{7-85}$$

and compute its transform $G_1(\omega)$.

3. *nth iteration.* We form the function (Fig. 7-12)

$$F_n(\omega) = G_{n-1}(\omega)p_\sigma(\omega) \tag{7-86}$$

compute its inverse transform,

$$f_n(t) \leftrightarrow F_n(\omega)$$

form the function

$$g_n(t) = f_n(t) + [f(t) - f_n(t)]p_\tau(t) \tag{7-87}$$

and compute its transform $G_n(\omega)$:

$$g_n(t) \leftrightarrow G_n(\omega) \tag{7-88}$$

We shall show that the function $F_n(\omega)$ tends to $F(\omega)$ as $n \to \infty$.

Note: From (7-86) it follows that the function $f_n(t)$ is BL, and

$$f_n(t) = g_{n-1}(t) * \frac{\sin \sigma t}{\pi t} \tag{7-89}$$

As we see from Eq. (7-87), the function $g_n(t)$ is obtained by replacing the segment of $f_n(t)$ in the interval $(-\tau, \tau)$ with the known segment $f_\tau(t)$ of $f(t)$. Thus, $F_n(\omega) = F(\omega) = 0$ for $|\omega| > \sigma$, and $g_n(t) = f_\tau(t) = f(t)$ for $|t| < \tau$. We maintain that by this use of the available information about $f(t)$ in the t and ω

[1] R. W. Gerchberg, Super-Resolution through Error Energy Reduction, *Opt. Acta*, vol. 21, no. 9, pp. 709–720, 1974.

[2] A. Papoulis, A New Algorithm in Spectral Analysis and Band-limited Extrapolation, *IEEE Trans. Circuits Syst.*, vol. CAS-22, no. 9, pp. 735–742, September, 1975.

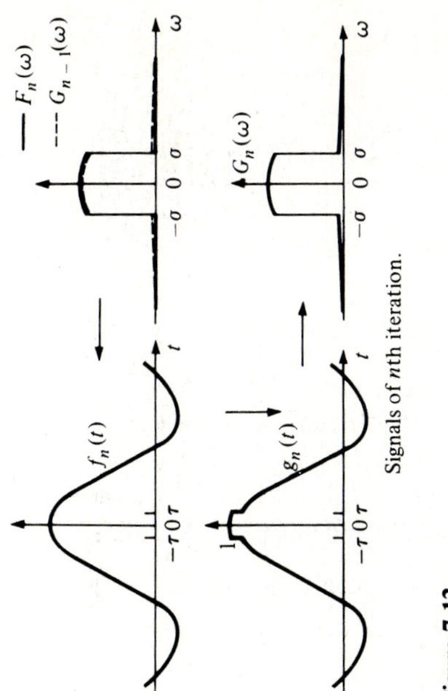

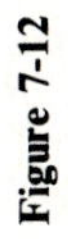

Figure 7-12 Signals of nth iteration.

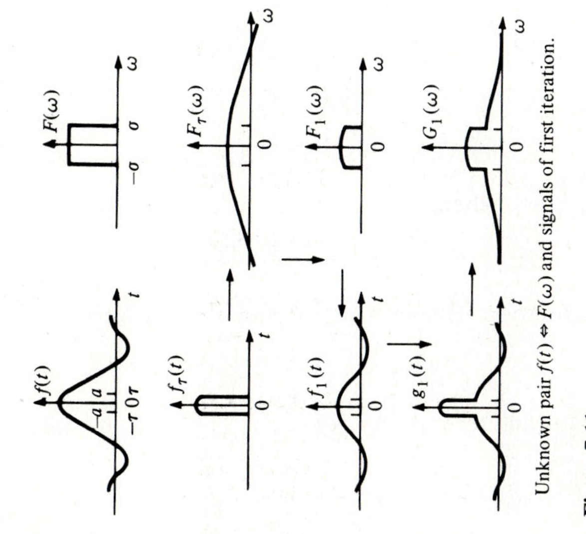

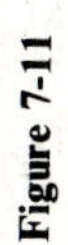

Figure 7-11 Unknown pair $f(t) \Leftrightarrow F(\omega)$ and signals of first iteration.

domains, we achieve a reduction of the mean-square value of the error $F(\omega) - F_n(\omega)$ twice in each iteration step.

Indeed, from Eqs. (7-86) and (7-87) and Parseval's formula, we conclude that

$$\frac{1}{2\pi}\int_{-\sigma}^{\sigma} |F(\omega) - F_n(\omega)|^2 \, d\omega = \int_{-\infty}^{\infty} |f(t) - f_n(t)|^2 \, dt > \int_{-\infty}^{\infty} |f(t) - g_n(t)|^2 \, dt$$

$$= \frac{1}{2\pi}\int_{-\infty}^{\infty} |F(\omega) - G_n(\omega)|^2 \, d\omega > \frac{1}{2\pi}\int_{-\sigma}^{\sigma} |F(\omega) - F_{n+1}(\omega)|^2 \, d\omega \tag{7-90}$$

Theorem 1 With a_k as in (7-81), the function $f_n(t)$ of the nth iteration is given by

$$f_n(t) = f(t) - \sum_{k=0}^{\infty} a_k(1 - \lambda_k)^n \varphi_k(t) \tag{7-91}$$

Proof Suppose, first, that $f(t)$ equals one of the prolate functions:

$$f(t) = \varphi_k(t) \tag{7-92}$$

In this case [see Eqs. (7-84) and (6-80)],

$$f_1(t) = \int_{-\tau}^{\tau} \varphi_k(x) \frac{\sin \sigma(t - x)}{\pi(t - x)} \, dx = \lambda_k \varphi_k(t)$$

We shall show by induction that

$$f_n(t) = A_n \varphi_k(t) \qquad A_n = 1 - (1 - \lambda_k)^n \tag{7-93}$$

Indeed, Eq. (7-93) is true for $n = 1$. Suppose that it is true for some $n > 1$. It then follows from (7-87) and (7-93) that

$$g_n(t) = A_n \varphi_k(t) + (1 - A_n)\varphi_k(t)p_\tau(t) \tag{7-94}$$

Inserting into (7-89), we obtain [see (6-80) and (6-91)]

$$f_{n+1}(t) = A_n \varphi_k(t) + (1 - A_n)\lambda_k \varphi_k(t) \tag{7-95}$$

Thus, $f_{n+1}(t) = A_{n+1}\varphi_k(t)$, where

$$A_{n+1} = A_n + (1 - A_n)\lambda_k \qquad n \geq 1 \tag{7-96}$$

This is a recursion equation with initial condition $A_1 = \lambda_k$; hence (Prob. 21),

$$A_n = 1 - (1 - \lambda_k)^n \tag{7-97}$$

Applying the above to each term in the expansion (7-81) of $f(t)$, we conclude that

$$f_n(t) = \sum_{k=0}^{\infty} a_k[1 - (1 - \lambda_k)^n]\varphi_k(t) \tag{7-98}$$

and (7-91) results.

Theorem 2 For any t,

$$f_n(t) \to f(t) \qquad \text{as } n \to \infty \tag{7-99}$$

Proof Denoting by $e_n(t)$ the error $f(t) - f_n(t)$ at the nth iteration, we have [see (7-91)]

$$e_n(t) = f(t) - f_n(t) = \sum_{k=0}^{\infty} a_k(1 - \lambda_k)^n \varphi_k(t) \tag{7-100}$$

From (6-102) it follows that the energy E_n of $e_n(t)$ is given by

$$E_n = \int_{-\infty}^{\infty} e_n^{\,2}(t)\,dt = \sum_{k=0}^{\infty} a_k^{\,2}(1 - \lambda_k)^{2n} \tag{7-101}$$

Lemma

$$E_n \to 0 \qquad \text{as } n \to \infty \tag{7-102}$$

Proof The energy E of $f(t)$ is finite:

$$E = \sum_{k=0}^{\infty} a_k^{\,2} < \infty$$

Hence, given $\epsilon > 0$, we can find an integer n such that

$$\sum_{k>N} a_k^{\,2} < \epsilon \tag{7-103}$$

The numbers λ_k are between 0 and 1, and $\lambda_k > \lambda_N$ for $k < N$. Hence,

$$1 - \lambda_k \le 1 - \lambda_N < 1 \qquad \text{for } k \le N$$

and

$$E_n = \sum_{k=0}^{N} + \sum_{k>N} a_k^{\,2}(1 - \lambda_k)^{2n} \le (1 - \lambda_N)^{2n} \sum_{k=0}^{N} a_k^{\,2} + \sum_{k>N} a_k^{\,2} \tag{7-104}$$

Since

$$(1 - \lambda_N)^{2n} \to 0 \qquad \text{as } n \to \infty$$

it follows from Eqs. (7-103) and (7-104) that

$$E_n < (1 - \lambda_N)^{2n} E + \epsilon \underset{n\to\infty}{\to} \epsilon$$

for any ϵ; hence, $E_n \to 0$.

Clearly, the error $e_n(t)$ is σ-BL because the functions $f(t)$ and $f_n(t)$ are σ-BL. Hence [see (6-53)],

$$|e_n(t)| \le \sqrt{\frac{E_n \sigma}{\pi}} \underset{n\to\infty}{\to} 0$$

and (7-99) results.

Numerical illustration We apply the iteration to the pair

$$f(t) = \frac{\sin \sigma t}{\pi t} \leftrightarrow p_\sigma(\omega) = F(\omega)$$

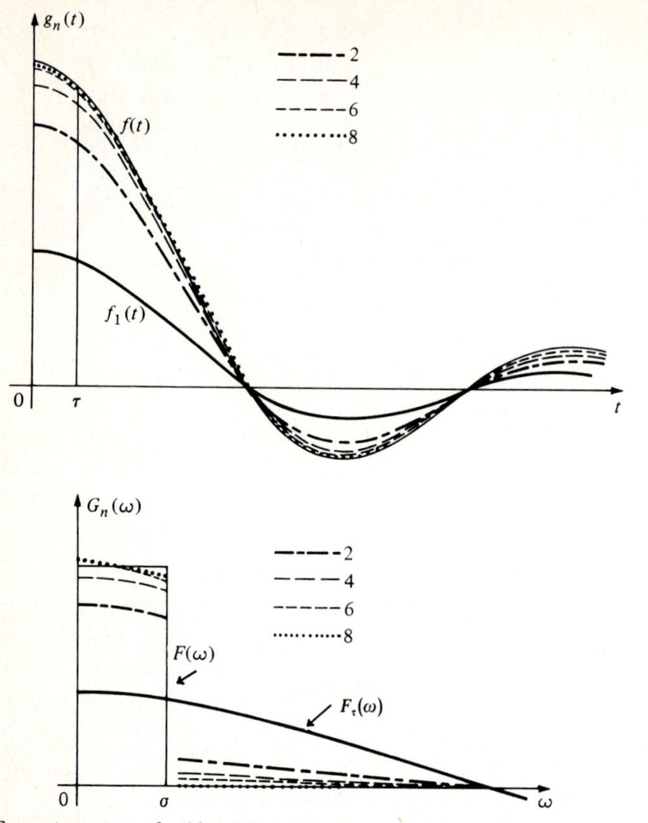

Computer output of $g_n(t)$ and $G_n(\omega)$ for $f(t) = \sin \sigma t/\pi t$ and $n = 2, 4, 6$, and 8.

Figure 7-13

choosing for τ the value $\pi/5\sigma$. The numerical results are shown in Fig. 7-13 for $n = 2, 4, 6, 8$. As we see from the figure, although the data interval is only $\frac{1}{5}$ of the main lobe of $f(t)$, the convergence of $F_n(\omega)$ to the unknown spectrum $p_\sigma(\omega)$ is remarkably rapid. With this value of τ, the method of windows would yield no satisfactory estimate of $F(\omega)$. We should point out, however, that the rapid convergence is due to the fact that, in the expansion (7-81) of the function $\sin \sigma t/\pi t$, only the first few terms are significant. If this is not the case, the convergence is not fast [see (7-101)].

Extrapolation of Digital Signals with Rational Transforms[1]

We shall discuss a method for determining a nonnegative periodic function in terms of a finite number of its Fourier-series coefficients under the assumption

[1] R. T. Lacoss, Data Adaptive Spectral Analysis Methods, *Geophysics*, vol. 86, pp. 661–675, 1971.

that the function is rational in $e^{-j\omega T}$ with constant numerator. The assumption will be justified at the end of this section (under "Maximum entropy").

This problem is of interest in spectral estimation, where the frequency function is denoted by $\bar{S}(\omega)$ (power spectrum), and its Fourier-series coefficients by $R[n]$ (discrete autocorrelation). Retaining this notation, we can phrase our problem as follows: Given $R[n]$ for $|n| \le M$, find $R[n]$ for $|n| > M$ subject to the condition that

$$\bar{S}(\omega) = \sum_{n=-\infty}^{\infty} R[n]e^{-jnT\omega} = \frac{1}{\sum_{n=-M}^{M} a_n e^{-jnT\omega}} \tag{7-105}$$

To solve the problem, it suffices to find the $2M + 1$ coefficients a_n.

Introducing the z transform $S(z)$ of $R[n]$, we can write Eq. (7-105) in the form

$$S(z) = \sum_{n=-\infty}^{\infty} R[n]z^{-n} = \frac{1}{\sum_{n=-M}^{M} a_n z^{-n}} \tag{7-106}$$

If we expand the right side into a Laurent series in z and equate the coefficients of z^{-n} for $|n| \le M$, we obtain $2M + 1$ equations relating the $2M + 1$ known values of $R[n]$ to the $2M + 1$ unknowns a_n. We can then, in principle, find a_n. However, the resulting system of equations is, in general, nonlinear and difficult to solve. As we shall presently see, the solution of the problem is considerably simplified if

$$\bar{S}(\omega) = S(e^{j\omega T}) > 0 \tag{7-107}$$

Indeed, from (7-105) and (7-107) it follows that $1/S(\omega)$ is a nonnegative trigonometric polynomial; therefore (Fejér-Riesz theorem), it can be written in the form

$$\frac{1}{\bar{S}(\omega)} = \frac{1}{p}\left|\sum_{n=0}^{M} b_n e^{-jnT\omega}\right|^2 \qquad b_0 = 1 \tag{7-108}$$

This follows from Eq. (7-33) with $p = |1/y_0|^2$ and $b_n = y_n/y_0$. In fact, as we have shown, the coefficients b_n can be chosen so that all zeros z_i of the polynomial

$$H(z) = \sum_{n=0}^{M} b_n z^{-n} \qquad b_0 = 1 \tag{7-109}$$

are inside the unit circle, and all zeros $1/z_i$ of the polynomial

$$H_{-}(z) = H^*\left(\frac{1}{z^*}\right) = \sum_{n=0}^{M} b_n^* z^n \tag{7-110}$$

are outside the unit circle. The inverse of the above sum is, therefore, an analytic function of z for $|z| \le 1$. Hence, it can be expanded into a power series in z

$$\frac{1}{H_{-}(z)} = 1 + c_1 z + c_2 z^2 + \cdots \qquad |z| \le 1 \tag{7-111}$$

The constant term is 1 because $b_0^* = 1$.

From the preceding equations, it follows that (elaborate)

$$S(z)H(z) = \frac{p}{H_{-}(z)} \tag{7-112}$$

We finally take inverse z transforms of both sides of (7-112). Since the inverse of the right side equals 0 for $n > 0$ [see (7-111)], we conclude from (7-106), (7-109), and the convolution theorem that

$$\sum_{k=0}^{M} R[n-k]b_k = \begin{cases} 0 & n > 0 \\ p & n = 0 \\ pc_n & n < 0 \end{cases} \tag{7-113}$$

With $n = 1, \ldots, M$, the above yields

$$\begin{aligned} R[1] + R[0]b_1 + R[-1]b_2 + \cdots + R[1-M]b_m &= 0 \\ R[2] + R[1]b_1 + R[0]b_2 + \cdots + R[2-M]b_m &= 0 \\ \cdots\cdots\cdots\cdots\cdots\cdots\cdots\cdots\cdots\cdots\cdots\cdots \\ R[M] + R[M-1]b_1 + R[M-2]b_2 + \cdots + R[0]b_m &= 0 \end{aligned} \tag{7-114}$$

because $b_0 = 1$. This is a system of M linear equations whose coefficients $R[n]$ are known. Solving, we find the M unknowns b_n. In fact, since the matrix of the coefficients

$$\begin{pmatrix} R[0] & R[-1] & \cdots & R[1-M] \\ R[1] & R[0] & \cdots & R[2-M] \\ \cdots & \cdots & \cdots & \cdots \\ R[M-1] & R[M-2] & \cdots & R[0] \end{pmatrix} \tag{7-115}$$

is Toeplitz, the solution of the system (7-114) is simple.

To complete the determination of $\bar{S}(\omega)$, we need the constant p. From (7-113) it follows, with $n = 0$, that

$$p = R[0] + R[-1]b_1 + \cdots + R[-M]b_m \tag{7-116}$$

Maximum entropy The assumption that $\bar{S}(\omega)$ is rational as in (7-105) can be justified as follows:

Suppose the coefficients $R[n]$ of $\bar{S}(\omega)$ are specified for $n \le M$; and, for $|n| > M$, they are to be chosen so as to maximize the integral

$$E = -\int_{-\sigma}^{\sigma} \log \bar{S}(\omega)\, d\omega \qquad \bar{S}(\omega) = \sum_{n=-\infty}^{\infty} R[n]e^{-jnT\omega} \tag{7-117}$$

We must then have

$$\frac{\partial E}{\partial R[n]} = 0 = \int_{-\sigma}^{\sigma} \frac{1}{\bar{S}(\omega)} \frac{\partial \bar{S}(\omega)}{\partial R[n]}\, d\omega = \int_{-\sigma}^{\sigma} \frac{e^{-jnT\omega}}{\bar{S}(\omega)}\, d\omega \qquad |n| > M \tag{7-118}$$

Hence, the Fourier-series coefficients of the periodic function $1/\bar{S}(\omega)$ must be zero for $|n| > M$.

An arbitrary $\bar{S}(\omega)$ need not, of course, be rational. However, in the absence of other a priori information, Eq. (7-105) suggests a reasonable form for $\bar{S}(\omega)$. Certainly, it is no less reasonable than the assumption that $R[n] = 0$ for $|n| > M$, implicit in the method of windows.

7-4 HILBERT TRANSFORMS

If a real or complex function $f(t)$ is causal, then the real and imaginary parts of its Fourier transform $F(\omega)$ are related. If a function $f(t)$ is real, then it is determined in terms of its transform on the positive frequency axis. The underlying relationships are known as *Hilbert transforms*. In this section, we examine these relationships and their parallels for periodic functions and discrete Fourier series. The results are stated in two forms. The first form deals with causal signals, and the second with real signals.

Fourier Transforms

We shall use the following Fourier pairs [see Eq. (3-50)]:

$$U(t) \leftrightarrow \pi\,\delta(\omega) + \frac{1}{j\omega} \tag{7-119}$$

$$\delta(t) + \frac{j}{\pi t} \leftrightarrow 2U(\omega) \tag{7-120}$$

Form 1 If $f(t) = 0$ for $t < 0$ and

$$F(\omega) = R(\omega) + jX(\omega) = \int_0^\infty f(t)e^{-j\omega t}\,dt \tag{7-121}$$

then

$$R(\omega) = \frac{1}{\pi}\int_{-\infty}^{\infty} \frac{X(y)}{\omega - y}\,dy \qquad X(\omega) = -\frac{1}{\pi}\int_{-\infty}^{\infty} \frac{R(y)}{\omega - y}\,dy \tag{7-122}$$

Thus, the real and imaginary parts of the Fourier transform of a causal function form a Hilbert transform pair.

Proof Clearly,

$$f(t) = f(t)U(t) \tag{7-123}$$

Taking transforms of both sides, we conclude from (7-119) and the frequency convolution theorem that

$$R(\omega) + jX(\omega) = \frac{1}{2\pi}[R(\omega) + jX(\omega)] * \left[\pi\,\delta(\omega) + \frac{1}{j\omega}\right]$$

Equating real and imaginary parts, we obtain

$$R(\omega) = \frac{1}{2} R(\omega) + \frac{1}{2\pi} X(\omega) * \frac{1}{\omega} \qquad X(\omega) = \frac{1}{2} X(\omega) - \frac{1}{2\pi} R(\omega) * \frac{1}{\omega}$$

and (7-122) results.

Notes:

1. As we know, the function $F(s/j)$ is analytic in the right half-plane Re $s \geq 0$. Equation (7-122) shows that its real and imaginary parts on the $s = j\omega$ axis are related.
2. We have tacitly assumed that $f(t)$ does not contain singularities at the origin, i.e., that $F(\pm\infty) = 0$ [see (7-123)].
3. If $f(t)$ is real, then (7-122) yields

$$R(\omega) = \int_0^\infty f(t) \cos \omega t \, dt \qquad X(\omega) = -\int_0^\infty f(t) \sin \omega t \, dt \quad \text{(7-124)}$$

Thus, the cosine and sine integrals of a real function form a Hilbert transform pair.

***Form* 2** If $f(t)$ is a real function with Fourier transform $F(\omega)$ and

$$z(t) = \frac{1}{\pi} \int_0^\infty F(\omega) e^{j\omega t} \, d\omega \qquad \text{(7-125)}$$

then

$$z(t) = f(t) + j\hat{f}(t) \qquad \text{where} \qquad \hat{f}(t) = \frac{1}{\pi} \int_{-\infty}^\infty \frac{f(\tau)}{t - \tau} \, d\tau \qquad \text{(7-126)}$$

Proof With $Z(\omega)$ the Fourier transform of $z(t)$, it follows from (7-125) that

$$Z(\omega) = 2U(\omega)F(\omega) \qquad \text{(7-127)}$$

Taking inverse transforms of both sides, we conclude from (7-120) and the convolution theorem that

$$z(t) = f(t) * \left[\delta(t) + \frac{j}{\pi t}\right] = f(t) + jf(t) * \frac{1}{\pi t} \qquad \text{(7-128)}$$

and (7-126) results.

Note: From the results of Sec. 7-1, it follows, with s changed to $-jt$, that $z(t)$ exists for every t in the upper half-plane Im $t \geq 0$, and in this region it is an analytic function of t.

Fourier Series

Unlike Fourier transforms and DFS, the correspondence between a periodic function $f(t)$ and its Fourier-series coefficients a_n is not symmetrical. Therefore,

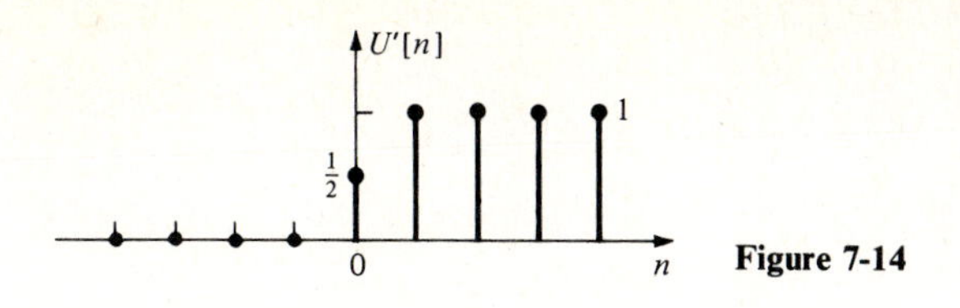

Figure 7-14

each form of Hilbert transforms has two versions. To establish them, we shall need the following:
For $|t| < T/2$:

$$U(t) = \sum_{n=-\infty}^{\infty} u_n e^{jn\omega_0 t} \qquad u_n = \frac{1}{T}\int_0^{T/2} e^{-jn\omega_0 t}\, dt = \begin{cases} \dfrac{1}{j\pi n} & n \text{ odd} \\ \frac{1}{2} & n = 0 \\ 0 & \text{otherwise} \end{cases} \tag{7-129}$$

For any t:

$$T \sum_{n=-\infty}^{\infty} \delta(t + nT) + j \cot \frac{\omega_0 t}{2} = \sum_{n=-\infty}^{\infty} 2U'[n] e^{jn\omega_0 t} \qquad U'[n] = \begin{cases} 1 & n > 0 \\ \frac{1}{2} & n = 0 \\ 0 & n < 0 \end{cases} \tag{7-130}$$

(Fig. 7-14) because [see Prob. 40 and Eq. (1-58)]

$$\cot \frac{\omega_0 t}{2} = 2\sum_{n=1}^{\infty} \sin n\omega_0 t \qquad T \sum_{n=-\infty}^{\infty} \delta(t + nT) = \sum_{n=-\infty}^{\infty} e^{jn\omega_0 t}$$

Form 1a If $f(t)$ is a periodic function vanishing in the interval $(-T/2, 0)$ as in Fig. 7-15, with Fourier-series coefficients

$$a_n = r_n + jx_n = \frac{1}{T}\int_0^{T/2} f(t) e^{-jn\omega_0 t}\, dt \tag{7-131}$$

then

$$r_n = \frac{2}{\pi} \sum_{k=-\infty}^{\infty} \frac{x_{n-2k-1}}{2k+1} \qquad x_n = -\frac{2}{\pi} \sum_{k=-\infty}^{\infty} \frac{r_{n-2k-1}}{2k+1} \tag{7-132}$$

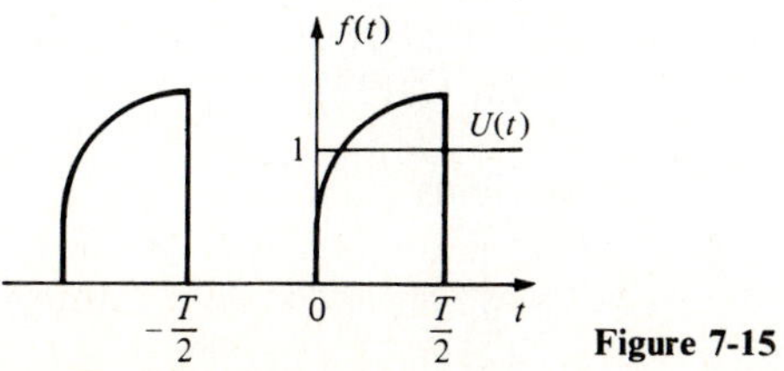

Figure 7-15

Proof For $|t| < T/2$,

$$f(t) = f(t)U(t)$$

Equating the Fourier-series coefficients of both sides, we conclude from the convolution theorem (3-73) that

$$r_n + jx_n = \sum_{k=-\infty}^{\infty} (r_{n-k} + jx_{n-k})u_k$$

From the above and Eq. (7-129), it follows that (elaborate)

$$r_n = \frac{r_n}{2} + \frac{1}{\pi}\sum_{k=-\infty}^{\infty} \frac{x_{n-2k-1}}{2k+1} \qquad x_n = \frac{x_n}{2} - \frac{1}{\pi}\sum_{k=-\infty}^{\infty} \frac{r_{n-2k-1}}{2k+1}$$

and (7-132) results.

Note: If $f(t)$ is real, then (7-131) yields

$$r_n = \frac{1}{T}\int_0^{T/2} f(t)\cos n\omega_0 t\, dt \qquad x_n = \frac{-1}{T}\int_0^{T/2} f(t)\sin n\omega_0 t\, dt \tag{7-133}$$

Thus, the cosine and sine coefficients of a real function satisfy (7-132).

Form 1b If $a_n = 0$ for $n \le 0$, that is, if

$$f(t) = r(t) + jx(t) = \sum_{n=1}^{\infty} a_n e^{jn\omega_0 t} \tag{7-134}$$

then

$$r(t) = -\frac{1}{T}\int_{-T/2}^{T/2} x(t-\tau)\cot\frac{\omega_0 \tau}{2}\, d\tau \qquad x(t) = \frac{1}{T}\int_{-T/2}^{T/2} r(t-\tau)\cot\frac{\omega_0 \tau}{2}\, d\tau \tag{7-135}$$

Proof With $U'[n]$ as in (7-130),

$$a_n = a_n U'[n]$$

Since a_n and $2U'[n]$ are the coefficients of the function $f(t)$ and the function in (7-130), respectively, we conclude from (3-72) that

$$r(t) + jx(t) = \frac{1}{2T}\int_{-T/2}^{T/2} [r(t-\tau) + jx(t-\tau)]\left[T\,\delta(\tau) + j\cot\frac{\omega_0 \tau}{2}\right] d\tau \tag{7-136}$$

because in the interval $(-T/2, T/2)$ the first sum in (7-130) equals $T\,\delta(t)$. Equating real and imaginary parts in (7-136), we obtain (7-135).

Note: If the coefficients a_n are real, then (7-134) yields

$$r(t) = \sum_{n=1}^{\infty} a_n \cos n\omega_0 t \qquad x(t) = \sum_{n=1}^{\infty} a_n \sin n\omega_0 t \tag{7-137}$$

Thus, the cosine and sine series of a real sequence satisfy Eqs. (7-135).

Form 2a If $f(t)$ is a real function with Fourier-series coefficients a_n and

$$z(t) = a_0 + 2\sum_{n=1}^{\infty} a_n e^{jn\omega_0 t} = 2\sum_{n=-\infty}^{\infty} U'[n]a_n e^{jn\omega_0 t} \tag{7-138}$$

then $z(t) = f(t) + j\hat{f}(t)$ where $\hat{f}(t) = \dfrac{1}{T}\displaystyle\int_{-T/2}^{T/2} f(t-\tau)\cot\frac{\omega_0 \tau}{2}\,d\tau$

(7-139)

Proof Denoting by z_n the Fourier-series coefficients of $z(t)$, we conclude from (7-138) that

$$z_n = 2U'[n]a_n$$

Hence, as in (7-136),

$$z(t) = \frac{1}{T}\int_{-T/2}^{T/2} f(t-\tau)\left[T\,\delta(\tau) + j\cot\frac{\omega_0\tau}{2}\right]d\tau$$

and (7-139) results.

Form 2b If the coefficients a_n of $f(t)$ are real and

$$z_n = \frac{2}{T}\int_0^{T/2} f(t)e^{-jn\omega_0 t}\,dt \tag{7-140}$$

then $z_n = a_n + j\hat{a}_n$ where $\hat{a}_n = -\dfrac{2}{\pi}\displaystyle\sum_{k=-\infty}^{\infty}\frac{a_{n-2k-1}}{2k+1}$ (7-141)

Proof With

$$z(t) = \sum_{n=-\infty}^{\infty} z_n e^{jn\omega_0 t}$$

it follows from (7-140) that, for $|t| < T/2$,

$$z(t) = 2f(t)U(t)$$

(Fig. 7-16). Hence [see Eqs. (7-129) and (3-73)],

$$z_n = 2\sum_{k=-\infty}^{\infty} a_{n-k}u_k = a_n + 2\sum_{k=-\infty}^{\infty}\frac{a_{n-2k-1}}{j\pi(2k+1)}$$

and (7-141) follows.

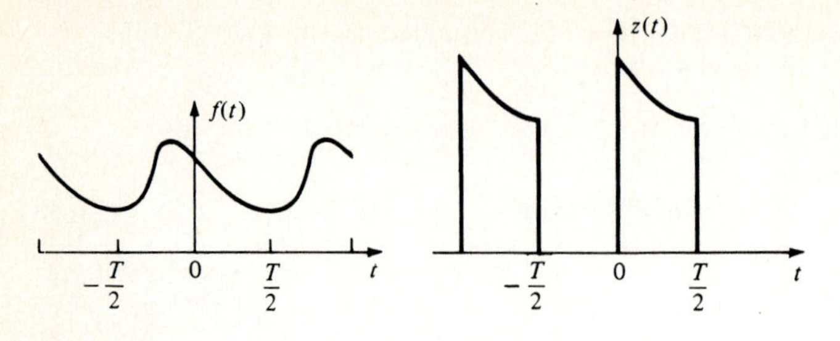

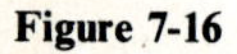
Figure 7-16

Discrete Fourier Series

We shall consider DFS of odd order $N = 2M + 1$ (the even case is similar):

$$A_m = \sum_{n=-M}^{M} a_n w^{mn} \qquad a_n = \frac{1}{N} \sum_{m=-M}^{M} A_m w^{-mn} \tag{7-142}$$

We shall need the following pairs: For $|n| \le M$ and $|m| \le M$,

$$U'[n] = \begin{cases} 1 & 0 < n \le M \\ \frac{1}{2} & n = 0 \\ 0 & -M \le n < 0 \end{cases} \leftrightarrow V[m] = \begin{cases} \dfrac{j}{2} \cot \dfrac{m\pi}{2N} & m \text{ odd} \\ M + \frac{1}{2} & m = 0 \\ \dfrac{-j}{2} \tan \dfrac{m\pi}{2N} & \text{otherwise} \end{cases} \tag{7-143}$$

$$\frac{1}{N} V[-n] \leftrightarrow U'[m] \tag{7-144}$$

To prove (7-143), we note that $V[0] = M + \frac{1}{2}$ and, for $m \neq 0$,

$$V[m] = \frac{1}{2} + \sum_{n=1}^{M} w^{mn} = \frac{1}{2} + \frac{w^{(M+1)m} - w^m}{w^m - 1} = -\frac{1}{2} \frac{w^{m/4} - (-1)^m w^{-m/4}}{w^{m/4} + (-1)^m w^{-m/4}} \tag{7-145}$$

because $w^{(M+1)m} = (-1)^m w^{m/2}$, and (7-143) follows with $w = e^{j2\pi/N}$. The proof of (7-144) follows from the symmetry of a_n and A_m (Prob. 51).

Form 1 If $a_n = 0$ for $-M \le n \le 0$ and

$$A_m = R_m + jX_m = \sum_{n=1}^{M} a_n w^{mn} \tag{7-146}$$

then

$$\begin{aligned} R_m &= -\frac{1}{N} \sum_k X_{m-k} \cot \frac{\pi}{2N} k + \frac{1}{N} \sum_k X_{m-k} \tan \frac{\pi}{2N} k \\ X_m &= \frac{1}{N} \sum_k R_{m-k} \cot \frac{\pi}{2N} k - \frac{1}{N} \sum_k R_{m-k} \tan \frac{\pi}{2N} k \end{aligned} \tag{7-147}$$

In the first sum in both equations, k varies over all odd integers from $-M$ to M; in the second, over all even integers except 0 from $-M$ to M.

Proof For $|n| \leq M$,

$$a_n = a_n U'[n] \tag{7-148}$$

Since the DFS of a_n and $U'[n]$ are the sequences A_m and $V[m]$, respectively, we conclude, taking DFS of both sides of Eq. (7-148), that [see (3-120)]

$$R_m + jX_m = \frac{1}{N} \sum_{k=-M}^{M} (R_{m-k} + jX_{m-k})V[k] \tag{7-149}$$

and (7-147) results (elaborate).

Form 2 If a_n is a real sequence with DFS A_m and

$$z_n = \frac{1}{N} A_0 + \frac{2}{N} \sum_{m=1}^{M} A_m w^{mn} \tag{7-150}$$

then

$$z_n = a_n + j\hat{a}_n \tag{7-151}$$

where

$$\hat{a}_n = -\frac{1}{N} \sum_k a_{n-k} \cot \frac{\pi}{2N} k + \frac{1}{N} \sum_k a_{n-k} \tan \frac{\pi}{2N} k \tag{7-152}$$

In the first sum, k is odd; in the second, it is even and different from 0 as in (7-147).

Proof Denoting by Z_m the DFS of z_m, we conclude from (7-150) that (Fig. 7-17)

$$Z_m = 2U'[m]A_m \tag{7-153}$$

The inverse transforms of both sides yield [see (7-144)]

$$z_n = \frac{2}{N} \sum_{k=-M}^{M} a_{n-k} V[-k]$$

as in (7-149), and (7-152) results (elaborate).

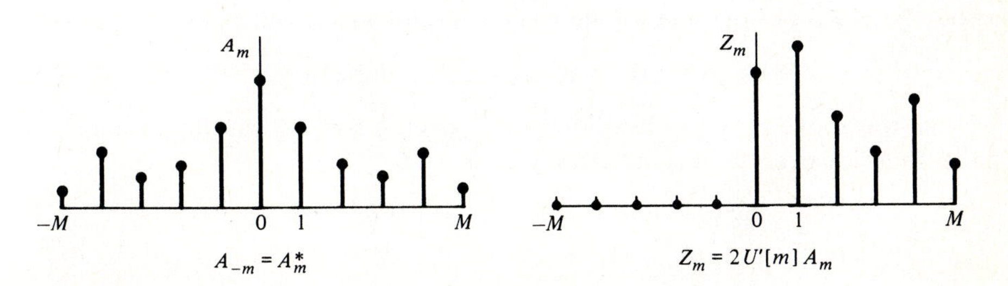

Figure 7-17

APPENDIX 7-A

The Paley-Wiener Theorem

The Paley-Wiener theorem is a rather advanced result in the theory of functions. We give below a proof starting not from first principles but from certain properties of functions of exponential type.[1] To do otherwise would require lengthy preparation.

Functions of exponential type A function $F(s)$ of the complex variable s is of *exponential type* (abbreviated ET) if there exist two constants A and τ such that

$$|F(s)| < Ae^{|s|\tau} \tag{7-A-1}$$

for every s.

If (7-A-1) holds for every τ greater than τ_0 but for no τ less than τ_0, then $F(s)$ is of type τ_0. A polynomial is of type zero; the function e^{2s} is of type 2; the function $(\sin 5s)/s$ is of type 5 because

$$\left|\frac{\sin 5s}{s}\right| = \left|\frac{e^{j5s} - e^{-j5s}}{2js}\right| < 2e^{5|s|}$$

The function e^{s^2} is not ET.

Thus, an ET function can grow as fast as, but no faster than, an exponential, as in (7-A-1). We note that the rate of growth of $F(s)$ need not be the same in all directions. In fact, in some directions, $F(s)$ might tend to zero as $|s| \to \infty$. For example, if $F(s) = (\sinh s)/s$, then $F(j\omega) = (\sin \omega)/\omega \to 0$ as $\omega \to \infty$.

In the following, we assume that $F(s)$ is ET as in (7-A-1).

Property 1: If $F(s)$ is analytic in the right half plane $\text{Re } s = \gamma \geq 0$ and bounded on the $j\omega$ axis:

$$|F(j\omega)| < B \qquad \text{for all real } \omega \tag{7-A-2}$$

then

$$|F(\gamma + j\omega)| < Me^{\gamma\tau} \qquad \text{for } \gamma \geq 0 \tag{7-A-3}$$

where M is the largest of the numbers A and B.

Thus, boundedness on the $j\omega$ axis permits us to replace the absolute value of s in (7-A-1) with its real part in the region $\gamma \geq 0$.

Property 2: If $F(s)$ is analytic for $\text{Re } s = \gamma \geq 0$ and $F(j\omega) \to 0$ as $\omega \to \infty$, then

$$F(\gamma + j\omega) \to 0 \qquad \text{as } \omega \to \infty \qquad \text{for every } \gamma \geq 0 \tag{7-A-4}$$

From the above property it follows that, given $\epsilon > 0$, we can find a constant ω_c such that, for every γ in a finite interval $(0, \gamma_1)$,

$$|F(\gamma + j\omega)| < \epsilon \qquad \text{for } |\omega| > \omega_c \tag{7-A-5}$$

[1] R. P. Boas, "Entire Functions," Academic Press, Inc., New York, 1954.

Property 3: If $F(s)$ is analytic for every s (entire function) and $\int_{-\infty}^{\infty} |F(j\omega)|^2 \, d\omega < \infty$, then

$$F(j\omega) \to 0 \qquad \text{as } |\omega| \to \infty$$

From the above it follows that $F(j\omega)$ is bounded on the imaginary axis, because it is certainly bounded in the finite interval $(-\omega_0, \omega_0)$ (since it is continuous) and it is less than ϵ for $|\omega| > \omega_0$. Combining properties 1, 2, and 3, we thus conclude that: An entire, ET function $F(s)$ with finite energy on the $j\omega$ axis satisfies (7-A-3) for some M; furthermore, $F(\gamma + j\omega) \to 0$ as $|\omega| \to \infty$ for every γ.

The Paley-Wiener theorem If $F(s)$ is an entire function of exponential type as in (7-A-1), and $F(j\omega)$ has finite energy (it belongs to L^2), then $F(s)$ has a time-limited inverse $f(t)$:

$$f(t) = 0 \qquad \text{for } |t| > \tau \tag{7-A-6}$$

Conversely, if $f(t)$ has finite energy in the interval $(-\tau, \tau)$ and

$$F(s) = \int_{-\tau}^{\tau} f(t) e^{-st} \, dt \tag{7-A-7}$$

then $F(s)$ is an entire function of exponential type, and $F(j\omega)$ belongs to L^2.

Proof Since $F(j\omega)$ has finite energy, it has an inverse

$$f(t) = \frac{1}{2\pi} \lim_{\omega_0 \to \infty} \int_{-\omega_0}^{\omega_0} F(j\omega) e^{j\omega t} \, d\omega \tag{7-A-8}$$

We shall show first that $f(t) = 0$ for $t < -\tau$.

From the analyticity of $F(s)$ and Cauchy's theorem, it follows that

$$\oint_{\Gamma} F(s) e^{st} \, ds = jI_1 + I_2 - jI_3 - I_4 = 0 \tag{7-A-9}$$

where Γ is the perimeter of the rectangle $ABCD$ of Fig. 7-18, and I_1, I_2, I_3, and I_4 are the line integrals along each side of this rectangle:

$$I_1 = \int_{-\omega_0}^{\omega_0} F(j\omega) e^{j\omega t} \, d\omega \qquad I_3 = \int_{-\omega_0}^{\omega_0} F(\gamma_0 + j\omega) e^{(\gamma_0 + j\omega)t} \, d\omega$$

$$I_2 = \int_0^{\gamma_0} F(\gamma + j\omega_0) e^{(\gamma + j\omega_0)t} \, d\gamma \qquad I_4 = \int_0^{\gamma_0} F(\gamma - j\omega_0) e^{(\gamma - j\omega_0)t} \, d\gamma$$

Since I_1 tends to $2\pi f(t)$, it suffices to show that the integrals I_2, I_3, and I_4 tend to zero as ω_0 and γ_0 tend to infinity.

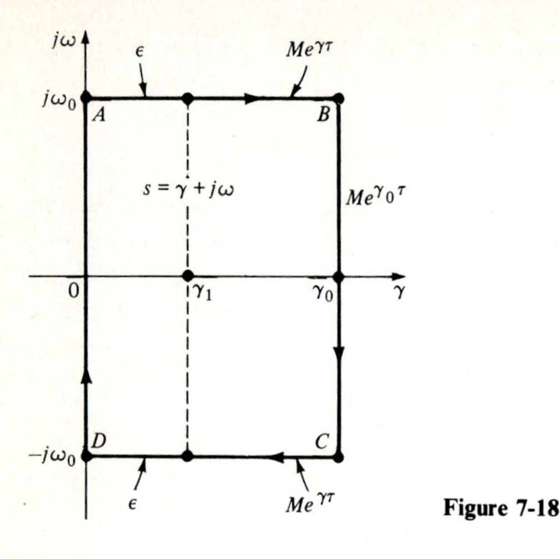

Figure 7-18

From property 1 it follows that $F(s)$ is bounded by $Me^{\gamma_0\tau}$ for every s on the line BC. Hence,

$$|I_3| \le e^{\gamma_0 t} \int_{-\omega_0}^{\omega_0} |F(\gamma_0 + j\omega)|\, d\omega \le 2M\omega_0\, e^{\gamma_0(t+\tau)} \underset{\gamma_0 \to \infty}{\to} 0$$

because $t + \tau < 0$ by assumption.

To show that $I_2 \to 0$, we express it as sum of two integrals, from 0 to γ_1 and from γ_1 to γ_0. Clearly [see (7-A-3)],

$$\left| \int_{\gamma_1}^{\gamma_0} F(\gamma + j\omega_0) e^{(\gamma + j\omega_0)t}\, d\gamma \right| \le M \int_{\gamma_1}^{\gamma_0} e^{\gamma(t+\tau)}\, d\gamma = M \frac{e^{(t+\tau)\gamma_0} - e^{(t+\tau)\gamma_1}}{t+\tau}$$

Since $t + \tau < 0$, the above can be made arbitrarily small if the constants γ_0 and γ_1 are sufficiently large. With γ_1 fixed, it follows from (7-A-5) that, if ω_0 is sufficiently large, then $|F(\gamma + j\omega)| < \epsilon$ for every γ in the interval $(0, \gamma_1)$. Hence,

$$\left| \int_0^{\gamma_1} F(\gamma + j\omega_0) e^{(\gamma + j\omega_0)t}\, dt \right| < \gamma_1 \epsilon e^{\gamma_1 |t|} \underset{\omega_0 \to \infty}{\to} 0$$

Therefore, $I_2 \to 0$ if γ_0 and ω_0 tend to infinity.

Reasoning similarly, we conclude that $I_4 \to 0$. In Fig. 7-18, we show the bounds of $F(s)$ in the relevant segments of Γ.

From the preceding it follows that $f(t) = 0$ for $t < -\tau$. The proof that $f(t) = 0$ for $t > \tau$ is the same (elaborate). The converse of the theorem was discussed in Sec. 6-1 (Theorems 1 and 2).

APPENDIX 7-B

Jordan's Lemma

If Γ is a circular arc AEB of radius r centered at the origin, and $F(s)$ is a function such that for every s on Γ

$$F(s) \to 0 \qquad \text{as } r \to \infty \tag{7-B-1}$$

then

$$\int_\Gamma e^{st} F(s)\, ds \to 0 \qquad \text{as } r \to \infty \tag{7-B-2}$$

1. For every $t < 0$, if $\Gamma = \Gamma_1$ is to the right of the vertical line AB (Fig. 7-1a)
2. For every $t > 0$, if $\Gamma = \Gamma_2$ is to the left of the vertical line AB (Fig. 7-1b)

Proof of part 1[1] Suppose, first, that $\gamma \geq 0$. Clearly [see (7-B-1)], given $\epsilon > 0$, we can find a constant r_0 such that $|F(s)| < \epsilon$ for $|s| > r_0$. With

$$s = re^{j\theta} \qquad ds = jre^{j\theta}\, d\theta$$

it follows that

$$\left| \int_\Gamma e^{st} F(s)\, ds \right| \leq \int_{-\pi/2}^{\pi/2} \left| e^{rt(\cos\theta + j\sin\theta)} F(re^{j\theta}) jre^{j\theta} \right| d\theta \leq \epsilon r \int_{-\pi/2}^{\pi/2} e^{rt\cos\theta}\, d\theta$$

Since $t < 0$ and $\cos\theta \geq 1 - 2|\theta|/\pi$ for every $|\theta| \leq \pi/2$, we conclude that

$$\int_{-\pi/2}^{\pi/2} e^{rt\cos\theta}\, d\theta < 2\int_0^{\pi/2} e^{tr(1-2\theta/\pi)}\, d\theta = \frac{\pi}{-tr}(1 - e^{rt}) < \frac{\pi}{-tr}$$

Therefore,

$$\left| \int_\Gamma e^{st} F(s)\, ds \right| < \frac{\pi\epsilon}{-t}$$

If $\gamma < 0$, then the arc Γ consists of a semicircle and two arcs whose length is less than $2|\gamma|$ for any r. Along these arcs, $|e^{st}| \leq e^{|\gamma t|}$ and $|F(s)| < \epsilon$ for $r > r_0$. Hence, their combined contribution does not exceed $4|\gamma|\epsilon e^{|\gamma t|}$ and, therefore, it tends to zero as $r \to \infty$.

Proof of part 2 The second part of the lemma follows from the first when t is changed to $-t$, and s to $-s$.

[1] E. T. Whittaker and G. N. Watson, "A Course of Modern Analysis," Cambridge University Press, New York, 1948.

CHAPTER

EIGHT

FREQUENCY MODULATION, UNCERTAINTY, AMBIGUITY

In this chapter, we concentrate on the properties of rapidly varying functions and their spectra.

In Sec. 8-1, we evaluate the transforms of FM signals with arbitrary envelopes and discuss the method of stationary phase. The results are approximate and are based on the moment expansion of the convolution integral (Sec. 4-1).

In Sec. 8-2, we present various forms of the uncertainty principle, which relates the duration of a signal to the duration of its transform. Applications include the principle of pulse compression and the properties of sophisticated *signals, i.e., of functions whose spectra have rapidly varying phase.*

In Sec. 8-3, we introduce briefly the elements of the theory of two-dimensional transforms and Hankel transforms.[1]

In Sec. 8-4, we define the ambiguity function[2] *and develop its main properties. We show that it remains invariant to a succession of operations involving FM modulation and quadratic phase filtering, subject only to coordinate transformations. This property has applications in radar, in optics, and in other areas.*

---

[1] A. Papoulis, "Systems and Transforms with Applications in Optics," McGraw-Hill Book Company, New York, 1968.

[2] P. M. Woodward, "Probability and Information Theory with Applications to Radar," p. 49, Pergamon Press, New York, 1953.

8-1 FREQUENCY MODULATION AND THE METHOD OF STATIONARY PHASE

Consider the normal integral

$$N(s) = \int_{-\infty}^{\infty} e^{-st^2}\, dt \tag{8-1}$$

We maintain that, if Re $s > 0$, then

$$\int_{-\infty}^{\infty} e^{-st^2}\, dt = \sqrt{\frac{\pi}{s}} \qquad \text{Re}\sqrt{\frac{\pi}{s}} > 0 \tag{8-2}$$

Proof

$$N^2(s) = \int_{-\infty}^{\infty} e^{-sx^2}\, dx \int_{-\infty}^{\infty} e^{-sy^2}\, dy = \int_{-\infty}^{\infty}\int_{-\infty}^{\infty} e^{-s(x^2+y^2)}\, dx\, dy$$

Changing into polar coordinates,

$$x = r\cos\theta \qquad y = r\sin\theta \qquad dx\, dy = r\, dr\, d\theta$$

we obtain

$$N^2(s) = \int_0^{\infty}\int_{-\pi}^{\pi} re^{-sr^2}\, dr\, d\theta = 2\pi\int_0^{\infty} re^{-sr^2}\, dr = \pi\int_0^{\infty} e^{-st}\, dt = \frac{\pi}{s}$$

Hence, $N(s) = \sqrt{\pi/s}$, and Eq. (8-2) results because the real part of $N(s)$ is positive as is easy to see from (8-1).

Corollary

$$\int_{-\infty}^{\infty} e^{j\beta t^2}\, dt = \sqrt{\frac{j\pi}{\beta}} = \sqrt{\frac{\pi}{\beta}}\, e^{j\pi/4} \tag{8-3}$$

The above follows from (8-2) with $s = \alpha - j\beta$ and $\alpha \to 0$. We can take the limit under the integral in (8-2) because the integral exists for $s = -j\beta$, although the integrand does not tend to zero as $t \to \infty$.

We shall now show that, for any $c = a + jb$,

$$\int_{-\infty}^{\infty} e^{-s(t+c)^2}\, dt = \sqrt{\frac{\pi}{s}} \qquad s = \alpha + j\beta \qquad \alpha > 0 \tag{8-4}$$

Proof The transformation $t + c = z$ maps the real t axis L into a line L_c passing through the point c and parallel to L (Fig. 8-1). If c is real, then L_c coincides with L, and (8-4) follows from (8-2). For c complex, however, the line integral

$$\int_{L_c} e^{-sz^2}\, dz$$

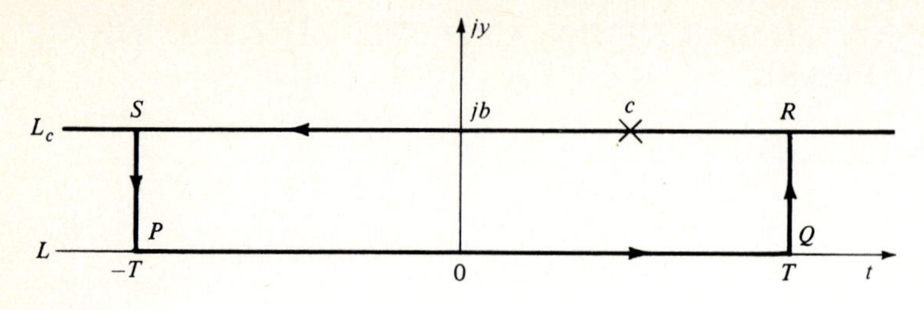

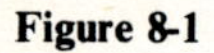
Figure 8-1

results. We shall reduce it to (8-2) using Cauchy's theorem (see footnote page 225).

The function e^{-sz^2} is obviously analytic in the entire z plane. Hence, the integral along the closed curve $PQRSP$ of Fig. 8-1 is zero:

$$\oint_{PQRSP} e^{-sz^2}\,dz = 0 \tag{8-5}$$

Along QR, $z = T + jy$; hence,

$$\text{Re}\, sz^2 = \alpha(T^2 - y^2) - 2\beta y T > \alpha(T^2 - b^2) - 2|\beta b| T \to \infty \quad \text{as} \quad T \to \infty$$

This leads to the conclusion that (elaborate)

$$\int_{QR} e^{-sz^2}\,dz \underset{T\to\infty}{\longrightarrow} 0$$

Similarly, the integral along SP tends to zero as $T \to \infty$. Furthermore,

$$\int_{SR} e^{-sz^2}\,dz \underset{T\to\infty}{\rightarrow} \int_{L_c} e^{-sz^2}\,dz \qquad \int_{PQ} e^{-sz^2}\,dz \underset{T\to\infty}{\rightarrow} \int_{-\infty}^{\infty} e^{-st^2}\,dt$$

From the above and (8-5), it follows, with $T \to \infty$, that

$$\int_{L_c} e^{-sz^2}\,dz - \int_{-\infty}^{\infty} e^{-st^2}\,dt = 0$$

Hence, Eq. (8-4) is true for any c. By a limiting argument, we can prove, as in (8-3), that (8-4) holds also if $\alpha = 0$.

Example 8-1 Using (8-4) we shall show that

$$e^{-st^2} \leftrightarrow \sqrt{\frac{\pi}{s}}\, e^{-\omega^2/4s} \qquad \text{Re}\, s \geq 0 \tag{8-6}$$

Proof The Fourier transform of e^{-st^2} is given by

$$F(\omega) = \int_{-\infty}^{\infty} e^{-st^2} e^{-j\omega t}\,dt$$

Since

$$st^2 + j\omega t = s\left(t + \frac{j\omega}{2s}\right)^2 + \frac{\omega^2}{4s} \tag{8-7}$$

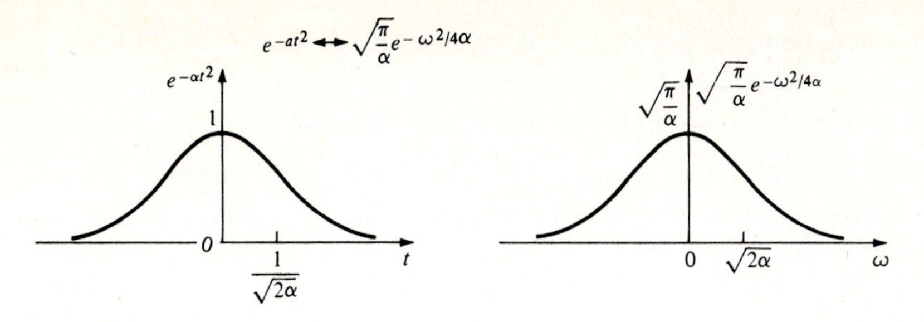

Figure 8-2

it follows that

$$F(\omega) = e^{-\omega^2/4s} \int_{-\infty}^{\infty} e^{-s(t + j\omega/2s)^2}\, dt$$

The above integral equals $\sqrt{\pi/s}$, as we see from (8-4) with $c = j\omega/2s$, and (8-6) results.

Special case If $s = \alpha$ is real, then (8-6) yields

$$e^{-\alpha t^2} \leftrightarrow \sqrt{\frac{\pi}{\alpha}}\, e^{-\omega^2/4\alpha} \tag{8-8}$$

Thus, a normal curve equals its own Fourier transform, properly scaled (Fig. 8-2).

Fresnel integrals A useful function in many applications is the integral

$$K(x) = \int_0^x e^{j\pi\tau^2/2}\, d\tau = C(x) + jS(x) \tag{8-9}$$

Its real and imaginary parts

$$C(x) = \int_0^x \cos\frac{\pi}{2}\tau^2\, d\tau \qquad S(x) = \int_0^x \sin\frac{\pi}{2}\tau^2\, d\tau \tag{8-10}$$

are the Fresnel integrals (Fig. 8-3).

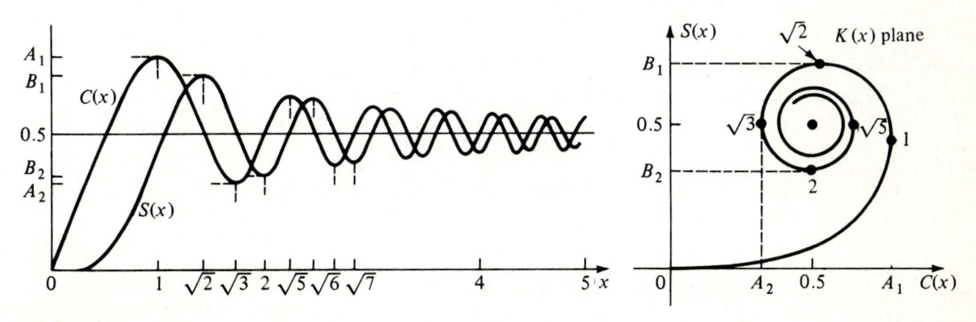

Figure 8-3

Since the function $e^{j\beta t^2}$ is even, it follows from (8-2), with $\beta = \pi/2$, that

$$K(\infty) = \sqrt{\frac{j}{2}} \qquad C(\infty) = \frac{1}{2} \qquad S(\infty) = \frac{1}{2} \tag{8-11}$$

Cornu spiral For each real x, $K(x)$ is a point in the complex plane. The locus of these points as x varies from 0 to ∞ is a curve known as the *Cornu spiral* (Fig. 8-3). The length of this curve from the origin to the point $K(x)$ equals x. This follows when (8-9) is differentiated:

$$\left|\frac{dK(x)}{dx}\right| = |e^{j\pi x^2/2}| = 1 \qquad \text{and} \qquad \int_0^x |dK(\tau)|\,d\tau = \int_0^x d\tau = x$$

Example 8-2 We shall show that the transform $F(\omega)$ of the signal

$$f(t) = r(t)e^{j\beta t^2} \qquad r(t) = \begin{cases} 1 & a < t < b \\ 0 & \text{otherwise} \end{cases}$$

is given by

$$F(\omega) = \sqrt{\frac{\pi}{2\beta}}\left[K\left(\frac{2\beta b - \omega}{\sqrt{2\beta\pi}}\right) - K\left(\frac{2\beta a - \omega}{\sqrt{2\beta\pi}}\right)\right]e^{-j\omega^2/4\beta} \tag{8-12}$$

Proof As in (8-7),

$$F(\omega) = \int_a^b e^{j\beta t^2}e^{-j\omega t}\,dt = e^{-j\omega^2/4\beta}\int_a^b e^{j\beta(t-\omega/2\beta)^2}\,dt$$

and with

$$\sqrt{\beta}\left(t - \frac{\omega}{2\beta}\right) = \tau\sqrt{\frac{\pi}{2}}$$

(8-12) follows from (8-9).

Frequency Modulation

Consider a monotone increasing function $\theta(t)$, as in Fig. 8-4. To every t we can associate an interval t_1 such that

$$\theta(t + t_1) = \theta(t) + 2\pi$$

Clearly, t_1 depends on t and is a constant only if $\theta(t) = \omega_0 t$.

We next form the function

$$y(t) = r(t)\cos\theta(t) \tag{8-13}$$

If, in any interval $(t, t + t_1)$, the envelope $r(t)$ of $y(t)$ and the slope

$$\omega(t) = \theta'(t) \tag{8-14}$$

of $\theta(t)$ have negligible variations, i.e., if

$$r(t + \tau) \simeq r(t) \qquad \omega(t + \tau) \simeq \omega(t) \qquad \theta(t + \tau) \simeq \theta(t) + \omega(t)\tau \qquad 0 \le \tau \le t_1$$

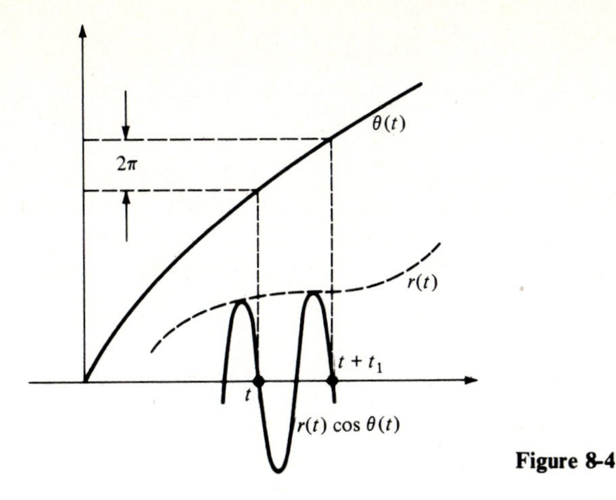

Figure 8-4

then, in this interval, $y(t)$ is approximately a sine wave in the variable τ

$$y(t+\tau) \simeq r(t)\cos[\omega(t)\tau + \theta(t)] \qquad 0 \le \tau \le t_1$$

and $\omega(t)$ is its *instantaneous frequency*. The signal $y(t)$ is amplitude- and frequency-modulated.

At the end of this section, we develop a method for the approximate evaluation of the transform $Y(\omega)$ of $y(t)$. As preparation, we consider an important special case.

Linear FM

$$\text{If} \qquad y(t) = r(t)\cos(\omega_0 t + \beta t^2) \qquad \text{then} \qquad \omega(t) = \omega_0 + 2\beta t \tag{8-15}$$

Thus, the instantaneous frequency of the above signal is linear in t. We shall discuss its spectral properties.

Consider the signal

$$f(t) = r(t)e^{j\beta t^2} \tag{8-16}$$

$$\text{With} \qquad F(\omega) = \int_{-\infty}^{\infty} r(t)e^{j\beta t^2}e^{-j\omega t}\,dt = e^{-j\omega^2/4\beta}\int_{-\infty}^{\infty} r(t)e^{j\beta(t-\omega/2\beta)^2}\,dt \tag{8-17}$$

its transform, it is easy to see that the transform of $y(t)$ is given by

$$Y(\omega) = \tfrac{1}{2}[F(\omega-\omega_0) + F^*(-\omega-\omega_0)] \tag{8-18}$$

It suffices, therefore, to find $F(\omega)$.

In general, the integral in (8-17) cannot be evaluated in closed form. However, if $r(t)$ is sufficiently smooth, then a simple approximate solution can be found. The result is based on the representation of the delta function as a limit [see Eq. (3-C-16)]:

$$\delta(t) = \lim_{c\to 0}\frac{1}{c\sqrt{j\pi}}\,e^{jt^2/c^2} \tag{8-19}$$

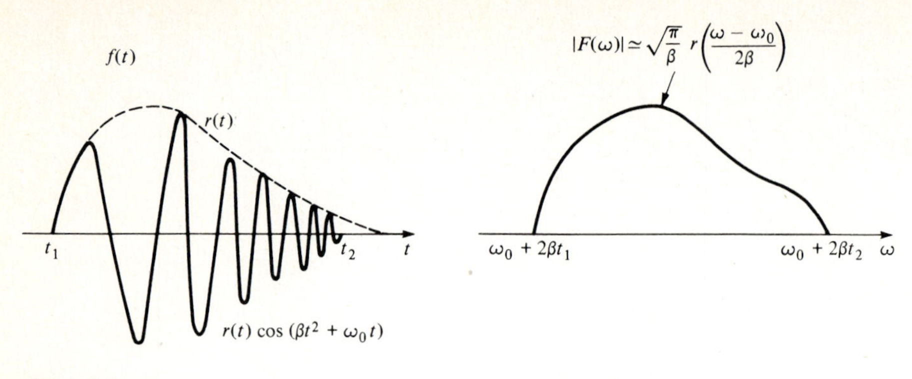

Figure 8-5

This leads to the approximation

$$\frac{1}{c\sqrt{j\pi}}\int_{-\infty}^{\infty} r(t)e^{j(t-t_0)^2/c^2}\,dt \simeq r(t_0) \tag{8-20}$$

for any $r(t)$ continuous at t_0 and for c sufficiently small.

Theorem 1 If the envelope $r(t)$ of the linear FM signal $f(t)$ is continuous, then for sufficiently large β (Fig. 8-5)

$$F(\omega) \simeq \sqrt{\frac{j\pi}{\beta}}\, e^{-j\omega^2/4\beta} r\left(\frac{\omega}{2\beta}\right) \tag{8-21}$$

Proof It follows from (8-20) with $t_0 = \omega/2\beta$ and $c = 1/\sqrt{\beta}$ [see Eq. (8-17)].

Notes:

1. The integral in (8-17) is a convolution

$$r(x) * e^{j\beta x^2}$$

evaluated at $x = \omega/2\beta$, and (8-21) is the first term in its moment expansion (4-9). Thus, $F(\omega)$ can be written as an asymptotic series [see (4-17)]

$$F(\omega) \simeq \sqrt{\frac{j\pi}{\beta}}\, e^{-j\omega^2/4\beta}\left[r\left(\frac{\omega}{2\beta}\right) + \frac{j}{4\beta} r''\left(\frac{\omega}{2\beta}\right) + \cdots\right] \tag{8-22}$$

2. As in any asymptotic expansion, it is difficult to give precise conditions for the validity of (8-21). However, the following gives some indication of the required size of β: If $t_0 = \omega/2\beta$ and

$$\left|\frac{r'(t)}{r(t_0)}\right| \ll \sqrt{\frac{\beta}{2\pi}} \qquad \text{for } |t - t_0| < \sqrt{\frac{2\pi}{\beta}} \tag{8-23}$$

then the error in the approximation (8-21) is small compared to $F(\omega)$.

To justify the above, we plot, in Fig. 8-6, the functions $\cos \beta(t - t_0)^2$ and $\sin \beta(t - t_0)^2$. As we see from the figure, these functions consist essen-

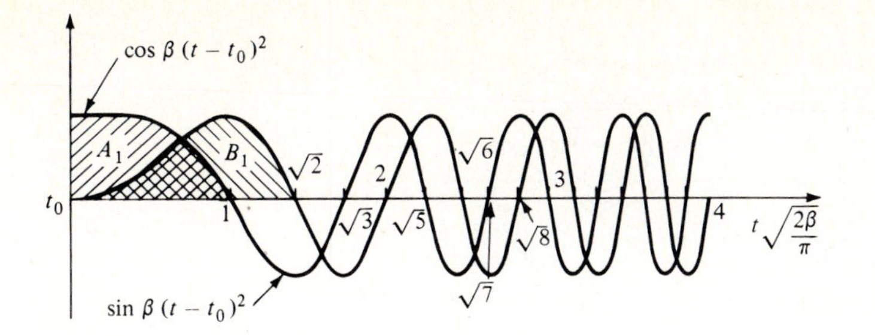

Figure 8-6

tially of pieces of sine waves for t outside the interval $(t_0 - \sqrt{2\pi/\beta}, t_0 + \sqrt{2\pi/\beta})$. Hence, the main contribution to the integral in Eq. (8-17) comes from this interval. If, therefore,

$$|r(t) - r(t_0)| = |r'(t_1)(t - t_0)| \ll r(t_0) \qquad \text{for } |t - t_0| < \sqrt{\frac{2\pi}{\beta}}$$

then $r(t)$ can be replaced with $r(t_0)$ in (8-17), and (8-21) results [see (8-3)].

3. If $r(t) = 0$ for $|t| > T$ and $\omega_0 > 2\beta T$, then the functions $F(\omega - \omega_0)$ and $F(-\omega - \omega_0)$ do not overlap. Therefore [see (8-18) and (8-21)],

$$Y(\omega) \simeq \frac{1}{2}\sqrt{\frac{j\pi}{\beta}}\, e^{-j(\omega - \omega_0)^2/2\beta}\, r\left(\frac{\omega - \omega_0}{2\beta}\right) \qquad \omega > 0$$

Thus, the amplitude of the transform $Y(\omega)$ of the linear FM signal $r(t)\cos(\omega_0 t + \beta t^2)$ equals the function $r(t)$, properly shifted and scaled, and its phase is a parabola centered at $\omega = \omega_0$.

4. The approximation (8-21) can be used to evaluate $F(\omega)$ even if $r(t)$ is not continuous. Indeed, denoting by t_i the discontinuity points of $r(t)$, and by A_i the corresponding jumps, we can write $r(t)$ as a sum

$$r(t) = r_c(t) + \sum_i A_i U(t - t_i)$$

where $r_c(t)$ is a continuous function. The part of $F(\omega)$ due to $r_c(t)$ can be found from (8-21). It suffices, therefore, to find the transform of the step-modulated FM signal $A_i U(t - t_i)e^{j\beta t^2}$. As we show in the next example, this transform can be expressed in terms of the Fresnel integral $K(x)$.

Example 8-3 From (8-12), it follows, with $a = 0$ and $b = \infty$, that

$$U(t)e^{j\beta t^2} \leftrightarrow \sqrt{\frac{\pi}{2\beta}}\left[\sqrt{\frac{j}{2}} + K\left(\frac{\omega}{\sqrt{2\beta\pi}}\right)\right]e^{-j\omega^2/4\beta} \tag{8-24}$$

because $K(\infty) = \sqrt{j/2}$ and $K(-x) = -K(x)$. The expression in brackets represents a vector $V(\omega)$ from the point $-\sqrt{j/2}$ to the point $K(\omega/\sqrt{2\beta\pi})$ of the Cornu spiral (Fig.

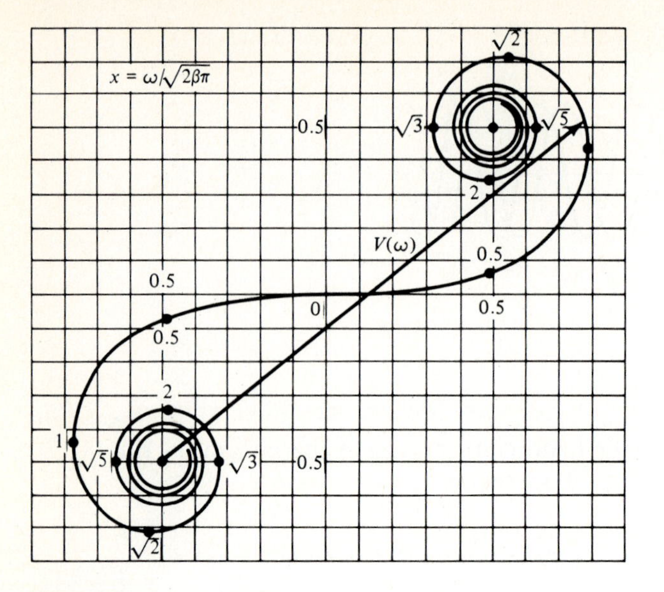

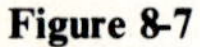
Figure 8-7

8-7). As we see from the figure,

$$V(\omega) \simeq \sqrt{2j}U(\omega) \qquad \text{for } |\omega| \gg \sqrt{2\beta\pi} \tag{8-25}$$

This shows that the approximation (8-21) also holds if $r(t)$ is discontinuous at t_i, provided that $|\omega - 2\beta t_i| \gg \sqrt{2\beta\pi}$ (elaborate).

In the next two examples, we compute $F(\omega)$ exactly and compare the results with the approximation (8-21).

Example 8-4 With $a = -T$ and $b = T$, Eq. (8-12) yields the transform of a pulse-modulated FM signal:

$$p_T(t)e^{j\beta t^2} \leftrightarrow \sqrt{\frac{\pi}{2\beta}}\left[K\left(\frac{2\beta T - \omega}{\sqrt{2\beta\pi}}\right) + K\left(\frac{2\beta T + \omega}{\sqrt{2\beta\pi}}\right)\right]e^{-j\omega^2/4\beta} \tag{8-26}$$

It is easy to see that if

$$|\omega \pm 2\beta T| \gg \sqrt{2\beta\pi}$$

then the expression in brackets in (8-26) equals $\sqrt{2j}\, p_T(\omega/2\beta)$ (elaborate). Thus, away from the discontinuity points, (8-26) agrees with (8-21).

In Fig. 8-8 we show the signal $p_T(t)\cos(\omega_0 t + \beta t^2)$ and its transform as obtained from (8-26) and (8-18).

Example 8-5 The pair

$$e^{-\alpha t^2}e^{j\beta t^2} \leftrightarrow \sqrt{\frac{j\pi}{\beta + j\alpha}}\, e^{-j\omega^2/4\beta_1}e^{-\alpha_1(\omega/2\beta)^2} \tag{8-27}$$

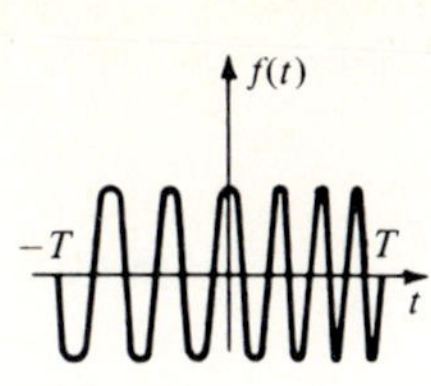

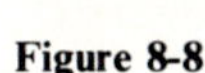

Figure 8-8

where
$$\alpha_1 = \frac{\alpha}{1 + \alpha^2/\beta^2} \qquad \beta_1 = \beta\left(1 + \frac{\alpha^2}{\beta^2}\right)$$

follows from (8-6) with $s = \alpha - j\beta$ (elaborate). In this example, $r(t) = e^{-\alpha t^2}$ is a normal curve. Clearly, if $\alpha \ll \beta$, then $\alpha_1 \simeq \alpha$ and $\beta_1 \simeq \beta$. Under this condition, (8-27) agrees with (8-21).

The Method of Stationary Phase[1]

The approximation (8-20) can be used to determine the transform of an FM signal with arbitrary phase. The result is based on the following.

Theorem 2 If the function $r(t)$ is continuous and the derivative of the function $\mu(t)$ vanishes at only a single point $t = t_0$ in the interval $(-\infty, \infty)$:

$$\mu'(t_0) = 0 \qquad \mu''(t_0) \neq 0$$

then, for sufficiently large k,

$$\int_{-\infty}^{\infty} r(t)e^{jk\mu(t)}\, dt \simeq e^{jk\mu(t_0)} r(t_0)\sqrt{\frac{2\pi j}{k\mu''(t_0)}} \tag{8-28}$$

Proof Since $\mu'(t) \neq 0$ in any interval that does not include the point t_0, we conclude, as in the proof of the Riemann lemma (3-B-4), that the integration over any such interval tends to zero as $k \to \infty$. Therefore, the major contribution to the integral in (8-28) comes from a small neighborhood near the point t_0. In this neighborhood,

$$\mu(t) \simeq \mu(t_0) + \frac{\mu''(t_0)}{2}(t - t_0)^2$$

Hence,
$$\int_{-\infty}^{\infty} r(t)e^{jk\mu(t)}\, dt \simeq e^{jk\mu(t_0)} \int_{-\infty}^{\infty} r(t)e^{jk\mu''(t_0)(t-t_0)^2/2}\, dt \tag{8-29}$$

and (8-29) follows from (8-20) with $k\mu''(t_0)/2 = 1/c^2$.

[1] F. T. Copson, "The Asymptotic Expansion of a Function Defined by a Definite Integral or Contour Integral," The Admiralty, London, S.R.E./ACS 106, 1946.

Notes:

1. Reasoning as in (8-23), we can show that the approximation (8-28) is satisfactory if

$$\left|\frac{r'(t)}{r(t_0)}\right| \ll \sqrt{\frac{k\mu''(t_0)}{4\pi}} \qquad \text{for } |t - t_0| < \epsilon = \sqrt{\frac{4\pi}{k\mu''(t_0)}} \tag{8-30}$$

If $r(t)$ is discontinuous at $t = t_c$, then (8-28) still holds, provided that $|t_0 - t_c| > \epsilon$ (elaborate).

2. If $\mu'(t)$ vanishes at several points $t = t_i$, then we must substitute the points t_i for the point t_0 in the right side of (8-28) and sum the results. If $\mu''(t_0) = 0$, then we must consider the next term in the asymptotic expansion of the integral in (8-29).

Example 8-6 We shall show that, for sufficiently large x,

$$J_0(x) = \frac{1}{2\pi}\int_{-\pi}^{\pi} e^{jx \sin t}\, dt \simeq \sqrt{\frac{2}{\pi x}} \cos\left(x - \frac{\pi}{4}\right) \tag{8-31}$$

Proof The above integral is of the form (8-28), where $r(t) = p_\pi(t)/2\pi$, $k = x$, and $\mu(t) = \sin t$. In the interval $(-\pi, \pi)$, $\mu'(t) = \cos t = 0$ for $t = t_1 = \pi/2$ and $t = t_2 = -\pi/2$. Since $\mu(t_1) = 1$, $\mu''(t_1) = -1$, $\mu(t_2) = -1$, and $\mu''(t_2) = 1$, we conclude that

$$J_0(x) \simeq \frac{e^{jx}}{2\pi}\sqrt{\frac{-2\pi j}{x}} + \frac{e^{-jx}}{2\pi}\sqrt{\frac{2\pi j}{x}}$$

and (8-31) results. From note 1 above, it follows that the approximation (8-31) is satisfactory if $|x| > 16/\pi$, because the distance from the stationary points $\pm\pi/2$ to the discontinuity points $\pm\pi$ is $\pi/2$.

Theorem 3 If the function $\theta(t)$ is monotone increasing or decreasing and the function $r(t)$ is sufficiently smooth, then

$$r(t)e^{j\theta(t)} \leftrightarrow e^{j[\theta(t_0) - \omega t_0]} r(t_0)\sqrt{\frac{2\pi j}{\theta''(t_0)}} \tag{8-32}$$

where t_0 is such that (Fig. 8-9)

$$\theta'(t_0) = \omega \tag{8-33}$$

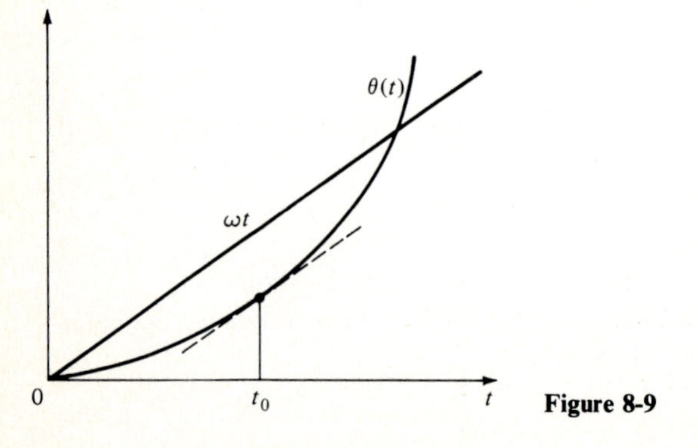

Figure 8-9

Proof The integral

$$\int_{-\infty}^{\infty} r(t)e^{j\theta(t)}e^{-j\omega t}\,dt$$

equals the integral in (8-28) if $k\mu(t) = \theta(t) - \omega t$. Hence, (8-32) follows from (8-28).

8-2 UNCERTAINTY PRINCIPLE AND SOPHISTICATED SIGNALS

If two functions $f(t)$ and $F(\omega)$ form a Fourier-integral pair, then they cannot both be of short duration. This is supported by the scaling theorem

$$af(at) \leftrightarrow F\left(\frac{\omega}{a}\right)$$

The above assertion, known as the *uncertainty principle*, can be given various interpretations, depending on the meaning of the term "duration."

Classical definition We use, as a measure of the duration of $f(t)$ and $F(\omega)$, the numbers d and D defined by

$$d^2 = \frac{1}{E}\int_{-\infty}^{\infty} t^2|f(t)|^2\,dt \qquad D^2 = \frac{1}{2\pi E}\int_{-\infty}^{\infty} \omega^2|F(\omega)|^2\,d\omega \tag{8-34}$$

where

$$E = \int_{-\infty}^{\infty} |f(t)|^2\,dt = \frac{1}{2\pi}\int_{-\infty}^{\infty} |F(\omega)|^2\,d\omega$$

Theorem 1 If

$$\sqrt{t}\,f(t) \to 0 \qquad \text{as } |t| \to \infty \tag{8-35}$$

then

$$Dd \geq \tfrac{1}{2} \tag{8-36}$$

The above is an equality only if $f(t)$ is a normal curve:

$$f(t) = Ae^{-\alpha t^2} \tag{8-37}$$

Proof We shall assume for simplicity that $f(t)$ is real and $E = 1$. From Schwarz' inequality (4-A-1), it follows that

$$\left|\int_{-\infty}^{\infty} tf\,\frac{df}{dt}\,dt\right|^2 \leq \int_{-\infty}^{\infty} t^2f^2\,dt \int_{-\infty}^{\infty} \left|\frac{df}{dt}\right|^2\,dt \tag{8-38}$$

Furthermore [see (8-35)],

$$\int_{-\infty}^{\infty} tf\,\frac{df}{dt}\,dt = \int_{-\infty}^{\infty} t\,\frac{df^2}{2} = t\,\frac{f^2}{2}\Bigg|_{-\infty}^{\infty} - \frac{1}{2}\int_{-\infty}^{\infty} f^2\,dt = -\frac{1}{2}$$

and

$$\int_{-\infty}^{\infty} \left|\frac{df}{dt}\right|^2\,dt = \frac{1}{2\pi}\int_{-\infty}^{\infty} \omega^2|F(\omega)|^2\,d\omega$$

because $f'(t) \leftrightarrow j\omega F(\omega)$. Inserting the last two equations into (8-38), we obtain (8-36).

If (8-36) is an equality, then (8-38) must also be an equality. This is possible only if $f'(t) = ktf(t)$, that is, if $f(t)$ is given by Eq. (8-37).

Energy concentration We next use, as a definition of the duration of $f(t)$ and $F(\omega)$, the numbers τ and σ defined by

$$\alpha = \frac{1}{E}\int_{-\tau}^{\tau} |f(t)|^2\, dt \qquad \beta = \frac{1}{2\pi E}\int_{-\sigma}^{\sigma} |F(\omega)|^2\, d\omega \tag{8-39}$$

where α and β are two *given constants*. Clearly, τ and σ depend on the form of $f(t)$. We shall determine the minimum value of the product

$$c = \tau\sigma$$

as $f(t)$ ranges over all functions.

As we have seen in Sec. 6-4, the maximum eigenvalue of the integral equation (6-80) is a function $\lambda_0(c)$ of c increasing monotonically from 0 to 1 as c increases from 0 to ∞. It has, therefore, an inverse which we shall denote by $c(\lambda_0)$ (Fig. 8-10).

Theorem 2

$$\tau\sigma \geq c(\lambda_0) \tag{8-40}$$

where $$\lambda_0 = \cos^2(\theta_1 + \theta_2) \qquad \cos^2\theta_1 = \alpha \qquad \cos^2\theta_2 = \beta \tag{8-41}$$

Equality holds only if $f(t) = a\varphi(t)p_\tau(t) + b\varphi(t)$ as in Eq. (6-120).

Proof As we have shown on page 212, if the product $c = \tau\sigma$ is fixed, then the point (α, β) must be in the region R_c of Fig. 6-13 bounded by the curve $\bar{\beta} = \cos^2(\theta_0 - \theta_1)$, where $\cos^2\theta_0 = \lambda_0$. Furthermore, if $c_1 \leq c_2$, then $R_{c_1} \subset R_{c_2}$. Therefore, for specified values of α and β, the product $c = \tau\sigma$ is minimum if the point (α, β) is on the boundary of the region R_c, that is, if $\beta = \bar{\beta} = \cos^2(\theta_0 - \theta_1)$.

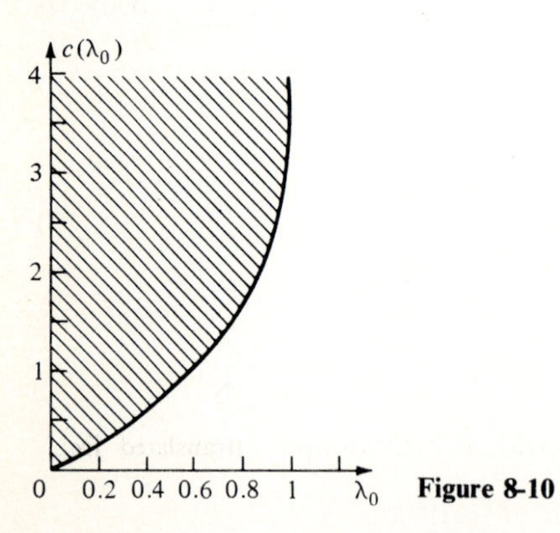

Figure 8-10

Example 8-7 We shall determine the minimum of the product $\tau\sigma$ for $\alpha = \beta = 0.9$.

Inserting into (8-41), we find $\theta_1 = \theta_2 = 18° \ 10'$ and $\lambda_0 = \cos^2 36° \ 20' = 0.65$. With this value of λ_0, the curve in Fig. 8-10 yields $c(\lambda_0) = 1.2$; hence, $\tau\sigma \geq 1.2$.

Time-limited signals If $f(t) = 0$ for $|t| > \tau$, then it is natural to consider the interval 2τ as a measure of the duration of $f(t)$. We shall reexamine the last two theorems using this measure.

Theorem 3 With D defined as in (8-34),

$$\tau D \geq \frac{\pi}{2} \tag{8-42}$$

Equality holds only if $f(t) = k \cos(\pi t/2\tau) p_\tau(t)$.

Proof It follows from Eq. (6-65) with t and ω interchanged.

Theorem 4 Given a number β, if σ is defined as in (8-39), then

$$\tau\sigma \geq c(\beta) \tag{8-43}$$

Equality holds only if $f(t) = a\varphi(t)p_\tau(t)$.

Proof Since $c(\lambda_0)$ is the inverse of $\lambda_0(c)$, where $c = \tau\sigma$, (8-43) is equivalent to $\beta \leq \lambda_0$. Hence, (8-43) is a consequence of (6-112).

Sophisticated Signals[1]

In the preceding definitions or in any other reasonable definitions of signal duration, the minimum m of the time-frequency duration product P is of the order of 1. The uncertainty principle states that P cannot be less than m, but it imposes no upper bound on P. In fact, as we shall presently see, we can have signals whose duration product P is arbitrarily large.

We shall say, loosely, that a signal $f(t)$ is *simple* if its product P is near the minimum m. All the examples in Chap. 3 are simple signals. We shall say that a signal is *sophisticated* if P is large compared to m. In Sec. 8-1 we dealt with such signals.

Consider a signal $f(t)$ with transform

$$F(\omega) = A(\omega)e^{j\varphi(\omega)}$$

As we know [see (3-39)],

$$\int_{-\infty}^{\infty} t^2 |f(t)|^2 \, dt = \frac{1}{2\pi} \int_{-\infty}^{\infty} \left[\left(\frac{dA}{d\omega} \right)^2 + A^2 \left(\frac{d\varphi}{d\omega} \right)^2 \right] d\omega \tag{8-44}$$

[1] D. E. Vackman, "Sophisticated Signals and the Uncertainty Principle" (translated from Russian), p. 49, Springer Publishing Co., Inc., New York, 1958.

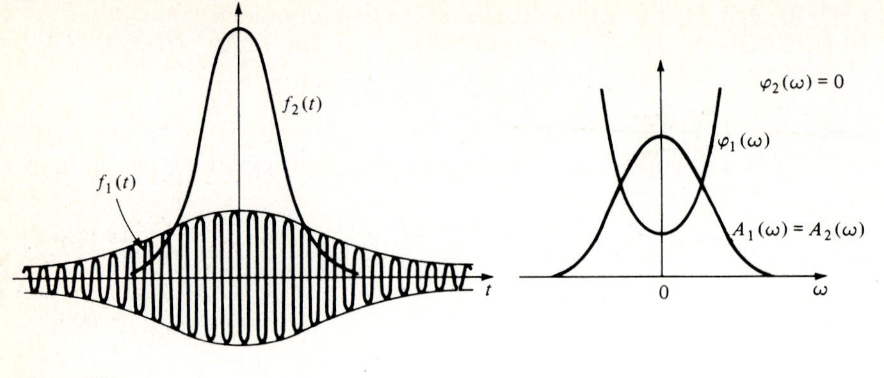

Figure 8-11

With d and D as in (8-34), it follows from the above that d depends on $A(\omega)$ and on $\varphi(\omega)$, whereas D depends only on $A(\omega)$. If, therefore, $f(t)$ is a simple signal, then its phase $\varphi(\omega)$ is smooth. Keeping $A(\omega)$ constant and choosing for $\varphi(\omega)$ a rapidly varying function, we obtain a sophisticated signal whose product Dd can be made arbitrarily large.

Example 8-8 From (8-27) it follows that, if $\beta \gg \alpha$, then

$$f_1(t) = e^{-\alpha t^2} e^{j\beta t^2} \leftrightarrow \sqrt{\frac{j\pi}{\beta}}\, e^{-\alpha\omega^2/4\beta^2} e^{-j\omega^2/4\beta} = F_1(\omega)$$

The computation in (8-34) yields $d^2 = 1/4\alpha$ and $D^2 = \beta^2/\alpha$. Hence, $f_1(t)$ is a sophisticated signal with

$$Dd = \frac{\beta}{2\alpha}$$

From (8-8), it follows that (Fig. 8-11)

$$f_2(t) = \sqrt{\frac{\beta}{\alpha}}\, e^{-\beta^2 t^2/\alpha} \leftrightarrow \sqrt{\frac{\pi}{\beta}}\, e^{-\alpha\omega^2/4\beta^2} = F_2(\omega)$$

Clearly, $f_2(t)$ is a simple signal with $Dd = \frac{1}{2}$ (see also pulse compression, page 294).

We determine next sophisticated signals with specified shapes.

Problem 1 Given a function $S(\omega) \geq 0$, find a linear FM signal

$$f(t) = r(t)e^{j\beta t^2}$$

of specified duration d and such that $|F(\omega)| = S(\omega)$.

Solution Since [see (8-21)]

$$r(t)e^{j\beta t^2} \leftrightarrow \sqrt{\frac{j\pi}{\beta}}\, e^{-j\omega^2/4\beta} r\left(\frac{\omega}{2\beta}\right)$$

it suffices to have

$$r(t) = \sqrt{\frac{\beta}{\pi}}\, S(2\beta t) \tag{8-45}$$

The constant β is determined from the duration requirement.

***Problem* 2** Given two functions $y(t) \geq 0$ and $S(\omega) \geq 0$ of equal energy, find an FM signal

$$f(t) = r(t)e^{j\theta(t)}$$

such that

$$|f(t)| = y(t) \qquad |F(\omega)| = S(\omega) \tag{8-46}$$

Solution With

$$\theta'(t) = w(t)$$

the instantaneous frequency of $f(t)$, we see from (8-32) that

$$r(t)e^{j\theta(t)} \leftrightarrow e^{j[\theta(t_0) - \omega t_0]} r(t_0)\sqrt{\frac{2\pi j}{w'(t_0)}} \qquad w(t_0) = \omega$$

Hence, Eq. (8-46) is satisfied if $r(t) = y(t)$ and

$$S(\omega) = y(t_0)\sqrt{\frac{2\pi}{w'(t_0)}} \qquad w(t_0) = \omega \tag{8-47}$$

This can be written in the form

$$\frac{dw}{dt_0} = 2\pi \frac{y^2(t_0)}{S^2(w)} \tag{8-48}$$

Integrating, we obtain

$$\frac{1}{2\pi}\int_{-\infty}^{w} S^2(\omega)\, d\omega = \int_{-\infty}^{t_0} y^2(t)\, dt \tag{8-49}$$

Thus, to find $w(t_0)$ for a specific $t = t_0$, we compute the area of $y^2(t)$ up to the point t_0 (Fig. 8-12) and then determine w from (8-49). The resulting function $w(t)$ is monotone increasing, and its integral yields $\theta(t)$.

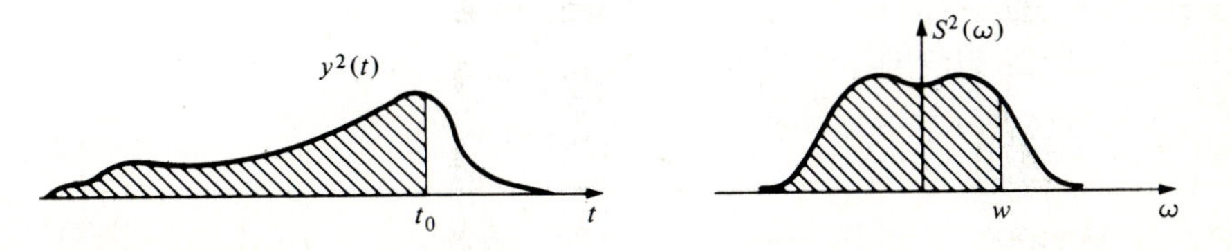

Figure 8-12

Note: The above holds only if the resulting $w(t)$ varies sufficiently rapidly. This is the case if the given functions $y(t)$ and $S(\omega)$ satisfy the condition

$$2S[w(t_0)] \ll \frac{y^2(t)}{y'(t_0)} \qquad \text{for } |t - t_0| < \frac{S[w(t_0)]}{y(t_0)}\sqrt{2} \tag{8-50}$$

This condition is a consequence of (8-30) and (8-48).

8-3 TWO-DIMENSIONAL TRANSFORMS AND HANKEL TRANSFORMS

The two-dimensional Fourier transform of a function $f(x, y)$ is the integral

$$F(u, v) = \int_{-\infty}^{\infty}\int_{-\infty}^{\infty} f(x, y)e^{-j(ux+vy)}\,dx\,dy \tag{8-51}$$

The function $F(u, v)$ can be expressed as a cascade of two one-dimensional transforms:

$$\Phi(u, y) = \int_{-\infty}^{\infty} f(x, y)e^{-jux}\,dx \qquad F(u, v) = \int_{-\infty}^{\infty} \Phi(u, y)e^{-jvy}\,dy \tag{8-52}$$

This representation permits us to derive various theorems for two-dimensional transforms in terms of the corresponding theorems for one-dimensional transforms. Thus, from the inversion formula (3-2), it follows that

$$\Phi(u, y) = \frac{1}{2\pi}\int_{-\infty}^{\infty} F(u, v)e^{jvy}\,dv \qquad f(x, y) = \frac{1}{2\pi}\int_{-\infty}^{\infty} \Phi(u, y)e^{jux}\,du$$

Hence,

$$f(x, y) = \frac{1}{4\pi^2}\int_{-\infty}^{\infty}\int_{-\infty}^{\infty} F(u, v)e^{j(ux+vy)}\,du\,dv \tag{8-53}$$

Equations (8-51) and (8-53) establish a one-to-one correspondence between the functions $f(x, y)$ and $F(u, v)$. This correspondence will be denoted by

$$f(x, y) \Leftrightarrow F(u, v)$$

or, in cascade form, by

$$f(x, y) \underset{x}{\leftrightarrow} \Phi(u, y) \underset{y}{\leftrightarrow} F(u, v)$$

Theorems The following theorems follow readily from (8-51) or (8-52) (elaborate):

$$f^*(x, y) \Leftrightarrow F^*(-u, -v) \qquad F(x, y) \Leftrightarrow 4\pi^2 f(-u, -v) \tag{8-54}$$

$$\frac{\partial f(x, y)}{\partial x} \Leftrightarrow juF(u, v) \qquad \frac{\partial f(x, y)}{\partial y} \Leftrightarrow jvF(u, v) \tag{8-55}$$

$$-jxf(x, y) \Leftrightarrow \frac{\partial F(u, v)}{\partial u} \qquad -jyf(x, y) \Leftrightarrow \frac{\partial F(u, v)}{\partial v} \tag{8-56}$$

$$\frac{\partial^2 f}{\partial x^2} + \frac{\partial^2 f}{\partial y^2} \Leftrightarrow -(u^2 + v^2)F(u, v) \tag{8-57}$$

$$f(ax, by) \Leftrightarrow \frac{1}{|ab|} F\left(\frac{u}{a}, \frac{v}{b}\right) \tag{8-58}$$

$$f(ax + by, cx + dy) \Leftrightarrow JF(Au + Cv, Bu + Dv) \tag{8-59}$$

$$\begin{pmatrix} A & B \\ C & D \end{pmatrix} = \begin{pmatrix} a & b \\ c & d \end{pmatrix}^{-1} \qquad J = |AD - BC|$$

If $f_1(x, y) \Leftrightarrow F_1(u, v)$ and $f_2(x, y) \Leftrightarrow F_2(u, v)$, then

$$\int_{-\infty}^{\infty} \int_{-\infty}^{\infty} f_1(x - \xi, y - \eta) f_2(\xi, \eta)\, d\xi\, d\eta \Leftrightarrow F_1(u, v)F_2(u, v) \tag{8-60}$$

$$\int_{-\infty}^{\infty} \int_{-\infty}^{\infty} f_1(x, y) f_2^*(x, y)\, dx\, dy = \frac{1}{4\pi^2} \int_{-\infty}^{\infty} \int_{-\infty}^{\infty} F_1(u, v) F_2^*(u, v)\, du\, dv \tag{8-61}$$

$$\int_{-\infty}^{\infty} \int_{-\infty}^{\infty} |f(x, y)|^2\, dx\, dy = \frac{1}{4\pi^2} \int_{-\infty}^{\infty} \int_{-\infty}^{\infty} |F(u, v)|^2\, du\, dv \tag{8-62}$$

If $f(x) \leftrightarrow F(u)$, then

$$f(x) \Leftrightarrow 2\pi F(u)\, \delta(v) \qquad f(x - y) \Leftrightarrow 2\pi F(u)\, \delta(u + v)$$

because

$$f(x) \underset{x}{\leftrightarrow} F(u) \underset{y}{\leftrightarrow} 2\pi F(u)\, \delta(v)$$

$$f(x - y) \underset{x}{\leftrightarrow} e^{-juy} F(u) \underset{y}{\leftrightarrow} 2\pi F(u)\, \delta(u + v)$$

If $f_1(x) \leftrightarrow F_1(u)$ and $f_2(y) \leftrightarrow F_2(v)$, then

$$f_1(x) f_2(y) \Leftrightarrow F_1(u) F_2(v) \tag{8-63}$$

Example 8-9 If $f(x, y)$ equals 1 in the rectangle $|x| < a$, $|y| < b$ and equals zero elsewhere:

$$f(x, y) = p_a(x) p_b(y)$$

then

$$F(u, v) = \frac{4 \sin au \sin bv}{uv}$$

Example 8-10 From (8-8) it follows that [see (8-63)]

$$e^{-\alpha(x^2 + y^2)} \Leftrightarrow \frac{\pi}{\alpha} e^{-(u^2 + v^2)/4\alpha} \tag{8-64}$$

Projection We shall call the function

$$g(x) = \int_{-\infty}^{\infty} f(x, y)\, dy \tag{8-65}$$

the *projection* of $f(x, y)$. With $G(u)$ the Fourier transform of $g(x)$, it follows from Eq. (8-51) that

$$F(u, 0) = G(u) \tag{8-66}$$

Hankel Transforms

Given a function $f(r)$, we form the integral

$$\bar{F}(w) = \int_0^\infty r f(r) J_0(wr)\, dr \tag{8-67}$$

where
$$J_0(x) = \frac{1}{2\pi} \int_{-\pi}^{\pi} e^{jx \sin\theta}\, d\theta \tag{8-68}$$

is the Bessel function of order zero (Fig. 8-13). If the integral in (8-67) exists for every $w \geq 0$, it defines a function $\bar{F}(w)$ known as the *Hankel transform* of $f(r)$:

$$f(r) \overset{h}{\leftrightarrow} \bar{F}(w)$$

The Hankel transform as Fourier transform We show next that the Hankel transform $\bar{F}(w)$ of $f(r)$ can be expressed as a one-dimensional and as a two-dimensional Fourier transform.

Theorem 1 If $f(\sqrt{x^2 + y^2}) \Leftrightarrow F(u, v)$, then

$$F(u, v) = 2\pi \bar{F}(\sqrt{u^2 + v^2}) \tag{8-69}$$

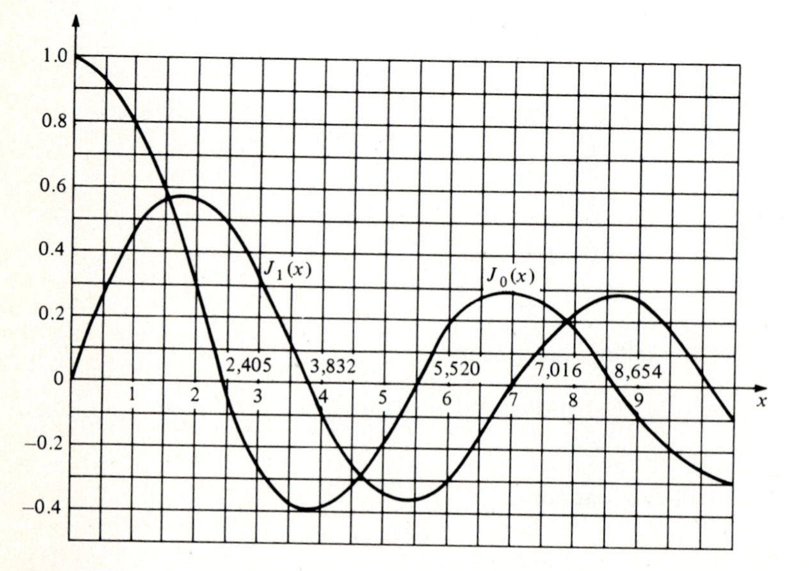

Figure 8-13

Proof With

$$x = r\cos\theta \qquad y = r\sin\theta \qquad r = \sqrt{x^2 + y^2}$$
$$u = w\cos\varphi \qquad v = w\sin\varphi \qquad w = \sqrt{u^2 + v^2} \tag{8-70}$$
$$ux + vy = wr\cos\theta\cos\varphi + wr\sin\theta\sin\varphi = wr\cos(\theta - \varphi)$$
$$dx\,dy = r\,dr\,d\theta$$

we have

$$F(u, v) = \int_{-\infty}^{\infty}\int_{-\infty}^{\infty} f(\sqrt{x^2 + y^2})e^{-j(ux+vy)}\,dx\,dy = \int_0^{\infty} rf(r)\int_{-\pi}^{\pi} e^{-jwr\cos(\theta-\varphi)}\,d\theta\,dr$$

It is easy to see that the last integral is independent of φ and equals $2\pi J_0(wr)$. Hence,

$$F(u, v) = 2\pi\int_0^{\infty} rf(r)J_0(wr)\,dr = 2\pi\bar{F}(w)$$

Example 8-11 From (8-69) and the pair (8-64), it follows that

$$e^{-\alpha r^2} \overset{h}{\leftrightarrow} \frac{1}{2\alpha}e^{-w^2/4\alpha} \tag{8-71}$$

Example 8-12 In this example, we shall use the well-known identity

$$\int_0^x tJ_0(t)\,dt = xJ_1(x) \tag{8-72}$$

where $J_1(x)$ is the Bessel function of order 1 (Fig. 8-13).
(*a*) If $f_1(r) = p_a(r)$ (Fig. 8-14), then

$$\bar{F}(w) = \int_0^a rJ_0(wr)\,dr = \frac{aJ_1(aw)}{w} \tag{8-73}$$

as we can see from Eq. (8-72) with $t = wr$ and $x = wa$.

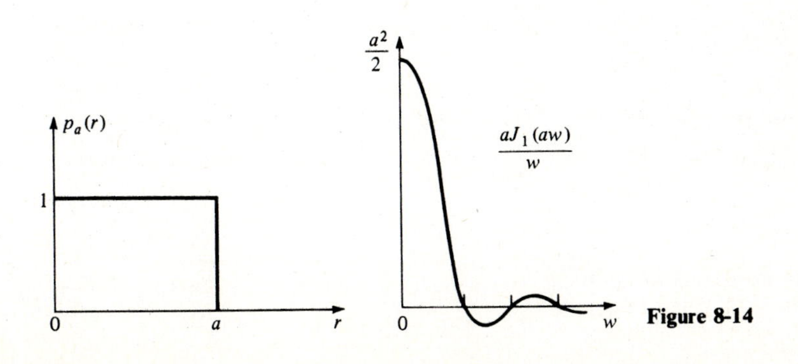

Figure 8-14

(*b*) From the above and (8-69), it follows that if

$$f(x, y) = \begin{cases} 1 & x^2 + y^2 \le a^2 \\ 0 & \text{otherwise} \end{cases}$$

then
$$F(u, v) = 2\pi a \frac{J_1(a\sqrt{u^2 + v^2})}{\sqrt{u^2 + v^2}} \tag{8-74}$$

Theorem 2 If $G(u)$ is the Fourier transform of the projection

$$g(x) = \int_{-\infty}^{\infty} f(\sqrt{x^2 + y^2})\, dy$$

of $f(r)$, then

$$G(u) = 2\pi \bar{F}(u) \tag{8-75}$$

Proof From (8-66) and (8-69), we conclude that $G(u) = F(u, 0) = 2\pi\bar{F}(u)$, and (8-75) results.

Corollary If a function $f(x, y) = f(r)$ has circular symmetry, then it is uniquely determined in terms of its projection $g(x)$.

Proof Given $g(x)$, we find $G(u)$ from Eq. (3-1), $\bar{F}(u)$ from (8-75), and $f(r)$ from (8-76).

Example 8-13 If $g(x) = e^{-\alpha x^2}$, then

$$G(u) = \sqrt{\frac{\pi}{\alpha}}\, e^{-u^2/4\alpha}$$

Hence,
$$\bar{F}(w) = \frac{1}{2\sqrt{\pi\alpha}} e^{-w^2/4\alpha} \qquad f(r) = \sqrt{\frac{\alpha}{\pi}}\, e^{-\alpha r^2}$$

Notes:

1. Except for a scale factor, the function $f(r)$ of the last example equals its projection $g(x)$. It can be shown (Prob. 89) that no other function has this property.
2. In Prob. 89 we show that[1] any function $f(r)$ can be expressed directly in terms of its projection $g(x)$.

Theorem 1 can be used to derive the properties of Hankel transforms in terms of the corresponding properties of Fourier transforms. We give next several illustrations.

[1] E. W. Marchand, Derivation of the Point Spread Function from the Line Spread Function, *J. Opt. Soc. Am.*, vol. 9, pp. 915–919, July, 1964.

Inversion formula

$$f(r) = \int_0^\infty w\bar{F}(w)J_0(rw)\,dw \tag{8-76}$$

Proof Inserting $F(u, v)$ as given by (8-69) into (8-53), we obtain

$$f(\sqrt{x^2+y^2}) = \frac{1}{2\pi}\int_{-\infty}^{\infty}\int_{-\infty}^{\infty} \bar{F}(\sqrt{u^2+v^2})e^{j(ux+vy)}\,du\,dv$$

$$= \frac{1}{2\pi}\int_0^\infty w\bar{F}(w)\int_{-\pi}^{\pi} e^{jrw\cos(\theta-\varphi)}\,d\varphi\,dw$$

and Eq. (8-76) results.

Parseval's formula If

$$f_1(r) \overset{h}{\leftrightarrow} \bar{F}_1(w), \qquad f_2(r) \overset{h}{\leftrightarrow} \bar{F}_2(w)$$

then

$$\int_0^\infty r f_1(r) f_2^*(r)\,dr = \int_0^\infty w\bar{F}_1(w)\bar{F}_2^*(w)\,dw \tag{8-77}$$

Proof It follows from (8-61) and (8-69) (elaborate).

Differentiation

$$\frac{d^2 f(r)}{dr^2} + \frac{1}{r}\frac{df(r)}{dr} \overset{h}{\leftrightarrow} -w^2\bar{F}(w) \tag{8-78}$$

Proof It follows from (8-69) because [see (8-57)]

$$\left(\frac{\partial}{\partial x^2} + \frac{\partial}{\partial y^2}\right) f(\sqrt{x^2+y^2}) = \frac{1}{r}\frac{d}{dr}\left[r\frac{df(r)}{dr}\right] \Leftrightarrow -(u^2+v^2)F(u, v)$$

Example 8-14 For any $a > 0$ [see (3-C-8)],

$$\int_0^\infty w\,\delta(w-a)\,J_0(rw)\,dw = aJ_0(ar)$$

Hence,

$$J_0(ar) \overset{h}{\leftrightarrow} \frac{1}{a}\,\delta(w-a) \tag{8-79}$$

Using the above, we shall show that the Bessel function $J_0(ar)$ satisfies the equation

$$\frac{d^2 J_0(ar)}{dr^2} + \frac{1}{r}\frac{dJ_0(ar)}{dr} = -a^2 J_0(ar) \tag{8-80}$$

Proof The Hankel transform of the left side equals $-w^2\,\delta(w-a)/a = -a\,\delta(w-a)$ [see (8-78)]. Equation (8-80) results because $a\,\delta(w-a)$ is the transform of $a^2 J_0(ar)$.

Fourier Series

A function $f(x, y)$ is doubly periodic if

$$f(x + mx_0, y + ny_0) = f(x, y) \tag{8-81}$$

for any integers m and n. The constants x_0 and y_0 are its periods.

We shall show that $f(x, y)$ can be written as a series

$$f(x, y) = \sum_{m=-\infty}^{\infty} \sum_{n=-\infty}^{\infty} a_{mn} e^{j(mu_0 x + nv_0 y)} \qquad u_0 = \frac{2\pi}{x_0} \qquad v_0 = \frac{2\pi}{y_0} \tag{8-82}$$

where

$$a_{mn} = \frac{1}{x_0 y_0} \int_0^{x_0} \int_0^{y_0} f(x, y) e^{-j(mu_0 x + nv_0 y)}\, dx\, dy \tag{8-83}$$

Proof From Eq. (8-81) it follows that $f(x + mx_0, y) = f(x, y)$. Thus, for any y, $f(x, y)$ is a periodic function of x. Therefore, it can be expanded into a series

$$f(x, y) = \sum_{m=-\infty}^{\infty} b_m(y) e^{jmu_0 x} \tag{8-84}$$

whose coefficients

$$b_m(y) = \frac{1}{x_0} \int_0^{x_0} f(x, y) e^{-jmu_0 x}\, dx \tag{8-85}$$

are functions of y. Since $f(x, y + ny_0) = f(x, y)$, it follows from (8-85) that the functions $b_m(y)$ are periodic. Hence,

$$b_m(y) = \sum_{n=-\infty}^{\infty} a_{mn} e^{jnv_0 y} \qquad a_{mn} = \frac{1}{y_0} \int_0^{y_0} b_m(y) e^{-jnv_0 y}\, dy \tag{8-86}$$

and (8-82) results.

8-4 THE AMBIGUITY FUNCTION

The ambiguity function of a signal $f(t)$ is the integral

$$\chi(u, \tau) = \frac{1}{E} \int_{-\infty}^{\infty} f\left(t + \frac{\tau}{2}\right) f^*\left(t - \frac{\tau}{2}\right) e^{-jut}\, dt \tag{8-87}$$

where E is the energy of $f(t)$. Thus, $\chi(u, \tau)$ is the Fourier transform in the variable t of the function

$$\gamma(t, \tau) = \frac{1}{E} f\left(t + \frac{\tau}{2}\right) f^*\left(t - \frac{\tau}{2}\right) \tag{8-88}$$

Example 8-15 If $f(t) = e^{-\alpha t^2}$, then $E = \sqrt{\pi/2\alpha}$ and

$$\gamma(t, \tau) = \sqrt{\frac{2\alpha}{\pi}}\, e^{-\alpha(t+\tau/2)^2} e^{-\alpha(t-\tau/2)^2} = \sqrt{\frac{2\alpha}{\pi}}\, e^{-2\alpha t^2} e^{-\alpha\tau^2/2}$$

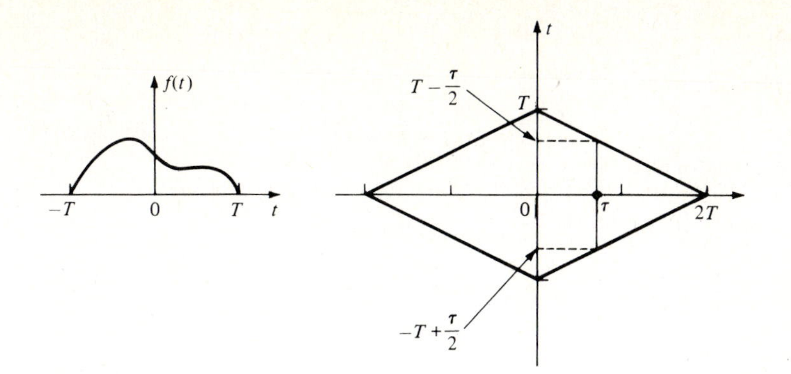

Figure 8-15

Hence [see (8-8)],

$$\chi(u, \tau) = e^{-u^2/8\alpha} e^{-\alpha\tau^2/2} \tag{8-89}$$

Time-limited functions If

$$f(t) = 0 \qquad \text{for } |t| > T$$

then $\gamma(t, \tau)$ is zero outside the rhombus of Fig. 8-15 bounded by the lines $t \pm \tau/2 = \pm T$. Hence,

$$\chi(u, \tau) = \frac{1}{E}\int_{-T+|\tau|/2}^{T-|\tau|/2} f\left(t + \frac{\tau}{2}\right) f^*\left(t - \frac{\tau}{2}\right) e^{-jut}\, dt \tag{8-90}$$

for $|\tau| < 2T$, and $\chi(u, \tau) = 0$ for $|\tau| > 2T$.

Example 8-16 If $f(t)$ is a pulse, $f(t) = p_T(t)$, then $E = 2T$ and Eq. (8-90) yields

$$\chi(u, \tau) = \frac{1}{2T} p_{2T}(\tau) \int_{-T+|\tau|/2}^{T-|\tau|/2} e^{-jut}\, dt = \frac{\sin (T - |\tau|/2)u}{Tu} p_{2T}(\tau)$$

We note that

$$\chi(0, \tau) = \left(1 - \frac{|\tau|}{2T}\right) p_{2T}(\tau) \qquad \chi(u, 0) = \frac{\sin Tu}{Tu}$$

Properties

We shall denote by $\Gamma(u, v)$ the Fourier transform of $\chi(u, \tau)$ in the variable τ:

$$\gamma(t, \tau) \underset{t}{\leftrightarrow} \chi(u, \tau) \underset{\tau}{\leftrightarrow} \Gamma(u, v) \tag{8-91}$$

Thus, $\Gamma(u, v)$ is the two-dimensional transform of $\gamma(t, \tau)$:

$$\gamma(t, \tau) \Leftrightarrow \Gamma(u, v) \tag{8-92}$$

We maintain that

$$\Gamma(u, v) = \frac{1}{E} F\left(v + \frac{u}{2}\right) F^*\left(v - \frac{u}{2}\right) \tag{8-93}$$

where $F(u)$ is the transform of $f(t)$.

Proof Clearly, $f(t)f^*(\tau) \Leftrightarrow F(u)F^*(-v)$. Hence [see (8-59)],

$$f\left(t + \frac{\tau}{2}\right) f^*\left(t - \frac{\tau}{2}\right) \Leftrightarrow F\left(v + \frac{u}{2}\right) F^*\left(v - \frac{u}{2}\right)$$

and (8-93) results.

Notes:

1. From (8-91) and (8-93), it follows that

$$\chi(u, \tau) = \frac{1}{2\pi E} \int_{-\infty}^{\infty} F\left(v + \frac{u}{2}\right) F^*\left(v - \frac{u}{2}\right) e^{jv\tau}\, dv \tag{8-94}$$

   Thus, $\chi(u, \tau)$ is also the ambiguity function of $F(u)$ if the variables u and τ are interchanged.

2. The functions

$$\begin{aligned} \chi(0, \tau) &= \frac{1}{E} \int_{-\infty}^{\infty} f\left(t + \frac{\tau}{2}\right) f^*\left(t - \frac{\tau}{2}\right) dt \\ \chi(u, 0) &= \frac{1}{2\pi E} \int_{-\infty}^{\infty} F\left(v + \frac{u}{2}\right) F^*\left(v - \frac{u}{2}\right) dv \end{aligned} \tag{8-95}$$

   are the normalized autocorrelations of $f(t)$ and $F(u)$, respectively (see Example 3-9).

Theorem 1 For any u and τ,

$$|\chi(u, \tau)| \le \chi(0, 0) = 1 \tag{8-96}$$

Proof Schwarz' inequality (4-A-1) yields

$$\left| \int_{-\infty}^{\infty} f\left(t + \frac{\tau}{2}\right) f^*\left(t - \frac{\tau}{2}\right) e^{-jut}\, dt \right|^2 \le \int_{-\infty}^{\infty} \left| f\left(t + \frac{\tau}{2}\right) \right|^2 dt \int_{-\infty}^{\infty} \left| f\left(t - \frac{\tau}{2}\right) \right|^2 dt$$

and (8-96) results [see Eq. (8-87)] because each integral on the right side of the above equals E.

Theorem 2

$$\frac{1}{2\pi} \int_{-\infty}^{\infty} \int_{-\infty}^{\infty} |\chi(u, \tau)|^2\, du\, d\tau = 1 \tag{8-97}$$

Proof Since $\chi(u, \tau)$ is the transform of $\gamma(t, \tau)$, (3-37) yields

$$\int_{-\infty}^{\infty} |\gamma(t, \tau)|^2 \, dt = \frac{1}{2\pi} \int_{-\infty}^{\infty} |\chi(u, \tau)|^2 \, du \tag{8-98}$$

With $t_1 = t + \tau/2$ and $t_2 = t - \tau/2$, it follows from (8-88) that

$$\int_{-\infty}^{\infty} \int_{-\infty}^{\infty} |\gamma(t, \tau)|^2 \, dt \, d\tau = \frac{1}{E^2} \int_{-\infty}^{\infty} \int_{-\infty}^{\infty} |f(t_1)|^2 |f(t_2)|^2 \, dt_1 \, dt_2 = 1$$

and Eq. (8-97) results when both sides of (8-98) are integrated.

Note: Equation (8-97) is known as the *radar uncertainty principle.*[1] It shows that the function $\chi(u, \tau)$ cannot be concentrated arbitrarily close to the origin.

Inversion formula and uniqueness We shall show that the function $f(t)$ is uniquely determined in terms of its ambiguity function within a constant factor. Indeed, from (8-87) and the Fourier inversion formula (3-2), it follows that

$$f\left(t + \frac{\tau}{2}\right) f^*\left(t - \frac{\tau}{2}\right) = \frac{E}{2\pi} \int_{-\infty}^{\infty} \chi(u, \tau) e^{jut} \, du$$

With $t_1 = t + \tau/2$ and $t_2 = t - \tau/2$, the above yields

$$f(t_1) f^*(t_2) = \frac{E}{2\pi} \int_{-\infty}^{\infty} \chi(u, t_1 - t_2) e^{ju(t_1 + t_2)/2} \, du \tag{8-99}$$

Setting $t_2 = 0$, we obtain $f(t_1)$ for any t_1.

Note: Equations (8-96) and (8-97) establish necessary but not sufficient conditions for a signal $\chi(u, \tau)$ to be an ambiguity function. The signal $\chi(u, \tau)$ must be such that if it is inserted into (8-99), the result is a product of the form $f(t_1) f^*(t_2)$. This condition is also sufficient because $\chi(u, \tau)$ is the ambiguity function of the signal $f(t)$ so determined.

Time and frequency shifts It is easy to see that if $f_1(t) = f(t + t_0)$, then $\chi_1(u, \tau) = \chi(u, \tau) e^{jt_0 u}$; and, if $F_2(\omega) = F(\omega - \omega_0)$, then $\chi_2(u, \tau) = \chi(u, \tau) e^{j\omega_0 \tau}$. Combining, we obtain the pair

$$f(t + t_0) e^{j\omega_0 t} \qquad \chi(u, \tau) e^{j(t_0 u + \omega_0 \tau)} \tag{8-100}$$

Coordinate Transformations[2]

Given a signal $f(t)$ with ambiguity function $\chi(u, \tau)$, we form the function

$$\chi_L(u, \tau) = \chi(au + b\tau, cu + d\tau) \tag{8-101}$$

[1] A. W. Rihaczek, "Principles of High-Resolution Radar," p. 118, McGraw-Hill Book Company, New York, 1969.

[2] A. Papoulis, Ambiguity Function in Fourier Optics, *JOSA*, vol. 64, no. 6, pp. 779–788, June, 1974.

obtained by a linear transformation of the coordinates u and τ of $\chi(u, \tau)$. If the function $\chi_L(u, \tau)$ so formed has an inverse $f_L(t)$, then the transformation matrix L must be unimodular:

$$L = \begin{pmatrix} a & b \\ c & d \end{pmatrix} \qquad ad - bc = 1 \tag{8-102}$$

because the functions $\chi(u, \tau)$ and $\chi_L(u, \tau)$ have the same volume [see Eq. (8-97)]. We shall show that this condition is also sufficient for $\chi_L(u, \tau)$ to have an inverse $f_L(t)$, and we shall express $f_L(t)$ in terms of $f(t)$.

As preparation, we discuss two special cases.

The P matrix We form the linear FM signal

$$f_P(t) = f(t)e^{j\alpha t^2/2} \tag{8-103}$$

Since $f\left(t + \frac{\tau}{2}\right)e^{j\alpha(t+\tau/2)^2/2} f^*\left(t - \frac{\tau}{2}\right)e^{-j\alpha(t-\tau/2)^2/2} = f\left(t + \frac{\tau}{2}\right)f^*\left(t + \frac{\tau}{2}\right)e^{j\alpha t\tau}$

and the signals $f_P(t)$ and $f(t)$ have the same energy, we conclude from (8-87) that

$$\chi_P(u, \tau) = \frac{1}{E}\int_{-\infty}^{\infty} f\left(t + \frac{\tau}{2}\right)f^*\left(t - \frac{\tau}{2}\right)e^{j\alpha t\tau}e^{-jut}\,dt = \chi(u - \alpha\tau, \tau) \tag{8-104}$$

Thus, $\chi_P(u, \tau)$ is obtained from $\chi(u, \tau)$ by a coordinate transformation as in (8-101). We shall denote the corresponding matrix by $P(\alpha)$:

$$P(\alpha) = \begin{pmatrix} 1 & -\alpha \\ 0 & 1 \end{pmatrix} \tag{8-105}$$

Example 8-17 Applying (8-104) to the pair

$$f(t) = e^{-\alpha t^2} \qquad \chi(u, \tau) = e^{-(u^2 + 4\alpha^2\tau^2)/8\alpha}$$

of Example 8-15, we obtain the pair

$$e^{-\alpha t^2}e^{j\beta t^2} \qquad e^{-[u^2 - 4\beta u\tau + 4(\alpha^2 + \beta^2)\tau^2]/8\alpha}$$

The Q matrix The signal

$$f_Q(t) = f(t) * e^{jt^2/2\beta} \tag{8-106}$$

is the output of a linear system with input $f(t)$ and impulse response $e^{jt^2/2\beta}$. Since [see (8-6)]

$$e^{jt^2/2\beta} \leftrightarrow \sqrt{2\pi j\beta}\, e^{-j\beta\omega^2/2}$$

we conclude that the Fourier transform of $f_Q(t)$ is given by

$$F_Q(\omega) = \sqrt{2\pi j\beta}\, F(\omega)e^{-j\beta\omega^2/2}$$

Hence [see Eq. (8-94)],

$$\chi_Q(u, \tau) = \frac{1}{2\pi E}\int_{-\infty}^{\infty} F\left(v + \frac{u}{2}\right)F^*\left(v - \frac{u}{2}\right)e^{-j\beta uv}e^{jv\tau}\,dv = \chi(u, \tau - \beta u)$$

We shall denote the corresponding matrix by $Q(\beta)$:

$$Q(\beta) = \begin{pmatrix} 1 & 0 \\ -\beta & 1 \end{pmatrix} \tag{8-107}$$

Canonical decomposition We show next that any unimodular matrix L can be written as a product of P and Q matrices. It will then follow that the corresponding $f_L(t)$ can be obtained by successively multiplying and convolving $f(t)$ with a linear FM signal, as in (8-103) and (8-106).

Case 1 $(c \neq 0)$ It is easy to see that

$$L = P(\alpha_1)Q(\beta_1)P(\alpha_2) \tag{8-108}$$

where
$$\alpha_1 = \frac{1-a}{c} \qquad \beta_1 = -c \qquad \alpha_2 = \frac{1-d}{c} \tag{8-109}$$

Hence,
$$f_L(t) = \{[f(t)e^{j\alpha_1 t^2/2}] * e^{jt^2/2\beta_1}\}e^{j\alpha_2 t^2/2}. \tag{8-110}$$

Case 2 $(b \neq 0)$ In this case,

$$L = Q(\beta_1)P(\alpha_1)Q(\beta_2) \tag{8-111}$$

where
$$\beta_1 = \frac{1-d}{b} \qquad \alpha_1 = -b \qquad \beta_2 = \frac{1-a}{b} \tag{8-112}$$

Hence,
$$f_L(t) = \{[f(t) * e^{jt^2/2\beta_1}]e^{j\alpha_1 t^2/2}\} * e^{jt^2/2\beta_2} \tag{8-113}$$

> **Note:** If $c \neq 0$ and $b \neq 0$, then $f_L(t)$ is given by Eqs. (8-110) and (8-113). From this, it follows that the functions on the right side of (8-110) and (8-113) must be proportional because $f_L(t)$ is uniquely determined from $\chi_L(u, \tau)$ within a factor.[1]

Case 3 $(c = 0$ and $b = 0)$ In this case [see (8-102)], $d = 1/a = k$; hence,

$$\chi_L(u, \tau) = \chi\left(\frac{u}{k}, k\tau\right) \tag{8-114}$$

The corresponding $f_L(t)$ is given by

$$f_L(t) = f\ (kt) \tag{8-115}$$

because the energy of $f_L(t)$ equals E/k and

$$\frac{k}{E}\int_{-\infty}^{\infty} f\left[k\left(t+\frac{\tau}{2}\right)\right]f^*\left[k\left(t-\frac{\tau}{2}\right)\right]e^{-jut}\,dt = \chi\left(\frac{u}{k}, k\tau\right)$$

as it is easy to see.

[1] A. Papoulis, Dual Optical Systems, *J. Opt. Soc. Am.*, vol. 68, no. 5, May, 1968.

We shall denote the matrix of the transformation (8-114) by $S(k)$:

$$S(k) = \begin{pmatrix} \frac{1}{k} & 0 \\ 0 & k \end{pmatrix} \tag{8-116}$$

If we multiply $S(k)$ by a P matrix, we obtain

$$S(k)P(\alpha) = \begin{pmatrix} \frac{1}{k} & -\frac{\alpha}{k} \\ 0 & k \end{pmatrix}$$

This product can be factored as in Eq. (8-111). And, since the inverse of $P(\alpha)$ equals $P(-\alpha)$, we conclude that

$$S(k) = Q(\beta_1)P(\alpha_1)Q(\beta_2)P(-\alpha) \tag{8-117}$$

where α is an arbitrary constant and

$$\beta_1 = \frac{k^2 - k}{\alpha} \qquad \alpha_1 = \frac{\alpha}{k} \qquad \beta_2 = \frac{1 - k}{\alpha}$$

Time scaling The ambiguity functions resulting from the two sides of (8-117) are equal; therefore, the corresponding time functions are proportional. This leads to the conclusion that for any k and α,

$$f(kt) = A(\{[f(t) * e^{jt^2/2\beta_1}]e^{j\alpha_1 t^2/2}\} * e^{-jt^2/2\beta_2})e^{-j\alpha t^2/2}$$

where A is a constant depending on k and α.

We have thus shown that a linear scaling of the time axis can be achieved by a combination of two linear FM modulators and two quadratic phase filters.

Real-time spectrum analyzer We shall, finally, show that the Fourier transform $F(\omega)$ of $f(t)$, expressed as a time function, has as its ambiguity function the function

$$\chi_L(u, \tau) = \chi(\tau, -u) \tag{8-118}$$

obtained from $\chi(u, \tau)$ with the coordinate transformation matrix

$$F = \begin{pmatrix} 0 & 1 \\ -1 & 0 \end{pmatrix} \tag{8-119}$$

Indeed, the energy of the signal $f_L(t) = F(t)$ equals $2\pi E$; hence,

$$\gamma_L(t, \tau) = \frac{1}{2\pi E} F\left(t + \frac{\tau}{2}\right)F^*\left(t - \frac{\tau}{2}\right) = \frac{1}{2\pi}\Gamma(\tau, t)$$

as we can see from Eqs. (8-88) and (8-93). Inserting into (8-87), we obtain

$$\chi_L(u, \tau) = \frac{1}{2\pi} \int_{-\infty}^{\infty} \Gamma(\tau, t) e^{-jut} \, dt = \chi(\tau, -u)$$

[see (8-94)], and (8-118) results.

The matrix F can be factored as in (8-108) or (8-111) because $b = 1 \neq 0$ and $c = -1 \neq 0$. Applying (8-108), we obtain

$$F = P(-1)Q(1)P(-1) \tag{8-120}$$

because $a = d = 0$, and (8-110) yields

$$F(t) = B\{[f(t)e^{-jt^2/2}] * e^{jt^2/2}\}e^{-jt^2/2} \tag{8-121}$$

Thus, the Fourier transform of a signal $f(t)$ can be obtained in real time by a combination of two linear FM generators and one quadratic phase filter. The result can also be written in the form (8-113).

Signal Duration

In Sec. 8-2, we used, as a measure of the width (duration) of a signal $f(t)$ and its transform $F(\omega)$, the moments

$$w = d^2 = \frac{1}{E} \int_{-\infty}^{\infty} t^2 |f(t)|^2 \, dt \qquad W = D^2 = \frac{1}{2\pi E} \int_{-\infty}^{\infty} \omega^2 |F(\omega)|^2 \, d\omega \tag{8-122}$$

In the following, we relate these moments to the derivatives of the ambiguity function of $f(t)$ at the origin. The result will give us a simple, exact expression for the effects of modulation and dispersion on the width of a signal.

Theorem 3a

$$\frac{-\partial^2 \chi(0, 0)}{\partial u^2} = w \qquad \frac{-\partial^2 \chi(0, 0)}{\partial \tau^2} = W \tag{8-123}$$

Proof Differentiating (8-87) twice with respect to u, and (8-94) twice with respect to τ, we obtain

$$\frac{\partial^2 \chi}{\partial u^2} = \frac{1}{E} \int_{-\infty}^{\infty} (-jt)^2 f\left(t + \frac{\tau}{2}\right) f^*\left(t - \frac{\tau}{2}\right) e^{-jut} \, dt$$

$$\frac{\partial^2 \chi}{\partial \tau^2} = \frac{1}{2\pi E} \int_{-\infty}^{\infty} (jv)^2 F\left(v + \frac{u}{2}\right) F^*\left(v - \frac{u}{2}\right) e^{jv\tau} \, dv$$

and (8-123) follows with $u = 0$ and $\tau = 0$.

For the next theorem, we shall need the mixed moment

$$\mu = \frac{j}{2E} \int_{-\infty}^{\infty} t[f'(t)f^*(t) - f(t)f'^*(t)] \, dt \tag{8-124}$$

Since $-jtf(t) \leftrightarrow F'(\omega)$ and $f'(t) \leftrightarrow j\omega F(\omega)$, we conclude from Parseval's formula that μ is also given by

$$\mu = \frac{-j}{4\pi E} \int_{-\infty}^{\infty} \omega[F'(\omega)F^*(\omega) - F(\omega)F'^*(\omega)]\, d\omega \tag{8-125}$$

Theorem 3b

$$\frac{-\partial^2 \chi(0, 0)}{\partial u\, \partial \tau} = \mu \tag{8-126}$$

Proof Differentiating Eq. (8-87) with respect to u and τ, we obtain

$$\frac{\partial^2 \chi}{\partial u\, \partial \tau} = \frac{1}{2E} \int_{-\infty}^{\infty} (-jt) \left[f'\left(t + \frac{\tau}{2}\right) f^*\left(t - \frac{\tau}{2}\right) - f\left(t + \frac{\tau}{2}\right) f'^*\left(t - \frac{\tau}{2}\right) \right] e^{-jut}\, dt$$

and (8-126) results with $u = 0$ and $\tau = 0$.

Notes:

1. We can assume, introducing a suitable shift in time or frequency [see (8-100)], that $\chi_u(0, 0) = 0$ and $\chi_\tau(0, 0) = 0$. Expanding $\chi(u, \tau)$ into a series and using (8-123) and (8-126), we obtain

$$\chi(u, \tau) = 1 - \tfrac{1}{2}(d^2u^2 + 2\mu u\tau + D^2\tau^2) + \cdots \tag{8-127}$$

Thus, near the origin, the contour lines of $\chi(u, \tau)$ are ellipses (Fig. 8-16). If $\mu = 0$, then the axes are along the coordinate lines, and their lengths are proportional to $1/d$ and $1/D$, respectively.

2. Writing $f(t)$ and $F(\omega)$ in polar form,

$$f(t) = r(t)e^{j\theta(t)} \qquad F(\omega) = A(\omega)e^{j\varphi(\omega)} \tag{8-128}$$

we conclude from (8-124) and (8-125) that (elaborate)

$$\mu = \frac{-1}{E} \int_{-\infty}^{\infty} tr^2(t)\theta'(t)\, dt = \frac{1}{2\pi E} \int_{-\infty}^{\infty} \omega A^2(\omega)\varphi'(\omega)\, d\omega \tag{8-129}$$

From the above, it follows that μ is a real number; it equals zero if $f(t)$ or $F(\omega)$ or both are real.

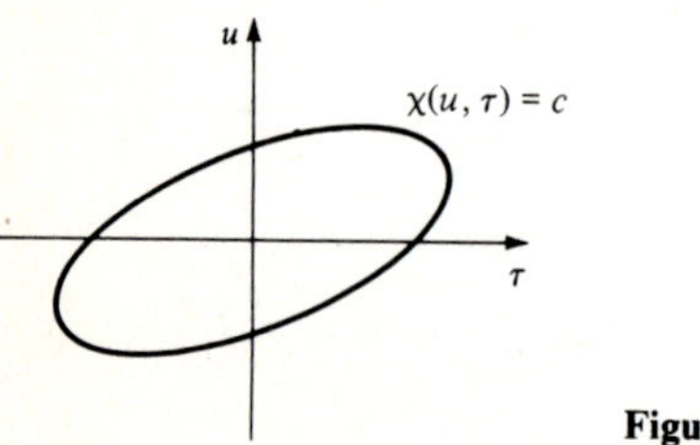

Figure 8-16

We denote by $f_L(t)$ the inverse of the function

$$\chi_L(u, \tau) = \chi(au + b\tau, cu + d\tau) \tag{8-130}$$

obtained by a linear transformation of the coordinates of the ambiguity function $\chi(u, \tau)$ of $f(t)$. We shall relate the moments w, W, and μ of $f(t)$ to the corresponding moments w_L, W_L, and μ_L of $f_L(t)$.

Theorem 4

$$\begin{aligned} w_L &= a^2 w + 2ac\mu + c^2 W \\ W_L &= b^2 w + 2bd\mu + d^2 W \\ \mu_L &= abw + (ad + bc)\mu + cdW \end{aligned} \tag{8-131}$$

Proof From Eq. (8-130), it follows that

$$\frac{\partial^2 \chi_L}{\partial u^2} = a^2 \frac{\partial^2 \chi}{\partial u^2} + 2ac \frac{\partial^2 \chi}{\partial u \, \partial \tau} + c^2 \frac{\partial^2 \chi}{\partial \tau^2}$$

$$\frac{\partial^2 \chi_L}{\partial \tau^2} = b^2 \frac{\partial^2 \chi}{\partial u^2} + 2bd \frac{\partial^2 \chi}{\partial u \, \partial \tau} + d^2 \frac{\partial^2 \chi}{\partial \tau^2}$$

$$\frac{\partial^2 \chi_L}{\partial u \, \partial \tau} = ab \frac{\partial^2 \chi}{\partial u^2} + (ad + bc) \frac{\partial^2 \chi}{\partial u \, \partial \tau} + cd \frac{\partial^2 \chi}{\partial \tau^2}$$

Setting $u = \tau = 0$ and using (8-123) and (8-126), we obtain (8-131).
The following special cases are of particular interest.

Width of the spectrum of an FM signal If $f_P(t) = f(t)e^{j\alpha t^2/2}$, then

$$\chi_P(u, \tau) = \chi(u - \alpha\tau, \tau)$$

[see (8-104)]. Hence, $a = 1$, $b = -\alpha$, $c = 0$, and $d = 1$, and (8-131) yields

$$W_P = \alpha^2 w - 2\alpha\mu + W \tag{8-132}$$

We have thus expressed the width W_P of the spectrum of a linear FM signal in terms of the moments of its envelope $f(t)$ and the slope α of the instantaneous frequency. This result is exact.

In Fig. 8-17 we show W_P as a function of α for $\mu < 0$, $\mu = 0$, and $\mu > 0$. We note that, if $\mu \leq 0$, then W_P increases monotonically. If $\mu > 0$, then W_P at first decreases, reaching a minimum for $\alpha = \mu/w$; then, for $\alpha > \mu/w$, it increases. If $f(t)$ is real, then $\mu = 0$ and

$$W_P = \alpha^2 w + W \tag{8-133}$$

In all cases, $W_P \simeq \alpha^2 w$ for large α.

Note: If W_P is determined from (8-21), then the result is $\alpha^2 w$ (elaborate) because (8-21) is an approximation valid only for large α.

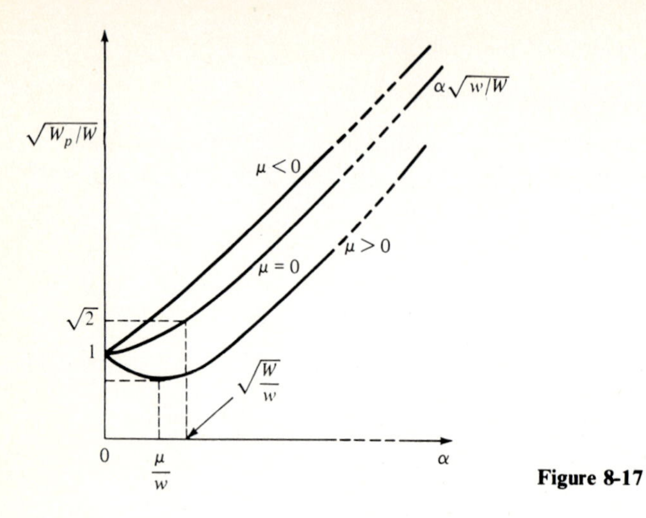

Figure 8-17

Quadratic phase filters and dispersion If $f(t)$ is the input to a quadratic phase filter

$$h(t) = e^{jt^2/2\beta} \qquad H(\omega) = \sqrt{2\pi j\beta}\, e^{-j\beta\omega^2/2}$$

then the resulting output is

$$f_Q(t) = f(t) * e^{jt^2/2\beta}$$

and its ambiguity function equals $\chi(u, \tau - \beta u)$ [see Eq. (8-107)]. In this case, $a = 1$, $b = 0$, $c = -\beta$, and $d = 1$, and (8-131) yields

$$w_Q = w - 2\beta\mu + \beta^2 W \tag{8-134}$$

If $\mu = 0$, then the signal spreads as it passes through the filter, and the increase in its width equals $\beta^2 W$.

Pulse compression The system of Fig. 8-18 consists of a quadratic phase filter as above and an FM modulator $e^{-j\alpha t^2/2}$. The input is a real signal $f(t)$, and the

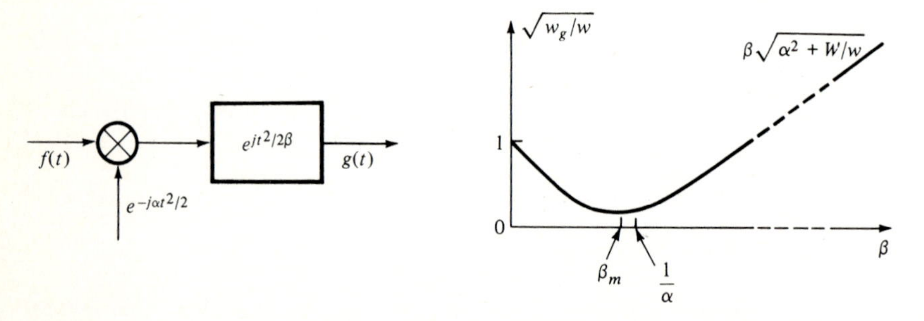

Figure 8-18

output $g(t)$ is given by

$$g(t) = [f(t)e^{-j\alpha t^2/2}] * e^{jt^2/2\beta} \tag{8-135}$$

We shall determine its width w_g.

Clearly, $g(t)$ is obtained by applying, to $\chi(u, \tau)$, the P and Q matrices in cascade [see (8-103) and (8-106)]. The resulting matrix equals

$$P(-\alpha)Q(\beta) = \begin{pmatrix} 1 - \alpha\beta & \alpha \\ -\beta & 1 \end{pmatrix}$$

Inserting its coefficients into (8-131), we obtain

$$w_g = (1 - \alpha\beta)^2 w + \beta^2 W \tag{8-136}$$

because the mixed moment μ of the real function $f(t)$ equals zero.

Compression ratio In Fig. 8-18, we show the width w_g of $g(t)$ as a function of β. As β increases w_g decreases, reaching a minimum for

$$\beta = \beta_m = \frac{1/\alpha}{1 + W/\alpha^2 w} \tag{8-137}$$

If $\alpha^2 w \gg W$, then $\beta_m \simeq 1/\alpha$.

For $\beta = 1/\alpha$, $w_g = W/\alpha^2$, resulting in a compression ratio

$$r \triangleq \sqrt{\frac{w}{w_g}} = \alpha\sqrt{\frac{w}{W}} \tag{8-138}$$

Thus, the compression ratio is proportional to the slope of the instantaneous frequency $\omega(t) = \alpha t$ of the modulator.

PART THREE

DATA SMOOTHING AND SPECTRAL ESTIMATION

The theory of stochastic processes is usually developed as a separate discipline. However, certain applications involve only second-order moments and can be discussed without special preparation. Such applications include various aspects of filtering, data smoothing, and spectral estimation. These important topics are closely related to the main ideas of this book and are presented in Part Three.

CHAPTER

NINE

STOCHASTIC PROCESSES

In this chapter, we discuss the theory of second-order moments and spectra in the context of linear systems.[1,2]

In Sec. 9-1, we concentrate on definitions and basic properties. In Sec. 9-2, we express the autocorrelation of the output of a system in terms of the autocorrelation of the input. In Sec. 9-3, we develop the spectral properties of system responses.

In Sec. 9-4, we present the discrete version of the earlier results.

---

[1] A. Papoulis, "Probability, Random Variables, and Stochastic Processes," McGraw-Hill Book Company, New York, 1965.

[2] Y. S. Pugachev, "Theory of Random Functions and Its Applications to Automatic Control Problems" (translated from Russian), Pergamon Press, New York, 1965.

9-1 CORRELATIONS AND SPECTRA

For an understanding of the spectral properties of random signals, various concepts from the theory of stochastic processes are needed. In this section, we develop these concepts, stressing second-order moments. The material is essentially self-contained. However, it is assumed that the reader is familiar with the elements of the theory of stochastic processes.

Definitions

A stochastic process is a family of functions $\mathbf{x}(t)$†, real or complex, defined in some probability space. At specified times $t_1, \ldots, t_n$, the quantities $\mathbf{x}(t_1), \ldots, \mathbf{x}(t_n)$ are random variables. We shall define their first- and second-order moments.

The *mean* or *expected value* of the process $\mathbf{x}(t)$, denoted by $\eta_x(t)$ or by $\eta(t)$, is the expected value of the random variable $\mathbf{x}(t)$:

$$E\{\mathbf{x}(t)\} = \eta(t) \tag{9-1}$$

The *autocorrelation* of the process $\mathbf{x}(t)$, denoted by $R_{xx}(t_1, t_2)$ or by $R(t_1, t_2)$, is the expected value of the product $\mathbf{x}(t_1)\,\mathbf{x}^*(t_2)$:

$$E\{\mathbf{x}(t_1)\mathbf{x}^*(t_2)\} = R(t_1, t_2) \tag{9-2}$$

The value of $R(t_1, t_2)$ for $t_1 = t_2 = t$ is the *average intensity* or *average power* of the process $\mathbf{x}(t)$:

$$E\{|\mathbf{x}(t)|^2\} = R(t, t) \tag{9-3}$$

Knowledge of $R(t_1, t_2)$ permits us to find the average intensity of any linear combination of the random variable $\mathbf{x}(t_i)$. For example,

$$\begin{aligned} E\{|\mathbf{x}(t_1) + \mathbf{x}(t_2)|^2\} &= E\{[\mathbf{x}(t_1) + \mathbf{x}(t_2)][\mathbf{x}^*(t_1) + \mathbf{x}^*(t_2)]\} \\ &= R(t_1, t_1) + R(t_1, t_2) + R(t_2, t_1) + R(t_2, t_2) \end{aligned}$$

This is a consequence of the linearity of expected values.

The *cross-correlation* of two processes $\mathbf{x}(t)$ and $\mathbf{y}(t)$, denoted by $R_{xy}(t_1, t_2)$, is the expected value of the product $\mathbf{x}(t_1)\mathbf{y}^*(t_2)$:

$$E\{\mathbf{x}(t_1)\mathbf{y}^*(t_2)\} = R_{xy}(t_1, t_2) \tag{9-4}$$

A related concept is the *cross-covariance* $C_{xy}(t_1, t_2)$, defined as the covariance of the random variables $\mathbf{x}(t_1)$ and $\mathbf{y}(t_2)$:

$$E\{[\mathbf{x}(t_1) - \eta_x(t_1)][\mathbf{y}^*(t_2) - \eta_y^*(t_2)]\} = C_{xy}(t_1, t_2) \tag{9-5}$$

From the above, it follows that

$$C_{xy}(t_1, t_2) = R_{xy}(t_1, t_2) - \eta_x(t_1)\eta_y^*(t_2) \tag{9-6}$$

† All random quantities will be represented by boldface letters.

The *autocovariance* $C_{xx}(t_1, t_2)$ or $C(t_1, t_2)$ of a process $\mathbf{x}(t)$ is defined similarly and is given by

$$C(t_1, t_2) = R(t_1, t_2) - \eta(t_1)\eta^*(t_2) \tag{9-7}$$

as we see by setting $\mathbf{y}(t) = \mathbf{x}(t)$ in Eq. (9-6). In particular,

$$C(t, t) = R(t, t) - |\eta(t)|^2 \tag{9-8}$$

is the variance of the random variable $\mathbf{x}(t)$.

Centering To any process $\mathbf{x}(t)$ we can associate another process $\mathbf{x}_c(t)$ obtained by subtracting from $\mathbf{x}(t)$ its mean $\eta(t)$:

$$\mathbf{x}_c(t) = \mathbf{x}(t) - \eta(t) \tag{9-9}$$

The process $\mathbf{x}_c(t)$ so formed has zero mean, and its autocovariance equals $C_{xx}(t_1, t_2)$:

$$C_{x_c x_c}(t_1, t_2) = R_{x_c x_c}(t_1, t_2) = C_{xx}(t_1, t_2) \tag{9-10}$$

Stationary processes A stochastic process is called *strict-sense stationary* if all its statistical properties are invariant to a shift of the time origin. A stochastic process is called *wide-sense stationary* if its mean is constant and its autocorrelation depends only on the difference $\tau = t_1 - t_2$:

$$E\{\mathbf{x}(t)\} = \eta = \text{constant} \tag{9-11}$$

$$R(t_1, t_2) = R(\tau) = E\{\mathbf{x}(t + \tau)\mathbf{x}^*(t)\} \tag{9-12}$$

From the above, it follows that

$$R(\tau) = E\left\{\mathbf{x}\left(t + \frac{\tau}{2}\right)\mathbf{x}^*\left(t - \frac{\tau}{2}\right)\right\} \tag{9-13}$$

If a process is strict-sense stationary, then it is also wide-sense stationary. The converse is not true in general. In the following, the term "stationary" will mean "wide-sense stationary."

Two processes $\mathbf{x}(t)$ and $\mathbf{y}(t)$ are jointly stationary if each is stationary and their cross-correlation $R_{xy}(t_1, t_2)$ depends only on $\tau = t_1 - t_2$:

$$R_{xy}(t_1, t_2) = R_{xy}(\tau) = E\{\mathbf{x}(t + \tau)\mathbf{y}^*(t)\} \tag{9-14}$$

From the above and Eq. (9-6), it follows that the cross-covariance $C_{xy}(t_1, t_2)$ depends only on τ:

$$C_{xy}(t_1, t_2) = C_{xy}(\tau) = R_{xy}(\tau) - \eta_x \eta_y^* \tag{9-15}$$

Similarly, if the process $\mathbf{x}(t)$ is stationary, then

$$C(t_1, t_2) = C(\tau) = R(\tau) - |\eta|^2 \tag{9-16}$$

In particular,

$$C(0) = E\{|\mathbf{x}(t) - \eta|^2\} = R(0) - |\eta|^2 \tag{9-17}$$

is the variance of the random variable $\mathbf{x}(t)$.

Notes:

1. If the autocovariance $C(t_1, t_2)$ of a process $\mathbf{x}(t)$ is of the form

$$C(t_1, t_2) = p(t_1)p^*(t_2)r(t_1 - t_2) \tag{9-18}$$

then the process

$$\mathbf{n}(t) = \frac{\mathbf{x}(t) - \eta(t)}{p(t)}$$

is stationary with zero mean and autocorrelation

$$R_{nn}(\tau) = r(\tau) \tag{9-19}$$

Thus, if $C(t_1, t_2)$ is given by (9-18), then the nonstationary process $\mathbf{x}(t)$ is a trivial extension of the stationary process $\mathbf{n}(t)$:

$$\mathbf{x}(t) = p(t)\mathbf{n}(t) + \eta(t)$$

2. The autocorrelation $R(\tau)$ of a stationary process $\mathbf{x}(t)$ can be defined as average intensity. Indeed, from (9-12), it follows readily that

$$E\{|\mathbf{x}(t)|^2\} = R(0)$$

and for any τ,

$$E\{[\mathbf{x}(t + \tau) \pm \mathbf{x}(t)]^2\} = 2[R(0) \pm R(\tau)] \tag{9-20}$$

where we assumed for simplicity that $\mathbf{x}(t)$ is real (see also Prob. 90).

3. *Quasistationary processes.* The autocorrelation $R(t_1, t_2)$ of a process $\mathbf{x}(t)$ can be written as a function $\bar{R}(t, \tau)$ of the variables

$$t = \frac{t_1 + t_2}{2} \qquad \tau = t_1 - t_2 \tag{9-21}$$

Thus,
$$\bar{R}(t, \tau) = R\left(t + \frac{\tau}{2}, t - \frac{\tau}{2}\right) = E\left\{\mathbf{x}\left(t + \frac{\tau}{2}\right)\mathbf{x}^*\left(t - \frac{\tau}{2}\right)\right\} \tag{9-22}$$

If $\mathbf{x}(t)$ is stationary, then $\bar{R}(t, \tau) = R(\tau)$ is independent of t. We shall say that the process $\mathbf{x}(t)$ is *quasistationary* if $\bar{R}(t, \tau)$ is a slowly varying function of t. This vague definition is best explained in terms of the following concept:

A process $\mathbf{x}(t)$ will be called *a-dependent*[1] if its autocorrelation $R(t_1, t_2)$ is zero for $|t_1 - t_2| > a$, that is, if

$$\bar{R}(t, \tau) = 0 \qquad \text{for } |\tau| > a \tag{9-23}$$

An a-dependent process is quasistationary if

$$\bar{R}(t + x, \tau) \simeq \bar{R}(t, \tau) \qquad \text{for } |x| < a \tag{9-24}$$

[1] The term "a-dependent," as it is used generally, means that the random variables $\mathbf{x}(t_1)$ and $\mathbf{x}(t_2)$ are independent for $|t_1 - t_2| > a$. This implies (9-23) if $\eta_x(t) = 0$, and it is equivalent to (9-23) if $\mathbf{x}(t)$ is also normal.

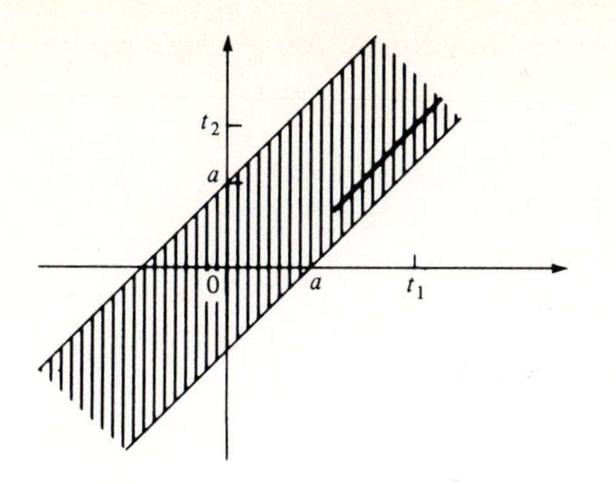

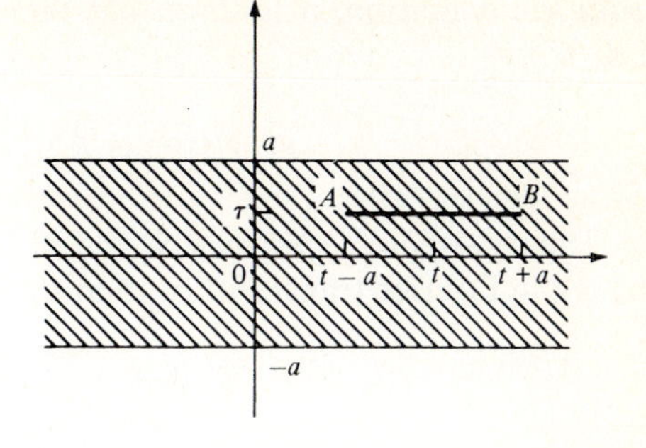

Figure 9-1

Thus, along any horizontal line AB of length $2a$ (Fig. 9-1), $\bar{R}(t, \tau)$ is approximately constant.

4. *White noise* is a process $\mathbf{x}(t)$ whose autocorrelation is given by

$$R(t_1, t_2) = I(t_1)\,\delta(t_1 - t_2) \tag{9-25}$$

Clearly, $R(t_1, t_2) = 0$ for $t_1 \neq t_2$; hence, $\mathbf{x}(t)$ is zero-dependent and quasi-stationary. From Eqs. (9-22) and (9-25), it follows that if $\mathbf{x}(t)$ is white noise, then [see (3-C-6)]

$$\bar{R}(t, \tau) = I\left(t + \frac{\tau}{2}\right)\delta(\tau) = I(t)\,\delta(\tau) \tag{9-26}$$

If $\mathbf{x}(t)$ is also stationary, then $I(t) = I =$ constant, and

$$\bar{R}(t, \tau) = R(\tau) = I\,\delta(\tau) \tag{9-27}$$

In Sec. 9-2 we show that, if the input to a linear system is an a-dependent process $\mathbf{x}(t)$ and a is small compared to the time constants of the system, then [see (9-51)], for the determination of the average intensity of the resulting response, $\mathbf{x}(t)$ can be replaced with a white-noise process with "average intensity"

$$I(t) = \int_{-a}^{a} \bar{R}(t, \tau)\,d\tau \tag{9-28}$$

Spectra

The *power spectrum* $S(\omega)$ of a stationary process $\mathbf{x}(t)$ is the Fourier transform of its autocorrelation:

$$S(\omega) = \int_{-\infty}^{\infty} R(\tau)e^{-j\omega\tau}\,d\tau \tag{9-29}$$

From the definition, it follows that the area of $S(\omega)$ equals the average power of $\mathbf{x}(t)$:

$$E\{|\mathbf{x}(t)|^2\} = R(0) = \frac{1}{2\pi}\int_{-\infty}^{\infty} S(\omega)\,d\omega \tag{9-30}$$

The *cross-power spectrum* $S_{xy}(\omega)$ of two jointly stationary processes $\mathbf{x}(t)$ and $\mathbf{y}(t)$ is the Fourier transform of their cross-correlation:

$$S_{xy}(\omega) = \int_{-\infty}^{\infty} R_{xy}(\tau)e^{-j\omega\tau}\,d\tau \tag{9-31}$$

It is easy to see from Eq. (9-12) that

$$R(-\tau) = R^*(\tau)$$

Hence, $S(\omega)$ is *real*. If the process $\mathbf{x}(t)$ is also real, then $R(-\tau) = R(\tau)$; hence, $S(\omega)$ is real and even.

Note: Power spectra are defined mainly for stationary processes. For nonstationary processes, the *time-varying spectrum* $\psi(t, \omega)$ is sometimes used. This is defined as the Fourier transform in the variable τ of the autocorrelation $\bar{R}(t, \tau)$ [see (9-22)]:

$$\psi(t, \omega) = \int_{-\infty}^{\infty} \bar{R}(t, \tau)e^{-j\omega\tau}\,d\tau \tag{9-32}$$

Its area in ω equals the average power of $\mathbf{x}(t)$, which is now time-dependent:

$$E\{|\mathbf{x}(t)|^2\} = \bar{R}(t, 0) = \frac{1}{2\pi}\int_{-\infty}^{\infty} \psi(t, \omega)\,d\omega \tag{9-33}$$

The function $\psi(t, \omega)$ is used mainly for quasistationary processes. In this case, its properties are essentially the same as the properties of $S(\omega)$ to be discussed in the next section.

Example 9-1 (See Fig. 9-2.)
(*a*) If $R(\tau) = I\,\delta(\tau)$, then $S(\omega) = I$.
(*b*) If $R(\tau) = e^{-\alpha|\tau|}$, then $S(\omega) = 2\alpha/(\alpha^2 + \omega^2)$.
(*c*) If $R(\tau) = e^{-\alpha|\tau|}\cos\beta\tau$, then

$$S(\omega) = \frac{\alpha}{\alpha^2 + (\omega - \beta)^2} + \frac{\alpha}{\alpha^2 + (\omega + \beta)^2}$$

Note: We describe a physical quantity with a stochastic process $\mathbf{x}(t)$ if we are interested not in its instantaneous values but only in certain averages. These averages can be given a statistical interpretation either as expected values of a family of functions (ensemble averages) or as time averages of a single function. In the next chapter we comment on the equivalence between these two interpretations. We note here, however, that a noncritical use of time averages might lead to wrong conclusions.

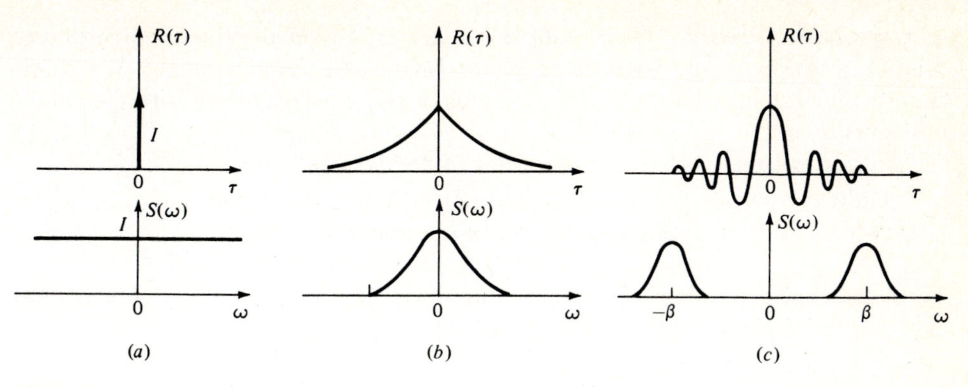

Figure 9-2

Suppose, for example, that we wish to determine the spectral properties of a sample $x(t)$ of a stationary process. We first note that the integral

$$\int_{-T}^{T} x(t)e^{-j\omega t}\,dt$$

does not converge but tends to ∞ as $\sqrt{T}$ as $T \to \infty$. To obtain a convergent spectrum, we form the integral

$$S_T(\omega) = \frac{1}{2T}\left|\int_{-T}^{T} x(t)e^{-j\omega t}\,dt\right|^2 \tag{9-34}$$

This integral tends to a limit, but the limit depends on the time origin. In other words, the limit of the integral

$$S_T{}^t(\omega) = \frac{1}{2T}\left|\int_{t-T}^{t+T} x(t)e^{-j\omega t}\,dt\right|^2 \tag{9-35}$$

is a function of t. Therefore, the spectral properties of a process represented by $x(t)$ cannot be determined from Eq. (9-35). In Chap. 11 we show that the "smoothed spectrum"

$$S_c{}^t(\omega) = \frac{1}{2c}\int_{\omega-c}^{\omega+c} S_T{}^t(y)\,dy \tag{9-36}$$

is not time-sensitive for large T and that it can be used to describe the spectral properties of the random signal $x(t)$.

9-2 LINEAR SYSTEMS WITH STOCHASTIC INPUTS

If the input to a linear system is a stochastic process $\mathbf{x}(t)$, then the resulting output

$$\mathbf{y}(t) = \int_{-\infty}^{\infty} \mathbf{x}(t-\alpha)h(\alpha)\,d\alpha = L[\mathbf{x}(t)] \tag{9-37}$$

is a stochastic process. Each sample of $\mathbf{y}(t)$ is obtained by convolving the sample of $\mathbf{x}(t)$ with the impulse response $h(t)$ of the given system.[1] We shall express the first- and second-order moments of $\mathbf{y}(t)$ in terms of the corresponding moments of $\mathbf{x}(t)$.

Taking expected values of both sides of (9-37), we obtain

$$E\{\mathbf{y}(t)\} = \int_{-\infty}^{\infty} E\{\mathbf{x}(t-\alpha)\}h(\alpha)\,d\alpha$$

Thus, the expected value $\eta_y(t)$ of the output $\mathbf{y}(t)$ is obtained by convolving the expected value $\eta_x(t)$ of the input $\mathbf{x}(t)$ with $h(t)$:

$$\eta_y(t) = \int_{-\infty}^{\infty} \eta_x(t-\alpha)h(\alpha)\,d\alpha = L[\eta_x(t)] \tag{9-38}$$

Note: The preceding basic result is a consequence of the linearity of expected values. In the context of linear systems it leads to the conclusion that

$$E\{L[\mathbf{x}(t)]\} = L[E\{\mathbf{x}(t)\}] \tag{9-39}$$

provided that the system L is deterministic, i.e., its impulse response $h(t)$ is a single function for any realization of the input.

Output Correlation

The autocorrelation $R_{yy}(t_1, t_2)$ of the output $\mathbf{y}(t)$ can be obtained similarly by taking expected values of the product

$$\mathbf{y}(t_1)\mathbf{y}^*(t_2) = \int_{-\infty}^{\infty}\int_{-\infty}^{\infty} \mathbf{x}(t_1-\alpha)\mathbf{x}^*(t_2-\beta)h(\alpha)h^*(\beta)\,d\alpha\,d\beta$$

We prefer to determine $R_{yy}(t_1, t_2)$ indirectly in terms of the cross-correlation $R_{xy}(t_1, t_2)$ between the input and the output.

Multiplying the conjugate of Eq. (9-37) by $\mathbf{x}(t_1)$ and taking expected values of both sides, we obtain

$$E\{\mathbf{x}(t_1)\mathbf{y}^*(t)\} = \int_{-\infty}^{\infty} E\{\mathbf{x}(t_1)\mathbf{x}^*(t-\alpha)\}h^*(\alpha)\,d\alpha$$

With $t = t_2$, the above yields

$$R_{xy}(t_1, t_2) = \int_{-\infty}^{\infty} R_{xx}(t_1, t_2-\alpha)h^*(\alpha)\,d\alpha = L_2^*[R_{xx}(t_1, t_2)] \tag{9-40}$$

The symbol L^* indicates the conjugate of the system L, that is, a system with impulse response $h^*(t)$. The subscript 2 in L_2^* means that the system operates

[1] All stochastic integrals will be interpreted in the mean-square sense. For their existence, the absolute integrability of the autocorrelation $R(t_1, t_2)$ of the integrand in the t_1t_2 plane is sufficient.

on $R_{xx}(t_1, t_2)$ with respect to the variable t_2; the variable t_1 is treated as a parameter.

Multiplying (9-37) by $\mathbf{y}^*(t_2)$ and taking expected values of both sides, we obtain

$$E\{\mathbf{y}(t)\mathbf{y}^*(t_2)\} = \int_{-\infty}^{\infty} E\{\mathbf{x}(t-\alpha)\mathbf{y}^*(t_2)\}h(\alpha)\,d\alpha$$

With $t = t_1$, the above yields

$$R_{yy}(t_1, t_2) = \int_{-\infty}^{\infty} R_{xy}(t_1 - \alpha, t_2)h(\alpha)\,d\alpha = L_1[R_{xy}(t_1, t_2)] \tag{9-41}$$

The function $R_{yy}(t_1, t_2)$ can be expressed directly in terms of $R_{xx}(t_1, t_2)$. Indeed, inserting (9-40) into (9-41), we conclude that

$$R_{yy}(t_1, t_2) = \int_{-\infty}^{\infty}\int_{-\infty}^{\infty} R_{xx}(t_1 - \alpha, t_2 - \beta)h(\alpha)h^*(\beta)\,d\alpha\,d\beta \tag{9-42}$$

Example 9-2 We wish to find the mean and autocorrelation of the derivative $\mathbf{x}'(t)$ of a process $\mathbf{x}(t)$. Clearly, $\mathbf{x}'(t)$ is the output of a differentiation L with input $\mathbf{x}(t)$; hence [see Eq. (9-39)],

$$E\{\mathbf{x}'(t)\} = L[\eta_x(t)] = \eta_x'(t)$$

Since L is real, it follows from (9-40) that

$$R_{xx'}(t_1, t_2) = L_2[R_{xx}(t_1, t_2)] = \frac{\partial R_{xx}(t_1, t_2)}{\partial t_2}$$

Similarly, (9-41) yields

$$R_{x'x'}(t_1, t_2) = L_1[R_{xx'}(t_1, t_2)] = \frac{\partial R_{xx'}(t_1, t_2)}{\partial t_1}$$

Combining, we obtain

$$R_{x'x'}(t_1, t_2) = \frac{\partial^2 R_{xx}(t_1, t_2)}{\partial t_1\,\partial t_2}$$

Note: If $\mathbf{n}(t)$ is a stationary process with autocorrelation $R(\tau)$, and $\mathbf{x}(t) = \mathbf{n}(t)U(t)$, then

$$R_{xx}(t_1, t_2) = E\{\mathbf{n}(t_1)U(t_1)\mathbf{n}^*(t_2)U(t_2)\} = R(t_1 - t_2)U(t_1)U(t_2) \tag{9-43}$$

This is useful in the study of linear systems with stationary inputs applied at $t = 0$.

Example 9-3 (Langevin's equation and brownian motion) If $\mathbf{y}(0) = 0$ and

$$\mathbf{y}'(t) + \alpha\mathbf{y}(t) = \mathbf{n}(t) \qquad t \geq 0$$

then $\mathbf{y}(t)$ can be considered as the output of a linear system with input $\mathbf{x}(t) = \mathbf{n}(t)U(t)$

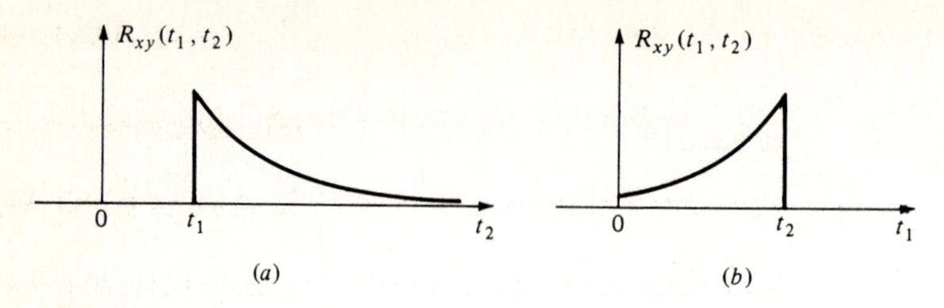

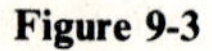
Figure 9-3

and impulse response

$$h(t) = e^{-\alpha t}U(t)$$

We shall determine $R_{yy}(t_1, t_2)$ under the assumption that $\mathbf{n}(t)$ is stationary white noise with $R(\tau) = I\,\delta(\tau)$.

From Eqs. (9-40) and (9-43), it follows that

$$R_{xy}(t_1, t_2) = e^{-\alpha t_2}U(t_2) * I\,\delta(t_2 - t_1)\,U(t_1)U(t_2) = Ie^{-\alpha(t_2 - t_1)}$$

for $t_2 > t_1 > 0$ and zero otherwise (Fig. 9-3). Considering the above as a function of t_1 and convolving with $h(t_1)$, we obtain

$$R_{yy}(t_1, t_2) = I\int_0^{t_1} e^{-\alpha\tau}e^{-\alpha(t_2 - t_1 + \tau)}\,d\tau = \frac{I}{2\alpha}(1 - e^{-2\alpha t_1})e^{-\alpha(t_2 - t_1)} \tag{9-44}$$

for $t_2 > t_1 > 0$. For $t_1 > t_2 > 0$, we interchange t_1 and t_2 because $R_{yy}(t_1, t_2) = R_{yy}(t_2, t_1)$.

We note that

$$E\{|\mathbf{y}(t)|^2\} = R_{yy}(t, t) = \frac{I}{2\alpha}(1 - e^{-2\alpha t})U(t)$$

This result can be derived simply from (9-46).

The following is an important consequence of Eqs. (9-40) and (9-41).

Theorem If the input $\mathbf{x}(t)$ to a linear system is nonstationary white noise with

$$R_{xx}(t_1, t_2) = I(t_1)\,\delta(t_1 - t_2) \tag{9-45}$$

then the average intensity of the resulting output $\mathbf{y}(t)$ is given by

$$E\{|\mathbf{y}(t)|^2\} = I(t) * |h(t)|^2 = \int_{-\infty}^{\infty} I(t - \alpha)|h(\alpha)|^2\,d\alpha \tag{9-46}$$

Proof Inserting Eq. (9-45) into (9-40), we obtain

$$R_{xy}(t_1, t_2) = I(t_1)\,\delta(t_1 - t_2) * h^*(t_2) = I(t_1)h^*(t_2 - t_1)$$

and (9-41) yields

$$R_{yy}(t_1, t_2) = \int_{-\infty}^{\infty} I(t_1 - \alpha)h^*[t_2 - (t_1 - \alpha)]h(\alpha)\,d\alpha$$

Setting $t_1 = t_2 = t$, we obtain (9-46) because $R_{yy}(t, t) = E\{|\mathbf{y}(t)|^2\}$.

Example 9-4 (Smoothing) If $h(t) = (1/2T)p_T(t)$ is a pulse, then

$$\mathbf{y}(t) = \mathbf{x}(t) * h(t) = \frac{1}{2T}\int_{t-T}^{t+T} \mathbf{x}(\tau)\,d\tau \tag{9-47}$$

Since $h^2(t) = (1/4T^2)p_T(t)$, it follows from (9-46) that

$$E\{|\mathbf{y}(t)|^2\} = \frac{1}{4T^2}\int_{t-T}^{t+T} I(\tau)\,d\tau \tag{9-48}$$

Corollary If $\mathbf{x}(t)$ is white noise as in (9-45), then

$$I(t) \geq 0 \tag{9-49}$$

Proof From (9-48), it follows that, for any a and b,

$$\int_a^b I(\tau)\,d\tau \geq 0$$

(elaborate). This is possible only if $I(t) \geq 0$ for every t.

Notes:

1. The nonstationary white noise $\mathbf{x}(t)$ can be written as a product

$$\mathbf{x}(t) = p(t)\mathbf{n}(t) \qquad \text{where} \qquad p(t) = \sqrt{I(t)}$$

and $\mathbf{n}(t)$ is stationary white noise with $R_{nn}(\tau) = \delta(\tau)$. Indeed,

$$R_{xx}(t_1, t_2) = p(t_1)p(t_2)R_{nn}(t_1, t_2) = p(t_1)p(t_2)\,\delta(t_1 - t_2) = I(t_1)\,\delta(t_1 - t_2).$$

2. Although $E\{|\mathbf{x}(t)|^2\} = \infty$, the function $I(t)$ is often called the *average intensity* of $\mathbf{x}(t)$. The reason is that, if $\mathbf{x}(t)$ is measured with a dynamic system whose impulse response is of sufficiently short duration, then [see Eq. (9-46)] the average intensity

$$E\{|\mathbf{y}(t)|^2\} \simeq I(t)\int_{-\infty}^{\infty} |h(t)|^2\,dt \tag{9-50}$$

of the resulting response is proportional to $I(t)$.

3. Suppose that $\mathbf{x}(t)$ is a-dependent and quasistationary as in (9-24), and a is sufficiently small so that $h(t)$ is approximately constant in any interval of length $2a$. It then follows that [see (9-42)]

$$E\{|\mathbf{y}(t)|^2\} = \int_{-\infty}^{\infty}\int_{-\infty}^{\infty} R_{xx}(t-\alpha, t-\beta)h(\alpha)h^*(\beta)\,d\alpha\,d\beta$$

$$\simeq \int_{-\infty}^{\infty} I(t-\alpha)|h(\alpha)|^2\,d\alpha \tag{9-51}$$

where

$$I(t) = \int_{-\infty}^{\infty} R_{xx}\left(t - \frac{\tau}{2}, t + \frac{\tau}{2}\right)d\tau$$

i.e., for the evaluation of $E\{|\mathbf{y}(t)|^2\}$ we can assume that $\mathbf{x}(t)$ is white noise with average intensity $I(t)$.

Example 9-5 (Wiener-Lévy process) The integral

$$\mathbf{y}(t) = \int_0^t \mathbf{n}(\tau)\, d\tau$$

can be considered as the output of a linear system with input $\mathbf{n}(t)U(t)$ and impulse response $U(t)$:

$$\mathbf{y}(t) = \mathbf{x}(t) * h(t) \qquad \mathbf{x}(t) = \mathbf{n}(t)U(t) \qquad h(t) = U(t)$$

Suppose that $\mathbf{n}(t)$ is stationary white noise with $R_{nn}(\tau) = I\,\delta(\tau)$. In this case,

$$R_{xx}(t_1, t_2) = U(t_1)U(t_2)I\,\delta(t_1 - t_2) = IU(t_1)\,\delta(t_1 - t_2)$$

Thus, $\mathbf{x}(t)$ is nonstationary white noise with $I(t) = IU(t)$, and Eq. (9-46) yields

$$E\{|\mathbf{y}(t)|^2\} = IU(t) * U^2(t) = I\int_0^t d\tau = It$$

Output covariance From (9-37) and (9-38), it follows that

$$\mathbf{y}(t) - \eta_y(t) = \int_{-\infty}^{\infty} [\mathbf{x}(t-\alpha) - \eta_x(t-\alpha)]h(\alpha)\, d\alpha \tag{9-52}$$

Thus, the response of the given system to the centered input $\mathbf{x}_c(t) = \mathbf{x}(t) - \eta_x(t)$ equals the centered output $\mathbf{y}_c(t) = \mathbf{y}(t) - \eta_y(t)$:

$$\mathbf{y}_c(t) = L[\mathbf{x}_c(t)] \tag{9-53}$$

Applying Eqs. (9-40) and (9-41) to the processes $\mathbf{x}_c(t)$ and $\mathbf{y}_c(t)$ and noting that their correlations equal the covariances of $\mathbf{x}(t)$ and $\mathbf{y}(t)$ [see (9-10)], we conclude that

$$\begin{aligned} C_{xy}(t_1, t_2) &= L_2^*[C_{xx}(t_1, t_2)] \\ C_{yy}(t_1, t_2) &= L_1[C_{xy}(t_1, t_2)] \end{aligned} \tag{9-54}$$

Combining, we obtain, as in (9-42),

$$C_{yy}(t_1, t_2) = \int_{-\infty}^{\infty}\int_{-\infty}^{\infty} C_{xx}(t_1 - \alpha, t_2 - \beta)h(\alpha)h^*(\beta)\, d\alpha\, d\beta \tag{9-55}$$

Higher-order moments can be determined similarly (see Appendix 11-A).

Stationary processes We now assume that the input $\mathbf{x}(t)$ is stationary with mean η_x and autocorrelation $R_{xx}(\tau)$. With $\mathbf{y}(t)$ the resulting output, we shall show that the processes $\mathbf{x}(t)$ and $\mathbf{y}(t)$ are jointly stationary. It suffices to prove that $\eta_y(t)$ is constant and that $R_{xy}(t_1, t_2)$ and $R_{yy}(t_1, t_2)$ depend on $t_1 - t_2 = \tau$ only.

The first condition is a consequence of (9-38):

$$\eta_y(t) = \eta_x \int_{-\infty}^{\infty} h(\alpha)\, d\alpha = \eta_y \tag{9-56}$$

Since $R_{xx}(t_1, t_2 - \alpha) = R_{xx}(t_1 - t_2 + \alpha)$, it follows from (9-40) that

$$R_{xy}(t_1, t_2) = \int_{-\infty}^{\infty} R_{xx}(\tau + \alpha)h^*(\alpha)\, d\alpha$$

Thus, $R_{xy}(t_1, t_2)$ depends only on $t_1 - t_2 = \tau$. Replacing α in the above integral with $-\alpha$, we conclude that

$$R_{xy}(\tau) = \int_{-\infty}^{\infty} R_{xx}(\tau - \alpha)h^*(-\alpha)\, d\alpha = R_{xx}(\tau) * h^*(-\tau) \tag{9-57}$$

Reasoning similarly, we deduce from (9-41) that $R_{yy}(t_1, t_2)$ depends on $t_1 - t_2 = \tau$ only and is given by

$$R_{yy}(\tau) = \int_{-\infty}^{\infty} R_{xy}(\tau - \alpha)h(\alpha)\, d\alpha = R_{xy}(\tau) * h(\tau) \tag{9-58}$$

The output autocorrelation $R_{yy}(\tau)$ can be expressed directly in terms of $R_{xx}(\tau)$. Inserting Eq. (9-57) into (9-58), we obtain

$$R_{yy}(\tau) = R_{xx}(\tau) * h^*(-\tau) * h(\tau) \tag{9-59}$$

This important result can be written in the form

$$R_{yy}(\tau) = R_{xx}(\tau) * \rho(\tau) = \int_{-\infty}^{\infty} R_{xx}(\tau - \alpha)\rho(\alpha)\, d\alpha \tag{9-60}$$

where
$$\rho(\tau) = h(\tau) * h^*(-\tau) = \int_{-\infty}^{\infty} h(\tau + \alpha)h^*(\alpha)\, d\alpha \tag{9-61}$$

We note that

$$E\{|\mathbf{y}(t)|^2\} = R_{yy}(0) = \int_{-\infty}^{\infty} R_{xx}(-\alpha)\rho(\alpha)\, d\alpha \tag{9-62}$$

White noise If $R_{xx}(\tau) = I\,\delta(\tau)$, then

$$R_{xy}(\tau) = Ih^*(-\tau) \qquad R_{yy}(\tau) = I\rho(\tau) \qquad E\{|\mathbf{y}(t)|^2\} = I\rho(0) \tag{9-63}$$

as we see from Eqs. (9-57) and (9-60).

Output covariance Applying (9-57) and (9-58) to the centered input $\mathbf{x}_c(t)$, we conclude as in (9-54) that

$$C_{xy}(\tau) = C_{xx}(\tau) * h^*(-\tau) \qquad C_{yy}(\tau) = C_{xy}(\tau) * h(\tau)$$
$$C_{yy}(\tau) = C_{xx}(\tau) * \rho(\tau) \tag{9-64}$$

Example 9-6 If $h(t) = (1/2T)p_T(t)$ is a pulse, then $\rho(t) = (1/4T^2)p_T(t) * p_T(t)$ is a triangle (Fig. 9-4). With

$$\mathbf{y}(t) = \frac{1}{2T}\int_{t-T}^{t+T} \mathbf{x}(\tau)\, d\tau$$

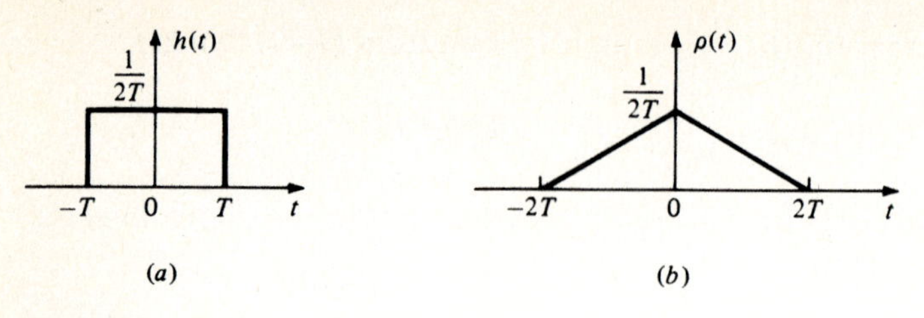

Figure 9-4

it follows from (9-64) that

$$C_{yy}(\tau) = C_{xx}(\tau) * \rho(\tau) = \frac{1}{2T}\int_{-2T}^{2T} C_{xx}(\tau - \alpha)\left(1 - \frac{|\alpha|}{2T}\right) d\alpha \tag{9-65}$$

Setting $\tau = 0$, we obtain the variance

$$C_{yy}(0) = \frac{1}{2T}\int_{-2T}^{2T} C_{xx}(-\alpha)\left(1 - \frac{|\alpha|}{2T}\right) d\alpha = \frac{1}{2T}\int_{-2T}^{2T} C_{xx}(\beta)\left(1 - \frac{|\beta|}{2T}\right) d\beta \tag{9-66}$$

of the process $\mathbf{y}(t)$.

9-3 SPECTRAL ANALYSIS

We shall express the power spectrum $S_{yy}(\omega)$ of the output $\mathbf{y}(t)$ of a linear system in terms of the power spectrum $S_{xx}(\omega)$ of the input $\mathbf{x}(t)$ and the system function $H(\omega)$.

Applying the convolution theorem to Eqs. (9-57) and (9-58) and using the Fourier transform pairs

$$h(t) \leftrightarrow H(\omega) \qquad h^*(-t) \leftrightarrow H^*(\omega) \qquad \rho(t) \leftrightarrow |H(\omega)|^2$$

$$R_{xx}(\tau) \leftrightarrow S_{xx}(\omega) \qquad R_{xy}(\tau) \leftrightarrow S_{xy}(\omega) \qquad R_{yy}(\tau) \leftrightarrow S_{yy}(\omega)$$

we obtain

$$S_{xy}(\omega) = S_{xx}(\omega)H^*(\omega) \qquad S_{yy}(\omega) = S_{xy}(\omega)H(\omega) \tag{9-67}$$

$$S_{yy}(\omega) = S_{xx}(\omega)|H(\omega)|^2 \tag{9-68}$$

Equation (9-68) is fundamental in applications of stochastic processes. We shall give several illustrations, using the system concept as a convenient device for simplifying the analysis.

Example 9-7 If

$$\mathbf{y}'(t) + \alpha\mathbf{y}(t) = \mathbf{x}(t)$$

then $\mathbf{y}(t)$ can be considered as the output of the system $H(\omega) = 1/(\alpha + j\omega)$. Suppose that

$\mathbf{x}(t)$ is stationary white noise with $R_{xx}(\tau) = I\,\delta(\tau)$. It then follows from (9-68) that

$$S_{yy}(\omega) = \frac{S_{xx}(\omega)}{\alpha^2 + \omega^2} = \frac{I}{\alpha^2 + \omega^2}$$

Hence,
$$R_{yy}(\tau) = \frac{I}{2\alpha} e^{-\alpha|\tau|} \qquad E\{\mathbf{y}^2(t)\} = \frac{I}{2\alpha}$$

Example 9-8 If

$$\mathbf{y}(t) = \mathbf{x}^{(n)}(t)$$

is the nth derivative of $\mathbf{x}(t)$, then $H(\omega) = (j\omega)^n$. Hence,

$$S_{yy}(\omega) = \omega^{2n} S_{xx}(\omega) \qquad R_{yy}(\tau) = (-1)^n \frac{d^{2n} R_{xx}(\tau)}{d\tau^{2n}}$$

Note: The preceding result (9-68) is the random version of the equation $|Y(\omega)|^2 = |X(\omega)|^2 |H(\omega)|^2$ relating the energy spectra of the input $x(t)$ and the output $y(t)$ of a linear system. In the deterministic case, the energy of the output is given by

$$\int_{-\infty}^{\infty} |y(t)|^2\, dt = \frac{1}{2\pi} \int_{-\infty}^{\infty} |X(\omega)|^2 |H(\omega)|^2\, d\omega$$

The stochastic version of the above is the average power of $\mathbf{y}(t)$:

$$E\{|\mathbf{y}(t)|^2\} = R_{yy}(0) = \frac{1}{2\pi} \int_{-\infty}^{\infty} S_{xx}(\omega) |H(\omega)|^2\, d\omega \tag{9-69}$$

as we can see from (9-68).

Example 9-9 (Tapped delay line) The process

$$\mathbf{y}(t) = a_0 \mathbf{x}(t) + a_1 \mathbf{x}(t - T) + \cdots + a_n \mathbf{x}(t - nT) \tag{9-70}$$

is the output of a system with input $\mathbf{x}(t)$ and system function

$$H(\omega) = a_0 + a_1 e^{-j\omega T} + \cdots + a_n e^{-jn\omega T} \tag{9-71}$$

This system can be realized by a delay line as in Fig. 9-5. Clearly,

$$|H(\omega)|^2 = H(\omega) H^*(\omega) = \sum_{k,\, r=0}^{n} a_k a_r^* e^{-j(k-r)\omega T}$$

Hence [see Eq. (9-69)],

$$E\{|\mathbf{y}(t)|^2\} = \frac{1}{2\pi} \int_{-\infty}^{\infty} S(\omega) \sum_{k,\, r=0}^{n} a_k a_r^* e^{-j(k-r)\omega T}\, d\omega = \sum_{k,\, r=0}^{n} a_k a_r^* R[(k-r)T] \tag{9-72}$$

nth difference A special case of (9-71) is the system

$$H(\omega) = (1 - e^{-j\omega T})^n = 1 - \binom{n}{1} e^{-j\omega T} + \cdots + (-1)^n \binom{n}{n} e^{-jn\omega T} \tag{9-73}$$

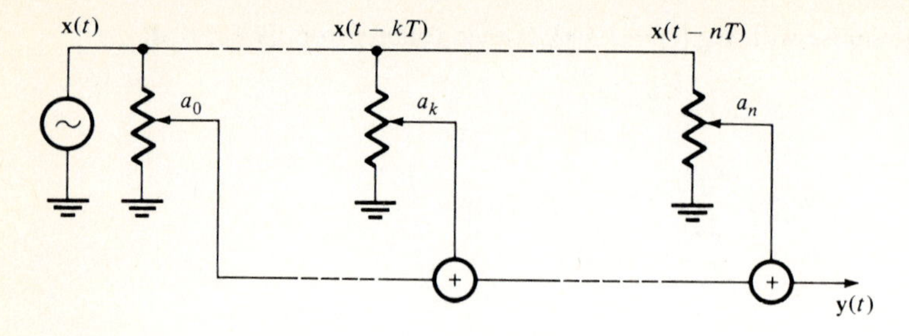

Figure 9-5

whose output

$$\mathbf{y}(t) = \mathbf{x}(t) - \binom{n}{1}\mathbf{x}(t - T) + \cdots + (-1)^n\mathbf{x}(t - nT) \tag{9-74}$$

is the nth difference of the input $\mathbf{x}(t)$. Hence [see (9-68)],

$$S_{yy}(\omega) = S_{xx}(\omega)\,|1 - e^{-j\omega T}|^{2n} = S_{xx}(\omega)\left(2 \sin\frac{\omega T}{2}\right)^{2n} \tag{9-75}$$

Power localization If (Fig. 9-6)

$$H(\omega) = \begin{cases} 1 & a < \omega < b \\ 0 & \text{otherwise} \end{cases}$$

then

$$S_{yy}(\omega) = \begin{cases} S_{xx}(\omega) & a < \omega < b \\ 0 & \text{otherwise} \end{cases}$$

Hence,

$$E\{|\mathbf{y}(t)|^2\} = \frac{1}{2\pi}\int_a^b S_{xx}(\omega)\, d\omega \tag{9-76}$$

Thus, the average power of $\mathbf{x}(t)$ (total area of its power spectrum) is localized on the frequency axis, in the sense that the area of $S_{xx}(\omega)$ in an interval (a, b) equals the average power of the output $\mathbf{y}(t)$ of the ideal bandpass filter $H(\omega)$, whose band is (a, b). This observation leads to the following important conclusion:

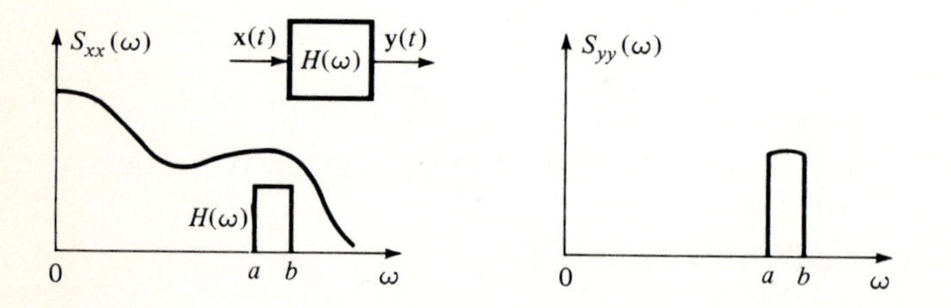

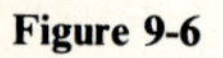
Figure 9-6

Theorem The power spectrum $S(\omega)$ of a process $\mathbf{x}(t)$ is nonnegative for all ω:

$$S(\omega) \geq 0 \tag{9-77}$$

Proof From Eq. (9-76), it follows that

$$\int_a^b S(\omega)\, d\omega \geq 0$$

and since this is true for any a and b, (9-77) results.

> **Note:** A function whose Fourier transform is positive is called *positive definite*. From (9-77), it follows that the autocorrelation $R(\tau)$ of $\mathbf{x}(t)$ is positive (nonnegative) definite; hence, it is continuous for every τ (see Prob. 76).

Corollary 1 For any τ,

$$|R(\tau)| \leq R(0) \tag{9-78}$$

Proof Since $S(\omega)$ is real and positive,

$$|R(\tau)| = \frac{1}{2\pi}\left|\int_{-\infty}^{\infty} S(\omega)e^{j\omega\tau}\, d\omega\right| \leq \frac{1}{2\pi}\int_{-\infty}^{\infty} S(\omega)\, d\omega = R(0)$$

Corollary 2 If, for some $\tau = \tau_1$, $|R(\tau_1)| = R(0)$, that is, if

$$R(\tau_1) = e^{j\alpha}R(0) \tag{9-79}$$

then

$$R(\tau) = e^{j\alpha\tau/\tau_1}w(\tau) \tag{9-80}$$

where $w(\tau) = w(\tau + \tau_1)$ is a periodic function with period τ_1.

Proof From (9-79), it follows that

$$0 = R(0) - e^{-j\alpha}R(\tau_1) = \frac{1}{2\pi}\int_{-\infty}^{\infty} S(\omega)(1 - e^{j(\omega\tau_1 - \alpha)})\, d\omega$$

$$= \frac{1}{2\pi}\int_{-\infty}^{\infty} S(\omega)[1 - \cos(\omega\tau_1 - \alpha)]\, d\omega$$

Since $S(\omega) \geq 0$ and $1 - \cos(\omega\tau_1 - \alpha) \geq 0$, the above can be true only if $S(\omega)$ consists of impulses at the zeros $\omega_n = (2\pi n + \alpha)/\tau_1$ of $1 - \cos(\omega\tau_1 - \alpha)$. Thus,

$$S(\omega) = \sum_{n=-\infty}^{\infty} a_n\, \delta\left(\omega - n\omega_0 - \frac{\alpha}{\tau_1}\right) \qquad \omega_0 = \frac{2\pi}{\tau_1}$$

and Eq. (9-80) results (elaborate).

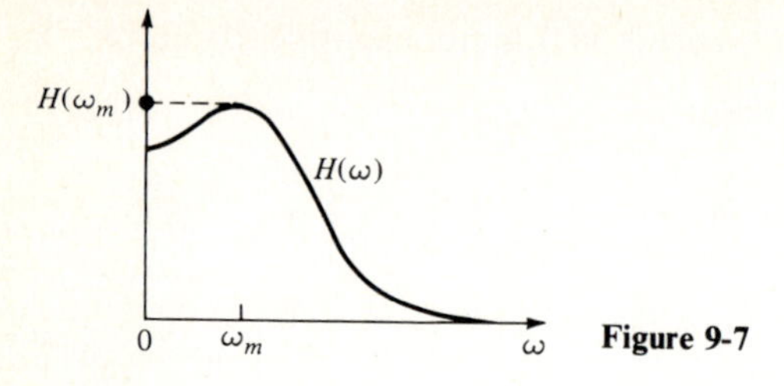

Figure 9-7

Stochastic resonance We denote by ω_m the frequency at which $|H(\omega)|$ is maximum (Fig. 9-7). From (9-69), it follows that

$$E\{|\mathbf{y}(t)|^2\} \leq \frac{|H(\omega_m)|^2}{2\pi} \int_{-\infty}^{\infty} S_{xx}(\omega)\, d\omega = |H(\omega_m)|^2 E\{|\mathbf{x}(t)|^2\} \tag{9-81}$$

Equality holds only if $S_{xx}(\omega)$ is an impulse at $\omega = \omega_m$, that is, if

$$R_{xx}(\tau) = Pe^{j\omega_m \tau} \tag{9-82}$$

A process whose autocorrelation is a harmonic function of τ, as above, is called *monochromatic*. We reach, then, the following conclusion:

Suppose that the input to a system is a process $\mathbf{x}(t)$ with given average power $E\{|\mathbf{x}(t)|^2\} = P$. It then follows that the average power $E\{|\mathbf{y}(t)|^2\}$ of the output is bounded by (9-81) and is maximum only if $\mathbf{x}(t)$ is monochromatic, as in (9-82).

The sampling theorem for stochastic processes We shall say that a stochastic process $\mathbf{x}(t)$ is σ-BL (σ-bandlimited) if its autocorrelation $R(\tau)$ is σ-BL in the sense of Sec. 6-1, that is, if

$$S(\omega) = 0 \qquad \text{for } |\omega| > \sigma \tag{9-83}$$

We shall show that if $\mathbf{x}(t)$ is σ-BL, and

$$\hat{\mathbf{x}}(t) = \sum_{n=-\infty}^{\infty} \mathbf{x}(nT) \frac{\sin \sigma(t - nT)}{\sigma(t - nT)} \qquad T = \frac{2\pi}{\sigma} \tag{9-84}$$

is its sampling expansion [see Eq. (5-9)], then $\mathbf{x}(t)$ equals $\hat{\mathbf{x}}(t)$ in the mean-square sense; i.e.,

$$E\{|\mathbf{x}(t) - \hat{\mathbf{x}}(t)|^2\} = 0 \tag{9-85}$$

To prove the above, we shall first determine the truncation error and show that it tends to zero with $N \to \infty$.

Truncated sum We shall use the sum

$$\hat{\mathbf{x}}_N(t, \tau) = \sum_{n=-N}^{N} \mathbf{x}(t + nT) \frac{\sin \sigma(\tau - nT)}{\sigma(\tau - nT)} \tag{9-86}$$

to approximate the given process $\mathbf{x}(t + \tau)$ for any t and τ. As we shall see, the shift in the time origin will permit us to apply the system results for the evaluation of the resulting error

$$\mathbf{y}_N(t, \tau) = \mathbf{x}(t+\tau) - \hat{\mathbf{x}}_N(t, \tau) \tag{9-87}$$

For a fixed τ, $\mathbf{y}_N(t, \tau)$ is a stochastic process in the variable t, and it equals the output of a linear system with input $\mathbf{x}(t)$ and system function

$$H_N(\omega, \tau) = e^{j\omega\tau} - \sum_{n=-N}^{N} \frac{\sin \sigma(\tau - nT)}{\sigma(\tau - nT)} e^{jnT\omega} \tag{9-88}$$

Hence [see Eqs. (9-69) and (9-83)],

$$E\{|\mathbf{y}_N(t, \tau)|^2\} = \frac{1}{2\pi}\int_{-\sigma}^{\sigma} S(\omega)|H_N(\omega, \tau)|^2\, d\tau \le |H_N(\omega_m, \tau)|E\{|\mathbf{x}(t)|^2\} \tag{9-89}$$

where $H_N(\omega_m, \tau)$ is the maximum of $H_N(\omega, \tau)$ *in the interval* $(-\sigma, \sigma)$.

If we expand the function $e^{j\omega\tau}$ into a Fourier series

$$e^{j\omega\tau} = \sum_{n=-\infty}^{\infty} a_n e^{jnT\omega} \qquad |\omega| \le \sigma \tag{9-90}$$

then

$$a_n = \frac{1}{2\sigma}\int_{-\sigma}^{\sigma} e^{j\omega\tau}e^{-jnT\omega}\, d\omega = \frac{\sin \sigma(\tau - nT)}{\sigma(\tau - nT)} \tag{9-91}$$

Hence, the sum in (9-88) is the truncation of the series (9-90), and $H_N(\omega, \tau)$ is the truncation error. As we know from the theory of Fourier series,

$$\lim_{N\to\infty} H_N(\omega, \tau) = 0 \qquad |\omega| \le \sigma$$

Hence, $E\{|\mathbf{y}_N(t)|^2\} \to 0$ with $N \to \infty$. Thus,

$$\mathbf{x}(t+\tau) = \sum_{n=-\infty}^{\infty} \mathbf{x}(t+nT)\frac{\sin \sigma(\tau - nT)}{\sigma(\tau - nT)} \tag{9-92}$$

in the mean-square sense.

Past samples We show next that, by increasing the sampling rate, we can express a bandlimited process in terms of its past samples only. For this purpose, we use as our estimator of $\mathbf{x}(t)$ the process

$$\mathbf{z}_n(t) = \binom{n}{1}\mathbf{x}(t - T_1) - \binom{n}{2}\mathbf{x}(t - 2T_1) + \cdots - (-1)^n\mathbf{x}(t - nT_1) \tag{9-93}$$

where

$$T_1 < \frac{\pi}{3\sigma}$$

As we have shown [see (9-75)], the power spectrum of the error $\mathbf{x}(t) - \mathbf{z}_n(t)$ equals $S(\omega)[2\sin(\omega T_1/2)]^{2n}$. Hence,

$$E\{|\mathbf{x}(t) - \mathbf{z}_n(t)|^2\} = \frac{1}{2\pi}\int_{-\sigma}^{\sigma} S(\omega)\left(2\sin\frac{\omega T_1}{2}\right)^{2n} d\omega \tag{9-94}$$

Clearly, $2 \sin (\omega T_1/2) < 1$ for $|\omega| \le \sigma$. Therefore,

$$\left(2 \sin \frac{\omega T_1}{2}\right)^{2n} \underset{n\to\infty}{\to} 0$$

and Eq. (9-94) yields

$$E\{|\mathbf{x}(t) - \mathbf{z}_n(t)|^2\} \underset{n\to\infty}{\to} 0$$

that is, $\mathbf{z}_n(t)$ tends to $\mathbf{x}(t)$ in the mean-square sense.

Factorization Consider a stationary process $\mathbf{x}(t)$ with power spectrum $S(\omega)$. With $\varphi(\omega)$ an arbitrary angle, we form a system with system function

$$H(\omega) = \sqrt{S(\omega)}\, e^{j\varphi(\omega)} \tag{9-95}$$

and apply to its input a white noise $\mathbf{n}(t)$ with power spectrum $S_{nn}(\omega) = 1$. From (9-68), it follows that the power spectrum $S_{yy}(\omega)$ of the resulting output $\mathbf{y}(t)$ equals the power spectrum of $\mathbf{x}(t)$:

$$S_{yy}(\omega) = S_{nn}(\omega)|H(\omega)|^2 = S(\omega) \tag{9-96}$$

Thus, in all applications involving only second-order moments, any process $\mathbf{x}(t)$ can be considered as the output of a suitable system driven by white noise.

In general, the system $H(\omega)$ so formed is not causal. However (see Sec. 7-2), if $S(\omega)$ satisfies the Paley-Wiener condition

$$\int_{-\infty}^{\infty} \frac{|\ln S(\omega)|}{1 + \omega^2}\, d\omega < \infty$$

then the angle $\varphi(\omega)$ can be chosen so that $H(\omega)$ has a causal inverse $h(t)$. In this case, $\mathbf{x}(t)$ is given by the moving average

$$\mathbf{x}(t) = \int_{-\infty}^{t} \mathbf{n}(\tau)h(t - \tau)\, d\tau \tag{9-97}$$

involving only past values of $\mathbf{n}(t)$.

Rational spectra If $S(\omega)$ is a rational function, then we can find a causal minimum-phase system (see Sec. 4-3)

$$H(s) = \frac{b_0 s^r + \cdots + b_r}{s^m + a_1 s^{m-1} + \cdots + a_m}$$

such that $|H(j\omega)| = \sqrt{S(\omega)}$. The resulting output $\mathbf{x}(t)$ satisfies the differential equation

$$\mathbf{x}^{(m)}(t) + a_1 \mathbf{x}^{(m-1)}(t) + \cdots + a_m \mathbf{x}(t) = b_0 \mathbf{n}^{(r)}(t) + \cdots + b_r \mathbf{n}(t)$$

Example 9-10 If $S(\omega) = 1/(1 + \omega^4)$, then

$$H(s) = \frac{1}{s^2 + \sqrt{2}\, s + 1}$$

(see Example 4-15). Hence,

$$\frac{d^2\mathbf{x}}{dt^2} + \sqrt{2}\,\frac{d\mathbf{x}(t)}{dt} + \mathbf{x}(t) = \mathbf{n}(t) \qquad S_{nn}(\omega) = 1$$

4 DISCRETE PROCESSES

In this section, we discuss briefly the discrete version of the preceding concepts. This involves a change from functions to sequences, from analog systems to digital systems, and from Fourier integrals to Fourier series.

Definitions[1]

A discrete process $\mathbf{x}[n]$ is a sequence of random variables, real or complex, defined for every integer n. The mean $\eta[n]$, autocorrelation $R[n_1, n_2]$, and autocovariance $C[n_1, n_2]$ of $\mathbf{x}(t)$ are, by definition,

$$E\{\mathbf{x}[n]\} = \eta[n] \tag{9-98}$$

$$E\{\mathbf{x}[n_1]\mathbf{x}^*[n_2]\} = R[n_1, n_2] \tag{9-99}$$

$$C[n_1, n_2] = R[n_1, n_2] - \eta[n_1]\eta^*[n_2] \tag{9-100}$$

The cross-correlation $R_{xy}[n_1, n_2]$ and cross-covariance $C_{xy}[n_1, n_2]$ of two discrete processes $\mathbf{x}[n]$ and $\mathbf{y}[n]$ are defined similarly:

$$E\{\mathbf{x}[n_1]\mathbf{y}^*[n_2]\} = R_{xy}[n_1, n_2] \tag{9-101}$$

$$C_{xy}[n_1, n_2] = R_{xy}[n_1, n_2] - \eta_x[n_1]\eta_y^*[n_2] \tag{9-102}$$

Stationary processes A process $\mathbf{x}[n]$ is called *stationary* (in the wide sense) if its mean is constant and its autocorrelation depends on the difference $m = n_1 - n_2$ only:

$$E\{\mathbf{x}[n+m]\mathbf{x}^*[n]\} = R[m] = C[m] + |\eta|^2 \tag{9-103}$$

Two processes $\mathbf{x}[n]$ and $\mathbf{y}[n]$ are jointly stationary if each is stationary and their cross-correlation depends on $n_1 - n_2$ only:

$$E\{\mathbf{x}[n+m]\mathbf{y}^*[n]\} = R_{xy}[m] = C_{xy}[m] + \eta_x\eta_y^* \tag{9-104}$$

Spectra The power spectrum of a stationary process $\mathbf{x}[n]$ is a periodic function $\overline{S}(\omega)$ with Fourier-series coefficients $R[m]$:

$$\overline{S}(\omega) = \sum_{m=-\infty}^{\infty} R[m]e^{-jmT\omega} \qquad R[m] = \frac{1}{2\sigma}\int_{-\sigma}^{\sigma} \overline{S}(\omega)e^{jmT\omega}\,d\omega \tag{9-105}$$

[1] M. Schwartz and L. Shaw, "Signal Processing," McGraw-Hill Book Company, New York, 1975.

In the above, $T = \pi/\sigma$ is an arbitrary constant. If $\mathbf{x}[n]$ is obtained by sampling an analog process, then T equals the sampling interval; otherwise, we can take $T = 1$ and $\sigma = \pi$.

We note that

$$E\{|\mathbf{x}[n]|^2\} = R[0] = \frac{1}{2\sigma}\int_{-\sigma}^{\sigma} \bar{S}(\omega)\, d\omega \tag{9-106}$$

The spectrum $\bar{S}(\omega)$ is real because $R[-m] = R^*[m]$, as is easy to see from Eq. (9-103). If $\mathbf{x}[n]$ is a real process, then $R[m]$ is a real, even sequence, and $\bar{S}(\omega)$ is a real, even function.

The cross-power spectrum $\bar{S}_{xy}(\omega)$ is defined similarly:

$$\bar{S}_{xy}(\omega) = \sum_{m=-\infty}^{\infty} R_{xy}[m]e^{-jmT\omega} \tag{9-107}$$

White noise We shall say that the process $\mathbf{x}[n]$ is *white noise* if $E\{\mathbf{x}[n_1]\mathbf{x}^*[n_2]\} = 0$ for $n_1 \neq n_2$. With $E\{|\mathbf{x}[n]|^2\} = I[n]$, it follows that the autocorrelation of $\mathbf{x}[n]$ is given by

$$R[n_1, n_2] = I[n_1]\,\delta[n_1 - n_2] \tag{9-108}$$

If the process $\mathbf{x}[n]$ is stationary, then $I[n] = I = \text{constant}$; hence,

$$R[m] = I\,\delta[m] \qquad \bar{S}(\omega) = I \tag{9-109}$$

Sampling Suppose that the discrete process $\mathbf{x}[n]$ is obtained by sampling an analog process $\mathbf{x}(t)$:

$$\mathbf{x}[n] = \mathbf{x}(nT)$$

With $\eta(t)$ and $R(t_1, t_2)$ the mean and autocorrelation of $\mathbf{x}(t)$, it follows that

$$\eta[n] = \eta(nT) \qquad R[n_1, n_2] = R(n_1T, n_2T) \tag{9-110}$$

If the analog process $\mathbf{x}(t)$ is stationary with autocorrelation $R(\tau)$ and power spectrum $S(\omega)$, then the discrete process $\mathbf{x}[n]$ is also stationary with autocorrelation $R[m] = R(mT)$ and power spectrum

$$\bar{S}(\omega) = \sum_{m=-\infty}^{\infty} R(mT)e^{-jmT\omega} = \frac{1}{T}\sum_{n=-\infty}^{\infty} S(\omega + 2\sigma n) \tag{9-111}$$

The last equality is a consequence of the Poisson sum formula (3-87).

Linear Systems

If the input to a discrete system is a discrete process $\mathbf{x}[n]$, then the resulting output

$$\mathbf{y}[n] = \sum_{k=-\infty}^{\infty} \mathbf{x}[n-k]h[k] \tag{9-112}$$

is a discrete process with mean

$$E\{\mathbf{y}[n]\} = \sum_{k=-\infty}^{\infty} E\{\mathbf{x}[n-k]\}h[k] \tag{9-113}$$

We shall determine its autocorrelation.
From Eq. (9-112), it follows that

$$\mathbf{x}[n_1]\mathbf{y}^*[n_2] = \sum_{k=-\infty}^{\infty} \mathbf{x}[n_1]\mathbf{x}^*[n_2-k]h^*[k]$$

$$\mathbf{y}[n_1]\mathbf{y}^*[n_2] = \sum_{k=-\infty}^{\infty} \mathbf{x}[n_1-k]\mathbf{y}^*[n_2]h[k]$$

Taking expected values, we obtain

$$\begin{aligned} R_{xy}[n_1, n_2] &= \sum_{k=-\infty}^{\infty} R_{xx}[n_1, n_2-k]h^*[k] \\ R_{yy}[n_1, n_2] &= \sum_{k=-\infty}^{\infty} R_{xy}[n_1-k, n_2]h[k] \end{aligned} \tag{9-114}$$

The covariances C_{xx}, C_{xy}, and C_{yy} satisfy the same equations.

White noise If $R_{xx}[n_1, n_2] = I[n_1]\,\delta[n_1-n_2]$, then

$$R_{xy}[n_1, n_2] = I[n_1]h^*[n_2-n_1]$$
$$R_{yy}[n_1, n_2] = \sum_{k=-\infty}^{\infty} I[n_1-k]h^*[n_2-n_1+k]h[k]$$

(elaborate). With $n_1 = n_2 = n$, the above yields

$$E\{|\mathbf{y}[n]|^2\} = R_{yy}[n, n] = \sum_{k=-\infty}^{\infty} I[n-k]|h[k]|^2 \tag{9-115}$$

as in (9-46).

Stationary processes If the input $\mathbf{x}[n]$ is stationary, then the output $\mathbf{y}[n]$ is also stationary, and (9-114) yields

$$R_{xy}[m] = \sum_{k=-\infty}^{\infty} R_{xx}[m+k]h^*[k] \qquad R_{yy}[m] = \sum_{k=-\infty}^{\infty} R_{xy}[m-k]h[k] \tag{9-116}$$

Combining, we obtain

$$R_{yy}[m] = R_{xx}[m] * \rho[m] \tag{9-117}$$

where

$$\rho[n] = \sum_{k=-\infty}^{\infty} h[n+k]h^*[k] \tag{9-118}$$

With

$$H(z) = \sum_{n=-\infty}^{\infty} h[n]z^{-n}$$

the system function of the given system, we conclude from Eqs. (9-116) and (9-117) that (elaborate)

$$\bar{S}_{xy}(\omega) = \bar{S}_{xx}(\omega)H^*(e^{j\omega T}) \qquad \bar{S}_{yy}(\omega) = \bar{S}_{xy}(\omega)H(e^{j\omega T}) \tag{9-119}$$

$$\bar{S}_{yy}(\omega) = \bar{S}_{xx}(\omega)|H(e^{j\omega T})|^2 \tag{9-120}$$

Notes:

1. From (9-120) it follows that

$$E\{|\mathbf{y}[n]|^2\} = \frac{1}{2\sigma}\int_{-\sigma}^{\sigma} \bar{S}_{xx}(\omega)|H(e^{j\omega T})|^2 \, d\omega \tag{9-121}$$

2. Reasoning as in the analog case, we conclude from (9-121) that

$$\bar{S}_{xx}(\omega) \geq 0 \tag{9-122}$$

Rational spectra Suppose that a discrete process $\mathbf{x}[n]$ satisfies the real recursion equation

$$\mathbf{x}[n] + a_1\mathbf{x}[n-1] + \cdots + a_r\mathbf{x}[n-r] = b_0\zeta[n] + \cdots + b_r\zeta[n-r] \tag{9-123}$$

where $\zeta[n]$ is white noise with $\bar{S}_{\zeta\zeta}(\omega) = 1$. Clearly, $\mathbf{x}[n]$ is the response of a linear system with input $\zeta[n]$ and system function $H(z)$, as in (2-37). Hence [see Eqs. (9-120) and (2-46)],

$$\bar{S}_{xx}(\omega) = |H(e^{j\omega T})|^2 = V(\cos \omega T) \tag{9-124}$$

where $V(\cos \omega T)$ is a rational function of $\cos \omega T$ of order r. With

$$S_{xx}(z) = \sum_{m=-\infty}^{\infty} R_{xx}[m]z^{-m} \tag{9-125}$$

the z transform of $R_{xx}[m]$, we conclude that [see Eq. (2-45)]

$$S_{xx}(z) = H(z)H\left(\frac{1}{z}\right) = V(w) \qquad w = \frac{1}{2}\left(z + \frac{1}{z}\right) \tag{9-126}$$

Suppose, finally, that $\mathbf{s}[n]$ is a process whose spectrum $S_{ss}(z)$ is a rational function of $z + z^{-1}$. It then follows that (see Sec. 2-3) $S_{ss}(z)$ can be factored as in (9-126), where $H(z)$ is a rational function of z. Therefore, $\mathbf{s}[n]$ is equivalent to a process $\mathbf{x}[n]$ satisfying a recursion equation of the form (9-123) driven by white noise (see also Sec. 10-4).

Example 9-11

$$\mathbf{x}[n] - a\mathbf{x}[n-1] = \zeta[n] \qquad \bar{S}_{\zeta\zeta}(\omega) = 1 \qquad H(z) = \frac{z}{z-a}$$

$$S_{xx}(z) = \frac{z}{(z-a)(1-az)} = \frac{-a^{-1}}{(z+z^{-1})-(a+a^{-1})}$$

$$R_{xx}[m] = \rho[m] = \frac{|a|^m}{1-a^2}$$

$$\bar{S}_{xx}(\omega) = S_{xx}(e^{j\omega T}) = \frac{1}{1-2a\cos\omega T + a^2}$$

CHAPTER
TEN

DATA SMOOTHING

One of the main reasons for introducing probabilistic considerations in a variety of applications is the problem of noise. In this chapter, we discuss various ways of reducing it. We assume that the noise is additive, the processing linear, and the optimality criterion the minimization of the mean-square error.[1]

In Sec. 10-1, we consider the problem of minimizing the signal-to-noise ratio at the output of the processor. This is useful mainly if the signal is of known form. The analysis includes the continuous and discrete forms of the matched-filter principle and the optimum tapped delay line.

In Sec. 10-2, the signal is of unknown form, but it is assumed that its band is narrow relative to the noise. The estimation involves running averages whose width is adaptively controlled to balance the bias due to smoothing of the signal with the variance due to noise.

In Sec. 10-3, we consider the problem of estimating random signals with known spectra (Wiener).[2] *The analysis is a simple extension of the orthogonality principle discussed in Sec. 5-2; the solution of the resulting integral equation for causal estimators (Wiener-Hopf) involves the factorization problem introduced in Sec. 4-3 for signals with rational spectra, and in Sec. 7-2 for arbitrary signals.*

In Sec. 10-4, we develop the discrete form of the Wiener filter and we show that, for wide-sense Markoff sequences, its causal realization is a recursive (Kalman) filter.

---

[1] E. Parzen, "Time Series Analysis Papers," Holden-Day, Inc., Publisher, San Francisco, 1967.

[2] N. Wiener, "Extrapolation, Interpolation, and Smoothing of Stationary Time Series," The Technology Press of the Massachusetts Institute of Technology, Cambridge, Mass., and John Wiley & Sons, Inc., New York, 1950.

10-1 KNOWN SIGNALS IN NOISE

A basic problem in the theory of stochastic processes is the estimation of a signal $f(t)$ in the presence of additive noise $\mathbf{n}(t)$. The available information (data) is the sum

$$\mathbf{x}(t) = f(t) + \mathbf{n}(t) \tag{10-1}$$

and the problem is to express $f(t)$ in terms of $\mathbf{x}(t)$. The solution depends on various assumptions about $f(t)$ and $\mathbf{n}(t)$. In this chapter, we consider the following three aspects of the problem:

1. The signal $f(t)$ is known.
2. The signal $f(t)$ is unknown.
3. The signal $\mathbf{f}(t)$ is a random process with known power spectrum.

In all three cases, the noise $\mathbf{n}(t)$ is a random process with known properties. The first case is of interest if our objective is to establish the presence or absence of $f(t)$ in the observed process $\mathbf{x}(t)$.

Linear estimation We apply $\mathbf{x}(t)$ to a linear system with impulse response $h(t)$. The resulting output $\mathbf{y}(t)$ is a sum

$$\mathbf{y}(t) = \mathbf{x}(t) * h(t) = y_f(t) + \mathbf{y}_n(t) \tag{10-2}$$

where $$y_f(t) = f(t) * h(t) \qquad \text{and} \qquad \mathbf{y}_n(t) = \mathbf{n}(t) * h(t) \tag{10-3}$$

are the components due to the signal $f(t)$ and to the noise $\mathbf{n}(t)$, respectively.

We wish to determine $h(t)$ such that, at a specified time $t = t_0$, the output *signal-to-noise* ratio

$$\frac{\mathrm{S}}{\mathrm{N}} = \frac{|y_f(t_0)|}{\sqrt{E\{|\mathbf{y}_n(t_0)|^2\}}} \tag{10-4}$$

is maximum. We shall give two solutions to this problem. In the first, no restrictions are imposed on the unknown system; in the second, it is assumed that the system is a tapped delay line, as in Eq. (10-11), whose coefficients a_k are to be found. For simplicity, we shall consider only *real* signals.

The Matched Filter

We assume that the noise $\mathbf{n}(t)$ is stationary with power spectrum $S(\omega)$. From (9-69) it follows that

$$E\{\mathbf{y}_n^2(t)\} = \frac{1}{2\pi}\int_{-\infty}^{\infty} S(\omega)|H(\omega)|^2\, d\omega \tag{10-5}$$

where $H(\omega)$ is the system function of the unknown system.

At $t = t_0$, the output due to $f(t)$ is given by

$$y_f(t_0) = \frac{1}{2\pi} \int_{-\infty}^{\infty} F(\omega)H(\omega)e^{j\omega t_0}\, d\omega \tag{10-6}$$

Applying Schwarz' inequality (4-A-1) to the above integral and using the identity

$$F(\omega)H(\omega) = \frac{F(\omega)}{\sqrt{S(\omega)}} H(\omega)\sqrt{S(\omega)}$$

we obtain

$$\left| \int_{-\infty}^{\infty} F(\omega)H(\omega)e^{j\omega t_0}\, d\omega \right|^2 \le \int_{-\infty}^{\infty} \frac{|F(\omega)|^2}{S(\omega)}\, d\omega \int_{-\infty}^{\infty} S(\omega)|H(\omega)|^2\, d\omega$$

Hence,
$$\left(\frac{\mathrm{S}}{\mathrm{N}}\right)^2 = \frac{y_f^{\,2}(t_0)}{E\{\mathbf{y}_n^{\,2}(t_0)\}} \le \frac{1}{2\pi} \int_{-\infty}^{\infty} \frac{|F(\omega)|^2}{S(\omega)}\, d\omega \tag{10-7}$$

The right side above is constant because $F(\omega)$ and $S(\omega)$ are specified. Therefore, the ratio S/N is maximum if (10-7) is an equality. This is the case only if [see (4-A-2)]

$$\sqrt{S(\omega)}\, H(\omega) = k \frac{F^*(\omega)}{\sqrt{S(\omega)}} e^{-jt_0\omega}$$

Thus, the optimum filter is given by

$$H(\omega) = k \frac{F^*(\omega)}{S(\omega)} e^{-jt_0\omega} \tag{10-8}$$

yielding
$$\left(\frac{\mathrm{S}}{\mathrm{N}}\right)^2_{\max} = \frac{1}{2\pi} \int_{-\infty}^{\infty} \frac{|F(\omega)|^2}{S(\omega)}\, d\omega \tag{10-9}$$

White noise If $S(\omega) = N_0$, then

$$H(\omega) = \frac{k}{N_0} F^*(\omega)e^{-jt_0\omega} \qquad h(t) = \frac{k}{N_0} f(t_0 - t) \tag{10-10}$$

(Fig. 10-1) and

$$\left(\frac{\mathrm{S}}{\mathrm{N}}\right)_{\max} = \sqrt{\frac{E}{N_0}}$$

where E is the energy of $f(t)$.

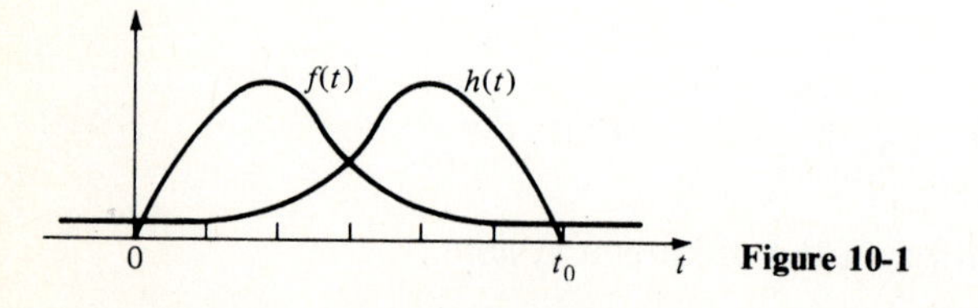

Figure 10-1

Optimum Tapped Delay Line

The matched filter is, in general, noncausal and difficult to realize. A simpler solution results if we use a tapped delay line

$$H(\omega) = a_0 + a_1 e^{-j\omega T} + \cdots + a_n e^{-jn\omega T} \tag{10-11}$$

as in Fig. 9-5. In the above, T is a given constant, and the coefficients a_k are to be determined so as to maximize the resulting signal-to-noise ratio. Clearly,

$$\mathbf{y}_n(t) = a_0\,\mathbf{n}(t) + a_1\mathbf{n}(t-T) + \cdots + a_n\,\mathbf{n}(t-nT) \tag{10-12}$$

Hence [see Eq. (9-72)],

$$E\{\mathbf{y}_n^2(t)\} = \sum_{k=0}^{n} a_k \sum_{r=0}^{n} a_r R(kT - rT) \tag{10-13}$$

where $R(\tau)$ is the autocorrelation of $\mathbf{n}(t)$.

If $H(\omega)$ is multiplied by a constant, the ratio S/N does not change. Therefore, we can assume that $y_f(t_0)$ is constant:

$$y_f(t_0) = a_0\, f(t_0) + a_1 f(t_0 - T) + \cdots + a_n f(t_0 - nT) = y_0 \tag{10-14}$$

Thus, our problem is to minimize (10-13) subject to the constraint (10-14).

With λ a constant (Lagrange multiplier), we form the difference

$$I(a_0, \ldots, a_n) = \sum_{k,\,r=0}^{n} a_k a_r R(kT - rT) - \lambda\left[\sum_{k=0}^{n} a_k f(t_0 - kT) - y_0\right]$$

As is known, the optimum a_k's must be such that

$$\frac{\partial I}{\partial a_k} = 0 = \sum_{r=0}^{n} a_r R(kT - rT) - \lambda f(t_0 - kT) \tag{10-15}$$

The solution of this system is obviously proportional to λ. And since $H(\omega)$ must be found within a factor, we can assume that $\lambda = 1$. With this choice of λ, (10-15) yields

$$\sum_{r=0}^{n} a_r R(kT - rT) = f(t_0 - kT) \qquad k = 0, 1, \ldots, n \tag{10-16}$$

To determine the resulting maximum S/N, we insert Eq. (10-16) into (10-13). This yields

$$E\{\mathbf{y}_n^2(t)\} = \sum_{k=0}^{n} a_k\, f(t_0 - kT) = y_f(t_0) \tag{10-17}$$

from which it follows that

$$\left(\frac{\mathrm{S}}{\mathrm{N}}\right)_{\max} = \sqrt{y_f(t_0)} \tag{10-18}$$

Notes:

1. The preceding results can be written in vector form. For this purpose, we form a square matrix R with elements

$$R_{kr} = R(kT - rT) \qquad k, r = 0, \ldots, n$$

and two column vectors f and a, with elements $f(t_0 - kT)$ and a_k, respectively. From (10-16) it follows that

$$Ra = f \qquad a = R^{-1}f \tag{10-19}$$

With f^T the transpose of f, we conclude from (10-14) and (10-18) that

$$y_f(t_0) = f^T a = f^T R^{-1} f \qquad \left(\frac{\text{S}}{\text{N}}\right)_{\max} = \sqrt{f^T R^{-1} f} \tag{10-20}$$

2. If the signal $f(t)$ is σ-BL and $\sigma = \pi/T$, then the tapped-delay-line solution approaches the matched-filter solution for large n, provided that sufficient delay is introduced and the data $\mathbf{x}(t)$ are filtered. If the noise $\mathbf{n}(t)$ is flat in the band $(-\sigma, \sigma)$ of the signal, i.e., if $R(\tau) = N_0 \sin \sigma\tau/\pi\tau$, then (10-16) yields

$$a_k = \frac{T}{N_0} f(t_0 - kT)$$

In this case, a_k equals the sample values of the white-noise matched-filter solution (10-10).

Discrete Processes

The discrete version of the preceding results can be readily derived. We discuss, briefly, the matched filter.

The input to a discrete system $H(z)$ is the *real* process

$$\mathbf{x}[n] = f[n] + \mathbf{n}[n] \tag{10-21}$$

where $f[n]$ is a known sequence with z transform $F(z)$, and $\mathbf{n}[n]$ is a discrete process with power spectrum $\bar{S}(\omega)$. The resulting output is

$$\mathbf{y}[n] = y_f[n] + \mathbf{y}_n[n]$$

where

$$y_f[n] = \frac{1}{2\sigma}\int_{-\sigma}^{\sigma} F(e^{j\omega T})H(e^{j\omega T})e^{jnT\omega}\, d\omega \tag{10-22}$$

and [see Eq. (9-121)]

$$E\{\mathbf{y}_n^2[n]\} = \frac{1}{2\sigma}\int_{-\sigma}^{\sigma} \bar{S}(\omega)|H(e^{j\omega T})|^2\, d\omega \tag{10-23}$$

We wish to find $H(z)$ such that the ratio

$$\frac{\mathrm{S}}{\mathrm{N}} = \frac{|y_f[n_0]|}{\sqrt{E\{\mathbf{y}_n^2[n_0]\}}} \tag{10-24}$$

is maximum.

Reasoning as in the analog case, we conclude that the optimum $H(z)$ must be such that

$$H(e^{j\omega T}) = k\,\frac{F^*(e^{j\omega T})}{\bar{S}(\omega)}\,e^{-jn_0 T\omega} \tag{10-25}$$

If $\bar{S}(\omega) = N_0$, then the above yields

$$H(z) = \frac{k}{N_0}\,F\left(\frac{1}{z}\right)z^{-n_0} \qquad h[n] = \frac{k}{N_0}\,f[n_0 - n] \tag{10-26}$$

as in (10-10) (elaborate).

10-2 UNKNOWN SIGNALS IN NOISE

Suppose, now, that

$$\mathbf{x}(t) = f(t) + \mathbf{n}(t) \tag{10-27}$$

where $f(t)$ is an unknown signal, and $\mathbf{n}(t)$ is a random process with zero mean and autocorrelation $C(t_1, t_2)$. Clearly,

$$E\{\mathbf{x}(t)\} = f(t) \qquad \sigma_x^2 = E\{\mathbf{n}^2(t)\} = C(t, t) \tag{10-28}$$

If, therefore, $C(t, t) \ll f^2(t)$, then the noise can be neglected. If, however, $C(t, t)$ is of the order of $f^2(t)$, then $\mathbf{x}(t)$ is not an adequate estimate of $f(t)$. We shall show that the effect of noise can be reduced by suitable smoothing.

The simplest form of smoothing is the moving average

$$\mathbf{y}(t) = \mathbf{x}(t) * \frac{1}{2T}\,p_T(t) = \frac{1}{2T}\int_{-T}^{T} \mathbf{x}(t - \tau)\,d\tau \tag{10-29}$$

Clearly, $\mathbf{y}(t)$ is a sum

$$\mathbf{y}(t) = y_f(t) + \mathbf{y}_n(t)$$

where

$$y_f(t) = \frac{1}{2T}\int_{-T}^{T} f(t - \tau)\,d\tau \qquad \mathbf{y}_n(t) = \frac{1}{2T}\int_{-T}^{T} \mathbf{n}(t - \tau)\,d\tau \tag{10-30}$$

We shall determine the size of T such that, if $\mathbf{y}(t)$ is used to estimate $f(t)$, then the mean-square value of the estimation error $\mathbf{y}(t) - f(t)$ is minimum.

Constant Signal

We assume first that the signal is constant,

$$f(t) = A = \text{constant}$$

and the noise stationary with autocorrelation $C(\tau)$. Since $E\{\mathbf{n}(t)\} = 0$ by assumption, it follows from Eq. (10-29) that

$$E\{\mathbf{y}(t)\} = y_f(t) = A \tag{10-31}$$

Thus, $\mathbf{y}(t)$ is an unbiased estimate of A. Its variance is given by [see (9-66)]

$$\sigma_y^2 = E\{\mathbf{y}_n^2(t)\} = \frac{1}{2T}\int_{-2T}^{2T} C(\tau)\left(1 - \frac{|\tau|}{2T}\right) d\tau \tag{10-32}$$

This quantity can be made arbitrarily small if T is large. In fact, if $C(\tau)$ is negligible for $|\tau| > a$ and $T \gg a$, then (see Fig. 10-2)

$$\sigma_y^2 \simeq \frac{1}{2T}\int_{-a}^{a} C(\tau)\, d\tau = \frac{I}{2T} \tag{10-33}$$

where I is the area of $C(\tau)$. Thus, if T is such that $I/2T \ll A^2$, then $\sigma_y^2 \ll A^2$. Hence, $\mathbf{y}(t) \simeq A$ with probability close to 1 (Tchebycheff's inequality). In other words, a constant can be estimated with any desired accuracy.

Notes:

1. The variance of $\mathbf{y}(t)$ can be expressed in terms of the power spectrum $S(\omega)$ of $\mathbf{n}(t)$ (elaborate):

$$\sigma_y^2 = \frac{1}{2\pi}\int_{-\infty}^{\infty} S(\omega)\,\frac{\sin^2 T\omega}{T^2\omega^2}\, d\omega \tag{10-34}$$

2. If $T \gg a$, then $\mathbf{n}(t)$ can be replaced with white noise with

$$C(\tau) = I\,\delta(\tau) \qquad I = \int_{-a}^{a} C(\tau)\, d\tau = S(0) \tag{10-35}$$

   In the frequency domain, this is valid if $S(\omega) \simeq S(0)$ for $|\omega| < 2\pi/T$ (Fig. 10-3).

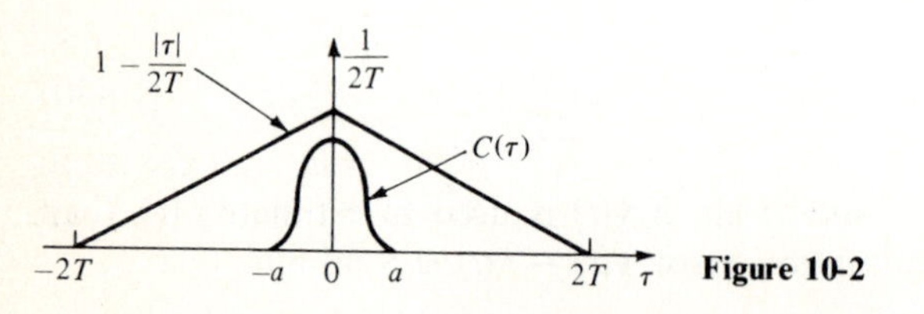

Figure 10-2

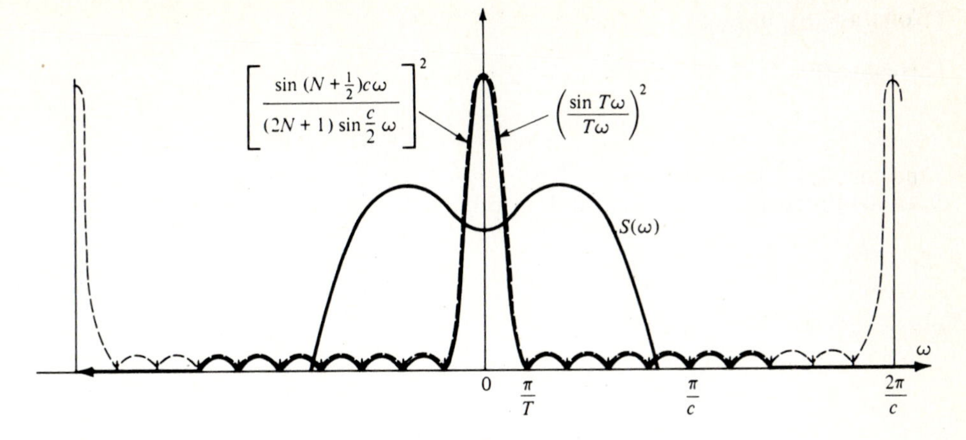

Figure 10-3

Sampling Suppose now that the constant A is estimated not by the integral $\mathbf{y}(t)$ but by the sum

$$\mathbf{z}(t) = \frac{1}{2N+1} \sum_{n=-N}^{N} \mathbf{x}(t - nc) \tag{10-36}$$

where c is a constant. Since $\mathbf{x}(t) = A + \mathbf{n}(t)$, it follows that

$$\mathbf{z}(t) = A + \mathbf{z}_n(t) \qquad \mathbf{z}_n(t) = \frac{1}{2N+1} \sum_{n=-N}^{N} \mathbf{n}(t - nc) \tag{10-37}$$

Clearly, $\mathbf{z}_n(t)$ is the output of a system with input $\mathbf{n}(t)$ and system function

$$H(\omega) = \frac{1}{2N+1} \sum_{n=-N}^{N} e^{-jnc\omega} = \frac{\sin(N+\frac{1}{2})c\omega}{(2N+1)\sin(c\omega/2)} \tag{10-38}$$

Hence [see Eq. (9-69)],

$$\sigma_z^2 = E\{\mathbf{z}_n^2(t)\} = \frac{1}{2\pi} \int_{-\infty}^{\infty} S(\omega) \frac{\sin^2(N+\frac{1}{2})c\omega}{(2N+1)^2 \sin^2(c\omega/2)} \, d\omega \tag{10-39}$$

If $S(\omega)$ is negligible for $|\omega| > \pi/c$, and N is large, then

$$\sigma_z^2 \simeq \frac{S(0)}{(2N+1)c}$$

(elaborate). If, also, $(2N+1)c = 2T$, then [see (10-33)] $\sigma_y = \sigma_z$. Under these assumptions, the integral estimator $\mathbf{y}(t)$ and the sample estimator $\mathbf{z}(t)$ are equivalent.

Time-varying Signal[1]

The estimator

$$\mathbf{y}(t) = \frac{1}{2T}\int_{-T}^{T} \mathbf{x}(t-\tau)\,d\tau = y_f(t) + \mathbf{y}_n(t)$$

of an arbitrary signal $f(t)$ is a random process with mean

$$E\{\mathbf{y}(t)\} = \frac{1}{2T}\int_{-T}^{T} f(t-\tau)\,d\tau \tag{10-40}$$

and variance

$$\sigma_y^2 = E\{\mathbf{y}_n^2(t)\} = \frac{1}{4T^2}\int_S\int C(\alpha, \beta)\,d\alpha\,d\beta \tag{10-41}$$

where S is a square of size $2T$ centered at the point (t, t) (Fig. 10-4). The last equality follows from Eq. (9-55) with $h(t) = p_T(t)/2T$ and $t_1 = t_2 = t$ (elaborate).

We shall assume that $\mathbf{n}(t)$ is white noise:

$$C(t_1, t_2) = I(t_1)\,\delta(t_1 - t_2) \tag{10-42}$$

Inserting into (10-41), we obtain [see (3-A-1)]

$$\sigma_y^2 = \frac{1}{4T^2}\int_{-T}^{T} I(t-\tau)\,d\tau = \frac{1}{2T} I(t+\tau_0) \qquad |\tau_0| < T \tag{10-43}$$

If we use the process $\mathbf{y}(t)$ as an estimate of $f(t)$, then the variance of the estimation is given by (10-43), and the bias

$$b = E\{\mathbf{y}(t)\} - f(t)$$

by [see (4-14)]

$$b = \frac{1}{2T}\int_{-T}^{T} f(t-\tau)\,d\tau - f(t) = \frac{T^2}{6} f''(t+\tau_1) \qquad |\tau_1| < T \tag{10-44}$$

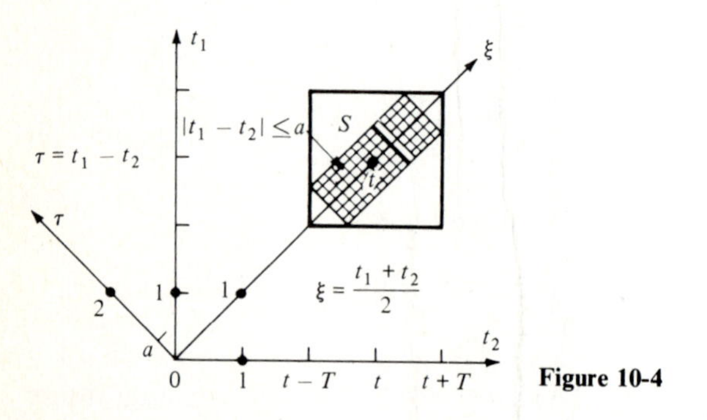

Figure 10-4

[1] A. Papoulis, Signal Processing in Nuclear Medicine, *J. Franklin Inst.*, December, 1973.

From (10-43) and (10-44) it follows that the choice of the smoothing interval T is dictated by the following conflicting requirements: For the variance σ_y^2 to be small, T should be large; for the bias b to be small, T should be small.

We determine next the optimum value of T such that the resulting mean-square estimation error

$$e = E\{[\mathbf{y}(t) - f(t)]^2\} = b^2 + \sigma_y^2 \tag{10-45}$$

is minimum.

Minimum mean-square estimation The dependence of the mean-square error e on the interval T is, in general, complicated, because the numbers τ_0 and τ_1 in Eqs. (10-43) and (10-44) also depend on T. However, if the functions $f''(t)$ and $I(t)$ are sufficiently smooth, then

$$I(t + \tau_0) \simeq I(t) \qquad f''(t + \tau_1) \simeq f''(t) \tag{10-46}$$

This leads to the approximation

$$e \simeq \frac{T^4}{36}[f''(t)]^2 + \frac{I(t)}{2T} \tag{10-47}$$

In Fig. 10-5, we plot the square of the bias b, the variance σ_y^2, and the mean-square error e as functions of T for a fixed t. It is easy to see that e is minimum if

$$T = T_m = \sqrt[5]{\frac{9I(t)}{2[f''(t)]^2}} \tag{10-48}$$

and its minimum value e_m is given by

$$e_m = \frac{5I(t)}{8T_m} \tag{10-49}$$

Notes:

1. If $T = T_m$, then $\sigma_y = 2b$. (10-50)

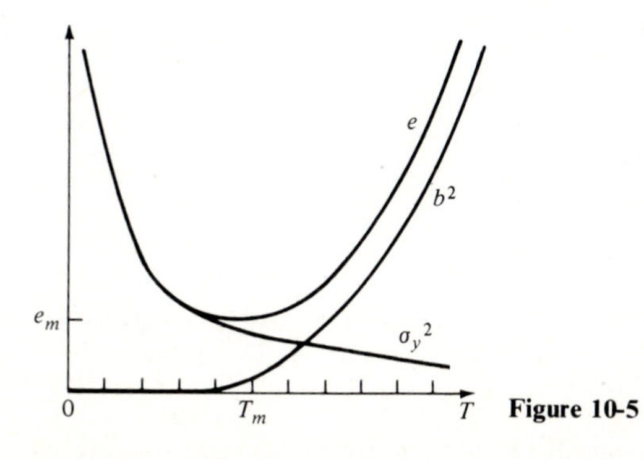

Figure 10-5

2. The white-noise assumption can be justified as follows: If the autocorrelation $C(t_1, t_2)$ of $\mathbf{n}(t)$ takes significant values in the entire square S of Fig. 10-4, then smoothing will not appreciably reduce the variance $C(t, t)$ of $\mathbf{x}(t)$, because the area of S equals $4T^2$. If, however, $C(t_1, t_2)$ takes significant values only for $|t_1 - t_2| < a$ and $a \ll T$, then with $\xi = (t_1 + t_2)/2$ and $\tau = t_1 - t_2$ we obtain from (10-41), neglecting end effects,

$$\sigma_y^2 = \frac{1}{4T^2}\int_{t-T}^{t+T}\int_{-a}^{a} C\left(\xi + \frac{\tau}{2}, \xi - \frac{\tau}{2}\right) d\tau\, d\xi = \frac{1}{4T^2}\int_{t-T}^{t+T} I(\xi)\, d\xi$$

where
$$I(\xi) = \int_{-a}^{a} C\left(\xi + \frac{\tau}{2}, \xi - \frac{\tau}{2}\right) d\tau$$

3. The optimum interval T_m depends on t. To determine it, we need to know or estimate the intensity $I(t)$ of the noise and the second derivative $f''(t)$ of $f(t)$.

The Two-to-One Rule in Data Smoothing[1]

We shall now use as the estimate of $f(t)$, the process

$$\mathbf{g}(t) = \int_{-\infty}^{\infty} \mathbf{x}(t - \tau)w(\tau)\, d\tau = g_f(t) + \mathbf{g}_n(t) \tag{10-51}$$

obtained by smoothing $\mathbf{x}(t)$ with the window $w(t)$. It will be assumed that $w(t)$ is an even, nonnegative function of unit area. Our objective is to find $w(t)$ such as to minimize the mean-square estimation error.

Variance. The component $\mathbf{g}_n(t)$ of $\mathbf{g}(t)$ is the output of a system with input $\mathbf{n}(t)$ and impulse response $w(t)$. Since $\mathbf{n}(t)$ is white noise [see Eq. (10-42)], it follows from (9-46) that the variance of $\mathbf{g}(t)$ is given by

$$\sigma_g^2 = E\{\mathbf{g}_n^2(t)\} = \int_{-\infty}^{\infty} I(t - \tau)w^2(\tau)\, d\tau = I(t + \tau_0)\int_{-\infty}^{\infty} w^2(\tau)\, d\tau \tag{10-52}$$

Bias. Clearly, $E\{\mathbf{g}(t)\} = g_f(t)$; hence, the bias $b = E\{\mathbf{g}(t)\} - f(t)$ of the estimation is given by [see also (7-66)]

$$b = \int_{-\infty}^{\infty} f(t - \tau)w(\tau)\, d\tau - f(t) = \frac{1}{2} f''(t + \tau_1)\int_{-\infty}^{\infty} \tau^2 w(\tau)\, d\tau \tag{10-53}$$

Mean-square error. With

$$E = \int_{-\infty}^{\infty} w^2(t)\, dt \qquad m_2 = \int_{-\infty}^{\infty} t^2 w(t)\, dt \tag{10-54}$$

[1] A. Papoulis, Two-to-One Rule in Data Smoothing, *IEEE Trans. Inf. Theory*, September, 1977.

the energy and second moment, respectively, of $w(t)$, we conclude, using the approximations (10-46), that the mean-square estimation error is given by

$$e = b^2 + \sigma_g{}^2 = \frac{m_2{}^2}{4}[f''(t)]^2 + EI(t) \tag{10-55}$$

Optimum scaling Suppose that $\bar{w}(t)$ is a given window of unit area, energy $\bar{E}$, and second moment $\bar{m}_2$. Scaling the time axis linearly, we obtain the function

$$w(t) = c\bar{w}(ct) \tag{10-56}$$

This function is, obviously, a window of unit area, and, as is easy to see, its energy E and second moment m_2 are given by

$$E = c\bar{E} \qquad m_2 = \frac{\bar{m}_2}{c^2} \tag{10-57}$$

Inserting into (10-55), we obtain

$$e = \frac{(\bar{m}_2)^2}{4c^4}[f''(t)]^2 + c\bar{E}I(t) \tag{10-58}$$

Clearly, e varies with c as in Fig. 10-5, and it is minimum if

$$c = c_m = \sqrt[5]{\frac{[\bar{m}_2 f''(t)]^2}{\bar{E}I(t)}} \tag{10-59}$$

yielding

$$e_m = \tfrac{5}{4}c_m \bar{E}I(t) \tag{10-60}$$

Since

$$b \simeq \frac{m_2}{2} f''(t) \qquad \sigma_g{}^2 \simeq EI(t)$$

it follows readily from Eqs. (10-57) and (10-59) that, if $c = c_m$, then

$$\sigma_g = 2b \tag{10-61}$$

This leads to the following conclusion: If the window $w(t)$ is of specified shape as in (10-56), and the scaling factor c is chosen so as to minimize the mean-square error, then the standard deviation σ_g of the resulting error equals twice the bias b, regardless of the form of $\bar{w}(t)$.

The parabolic window The minimum mean-square error e_m depends on the shape of $\bar{w}(t)$. We shall determine the optimum $\bar{w}(t)$ so as to minimize e_m. Since $\bar{w}(t)$ need be specified only within a scaling factor, it suffices to assume that $\bar{E}$ is fixed. Our problem is thus to find a nonnegative function $\bar{w}(t)$ of specified energy $\bar{E}$, such that its second moment $\bar{m}_2$ is minimum. As we show in Prob. 78, the optimum $\bar{w}(t)$ is a truncated parabola

$$\bar{w}(t) = \frac{3}{4T}\left[1 - \left(\frac{t}{T}\right)^2\right]p_T(t) \qquad \bar{E} = \frac{3}{5T} \qquad \bar{m}_2 = \frac{T^2}{5}$$

as in Fig. 10-6.

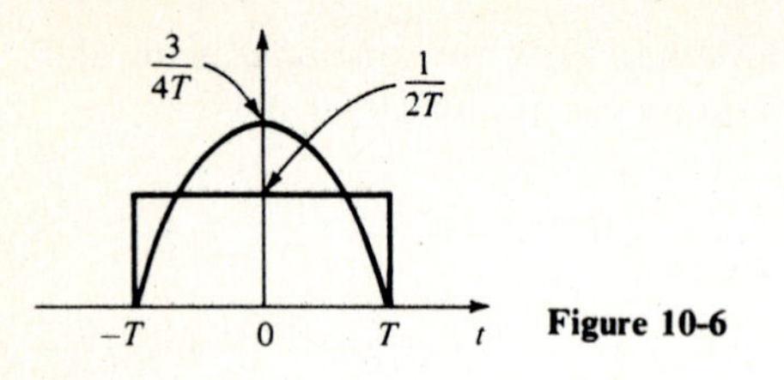

Figure 10-6

10-3 RANDOM SIGNALS AND WIENER FILTERS

We finally assume that

$$\mathbf{x}(t) = \mathbf{f}(t) + \mathbf{n}(t) \tag{10-62}$$

where now the signal $\mathbf{f}(t)$ is an unknown sample of a stochastic process. The processes $\mathbf{f}(t)$ and $\mathbf{n}(t)$ are jointly stationary with known power spectra. Our objective is to estimate $\mathbf{f}(t)$ in terms of the observed data $\mathbf{x}(t)$. As preparation, we discuss, briefly, the discrete estimation problem.

Discrete Linear Mean-Square Estimation

We wish to estimate an arbitrary random variable $\mathbf{f}$ with a linear combination

$$\hat{\mathbf{f}} = \sum_{k=1}^{n} a_k \mathbf{y}_k \tag{10-63}$$

of n given random variables $\mathbf{y}_k$ (data). The constants a_k are to be chosen so as to minimize the resulting mean-square error

$$e = E\{[\mathbf{f} - (a_1\mathbf{y}_1 + \cdots + a_n\mathbf{y}_n)]^2\} \tag{10-64}$$

This problem is identical to the problem considered in Sec. 5-2 if we define the inner product of two random variables $\mathbf{x}$ and $\mathbf{y}$ as the expected value of their product:

$$\langle \mathbf{x}, \mathbf{y} \rangle = E\{\mathbf{xy}\}$$

The solution is, therefore, given by the system (5-32). We shall, however, rederive this fundamental result for easy reference.

The orthogonality principle The optimum constants a_k must be such that the estimation error

$$\mathbf{f} - (a_1\mathbf{y}_1 + \cdots + a_n\mathbf{y}_n)$$

is orthogonal to the data $\mathbf{y}_i$; that is,

$$E\{[\mathbf{f} - (a_1\mathbf{y}_1 + \cdots + a_n\mathbf{y}_n)]\mathbf{y}_i\} = 0 \qquad i = 1, \ldots, n \tag{10-65}$$

Proof Clearly, the mean-square error e is a function of the coefficients a_i and is minimum if

$$\frac{\partial e}{\partial a_i} = E\{-2[\mathbf{f} - (a_1\mathbf{y}_1 + \cdots + a_n\mathbf{y}_n)]\mathbf{y}_i\} = 0$$

i.e., if (10-65) holds. The interchange between differentiation and expected value is a simple consequence of the linearity of expected values (elaborate).

Thus, to determine the n constants a_i it suffices to solve the system

$$E\{\mathbf{y}_1\mathbf{y}_i\}a_1 + E\{\mathbf{y}_2\mathbf{y}_i\}a_2 + \cdots + E\{\mathbf{y}_n\mathbf{y}_i\}a_n = E\{\mathbf{f}\mathbf{y}_i\} \tag{10-66}$$

The resulting mean-square error e is given by

$$e = E\{(\mathbf{f} - \hat{\mathbf{f}})^2\} = E\{(\mathbf{f} - \hat{\mathbf{f}})\mathbf{f}\} = E\{\mathbf{f}^2\} = E\{\hat{\mathbf{f}}^2\} \tag{10-67}$$

because, as we can see from Eq. (10-65), the estimator $\hat{\mathbf{f}}$ is orthogonal to the estimation error $\mathbf{f} - \hat{\mathbf{f}}$. [See also (5-40).]

Example 10-1 Given a stationary process $\mathbf{x}(t)$ and a constant λ, we wish to find a such that, if $\mathbf{x}(t + \lambda)$ is estimated by $a\mathbf{x}(t)$, the error

$$e = E\{[\mathbf{x}(t + \lambda) - a\mathbf{x}(t)]^2\}$$

is minimum.

This is a special case of (10-64) if $\mathbf{f} = \mathbf{x}(t + \lambda)$ and $\mathbf{y}_1 = \mathbf{x}(t)$. From the orthogonality principle (10-65), it follows that the constant a must be such that

$$E\{[\mathbf{x}(t + \lambda) - a\mathbf{x}(t)]\mathbf{x}(t)\} = 0$$

Solving, we find $a = R(\lambda)/R(0)$, where $R(\tau)$ is the autocorrelation of $\mathbf{x}(t)$. The mean-square error is given by

$$e = E\{[\mathbf{x}(t + \lambda) - a\mathbf{x}(t)]\mathbf{x}(t + \lambda)\} = R(0) - \frac{R^2(\lambda)}{R(0)}$$

The Wiener Filter[1]

We wish to estimate the process $\mathbf{f}(t)$ with a linear combination of past and future values of the data. This means that we seek the impulse response $h(t)$ of a linear, noncausal filter such that, if $\mathbf{x}(t)$ is its input, then the resulting response

$$\hat{\mathbf{f}}(t) = \int_{-\infty}^{\infty} \mathbf{x}(t - \alpha)h(\alpha)\, d\alpha \tag{10-68}$$

is the best mean-square estimate of $\mathbf{f}(t)$; that is, the mean-square error

$$e = E\{[\mathbf{f}(t) - \hat{\mathbf{f}}(t)]^2\} \tag{10-69}$$

is minimum.

[1] L. A. Wainstein and V. D. Zubakov, "Extraction of Signals from Noise," Prentice-Hall, Inc., Englewood Cliffs, N.J., 1962.

Clearly, $\hat{\mathbf{f}}(t)$ is a linear combination of the data $\mathbf{x}(t-\tau)$, where τ takes all values from $-\infty$ to ∞. Extending the orthogonality principle to infinite sums, and by a limiting argument to integrals, we can show that the optimum weight $h(t)$ must be such that the error $\mathbf{f}(t) - \hat{\mathbf{f}}(t)$ is orthogonal to the data $\mathbf{x}(t-\tau)$:

$$E\left\{\left[\mathbf{f}(t) - \int_{-\infty}^{\infty} \mathbf{x}(t-\alpha)h(\alpha)\,d\alpha\right]\mathbf{x}(t-\tau)\right\} = 0 \qquad \text{for all } \tau \tag{10-70}$$

The above yields

$$R_{fx}(\tau) = \int_{-\infty}^{\infty} R_{xx}(\tau-\alpha)h(\alpha)\,d\alpha \qquad \text{for all } \tau \tag{10-71}$$

Thus, the impulse response $h(t)$ of the optimum system is obtained by solving the integral equation (10-71).

Solution With

$$R_{fx}(\tau) \leftrightarrow S_{fx}(\omega) \qquad R_{xx}(\tau) \leftrightarrow S_{xx}(\omega) \qquad h(t) \leftrightarrow H(\omega)$$

it follows from Eq. (10-71) and the convolution theorem that

$$S_{fx}(\omega) = S_{xx}(\omega)H(\omega)$$

Hence, the system function of the optimum filter is given by

$$H(\omega) = \frac{S_{fx}(\omega)}{S_{xx}(\omega)} \tag{10-72}$$

Mean-square error Since $\mathbf{f}(t) - \hat{\mathbf{f}}(t)$ is orthogonal to $\mathbf{x}(t-\tau)$ for every τ, we conclude that

$$E\{[\mathbf{f}(t) - \hat{\mathbf{f}}(t)]\hat{\mathbf{f}}(t)\} = 0$$

and (10-69) yields

$$e = E\left\{\left[\mathbf{f}(t) - \int_{-\infty}^{\infty} \mathbf{x}(t-\alpha)h(\alpha)\,d\alpha\right]\mathbf{f}(t)\right\}$$

$$= R_{ff}(0) - \int_{-\infty}^{\infty} R_{fx}(\alpha)h(\alpha)\,d\alpha \tag{10-73}$$

Expressing $R_{ff}(0)$ in terms of $S_{ff}(\omega)$ and using Parseval's formula for the integral in (10-73), we obtain

$$e = \frac{1}{2\pi}\int_{-\infty}^{\infty} [S_{ff}(\omega) - S_{fx}^*(\omega)H(\omega)]\,d\omega \tag{10-74}$$

Special case If the signal $\mathbf{f}(t)$ and the noise $\mathbf{n}(t)$ are orthogonal, i.e., if $R_{fn}(\tau) = 0$, then

$$S_{fx}(\omega) = S_{ff}(\omega) \qquad S_{xx}(\omega) = S_{ff}(\omega) + S_{nn}(\omega)$$

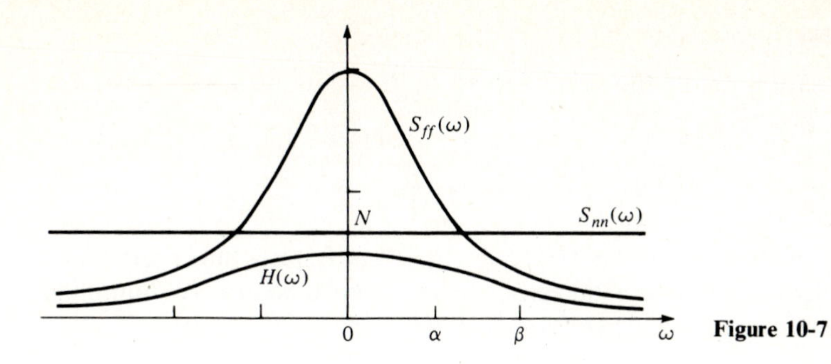

Figure 10-7

Hence,
$$H(\omega) = \frac{S_{ff}(\omega)}{S_{ff}(\omega) + S_{nn}(\omega)} \tag{10-75}$$

$$e = \frac{1}{2\pi}\int_{-\infty}^{\infty} \frac{S_{ff}(\omega)S_{nn}(\omega)}{S_{ff}(\omega) + S_{nn}(\omega)}\, d\omega \tag{10-76}$$

Note: If $S_{nn}(\omega) = N \gg S_{ff}(\omega)$, then

$$H(\omega) \simeq \frac{1}{N} S_{ff}(\omega) \qquad h(t) \simeq \frac{1}{N} R_{ff}(\tau)$$

$$e \simeq \frac{1}{2\pi}\int_{-\infty}^{\infty} S_{ff}(\omega)\, d\omega = E\{\mathbf{f}^2(t)\}$$

Example 10-2 Suppose that (Fig. 10-7)

$$S_{ff}(\omega) = \frac{k}{\alpha^2 + \omega^2} \qquad S_{nn}(\omega) = N \qquad S_{fn}(\omega) = 0$$

It then follows from (10-75) that

$$H(\omega) = \frac{k}{k + N(\alpha^2 + \omega^2)} = \frac{k/N}{\beta^2 + \omega^2} \qquad \beta^2 = \alpha^2 + \frac{k}{N}$$

Hence,
$$h(t) = \frac{k}{2N\beta} e^{-\beta|t|}$$

and [see (10-76)]

$$e = \frac{1}{2\pi}\int_{-\infty}^{\infty} \frac{k}{\beta^2 + \omega^2}\, d\omega = \frac{k}{2\beta}$$

Special case If $k/N = 3\alpha^2$, then $\beta = 2\alpha$ and

$$h(t) = \tfrac{3}{4}\alpha e^{-2\alpha|t|} \qquad e = \frac{k}{4\alpha} \tag{10-77}$$

Causal Estimation

In real-time estimation, only the past values of the data can be used. The estimator is, then, given by

$$\hat{\mathbf{f}}(t) = \int_0^\infty \mathbf{x}(t-\alpha)h(\alpha)\,d\alpha \tag{10-78}$$

i.e., it is the output of a causal filter with input $\mathbf{x}(t)$. Thus, $\mathbf{f}(t)$ is a linear combination of the data $\mathbf{x}(t-\tau)$, where τ takes all values from 0 to ∞. To minimize the mean-square error, we must, therefore, find $h(t)$ such that

$$E\left\{\left[\mathbf{f}(t) - \int_0^\infty \mathbf{x}(t-\alpha)h(\alpha)\,d\alpha\right]\mathbf{x}(t-\tau)\right\} = 0 \qquad \tau > 0 \tag{10-79}$$

This leads to the *Wiener-Hopf* integral equation,

$$R_{fx}(\tau) - \int_0^\infty R_{xx}(\tau-\alpha)h(\alpha)\,d\alpha = 0 \qquad \tau > 0 \tag{10-80}$$

which, unlike Eq. (10-71), is valid for $\tau > 0$ only.

Solution We denote by $y(\tau)$ the left side of (10-80):

$$y(\tau) = R_{fx}(\tau) - \int_0^\infty R_{xx}(\tau-\alpha)h(\alpha)\,d\alpha \tag{10-81}$$

Clearly, $\qquad y(\tau) = 0 \qquad \text{for } \tau > 0$

and $\qquad h(\tau) = 0 \qquad \text{for } \tau < 0 \qquad$ (10-82)

It suffices, therefore, to find a causal function $h(\tau)$ and an anticausal function $y(\tau)$ satisfying (10-81).

We denote by $Y(s)$ and $H(s)$ the Laplace transforms of $y(\tau)$ and $h(\tau)$, respectively. As we know (Sec. 7-1), $Y(s)$ is analytic for $\text{Re } s < 0$, and $H(s)$ is analytic for $\text{Re } s > 0$.

The integral in (10-81) is a convolution of $R_{xx}(\tau)$ with $h(\tau)$. Taking Laplace transforms of both sides, we conclude, therefore, that

$$Y(s) = S_{fx}(-js) - S_{xx}(-js)H(s) \tag{10-83}$$

The functions $S_{fx}(-js)$ and $S_{xx}(-js)$ are the Laplace transforms of $R_{fx}(\tau)$ and $R_{xx}(\tau)$, respectively, and they are uniquely determined in terms of the spectra $S_{fx}(\omega)$ and $S_{xx}(\omega)$. As we know, they are analytic in a vertical strip containing the $j\omega$ axis. Furthermore, since $R_{xx}(\tau)$ is real and even, its transform $S_{xx}(-js)$ is real for real s and even.

Our problem is to find two functions $Y(s)$ and $H(s)$ satisfying (10-83) and the stated analyticity conditions:

$$Y(s) \text{ for Re } s < 0 \qquad H(s) \text{ for Re } s > 0 \tag{10-84}$$

This problem can be solved as follows:

Step 1. We express the transform $S_{xx}(-js)$ as a product

$$S_{xx}(-js) = A^+(s)A^-(s) \tag{10-85}$$

where the function $A^+(s)$ and its inverse $1/A^+(s)$ are analytic for $\text{Re}\, s > 0$, and the function $A^-(s)$ and its inverse $1/A^-(s)$ are analytic for $\text{Re}\, s < 0$ (see Sec. 7-2). If the spectrum $S_{xx}(\omega)$ is rational in ω, then, to determine $A^+(s)$ and $A^-(s)$, we factor $S_{xx}(-js)$. We assign the left-hand-plane poles and zeros to $A^+(s)$, and the right-hand-plane poles and zeros to $A^-(s)$ (see Sec. 4-3).

Step 2. We form the ratio $S_{fx}(-js)/A^-(s)$ and express it as a sum

$$\frac{S_{fx}(-js)}{A^-(s)} = B^+(s) + B^-(s) \tag{10-86}$$

where the function $B^+(s)$ is analytic for $\text{Re}\, s > 0$, and the function $B^-(s)$ is analytic for $\text{Re}\, s < 0$ [see Eq. (10-90)]. If the spectrum $S_{fx}(\omega)$ is rational in ω, then, to determine $B^+(s)$ and $B^-(s)$, we expand the left side of (10-86) into a sum of linear terms (partial-fraction expansion). We assign the terms whose poles are on the left-hand plane to $B^+(s)$, and the remaining terms to $B^-(s)$.

Step 3. The unknown $H(s)$ is given by

$$H(s) = \frac{B^+(s)}{A^+(s)} \tag{10-87}$$

Proof Clearly, $H(s)$ is analytic for $\text{Re}\, s > 0$, because the functions $B^+(s)$ and $1/A^+(s)$ are, by construction, analytic for $\text{Re}\, s > 0$. To complete the proof, it suffices to show that if $H(s)$ is inserted into Eq. (10-83), the resulting $Y(s)$ is analytic for $\text{Re}\, s < 0$.

Using (10-87) and the expansions (10-85) and (10-86), we obtain

$$Y(s) = [B^+(s) + B^-(s)]A^-(s) - \frac{A^+(s)A^-(s)B^+(s)}{A^+(s)} = A^-(s)B^-(s)$$

and since the functions $A^-(s)$ and $B^-(s)$ are analytic for $\text{Re}\, s < 0$ by construction, the proof is complete.

Mean-square error Reasoning as in the noncausal case, we conclude that the mean-state error is given by (10-73), provided that the integrations are from 0 to ∞:

$$e = R_{ff}(0) - \int_0^\infty R_{fx}(\alpha)h(\alpha)\, d\alpha \tag{10-88}$$

Example 10-3 Suppose that

$$S_{ff}(\omega) = \frac{k}{\alpha^2 + \omega^2} \qquad S_{nn}(\omega) = N \qquad S_{fn}(\omega) = 0$$

as in Example 10-2. In this case, $S_{xx}(\omega) = S_{ff}(\omega) + N$ and $S_{fx}(\omega) = S_{ff}(\omega)$.

Step 1

$$S_{xx}(-js) = \frac{k}{\alpha^2 - s^2} + N = N\frac{\beta^2 - s^2}{\alpha^2 - s^2} \qquad \beta^2 = \alpha^2 + \frac{k}{N}$$

$$A^+(s) = \sqrt{N}\frac{\beta + s}{\alpha + s} \qquad A^-(s) = \sqrt{N}\frac{\beta - s}{\alpha - s}$$

Step 2

$$\frac{S_{fx}(-js)}{A^-(s)} = \frac{k(\alpha - s)}{(\alpha^2 - s^2)\sqrt{N}(\beta - s)} = \frac{(\beta - \alpha)\sqrt{N}}{\alpha + s} + \frac{(\beta - \alpha)\sqrt{N}}{\beta - s}$$

$$B^+(s) = \frac{(\beta - \alpha)\sqrt{N}}{\alpha + s} \qquad B^-(s) = \frac{(\beta - \alpha)\sqrt{N}}{\beta - s}$$

Step 3

$$H(s) = \frac{\beta - \alpha}{\beta + s} \qquad h(t) = (\beta - \alpha)e^{-\beta t}U(t)$$

and [see Eq. (10-88)]

$$e = \frac{k}{2\alpha} - \int_0^\infty \frac{k}{2\alpha} e^{-\alpha t}(\beta - \alpha)e^{-\beta t}\,dt = \frac{k}{\alpha + \beta}$$

Special case If $k/N = 3\alpha^2$, then $\beta = 2\alpha$ and

$$h(t) = \alpha e^{-2\alpha t}U(t) \qquad e = \frac{k}{3\alpha}$$

In Fig. 10-8 we show the impulse responses of the causal and noncausal filters. As we see from Example 10-2, the causality requirement increases the mean-square error by the factor 4/3.

Notes:

1. The spectrum $S_{xx}(\omega)$ is real and even; therefore, $A^-(s) = A^+(-s)$ (elaborate).

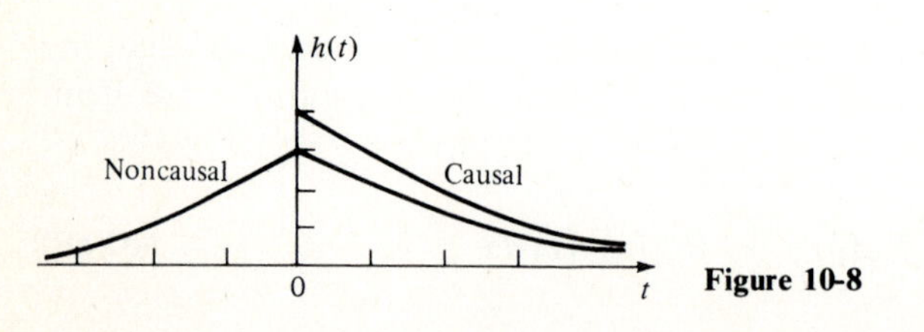

Figure 10-8

2. The factorization (10-85), expressed in the time domain, states that the autocorrelation $R_{xx}(\tau)$ of $\mathbf{x}(t)$ can be written as a convolution

$$R_{xx}(\tau) = a^+(\tau) * a^+(-\tau) \tag{10-89}$$

where $a^+(\tau)$ is a causal function. Clearly, $a^+(\tau)$ is the inverse transform of $A^+(s)$.

The time version of (10-86) states that the inverse transform $z(t)$ of the left side can be written as a sum

$$z(t) = b^+(t) + b^-(t) \tag{10-90}$$

where $b^+(t) = z(t)U(t) = 0$ for $t < 0$, and $b^-(t) = z(t)U(-t) = 0$ for $t > 0$.

Filtering by a Moving Average

In Sec. 10-2, we showed that mere averaging can lead to a satisfactory estimate of an unknown signal if the noise is sufficiently small. We shall develop, next, similar results for random signals.

We assume that the noise is white and orthogonal to the signal

$$S_{nn}(\omega) = N \qquad S_{fn}(\omega) = 0 \tag{10-91}$$

Our objective is to find the optimum interval T such that, if the integral

$$\mathbf{y}(t) = \frac{1}{2T}\int_{-T}^{T} \mathbf{x}(t - \tau)\, d\tau = \mathbf{y}_f(t) + \mathbf{y}_n(t) \tag{10-92}$$

is used to estimate $\mathbf{f}(t)$, the mean-square error e is minimum.

We maintain that the optimum T is given by

$$T = \sqrt[5]{\frac{9N}{2D}} \tag{10-93}$$

where

$$D = E\{[\mathbf{f}''(t)]^2\} = \frac{1}{2\pi}\int_{-\infty}^{\infty} \omega^4 S_{ff}(\omega)\, d\omega \tag{10-94}$$

Thus, to determine T, it suffices to know the constant N and the average power of $\mathbf{f}''(t)$.

Proof Since $\mathbf{n}(t)$ is orthogonal to $\mathbf{f}(t)$, $\mathbf{y}_n(t)$ is orthogonal to $\mathbf{y}_f(t)$ and to $\mathbf{y}_f(t) - \mathbf{f}(t)$. Hence,

$$\begin{aligned} e = E\{[\mathbf{y}(t) - \mathbf{f}(t)]^2\} &= E\{[\mathbf{y}_f(t) - \mathbf{f}(t) + \mathbf{y}_n(t)]^2\} \\ &= E\{[\mathbf{y}_f(t) - \mathbf{f}(t)]^2\} + E\{\mathbf{y}_n^2(t)\} \end{aligned} \tag{10-95}$$

Clearly, $\mathbf{y}_n(t)$ is the response of the system $\sin T\omega/T\omega$ to the input $\mathbf{n}(t)$. Hence,

$$E\{\mathbf{y}_n^2(t)\} = \frac{N}{2\pi}\int_{-\infty}^{\infty} \frac{\sin^2 T\omega}{T^2\omega^2}\, d\omega = \frac{N}{2T} \tag{10-96}$$

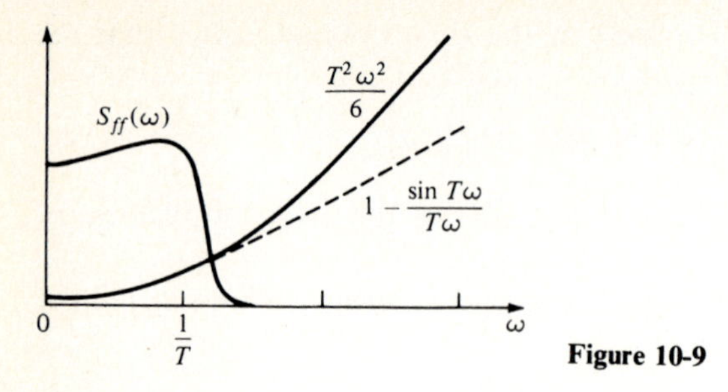

Figure 10-9

Furthermore, $\mathbf{y}_f(t) - \mathbf{f}(t)$ is the response of the system

$$\frac{\sin T\omega}{T\omega} - 1 = -\frac{T^2}{6}\omega^2 + \cdots$$

to the input $\mathbf{f}(t)$. Therefore,

$$E\{[\mathbf{y}_f(t) - \mathbf{f}(t)]^2\} = \frac{1}{2\pi}\int_{-\infty}^{\infty} S_{ff}(\omega)\left(\frac{\sin T\omega}{T\omega} - 1\right)^2 d\omega$$

$$\simeq \frac{T^4}{72\pi}\int_{-\infty}^{\infty} \omega^4 S_{ff}(\omega)\, d\omega \tag{10-97}$$

The last approximation is based on the assumption that the band of $S_{ff}(\omega)$ is of the order of $1/T$ (Fig. 10-9).

Inserting (10-97) and (10-96) into Eq. (10-95), we obtain

$$e \simeq \frac{T^4}{36}D + \frac{N}{2T} \tag{10-98}$$

This is obviously minimum if T equals the radical in (10-93).

Note: From (10-93), (10-96), and (10-98), it follows that

$$E\{\mathbf{y}_n^{\,2}(t)\} = \frac{N}{2T} = \frac{T^4 D}{9} = 4E\{[\mathbf{y}_f(t) - \mathbf{f}(t)]^2\} \tag{10-99}$$

Thus, for optimum estimation, the mean-square error due to the noise must equal four times the mean-square error due to the smoothing of the signal.

10-4 DISCRETE PROCESSES

The preceding results can be readily extended to random sequences. We shall discuss only the discrete form of the Wiener filter. It will be assumed that all processes are stationary and real.

We wish to estimate a discrete process $\mathbf{f}[n]$ in terms of the data

$$\mathbf{x}[n] = \mathbf{f}[n] + \mathbf{v}[n] \tag{10-100}$$

The estimator of $\mathbf{f}[n]$ is the sum

$$\hat{\mathbf{f}}[n] = \sum_{k=-\infty}^{\infty} \mathbf{x}[n-k]h[k] \tag{10-101}$$

and our problem is to find $h[k]$ such that the mean-square error

$$e = \sum_{n=-\infty}^{\infty} (\mathbf{f}[n] - \hat{\mathbf{f}}[n])^2 \tag{10-102}$$

is minimum.

From the orthogonality principle, Eq. (10-65), it follows that the error $\mathbf{f}[n] - \hat{\mathbf{f}}[n]$ must be orthogonal to the data $\mathbf{x}[n-m]$ for every m:

$$E\left\{\left(\mathbf{f}[n] - \sum_{k=-\infty}^{\infty} \mathbf{x}[n-k]h[k]\right)\mathbf{x}[n-m]\right\} = 0 \tag{10-103}$$

The above yields

$$R_{fx}[m] = \sum_{k=-\infty}^{\infty} R_{xx}[m-k]h[k] \qquad \text{for all } m \tag{10-104}$$

Thus, the optimum $h[n]$ is obtained by solving the system (10-104).

Solution Since (10-104) holds for every m, we conclude taking z transforms of both sides that [see (9-125)] $S_{fx}(z) = S_{xx}(z)H(z)$. Hence, the system function of the optimum filter is given by

$$H(z) = \frac{S_{fx}(z)}{S_{xx}(z)} \tag{10-105}$$

Mean-square error With $h[n]$ so determined, (10-102) yields

$$e = R_{ff}(0) - \sum_{k=-\infty}^{\infty} R_{fx}[k]h[k] \tag{10-106}$$

as in (10-73).

Special case If $R_{f\nu}[m] = 0$, then

$$S_{fx}(z) = S_{ff}(z) \qquad S_{xx}(z) = S_{ff}(z) + S_{\nu\nu}(z)$$

$$H(z) = \frac{S_{ff}(z)}{S_{ff}(z) + S_{\nu\nu}(z)} \tag{10-107}$$

Example 10-4 Suppose that $\mathbf{v}[n]$ is white noise

$$R_{vv}[m] = N\,\delta[m] \qquad S_{vv}(z) = N$$

and $\mathbf{f}[n]$ is a *first-order* process

$$R_{ff}[m] = a^{|m|} \qquad 0 < a < 1 \tag{10-108}$$

In this case, (see Example 9-11)

$$S_{ff}(z) = \frac{a - a^{-1}}{(z + z^{-1}) - (a + a^{-1})} \tag{10-109}$$

Substituting into (10-107), we obtain

$$H(z) = \frac{a - a^{-1}}{(a - a^{-1}) + N(z + z^{-1}) - N(a + a^{-1})} = \frac{(a - a^{-1})/N}{(z + z^{-1}) - (b + b^{-1})} \tag{10-110}$$

where

$$b + b^{-1} = a + a^{-1} + \frac{1}{N}(a^{-1} - a) \qquad 0 < b < 1 \tag{10-111}$$

Hence,

$$h[n] = \frac{a^{-1} - a}{N(b^{-1} - b)}\, b^{|n|} \tag{10-112}$$

From (10-106) it follows that (elaborate)

$$e = 1 - \frac{a^{-1} - a}{N(b^{-1} - b)} \sum_{k=-\infty}^{\infty} (ab)^{|k|} = \frac{a^{-1} - a}{b^{-1} - b} \tag{10-113}$$

Special case If $a = 0.8$ and $N = 1$, then

$$b = 0.5 \qquad h[n] = 0.3 \times 2^{-|n|} \qquad e = 0.3$$

Causal Estimation

If only the past values of the data are available, then

$$\hat{\mathbf{f}}[n] = \sum_{k=0}^{\infty} \mathbf{x}[n - k]h[k] \tag{10-114}$$

In this case, $h[n]$ is optimum if the error $\mathbf{f}[n] - \hat{\mathbf{f}}[n]$ is orthogonal to the data $\mathbf{x}[n - m]$ for $m \geq 0$ only:

$$E\left\{\left(\mathbf{f}[n] - \sum_{k=0}^{\infty} \mathbf{x}[n - k]h[k]\right)\mathbf{x}[n - m]\right\} = 0$$

This leads to the discrete form

$$R_{fx}[m] - \sum_{k=0}^{\infty} R_{xx}[m - k]h[k] \qquad m \geq 0 \tag{10-115}$$

of the Wiener-Hopf equation (10-80).

Solution We form the sequence

$$y[m] = R_{fx}[m] - \sum_{k=0}^{\infty} R_{xx}[m-k]h[k] \tag{10-116}$$

Clearly,

$$y[m] = 0 \qquad \text{for } m \geq 0 \qquad \text{and} \qquad h[n] = 0 \qquad \text{for } n < 0 \tag{10-117}$$

Hence, the z transform $Y(z)$ of $y[m]$ is analytic for $|z| < 1$ and the z transform $H(z)$ of $h[n]$ is analytic for $|z| > 1$. From (10-116) it follows that

$$Y(z) = S_{fx}(z) - S_{xx}(z)H(z) \tag{10-118}$$

Thus, our problem is to find two functions $Y(z)$ and $H(z)$ satisfying (10-118) and the stated analyticity conditions. This problem can be solved as follows:

Step 1. Since

$$S_{xx}(e^{j\omega T}) = \bar{S}_{xx}(\omega) \geq 0$$

[see (9-122)], we conclude from the discussion of Sec. 7-2 that $S_{xx}(z)$ can be factored into a product

$$S_{xx}(z) = A^{+}(z)A^{-}(z) \tag{10-119}$$

where the function $A^{+}(z)$ and its inverse are analytic for $|z| > 1$ and the function $A^{-}(z)$ and its inverse are analytic for $|z| < 1$. If $S_{xx}(z)$ is a rational function, then to determine $A^{+}(z)$ and $A^{-}(z)$, we factor $S_{xx}(z)$ and assign its poles and zeros that are outside the unit circle to $A^{+}(z)$ and the poles and zeros that are inside the unit circle to $A^{-}(z)$. The constant factor can be chosen arbitrarily.

Step 2. We form the ratio $S_{fx}(z)/A^{-}(z)$ and express it as a sum:

$$\frac{S_{fx}(z)}{A^{-}(z)} = B^{+}(z) + B^{-}(z) \tag{10-120}$$

where the function $B^{+}(z)$ is analytic for $|z| > 1$ and the function $B^{-}(z)$ is analytic for $|z| < 1$. This can be done by determining the inverse transform $b[n]$ of $S_{fx}(z)/A^{-}(z)$. The function $B^{+}(z)$ is then the transform of the causal part $b[n]U[n]$ of $b[n]$. If $S_{yx}(z)/A^{-}(z)$ is rational, then we expand it into partial fractions and assign all terms with poles in the region $|z| < 1$ to the function $B^{+}(z)$, and the remaining terms to $B^{-}(z)$.

Step 3. The unknown $H(z)$ is given by

$$H(z) = \frac{B^{+}(z)}{A^{+}(z)} \tag{10-121}$$

Proof Clearly, $H(z)$ is analytic for $|z| > 1$. It suffices, therefore, to show that if it is inserted into Eq. (10-118), the resulting $Y(z)$ is analytic for $|z| < 1$. Using (10-121) and the expansions (10-120) and (10-119), we obtain

$$Y(z) = A^-(z)B^-(z)$$

This completes the proof because the functions $A^-(z)$ and $B^-(z)$ are analytic to $|z| < 1$ by construction.

Mean-square error It is easy to see that the mean-square error e is given by

$$e = R_{ff}[0] - \sum_{k=0}^{\infty} R_{fx}[k]h[k] \tag{10-122}$$

Example 10-5 We assume as in Example 10-4 that

$$S_{ff}(z) = \frac{a^{-1} - a}{(a + a^{-1}) - (z + z^{-1})} \qquad S_{vv}(z) = N \qquad S_{fv}(z) = 0$$

Step 1

$$S_{xx}(z) = \frac{a^{-1} - a}{(a + a^{-1}) - (z + z^{-1})} + N = N\frac{(b + b^{-1}) - (z + z^{-1})}{(a + a^{-1}) - (z + z^{-1})}$$

$$= N\frac{(b - z)(1 - b^{-1}z^{-1})}{(a - z)(1 - a^{-1}z^{-1})}$$

$$b + b^{-1} = a + a^{-1} + \frac{1}{N}(a^{-1} - a)$$

Hence,

$$A^+(z) = \frac{z - b}{z - a} \qquad A^-(z) = N\frac{z - b^{-1}}{z - a^{-1}}$$

Step 2

$$\frac{S_{fx}(z)}{A^-(z)} = \frac{(a^{-1} - a)z}{N(a - z)(z - b^{-1})} = \frac{cz}{z - a} - \frac{cz}{z - b^{-1}} \qquad c = 1 - ba^{-1}$$

Hence,

$$B^+(z) = \frac{cz}{z - a} \qquad B^-(z) = \frac{-cz}{z - b^{-1}}$$

Step 3

$$H(z) = \frac{cz}{z - b} \qquad h[n] = cb^nU[n]$$

Thus, the causal Wiener filter is a first-order system, as in Fig. 10-10, and the estimator $\hat{\mathbf{f}}[n]$ of $\mathbf{f}[n]$ satisfies the recursion equation

$$\hat{\mathbf{f}}[n] - b\hat{f}[n - 1] = c\mathbf{x}[n] \tag{10-123}$$

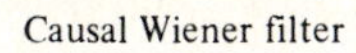

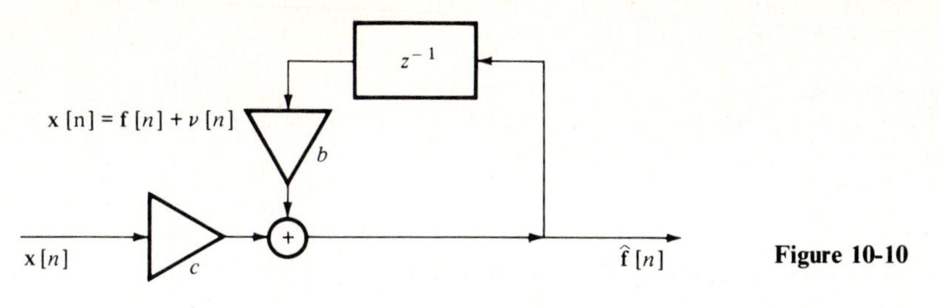

Figure 10-10

The resulting mean-square error is given by

$$e = 1 - c \sum_{k=0}^{\infty} (ab)^k = \frac{a^{-1} - a}{b^{-1} - a}$$

This follows from (10-122) because $R_{ff}[m] = a^{|m|}$

Special case If $a = 0.8$ and $N = 1$, then $b = 0.5$ and $c = 0.375$.

$$h[n] = 0.375 \times 2^{-n} \qquad e = 0.375$$

Comparing with the results in Example 10-4, we see that the causality requirement resulted in an increase of the mean-square error from 0.3 to 0.375.

Wide-sense Markoff sequences and recursive filtering (Kalman) The sequence $h[n]$ has, in general, infinitely many nonzero terms; therefore, for the realization of the Wiener filter, we must either truncate $h[n]$ (nonrecursive) or approximate $H(z)$ by a rational function (recursive). For a recursive realization, approximation is not necessary if the signal $\mathbf{f}[n]$ and the noise $\mathbf{v}[n]$ have rational spectra. In this case, $H(z)$ is a rational function of order $r_1 + r_2$, where r_1 and r_2 are the orders of $S_{ff}(z)$ and $S_{vv}(z)$, respectively, in the variable $z + z^{-1}$ (elaborate). This is the essence of the Kalman filter for linear data smoothing. We shall discuss briefly the properties of signals that can be so smoothed. For simplicity, we shall consider only the first-order case.

Suppose that $\mathbf{s}[n]$ is a first-order process with autocorrelation

$$R[m] = a^{|m|}$$

We form the sequence

$$\zeta[n] = \mathbf{s}[n] - a\mathbf{s}[n-1] \tag{10-125}$$

We maintain that $\zeta[n]$ is white noise with autocorrelation

$$R_{\zeta\zeta}[m] = (1 - a^2)\delta[m] \tag{10-126}$$

Proof Clearly,

$$E\{\zeta^2[n]\} = R[0] - 2aR[1] - a^2R[0] = 1 + a^2$$

Furthermore, for any $m > 0$,

$$E\{\zeta[n]\mathbf{s}[n-m]\} = R[m] - aR[m-1] = 0 \qquad (10\text{-}127)$$

Hence, for any $k > 0$,

$$E\{\zeta[n]\zeta[n-k]\} = 0 \qquad (10\text{-}128)$$

and (10-126) results.

Definition A process $\mathbf{s}[n]$ is called wide-sense Markoff[1] if the optimum linear mean-square estimate of $\mathbf{s}[n]$ in terms of its past values $\mathbf{s}[n-1]$, $\mathbf{s}[n-2]$, ..., equals $a\mathbf{s}[n-1]$, where $a = R[1]/R[0]$.

From the orthogonality principle [see Eq. (10-65)], it follows that if $\mathbf{s}[n]$ is a wide-sense Markoff process, then the difference $\mathbf{s}[n] - a\mathbf{s}[n-1]$ is orthogonal to $\mathbf{s}[n-1]$, $\mathbf{s}[n-2]$, ..., that is,

$$E\{(\mathbf{s}[n] - a\mathbf{s}[n-1])\mathbf{s}[n-m]\} = 0 \qquad m > 0 \qquad (10\text{-}129)$$

This leads to the conclusion that

$$R[m] - aR[m-1] = 0 \qquad m > 0 \qquad (10\text{-}130)$$

Thus, $R[m]$ satisfies a first-order recursion. Solving, we obtain $R[m] = R[0]a^{|m|}$ because $R[m]$ is even. From the above we conclude that a wide-sense Markoff process is of first order.

Conversely, if $\mathbf{s}[n]$ is a first-order process, as in Eq. (10-124), then [see (10-127)] the difference $\mathbf{s}[n] - a\mathbf{s}[n-1]$ is orthogonal to $\mathbf{s}[n-m]$ for every $m > 0$. Therefore, $a\mathbf{s}[n-1]$ is the optimum estimate of $\mathbf{s}[n]$ in terms of its past; that is, $\mathbf{s}[n]$ is wide-sense Markoff.

Suppose, finally, that a first-order process $\mathbf{s}[n]$ is to be estimated in terms of the data $\mathbf{x}[n] = \mathbf{s}[n] + \mathbf{v}[n]$, where $\mathbf{v}[n]$ is white noise. As we have shown in Example 10-5, the resulting causal Wiener filter is a first-order system.

We thus reach the conclusion that the following four statements are equivalent:

1. A process $\mathbf{s}[n]$ is of first order with autocorrelation $R[m] = R[0]a^{|m|}$.
2. It satisfies a first-order recursion $\mathbf{s}[n] - a\mathbf{s}[n-1] = \zeta[n]$, where $\zeta[n]$ is white noise.
3. If it is to be estimated in terms of the data $\mathbf{s}[n] + \mathbf{v}[n]$ containing the white-noise process $\mathbf{v}[n]$, then the resulting causal Wiener filter can be realized by a first-order recursive system, as in Fig. 10-10.
4. It is wide-sense Markoff.

Note: A wide-sense Markoff process is not, in general, Markoff. In fact, a Markoff process need not be wide-sense Markoff unless it is also normal.[2]

[1] J. L. Doob, "Stochastic Processes," John Wiley & Sons, Inc., New York, 1953.

[2] A. Papoulis, Markoff and Wide-Sense Markoff Sequences, *Proc. IEEE*, vol. 53, no. 10, pp. 1661, October, 1965.

CHAPTER

ELEVEN

ERGODICITY, CORRELATION ESTIMATORS, FOURIER TRANSFORMS

In this chapter, we consider the problem of estimating the mean, the autocorrelation, and the energy spectrum of a stochastic process in terms of a single sample.[1]

In Sec. 11-1, we investigate the conditions under which time averages equal ensemble averages (ergodicity). As preparation, we evaluate the second-order moments of random sums and integrals. In Sec. 11-2, we apply the theory to the estimation of the autocorrelation of a stationary process $\mathbf{x}(t)$. *In Sec. 11-3, we determine the variance of the amplitude spectrum and the energy spectrum of* $\mathbf{x}(t)$, *and we examine the effect of noise in the evaluation of Fourier series and integrals.*

The analysis is based on the assumption that the process $\mathbf{x}(t)$ *is normal. However, as we show in Appendix 11-A, the results are valid also for more general processes if the sample size is sufficiently large.*[2]

---

[1] J. S. Bendat and A. G. Piersol, "Random Data: Analysis and Measurement Procedures," Interscience Publishers, a division of John Wiley & Sons, Inc., New York, 1971.

[2] A. Papoulis, Spectral Estimation of a Certain Class of Non-stationary Processes, *Trans. Sixth Prague Conf. Inf. Theory and Autom.*, Academia, Prague, 1973.

11-1 ERGODICITY

Given a random process $\mathbf{x}(t)$, we form the integral

$$\boldsymbol{\eta}_T = \frac{1}{2T}\int_{-T}^{T} \mathbf{x}(t)\, dt \tag{11-1}$$

Clearly, $\boldsymbol{\eta}_T$ varies with each sample of $\mathbf{x}(t)$, that is, it is a random variable. We shall examine the conditions under which it tends to the constant $E\{\mathbf{x}(t)\}$ as $T \to \infty$:

$$\lim \boldsymbol{\eta}_T = E\{\mathbf{x}(t)\} \qquad T \to \infty \tag{11-2}$$

If the limit of $\boldsymbol{\eta}_T$ equals $E\{\mathbf{x}(t)\}$, then, for sufficiently large T, the random variable $\boldsymbol{\eta}_T$ is close to $E\{\mathbf{x}(t)\}$ for most samples of $\mathbf{x}(t)$. Hence, the ensemble average $E\{\mathbf{x}(t)\}$ of the given process can be determined from a time average of a single sample. Ergodicity is the theory dealing with this topic. The term has the following three interpretations:

1. Under what conditions does the limit of $\boldsymbol{\eta}_T$ exist as $T \to \infty$? This question is of theoretical interest, and it is answered by Birkhoff's ergodic theorem: If $\mathbf{x}(t)$ is a stationary process with finite mean, then the time average $\boldsymbol{\eta}_T$ tends to a limit $\boldsymbol{\eta}$ for almost all samples of $\mathbf{x}(t)$.
2. The limit $\boldsymbol{\eta}$, if it exists, is, in general, a random variable whose value varies with each sample. Under what conditions is it a constant equal to $E\{\mathbf{x}(t)\}$? In our treatment of ergodicity, we deal with interpretation 2.
3. Suppose that $x(t)$ is a single time signal representing a physical quantity. Under what conditions can it be considered as a sample of an ergodic process? If it can, then averages based on past observations can be used to determine the statistics of the process which, in turn, leads to predictions of future averages.

Interpretations 1 and 2 are mathematical; i.e., they involve deduction within an axiomatic framework. Interpretation 3 is physical and, like any physical theory, it must be given an inductive interpretation.

Stochastic Integrals

The integral

$$\mathbf{p} = \int_a^b \mathbf{z}(t)\, dt \tag{11-3}$$

is a random variable with mean

$$\eta_p = E\{\mathbf{p}\} = \int_a^b E\{\mathbf{z}(t)\}\, dt = \int_a^b \eta_z(t)\, dt \tag{11-4}$$

and second moment

$$E\{|\mathbf{p}|^2\} = \int_a^b \int_a^b R_{zz}(t_1, t_2)\, dt_1\, dt_2 \tag{11-5}$$

because $$|\mathbf{p}|^2 = \mathbf{p}\mathbf{p}^* = \int_a^b \int_a^b \mathbf{z}(t_1)\mathbf{z}^*(t_2)\, dt_1\, dt_2$$

Clearly, $$\mathbf{p} - \eta_p = \int_a^b [\mathbf{z}(t) - \eta_z(t)]\, dt \tag{11-6}$$

as we see by subtracting Eq. (11-4) from (11-3). Applying (11-5) to (11-6), we conclude that the variance of $\mathbf{p}$ is given by

$$\sigma_p^2 = E\{|\mathbf{p} - \eta_p|^2\} = \int_a^b \int_a^b C_{zz}(t_1, t_2)\, dt_1\, dt_2 \tag{11-7}$$

because the autocorrelation of the centered process $\mathbf{z}(t) - \eta_z(t)$ equals the autocovariance $C_{zz}(t_1, t_2)$ of $\mathbf{z}(t)$.

The joint moment of the random variables $\mathbf{p}$ and

$$\mathbf{q} = \int_c^d \mathbf{w}(t)\, dt \tag{11-8}$$

is given by

$$E\{\mathbf{p}\mathbf{q}^*\} = \int_a^b \int_c^d R_{zw}(t_1, t_2)\, dt_1\, dt_2 \tag{11-9}$$

because $$\mathbf{p}\mathbf{q}^* = \int_a^b \int_c^d \mathbf{z}(t_1)\mathbf{w}^*(t_2)\, dt_1\, dt_2$$

Applying (11-9) to the integrals $\mathbf{p} - \eta_p$ and $\mathbf{q} - \eta_q$ of the centered processes $\mathbf{z}(t) - \eta_z(t)$ and $\mathbf{w}(t) - \eta_w(t)$, we obtain, as in (11-7), the covariance of the random variables $\mathbf{p}$ and $\mathbf{q}$:

$$\operatorname{cov}(\mathbf{p}, \mathbf{q}) = E\{(\mathbf{p} - \eta_p)(\mathbf{q}^* - \eta_q^*)\} = \int_a^b \int_c^d C_{zw}(t_1, t_2)\, dt_1\, dt_2 \tag{11-10}$$

Example 11-1 If $\mathbf{z}(t)$ is white noise with zero mean and autocorrelation

$$R_{zz}(t_1, t_2) = I(t_1)\, \delta(t_2 - t_1)$$

then $\eta_p = 0$, and (11-5) yields

$$\sigma_p^2 = E\{|\mathbf{p}|^2\} = \int_a^b \int_a^b I(t_1)\, \delta(t_2 - t_1)\, dt_2\, dt_1 = \int_a^b I(t_1)\, dt_1 \tag{11-11}$$

because $$\int_a^b \delta(t_2 - t_1)\, dt_2 = 1 \qquad \text{for } a < t_1 < b$$

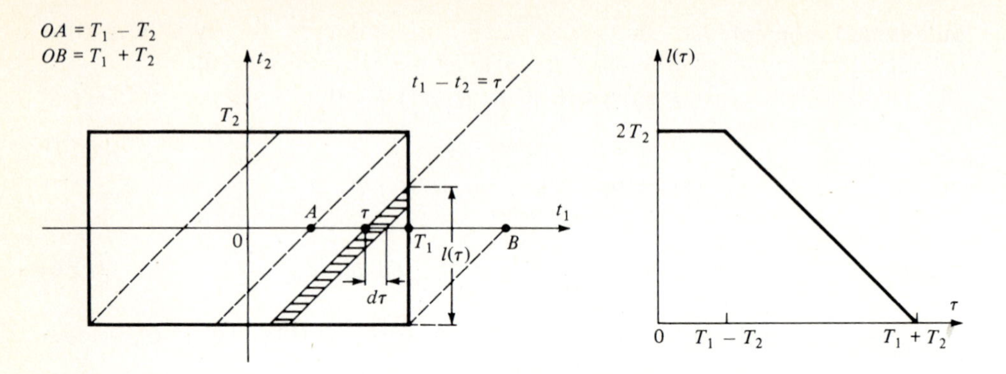

Figure 11-1

Stationary processes If $\mathbf{z}(t)$ and $\mathbf{w}(t)$ are two jointly stationary processes and

$$\mathbf{p} = \int_{-T_1}^{T_1} \mathbf{z}(t)\,dt \qquad \mathbf{q} = \int_{-T_2}^{T_2} \mathbf{w}(t)\,dt \tag{11-12}$$

then [see Eq. (11-9)]

$$E\{\mathbf{pq}^*\} = \int_{-T_1}^{T_1} \int_{-T_2}^{T_2} R_{zw}(t_1 - t_2)\,dt_1\,dt_2 \tag{11-13}$$

The region of integration is a rectangle (Fig. 11-1). Along the line $t_1 - t_2 = \tau$, the integrand $R_{zw}(\tau)$ is constant; hence, the integral over the shaded differential parallelogram equals $R_{zw}(\tau)$ times its area $l(\tau)\,d\tau$. As we see from the figure, if $T_1 > T_2$, then

$$l(\tau) = \begin{cases} T_1 + T_2 - |\tau| & T_1 - T_2 < |\tau| < T_1 + T_2 \\ 2T_2 & |\tau| < T_1 - T_2 \end{cases} \tag{11-14}$$

Hence,

$$E\{\mathbf{pq}^*\} = \int_{-(T_1+T_2)}^{T_1+T_2} l(\tau) R_{zw}(\tau)\,d\tau \tag{11-15}$$

Similarly, (11-10) yields

$$\text{cov}(\mathbf{p}, \mathbf{q}) = \int_{-(T_1+T_2)}^{T_1+T_2} l(\tau) C_{zw}(\tau)\,d\tau \tag{11-16}$$

Ergodicity of the Mean

A stochastic process $\mathbf{x}(t)$ is called *ergodic* if all its statistical properties can be determined from a single sample. An ergodic process must be stationary in the strict sense.

Ergodicity is also defined in a more limited sense: If certain statistical parameters P of $\mathbf{x}(t)$ can be determined from a single sample, then the process is called *ergodic in P*. In this case, stationarity with respect to P only is necessary. In this section, we establish conditions for the ergodicity of $E\{\mathbf{x}(t)\}$.

Suppose that $\mathbf{x}(t)$ is a stochastic process with constant mean

$$E\{\mathbf{x}(t)\} = \eta \tag{11-17}$$

We wish to see under what conditions the time average

$$\boldsymbol{\eta}_T = \frac{1}{2T}\int_{-T}^{T} \mathbf{x}(t)\, dt \tag{11-18}$$

is close to η. Clearly,

$$E\{\boldsymbol{\eta}_T\} = \frac{1}{2T}\int_{-T}^{T} E\{\mathbf{x}(t)\}\, dt = \eta \tag{11-19}$$

If the process $\mathbf{x}(t)$ is covariance stationary, i.e., if $C(t_1, t_2) = C(t_1 - t_2)$, then the variance of $\boldsymbol{\eta}_T$ is given by

$$\sigma_{\eta_T}^{\ 2} = \frac{1}{2T}\int_{-2T}^{2T} C(\tau)\left(1 - \frac{|\tau|}{2T}\right) d\tau \tag{11-20}$$

This follows from (11-16) with $\mathbf{z}(t) = \mathbf{w}(t) = \mathbf{x}(t)/2T$ and $T_1 = T_2 = T$ [see also Eq. (9-66)]. If, therefore,

$$\frac{1}{2T}\int_{-2T}^{2T} C(\tau)\left(1 - \frac{|\tau|}{2T}\right) d\tau \xrightarrow[T\to\infty]{} 0 \tag{11-21}$$

then,

$$\sigma_{\eta_T}^{\ 2} \xrightarrow[T\to\infty]{} 0 \tag{11-22}$$

and since the mean of $\boldsymbol{\eta}_T$ equals η, we conclude that (Tchebycheff's inequality)

$$\boldsymbol{\eta}_T \xrightarrow[T\to\infty]{} \eta \qquad \text{with probability 1} \tag{11-23}$$

Thus, if the autocovariance $C(\tau)$ of $\mathbf{x}(t)$ satisfies (11-21), then, for sufficiently large T, the time average $\boldsymbol{\eta}_T$ computed from a single sample is close to the mean η of $\mathbf{x}(t)$. Condition (11-21) is, therefore, necessary and sufficient for the process $\mathbf{x}(t)$ to be mean-ergodic.

Sufficient conditions

1. If

$$\int_{-\infty}^{\infty} |C(\tau)|\, d\tau < \infty \tag{11-24}$$

then the process $\mathbf{x}(t)$ is mean-ergodic.

Proof Condition (11-21) follows from (11-24) because

$$\left|\int_{-2T}^{2T} C(\tau)\left(1 - \frac{|\tau|}{2T}\right) d\tau\right| \le \int_{-2T}^{2T} |C(\tau)|\, d\tau < \infty \tag{11-25}$$

2. If $C(0) < \infty$ and

$$C(\tau) \underset{\tau\to\infty}{\longrightarrow} 0 \tag{11-26}$$

then the process $\mathbf{x}(t)$ is mean-ergodic.

Proof If $C(\tau)$ tends to zero with $\tau \to \infty$, then, given $\epsilon > 0$, we can find a constant α such that

$$|C(\tau)| < \epsilon \qquad \text{for } |\tau| > \alpha \tag{11-27}$$

Clearly, $$\left|\int_{-2T}^{2T} C(\tau)\left(1 - \frac{|\tau|}{2T}\right) d\tau\right| \le \int_{-\alpha}^{\alpha} |C(\tau)|\, d\tau + \int_{\alpha \le |\tau| \le 2T} |C(\tau)|\, d\tau$$

Applying (9-78) to the autocorrelation $C(\tau)$ of the centered process $\mathbf{x}(t) - \eta$, we obtain $|C(\tau)| \le C(0)$; hence, $\int_{-\alpha}^{\alpha} |(C(\tau))|\, d\tau \le 2\alpha C(0)$. Furthermore [see (11-27)],

$$\int_{|\alpha| \le |\tau| \le 2T} |C(\tau)| d\tau < 4\epsilon T$$

Hence, $$\frac{1}{2T}\int_{-2T}^{2T} C(\tau)\left(1 - \frac{|\tau|}{2T}\right) d\tau \le \frac{\alpha}{T} C(0) + 2\epsilon \underset{T\to\infty}{\longrightarrow} 2\epsilon$$

And since this is true for any ϵ, (11-21) results.

Corollary If

$$C(\tau) = 0 \qquad \text{for } |\tau| > a \tag{11-28}$$

i.e., if the process $\mathbf{x}(t) - \eta$ is a-dependent, then $\mathbf{x}(t)$ is mean-ergodic. Indeed, if $C(\tau)$ satisfies Eq. (11-28), then (11-26) is also true.

Notes:

1. With

$$S_c(\omega) = \int_{-\infty}^{\infty} C(\tau) e^{-j\omega\tau}\, d\tau$$

the power spectrum of the centered process $\mathbf{x}(t) - \eta$, it follows from (11-20) and the convolution theorem that

$$\sigma_{\eta_T}{}^2 = \frac{1}{2\pi}\int_{-\infty}^{\infty} S_c(\omega) \frac{\sin^2 \omega T}{\omega^2 T^2}\, d\omega \tag{11-29}$$

[see also (10-34)].

2. For ergodicity of the mean, it is necessary that $E\{\mathbf{x}(t)\}$ be constant; however, the autocovariance $C(t_1, t_2)$ of $\mathbf{x}(t)$ need not be a function of

$t_1 - t_2$ only. If it is not, then condition (11-21) must be replaced by [see (11-7)]

$$\sigma_{\eta_2}{}^2 = \frac{1}{4T^2}\int_{-T}^{T}\int_{-T}^{T} C(t_1, t_2)\, dt_1\, dt_2 \xrightarrow[T\to\infty]{} 0 \tag{11-30}$$

In Prob. 93 we discuss sufficient conditions for the above to be true.

Stochastic Sums

Given a discrete process $\mathbf{x}[n]$ and two integers a and b, we form the sum

$$\mathbf{p} = \sum_{n=a}^{b} \mathbf{x}[n] \tag{11-31}$$

Clearly, $\mathbf{p}$ is a random variable with mean

$$E\{\mathbf{p}\} = \eta_p = \sum_{n=a}^{b} E\{\mathbf{x}[n]\} = \sum_{n=a}^{b} \eta_x[n] \tag{11-32}$$

To find its variance, we note that

$$\mathbf{pp}^* = \sum_{k=a}^{b}\sum_{r=a}^{b} \mathbf{x}[k]\mathbf{x}^*[r] \tag{11-33}$$

Hence,

$$E\{|\mathbf{p}|^2\} = \sum_{k=a}^{b}\sum_{r=a}^{b} R_{xx}[k, r] \tag{11-34}$$

Applying the above to the centered sum

$$\mathbf{p} - \eta_p = \sum_{n=a}^{b} (\mathbf{x}[n] - \eta_x[n])$$

we obtain

$$\sigma_p{}^2 = E\{|\mathbf{p} - \eta_p|^2\} = \sum_{k=a}^{b}\sum_{r=a}^{b} C_{xx}[k, r] \tag{11-35}$$

because the autocorrelation of $\mathbf{x}[n] - \eta_x[n]$ equals the autocovariance $C_{xx}[n_1, n_2]$ of $\mathbf{x}[n]$.

Similarly, if

$$\mathbf{q} = \sum_{n=c}^{d} \mathbf{y}[n] \tag{11-36}$$

then

$$E\{\mathbf{pq}^*\} = \sum_{k=a}^{b}\sum_{r=c}^{d} R_{xy}[k, r] \qquad \text{cov}(\mathbf{p}, \mathbf{q}) = \sum_{k=a}^{b}\sum_{r=c}^{d} C_{xy}[k, r] \tag{11-37}$$

Example 11-2 If $\mathbf{x}[n]$ is white noise with zero mean and autocorrelation

$$R_{xx}[n_1, n_2] = I[n_1]\,\delta[n_1 - n_2]$$

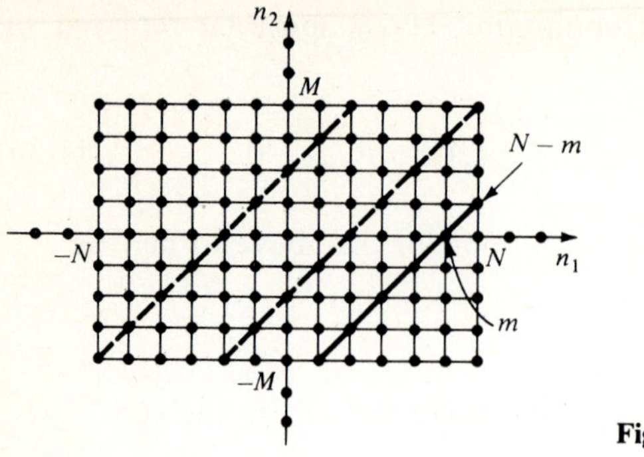

Figure 11-2

then $\eta_p = 0$, and Eq. (11-35) yields

$$\sigma_p^2 = \sum_{k=a}^{b} \sum_{r=a}^{b} I[k]\, \delta[k - r] = \sum_{k=a}^{b} I[k]$$

Stationary processes If $\mathbf{x}[n]$ and $\mathbf{y}[n]$ are two jointly stationary processes and

$$\mathbf{p} = \sum_{n=-N}^{N} \mathbf{x}[n] \qquad \mathbf{q} = \sum_{n=-M}^{M} \mathbf{y}[n] \qquad N \geq M$$

then
$$E\{\mathbf{pq}^*\} = \sum_{k=-N}^{N} \sum_{r=-M}^{M} R_{xy}[k - r] = \sum_{m=-M-N}^{M+N} l[m] R_{xy}[m] \tag{11-38}$$

where
$$l[m] = \begin{cases} M + N + 1 - |m| & N - M \leq |m| \leq M + N \\ 2M + 1 & |m| < N - M \end{cases} \tag{11-39}$$

This follows readily with $k - r = m$ (Fig. 11-2).

Setting $\mathbf{x}[n] = \mathbf{y}[n]$ and $N = M$ in (11-38), we obtain

$$E\{|\mathbf{p}|^2\} = \sum_{m=-2N}^{2N} (2N + 1 - |m|) R_{xx}[m] \tag{11-40}$$

and with $\mathbf{p}$ replaced by $\mathbf{p} - \eta_p$,

$$\sigma_p^2 = \sum_{m=-2N}^{2N} (2N + 1 - |m|) C_{xx}[m] \tag{11-41}$$

Ergodicity of the mean We wish to determine the mean η of a discrete process $\mathbf{x}[n]$ from a segment $|n| \leq N$ of a single realization of $\mathbf{x}[n]$. For this purpose, we form the sum

$$\boldsymbol{\eta}_N = \frac{1}{2N + 1} \sum_{n=-N}^{N} \mathbf{x}[n] \tag{11-42}$$

This sum is a random variable with mean η and variance [see (11-41)]

$$\sigma_{\eta_N}{}^2 = \frac{1}{2N+1} \sum_{m=-2N}^{2N} C[m]\left(1 - \frac{|m|}{2N+1}\right) \tag{11-43}$$

If therefore, the process $\mathbf{x}[n]$ is such that

$$\frac{1}{2N+1} \sum_{m=-2N}^{2N} C[m]\left(1 - \frac{|m|}{2N+1}\right) \xrightarrow[N\to\infty]{} 0 \tag{11-44}$$

then it is mean ergodic; i.e., its sample mean $\boldsymbol{\eta}_N$ tends to η as $N \to \infty$.

We can show, as in the analog case (page 356), that if

$$\sum_{m=-\infty}^{\infty} |C[m]| < \infty \qquad \text{or if } C[m] \xrightarrow[m\to\infty]{} 0 \tag{11-45}$$

then the ergodicity condition (11-44) holds (elaborate).

11-2 CORRELATION ESTIMATES

Consider a real, stationary process $\mathbf{x}(t)$ with autocorrelation

$$R(\tau) = E\{\mathbf{x}(t+\tau)\mathbf{x}(t)\}$$

Clearly, $R(\tau)$ is the expected value of the process

$$\boldsymbol{\varphi}(t) = \mathbf{x}(t+\tau)\mathbf{x}(t) \tag{11-46}$$

where τ is considered as a parameter. To establish conditions for the ergodicity of $\mathbf{x}(t)$ with respect to its autocorrelation, we can apply, therefore, the results of the last section to the process $\boldsymbol{\varphi}(t)$.

Ergodicity The integral

$$\mathbf{R}_c(\tau) = \frac{1}{2c}\int_{-c}^{c} \boldsymbol{\varphi}(t)\, dt \tag{11-47}$$

is the time average of $\boldsymbol{\varphi}(t)$. Therefore, its mean is given by

$$E\{\mathbf{R}_c(\tau)\} = E\{\boldsymbol{\varphi}(t)\} = R(\tau) \tag{11-48}$$

and its variance by

$$\sigma_c{}^2 = \frac{1}{2c}\int_{-2c}^{2c} C_{\varphi\varphi}(\alpha)\left(1 - \frac{|\alpha|}{2c}\right) d\alpha \tag{11-49}$$

where $C_{\varphi\varphi}(\alpha)$ is the autocovariance of $\boldsymbol{\varphi}(t)$. The above follows from Eq. (11-20) if $\mathbf{x}(t)$ is replaced with $\boldsymbol{\varphi}(t)$ in (11-18).

The process $\mathbf{x}(t)$ is thus correlation-ergodic if the right side of (11-49) tends to zero as $c \to \infty$. If this is the case, then for sufficiently large c, the integral $\mathbf{R}_c(\tau)$

is close to $R(\tau)$. With

$$R_{\varphi\varphi}(\alpha) = E\{\boldsymbol{\varphi}(t+\alpha)\boldsymbol{\varphi}(t)\} = E\{\mathbf{x}(t+\alpha+\tau)\mathbf{x}(t+\alpha)\mathbf{x}(t+\tau)\mathbf{x}(t)\} \quad (11\text{-}50)$$

the autocorrelation of $\boldsymbol{\varphi}(t)$, we have

$$C_{\varphi\varphi}(\alpha) = R_{\varphi\varphi}(\alpha) - E^2\{\boldsymbol{\varphi}(t)\} = R_{\varphi\varphi}(\alpha) - R^2(\tau) \quad (11\text{-}51)$$

Hence, to establish the ergodicity of $\mathbf{x}(t)$ with respect to its correlation, we need its fourth-order moment $R_{\varphi\varphi}(\alpha)$.

Normal process If $\mathbf{x}(t)$ is normal, then this fourth-order moment can be expressed in terms of $R(\tau)$. Indeed, assuming for simplicity that $E\{\mathbf{x}(t)\} = 0$, we conclude from (11-A-5) that

$$\begin{aligned} E\{\mathbf{x}(t+\tau+\alpha)\mathbf{x}(t+\alpha)\mathbf{x}(t+\tau)\mathbf{x}(t)\} &= E\{\mathbf{x}(t+\tau+\alpha)\mathbf{x}(t+\alpha)\}E\{\mathbf{x}(t+\tau)\mathbf{x}(t)\} \\ &+ E\{\mathbf{x}(t+\tau+\alpha)\mathbf{x}(t+\tau)\}E\{\mathbf{x}(t+\alpha)\mathbf{x}(t)\} \\ &+ E\{\mathbf{x}(t+\tau+\alpha)\mathbf{x}(t)\}E\{\mathbf{x}(t+\alpha)\mathbf{x}(t+\tau)\} \end{aligned}$$

Hence,

$$R_{\varphi\varphi}(\alpha) = R(\tau)R(\tau) + R(\alpha)R(\alpha) + R(\tau+\alpha)R(\tau-\alpha) \quad (11\text{-}52)$$

and (11-51) yields

$$C_{\varphi\varphi}(\alpha) = R(\tau+\alpha)R(\tau-\alpha) + R^2(\alpha) \quad (11\text{-}53)$$

From the above and (11-49), it follows that a normal process with zero mean is autocorrelation-ergodic if

$$\frac{1}{2c}\int_{-2c}^{2c}[R(\tau+\alpha)R(\tau-\alpha) + R^2(\alpha)]\left(1 - \frac{|\alpha|}{2c}\right)d\alpha \xrightarrow[c\to\infty]{} 0 \quad (11\text{-}54)$$

Sufficient condition If $\mathbf{x}(t)$ is a-dependent, then it is autocorrelation-ergodic. Indeed, if $R(\tau) = 0$ for $|\tau| > a$, then [see Eq. (11-53)] $C_{\varphi\varphi}(\alpha) = 0$ for $|\alpha| > a$ as in (11-28).

> **Note:** It can be shown that this condition is sufficient even if the process $\mathbf{x}(t)$ is not normal (see Appendix 11-A).

Finite Sample

We now assume that the sample $\mathbf{x}(t)$ is available only in the interval $(-T, T)$. In this case, the product $\mathbf{x}(t+\tau)\mathbf{x}(t)$ can be determined only for values of t in the interval $(-T, T-\tau)$ if $\tau > 0$, and in the interval $(-T+|\tau|, T)$ if $\tau < 0$. To avoid separate formulas for $\tau > 0$ and $\tau < 0$, we introduce the symmetrical function

$$\mathbf{R}^T(\tau) = \frac{1}{2T - |\tau|}\int_{-T+|\tau|/2}^{T-|\tau|/2} \mathbf{x}\left(t + \frac{\tau}{2}\right)\mathbf{x}\left(t - \frac{\tau}{2}\right)dt \quad (11\text{-}55)$$

Clearly, $\mathbf{R}^T(\tau) = \mathbf{R}_c(\tau)$ as in (11-47), provided that $c = T - |\tau|/2$ and the integrand $\boldsymbol{\varphi}(t)$ is replaced with $\boldsymbol{\varphi}(t - \tau/2)$. Hence,

$$E\{\mathbf{R}^T(\tau)\} = R(\tau) \tag{11-56}$$

That is, $\mathbf{R}^T(\tau)$ is an unbiased estimator of $R(\tau)$ with variance

$$\sigma_{R^T}{}^2 = \sigma_c{}^2 \qquad c = T - \frac{|\tau|}{2} \tag{11-57}$$

If, therefore, the process $\mathbf{x}(t)$ is correlation-ergodic, then

$$\mathbf{R}^T(\tau) \xrightarrow[T \to \infty]{} R(\tau)$$

and the time average $\mathbf{R}^T(\tau)$ can be used to estimate $R(\tau)$, provided that c is sufficiently large. Since $c = T - |\tau|/2$, the required value of T for a certain accuracy is large if τ is large.

Biased estimator We shall also use the estimator

$$\mathbf{R}_T(\tau) = \frac{1}{2T} \int_{-T+|\tau|/2}^{T-|\tau|/2} \mathbf{x}\left(t + \frac{\tau}{2}\right) \mathbf{x}\left(t - \frac{\tau}{2}\right) dt \qquad |\tau| < 2T \tag{11-58}$$

whose mean is no longer $R(\tau)$:

$$E\{\mathbf{R}_T(\tau)\} = \left(1 - \frac{|\tau|}{2T}\right) R(\tau) \tag{11-59}$$

Assuming that $\mathbf{R}_T(\tau)$ is defined for $|\tau| > 2T$ such that

$$\mathbf{R}_T(\tau) = 0 \qquad \text{for } |\tau| > 2T \tag{11-60}$$

we can easily see that the integral in (11-58) is a convolution

$$\mathbf{R}_T(\tau) = \frac{1}{2T} \mathbf{x}_T(\tau) * \mathbf{x}_T(-\tau) \tag{11-61}$$

where

$$\mathbf{x}_T(t) = \mathbf{x}(t) p_T(t) = \begin{cases} \mathbf{x}(t) & |t| \le T \\ 0 & |t| > T \end{cases} \tag{11-62}$$

equals the given segment of $\mathbf{x}(t)$.

With $\mathbf{S}_T(\omega)$ the Fourier transform of $\mathbf{R}_T(\tau)$, it follows from Eq. (11-61) and the convolution theorem that

$$\mathbf{S}_T(\omega) = \int_{-2T}^{2T} \mathbf{R}_T(\tau) e^{-j\omega\tau} \, d\tau = \frac{1}{2T} \left| \int_{-T}^{T} \mathbf{x}(t) e^{-j\omega t} \, dt \right|^2 \tag{11-63}$$

Thus, $\mathbf{S}_T(\omega)$ can be determined directly from $\mathbf{x}_T(t)$.

Fundamental remark If the process $\mathbf{x}(t)$ is correlation-ergodic, then $\mathbf{R}^T(\tau) \simeq R(\tau)$, provided that the interval of integration $2T - |\tau|$ in (11-55) is sufficiently large. If, however, T is fixed, no matter how large, the approximation $\mathbf{R}^T(\tau) \simeq R(\tau)$

holds not for every τ, but only for $\tau \ll T$. For each τ, we can choose T such that $\mathbf{R}^T(\tau) \simeq R(\tau)$, but the required T will depend on τ.

The same holds for the biased estimator

$$\mathbf{R}_T(\tau) = \left(1 - \frac{|\tau|}{2T}\right)\mathbf{R}^T(\tau)$$

because its bias

$$b = E\{\mathbf{R}_T(\tau)\} - R(\tau) = -\frac{|\tau|}{2T}R(\tau)$$

and its variance

$$\sigma_{R_T}{}^2 = \left(1 - \frac{|\tau|}{2T}\right)^2 \sigma_{R^T}{}^2 \tag{11-64}$$

tend to zero with $T \to \infty$. Thus, although

$$\mathbf{R}_T(\tau) \xrightarrow[T\to\infty]{} R(\tau) \tag{11-65}$$

the convergence is not uniform in τ. Hence, we cannot conclude that the transform $\mathbf{S}_T(\omega)$ of $\mathbf{R}_T(\tau)$ tends to the transform $S(\omega)$ of $R(\tau)$. In fact, as we show in the next chapter, $\mathbf{S}_T(\omega)$ tends to a random variable whose variance equals $S^2(\omega)$. The explanation is that $\mathbf{S}_T(\omega)$ is given by the integral in Eq. (11-63) involving $\mathbf{R}_T(\tau)$ for *every* $|\tau| < 2T$; and, for τ near $2T$, $\mathbf{R}_T(\tau)$ is not a reliable estimate of $R(\tau)$.

Autocovariance of the estimator The sample autocorrelation $\mathbf{R}_T(\tau)$, considered as a function of τ, is a nonstationary process with mean $(1 - |\tau|/2T)R(\tau)$. We shall determine its autocorrelation and autocovariance. With

$$\mathbf{z}(t) = \frac{1}{2T}\mathbf{x}\left(t + \frac{u}{2}\right)\mathbf{x}\left(t - \frac{u}{2}\right) \qquad \mathbf{w}(t) = \frac{1}{2T}\mathbf{x}\left(t + \frac{v}{2}\right)\mathbf{x}\left(t - \frac{v}{2}\right)$$

it follows from (11-58) that

$$\begin{aligned} \mathbf{R}_T(u) &= \int_{-T_1}^{T_1} \mathbf{z}(t)\,dt \qquad T_1 = T - \frac{|u|}{2} \\ \mathbf{R}_T(v) &= \int_{-T_2}^{T_2} \mathbf{w}(t)\,dt \qquad T_2 = T - \frac{|v|}{2} \end{aligned} \tag{11-66}$$

The above integrals are of the form (11-12) if we view u and v as two parameters. Hence [see (11-15)],

$$E\{\mathbf{R}_T(u)\mathbf{R}_T(v)\} = \frac{1}{4T^2}\int_{-\infty}^{\infty} l(\alpha)E\left\{\mathbf{x}\left(t + \alpha + \frac{u}{2}\right)\mathbf{x}\left(t + \alpha - \frac{u}{2}\right)\mathbf{x}\left(t + \frac{v}{2}\right)\mathbf{x}\left(t - \frac{v}{2}\right)\right\} d\alpha \tag{11-67}$$

where
$$l(\alpha) = \begin{cases} 2T_2 & |\alpha| < T_1 - T_2 \\ T_1 + T_2 - |\alpha| & T_1 - T_2 < |\alpha| < T_1 + T_2 \\ 0 & |\alpha| > T_1 + T_2 \end{cases} \tag{11-68}$$

as in (11-14). We have assumed in (11-68) that $T_1 > T_2$, that is, that $|u| < |v|$. If $|u| > |v|$, then we must interchange T_1 and T_2.

To determine the autocovariance of $\mathbf{R}_T(\tau)$, we subtract from (11-67) the product

$$E\{\mathbf{R}_T(u)\}E\{\mathbf{R}_T(v)\} = \frac{T_1 T_2}{T^2} R(u)R(v)$$

[see (11-59)]. Since the area of $l(\alpha)$ equals $4T_1T_2$, it suffices to subtract, from the term $E\{\ \}$ in the integrand of (11-67), the product $R(u)R(v)$ (elaborate).

Normal processes Suppose, finally, that $\mathbf{x}(t)$ is a normal process with zero mean. Applying (11-A-5) to (11-67), we conclude, as in (11-52), that

$$\begin{aligned} E\{\mathbf{R}_T(u)\mathbf{R}_T(v)\} = \frac{1}{4T^2}\int_{-\infty}^{\infty} l(\alpha)\bigg[R(u)R(v) + R\left(\alpha + \frac{u}{2} - \frac{v}{2}\right)R\left(\alpha - \frac{u}{2} + \frac{v}{2}\right) \\ + R\left(\alpha + \frac{u}{2} + \frac{v}{2}\right)R\left(\alpha - \frac{u}{2} - \frac{v}{2}\right)\bigg]\,d\alpha \end{aligned} \tag{11-69}$$

Hence,

$$\begin{aligned} \operatorname{cov}[\mathbf{R}_T(u), \mathbf{R}_T(v)] = \frac{1}{4T^2}\int_{-\infty}^{\infty} l(\alpha)\bigg[R\left(\alpha + \frac{u}{2} - \frac{v}{2}\right)R\left(\alpha - \frac{u}{2} + \frac{v}{2}\right) \\ + R\left(\alpha + \frac{u}{2} + \frac{v}{2}\right)R\left(\alpha - \frac{u}{2} - \frac{v}{2}\right)\bigg]\,d\alpha \end{aligned} \tag{11-70}$$

If $u = v = \tau$, then $l(\alpha) = 2T - |\tau| - |\alpha|$ for $|\alpha| < 2T - |\tau|$, and Eq. (11-70) yields the following expression for the variance of $\mathbf{R}_T(\tau)$:

$$\sigma_{R_T}{}^2 = \frac{1}{2T}\int_{-2T+|\tau|}^{2T-|\tau|} [R^2(\alpha) + R(\alpha + \tau)R(\alpha - \tau)]\left(1 - \frac{|\tau| + |\alpha|}{2T}\right) d\alpha \tag{11-71}$$

Note: The autocorrelation (11-67) of $\mathbf{R}_T(\tau)$ is needed to evaluate the variance of the sample spectrum $\mathbf{S}_T(\omega)$ if $\mathbf{S}_T(\omega)$ is expressed as the Fourier transform of $\mathbf{R}_T(\tau)$ as in (11-63) (see Sec. 11-3). If however, $\mathbf{S}_T(\omega)$ is determined directly in terms of $\mathbf{x}_T(t)$. then, as we show in the next chapter, (11-67) need not be used.

11-3 FOURIER TRANSFORMS OF RANDOM SIGNALS

In this section, we develop the second-order moments of the Fourier transform $\mathbf{X}(\omega)$ and the energy spectrum $|\mathbf{X}(\omega)|^2$ of a random process $\mathbf{x}(t)$. The results are used in the study of Fourier integrals of noisy signals and in the estimation of power spectra of stochastic processes.

As preparation, we introduce the two-dimensional transform $\Gamma(u, v)$ of the autocorrelation $R(t_1, t_2)$ of $\mathbf{x}(t)$:

$$\Gamma(u, v) = \int_{-\infty}^{\infty} \int_{-\infty}^{\infty} R(t_1, t_2)e^{-j(ut_1 + vt_2)}\, dt_1\, dt_2 \tag{11-72}$$

From the inversion formula (8-53), it follows that

$$R(t_1, t_2) = \frac{1}{4\pi^2} \int_{-\infty}^{\infty} \int_{-\infty}^{\infty} \Gamma(u, v)e^{j(ut_1 + vt_2)}\, dt_1\, dt_2 \tag{11-73}$$

Linear systems Suppose that $\mathbf{y}(t)$ is the output of a linear system with input $\mathbf{x}(t)$:

$$\mathbf{y}(t) = \mathbf{x}(t) * h(t) \tag{11-74}$$

As we know [see Eqs. (9-40) and (9-41)],

$$\begin{aligned} R_{xy}(t_1, t_2) &= R_{xx}(t_1, t_2) * h^*(t_2) \\ R_{yy}(t_1, t_2) &= R_{xy}(t_1, t_2) * h(t_1) \end{aligned} \tag{11-75}$$

Denoting by $\Gamma_{xx}(u, v)$, $\Gamma_{xy}(u, v)$, and $\Gamma_{yy}(u, v)$ the transforms of $R_{xx}(t_1, t_2)$, $R_{xy}(t_1, t_2)$, and $R_{yy}(t_1, t_2)$, respectively, we conclude from (11-75) and the convolution theorem that (elaborate)

$$\Gamma_{xy}(u, v) = \Gamma_{xx}(u, v)H^*(-v) \qquad \Gamma_{yy}(u, v) = \Gamma_{xy}(u, v)H(u) \tag{11-76}$$

where $H(\omega)$ is the transform of $h(t)$. Hence,

$$\Gamma_{yy}(u, v) = \Gamma_{xx}(u, v)H(u)H^*(-v) \tag{11-77}$$

Stationary processes If the process $\mathbf{x}(t)$ is stationary with power spectrum $S(\omega)$, then

$$\Gamma(u, v) = 2\pi S(u)\, \delta(u + v) \tag{11-78}$$

Proof As we know, $S(\omega)$ is the transform of the autocorrelation $R(\tau)$ of $\mathbf{x}(t)$. Therefore [shifting theorem (3-20)],

$$R(t_1 - t_2) \underset{t_1}{\leftrightarrow} S(u)e^{-jt_2u} \tag{11-79}$$

and (11-78) results because [see (3-52)]

$$S(u)e^{-jt_2u} \underset{t_2}{\leftrightarrow} 2\pi S(u)\, \delta(u + v) \tag{11-80}$$

Windows We form the product

$$\mathbf{x}_w(t) = w(t)\mathbf{x}(t) \tag{11-81}$$

where $\mathbf{x}(t)$ is a stationary process, and $w(t)$ a given signal with transform $W(\omega)$. Clearly, $\mathbf{x}_w(t)$ is a nonstationary process with autocorrelation

$$R_w(t_1, t_2) = E\{w(t_1)\mathbf{x}(t_1)w^*(t_2)\mathbf{x}^*(t_2)\} = w(t_1)w^*(t_2)R(t_1 - t_2) \tag{11-82}$$

We maintain that the two-dimensional transform $\Gamma_w(u, v)$ of $R_w(t_1, t_2)$ is given by

$$\Gamma_w(u, v) = \frac{1}{2\pi}\int_{-\infty}^{\infty} W(u - y)W^*(-v - y)S(y)\,dy \tag{11-83}$$

Proof Since $w(t) \leftrightarrow W(\omega)$ and $R(\tau) \leftrightarrow S(\omega)$, it follows that [see (3-30)]

$$w(t_1)w^*(t_2)R(t_1 - t_2) \underset{t_1}{\leftrightarrow} \frac{1}{2\pi}\int_{-\infty}^{\infty} W(u - y)S(y)e^{-jt_2y}w^*(t_2)\,dy \tag{11-84}$$

But $w^*(t_2) \leftrightarrow W^*(-v)$; hence [see (3-20)],

$$e^{-jt_2y}w^*(t_2) \underset{t_2}{\leftrightarrow} W^*[-(v + y)] \tag{11-85}$$

The transform of the right side of (11-84) with respect to t_2 yields Eq. (11-83).

Fourier Transforms of Stochastic Processes

The Fourier transform

$$\mathbf{X}(\omega) = \int_{-\infty}^{\infty} \mathbf{x}(t)e^{-j\omega t}\,dt \tag{11-86}$$

of a stochastic process $\mathbf{x}(t)$ is a stochastic process with mean

$$E\{\mathbf{X}(\omega)\} = \int_{-\infty}^{\infty} E\{\mathbf{x}(t)\}e^{-j\omega t}\,dt = \int_{-\infty}^{\infty} \eta(t)e^{-j\omega t}\,dt \tag{11-87}$$

the Fourier transform of the mean $\eta(t)$ of $\mathbf{x}(t)$.

To find its autocorrelation, we form the product

$$\mathbf{X}(u)\mathbf{X}^*(v) = \int_{-\infty}^{\infty}\int_{-\infty}^{\infty} \mathbf{x}(t_1)e^{-jut_1}\mathbf{x}^*(t_2)e^{jvt_2}\,dt_1\,dt_2 \tag{11-88}$$

and take expected values of both sides:

$$E\{\mathbf{X}(u)\mathbf{X}^*(v)\} = \int_{-\infty}^{\infty}\int_{-\infty}^{\infty} R(t_1, t_2)e^{-j(ut_1 - vt_2)}\,dt_1\,dt_2 \tag{11-89}$$

Hence [see (11-72)],

$$E\{\mathbf{X}(u)\mathbf{X}^*(v)\} = \Gamma(u, -v) \tag{11-90}$$

The average power of $\mathbf{X}(\omega)$ is obtained from the above with $u = v = \omega$:

$$E\{|\mathbf{X}(\omega)|^2\} = \Gamma(\omega, -\omega) \tag{11-91}$$

White noise Suppose that $\mathbf{x}(t)$ is nonstationary white noise:

$$R(t_1, t_2) = I(t_2)\,\delta(t_1 - t_2)$$

With
$$J(y) = \int_{-\infty}^{\infty} I(t)e^{-jyt}\,dt$$

the Fourier transform of $I(t)$, we have

$$I(t_2)\,\delta(t_1 - t_2) \underset{t_1}{\leftrightarrow} I(t_2)e^{-jt_2u} \underset{t_2}{\leftrightarrow} J(u + v)$$

Hence,
$$\Gamma(u, v) = J(u + v)$$

Inserting into Eq. (11-90), we obtain, with $u = \omega + y$ and $u = \omega$,

$$E\{\mathbf{X}(\omega + y)\mathbf{X}^*(\omega)\} = J(y) \tag{11-92}$$

Thus, the Fourier transform of nonstationary white noise with average intensity $I(t)$ is a stationary process $\mathbf{X}(\omega)$ whose autocorrelation equals the Fourier transform $J(y)$ of $I(t)$.

Stationary processes From (11-90) and (11-78) it follows that, if $\mathbf{x}(t)$ is stationary, then

$$E\{\mathbf{X}(u)\mathbf{X}^*(v)\} = 2\pi S(u)\,\delta(u - v) \tag{11-93}$$

Thus, the Fourier transform $\mathbf{X}(\omega)$ of a stationary process $\mathbf{x}(t)$ with power spectrum $S(\omega)$ is nonstationary white noise with average intensity $2\pi S(\omega)$.

Windows If

$$\mathbf{X}_w(\omega) = \int_{-\infty}^{\infty} w(t)\mathbf{x}(t)e^{-j\omega t}\,dt \tag{11-94}$$

is the transform of the product $\mathbf{x}_w(t) = w(t)\mathbf{x}(t)$, then (11-83) yields (elaborate)

$$E\{\mathbf{X}_w(u)\mathbf{X}_w^*(v)\} = \frac{1}{2\pi}\int_{-\infty}^{\infty} W(u - y)W^*(v - y)S(y)\,dy \tag{11-95}$$

Hence,

$$E\{|\mathbf{X}_w(\omega)|^2\} = \frac{1}{2\pi}\int_{-\infty}^{\infty} |W(\omega - y)|^2 S(y)\,dy = \frac{1}{2\pi}|W(\omega)|^2 * S(\omega) \tag{11-96}$$

Truncation The integral

$$\mathbf{X}_T(\omega) = \int_{-T}^{T} \mathbf{x}(t)e^{-j\omega t}\,dt$$

is the Fourier transform of the segment

$$\mathbf{x}_T(t) = p_T(t)\mathbf{x}(t)$$

of $\mathbf{x}(t)$. This is a special case of Eq. (11-94) with

$$w(t) = p_T(t) \qquad W(\omega) = \frac{2 \sin T\omega}{\omega}$$

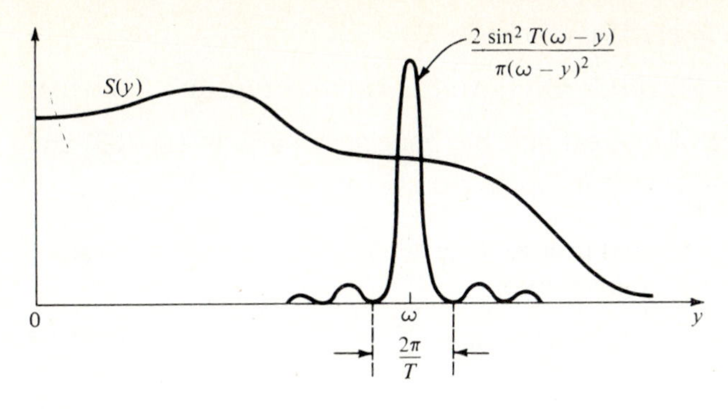

Figure 11-3

Therefore,

$$E\{\mathbf{X}_T(u)\mathbf{X}_T^*(v)\} = \frac{1}{2\pi}\int_{-\infty}^{\infty} \frac{2\sin T(u-y)}{u-y}\,\frac{2\sin T(v-y)}{v-y}\,S(y)\,dy \tag{11-97}$$

and

$$E\{|\mathbf{X}_T(\omega)|^2\} = S(\omega) * \frac{2\sin^2 T\omega}{\pi\omega^2} \tag{11-98}$$

Large T If T is so large that $S(\omega)$ is approximately constant in any interval of length $2\pi/T$ (Fig. 11-3), then (11-98) yields

$$E\{|\mathbf{X}_T(\omega)|^2\} \simeq S(\omega)\int_{-\infty}^{\infty} \frac{2\sin^2 T(\omega-y)}{\pi(\omega-y)^2}\,dy = 2TS(\omega) \tag{11-99}$$

Covariance of energy spectrum† In general, the transform $\Gamma(u, v)$ of the autocorrelation $R(t_1, t_2)$ of the process $\mathbf{x}(t)$ is complex:

$$\Gamma(u, v) = \Gamma_r(u, v) + j\Gamma_i(u, v) \tag{11-100}$$

We maintain that, if $\mathbf{A}(\omega)$ and $\mathbf{B}(\omega)$ are the real and imaginary parts of the Fourier transform $\mathbf{X}(\omega)$ of $\mathbf{x}(t)$:

$$\mathbf{X}(\omega) = \mathbf{A}(\omega) + j\mathbf{B}(\omega) \tag{11-101}$$

then

$$\begin{aligned} E\{\mathbf{A}(u)\mathbf{A}(v)\} &= \tfrac{1}{2}[\Gamma_r(u, -v) + \Gamma_r(u, v)] \\ E\{\mathbf{B}(u)\mathbf{B}(v)\} &= \tfrac{1}{2}[\Gamma_r(u, -v) - \Gamma_r(u, v)] \\ E\{\mathbf{A}(u)\mathbf{B}(v)\} &= \tfrac{1}{2}[\Gamma_i(u, v) - \Gamma_i(u, -v)] \\ E\{\mathbf{A}(v)\mathbf{B}(u)\} &= \tfrac{1}{2}[\Gamma_i(u, v) + \Gamma_i(u, -v)] \end{aligned} \tag{11-102}$$

Proof From Eq. (11-90), it follows that

$$E\{\mathbf{X}(u)\mathbf{X}^*(v)\} = E\{[\mathbf{A}(u) + j\mathbf{B}(u)][\mathbf{A}(v) - j\mathbf{B}(v)]\} = \Gamma(u, -v) \tag{11-103}$$

† A. Papoulis, Identification of Systems Driven by Nonstationary Noise, *IEEE Trans. Inf. Theory*, January, 1978.

Replacing $\mathbf{X}^*(v)$ with $\mathbf{X}(v)$ in (11-89), we obtain

$$E\{\mathbf{X}(u)\mathbf{X}(v)\} = E\{[\mathbf{A}(u) + j\mathbf{B}(u)][\mathbf{A}(v) + j\mathbf{B}(v)]\} = \Gamma(u, v) \tag{11-104}$$

The sum and the difference of the real and the imaginary parts of (11-103) and (11-104) yield (11-102).

Normal processes If $\mathbf{x}(t)$ is a normal process with zero mean, then the covariance of its energy spectrum $|\mathbf{X}(\omega)|^2$ is given by

$$\text{cov}[|\mathbf{X}(u)|^2, |\mathbf{X}(v)|^2] = |\Gamma(u, -v)|^2 + |\Gamma(u, v)|^2 \tag{11-105}$$

Proof As is well known, a linear combination of normal random variables is normal. This property, extended to integrals, leads to the conclusion that the Fourier transform $\mathbf{X}(\omega)$ of a normal process $\mathbf{x}(t)$ is also normal. Hence, its real part $\mathbf{A}(\omega)$ and imaginary part $\mathbf{B}(\omega)$ are jointly normal. Furthermore, since $E\{\mathbf{x}(t)\} = 0$, it follows from (11-87) that

$$E\{\mathbf{X}(\omega)\} = 0 \qquad E\{\mathbf{A}(\omega)\} = 0 \qquad E\{\mathbf{B}(\omega)\} = 0$$

Thus, the random variables $\mathbf{A}(u)$, $\mathbf{A}(v)$, $\mathbf{B}(u)$, and $\mathbf{B}(v)$ are jointly normal with zero mean. Hence [see (11-A-2) and (11-102)],

$$E\{\mathbf{A}^2(u)\mathbf{A}^2(v)\} = E\{\mathbf{A}^2(u)\}E\{\mathbf{A}^2(v)\} + 2E^2\{\mathbf{A}(u)\mathbf{A}(v)\}$$
$$E\{\mathbf{B}^2(u)\mathbf{B}^2(v)\} = E\{\mathbf{B}^2(u)\}E\{\mathbf{B}^2(v)\} + 2E^2\{\mathbf{B}(u)\mathbf{B}(v)\}$$
$$E\{\mathbf{A}^2(u)\mathbf{B}^2(v)\} = E\{\mathbf{A}^2(u)\}E\{\mathbf{B}^2(v)\} + 2E^2\{\mathbf{A}(u)\mathbf{B}(v)\}$$
$$E\{\mathbf{A}^2(v)\mathbf{B}^2(u)\} = E\{\mathbf{A}^2(v)\}E\{\mathbf{B}^2(u)\} + 2E^2\{\mathbf{A}(v)\mathbf{B}(u)\}$$

This leads to the conclusion that

$$E\{[\mathbf{A}^2(u) + \mathbf{B}^2(u)][\mathbf{A}^2(v) + \mathbf{B}^2(v)]\} - E\{\mathbf{A}^2(u) + \mathbf{B}^2(u)\}E\{\mathbf{A}^2(v) + \mathbf{B}^2(v)\}$$
$$= 2E^2\{\mathbf{A}(u)\mathbf{A}(v)\} + 2E^2\{\mathbf{B}(u)\mathbf{B}(v)\} + 2E^2\{\mathbf{A}(u)\mathbf{B}(v)\} + 2E^2\{\mathbf{A}(v)\mathbf{B}(u)\}$$

and since $|\mathbf{X}(\omega)|^2 = \mathbf{A}^2(\omega) + \mathbf{B}^2(\omega)$, the above yields [see (11-102)]

$$\text{cov}[|\mathbf{X}(u)|^2, |\mathbf{X}(v)|^2] = \tfrac{1}{2}[\Gamma_r(u, -v) + \Gamma_r(u, v)]^2 + \tfrac{1}{2}[\Gamma_r(u, -v) - \Gamma_r(u, v)]^2$$
$$+ \tfrac{1}{2}[\Gamma_i(u, v) - \Gamma_i(u, -v)]^2 + \tfrac{1}{2}[\Gamma_i(u, v) + \Gamma_i(u, v)]^2$$

and Eq. (11-105) results.

We note that, if $R(t_1, t_2) = R(-t_1, -t_2)$, then $\Gamma(u, v)$ is real. This is the case if $\mathbf{x}(t) = w(t)\mathbf{n}(t)$, where $w(t)$ is a real, even signal and $\mathbf{n}(t)$ is a real stationary process.

Fourier Transforms of Noisy Signals

We wish to find the Fourier transform

$$F(\omega) = \int_{-\infty}^{\infty} f(t)e^{-j\omega t}\, dt$$

of a deterministic signal $f(t)$ contaminated by additive noise. The available information is the sum

$$\mathbf{x}(t) = f(t) + \mathbf{n}(t)$$

where it is assumed that the noise $\mathbf{n}(t)$ is a stationary process with zero mean and power spectrum $S(\omega)$. Clearly,

$$E\{\mathbf{x}(t)\} = f(t)$$

Hence, the transform

$$\mathbf{X}(\omega) = \int_{-\infty}^{\infty} \mathbf{x}(t)e^{-j\omega t}\,dt = F(\omega) + \mathbf{N}(\omega) \tag{11-106}$$

is a process with mean

$$E\{\mathbf{X}(\omega)\} = F(\omega) \tag{11-107}$$

and autocovariance

$$\operatorname{cov}[\mathbf{X}(u), \mathbf{X}(u)] = E\{\mathbf{N}(u)\mathbf{N}^*(v)\} = 2\pi S(u)\,\delta(u - v) \tag{11-108}$$

The last equality follows from Eq. (11-93) because $\mathbf{N}(\omega)$ is the transform of the stationary process $\mathbf{n}(t)$.

As we see from (11-108), the variance of $\mathbf{X}(\omega)$ is infinite. Hence, $\mathbf{X}(\omega)$ cannot be used to estimate $F(\omega)$. For a satisfactory estimate, we must either truncate $\mathbf{x}(t)$ or smooth $\mathbf{X}(\omega)$.

Truncation We form the integral of the segment $p_T(t)\mathbf{x}(t)$ of $\mathbf{x}(t)$,

$$\mathbf{X}_T(\omega) = \int_{-T}^{T} \mathbf{x}(t)e^{j\omega t}\,dt = F_T(\omega) + \mathbf{N}_T(\omega) \tag{11-109}$$

where [see (3-30)]

$$F_T(\omega) = \int_{-T}^{T} f(t)e^{-j\omega t}\,dt = \int_{-\infty}^{\infty} F(\omega - y)\frac{\sin Ty}{\pi y}\,dy \tag{11-110}$$

Clearly, $\mathbf{X}_T(\omega)$ is a biased estimator of $F(\omega)$ with mean

$$E\{\mathbf{X}_T(\omega)\} = F_T(\omega)$$

and variance [see (11-98)]

$$\operatorname{var}[\mathbf{X}_T(\omega)] = E\{|\mathbf{N}_T(\omega)|^2\} = S(\omega) * \frac{2\sin^2 T\omega}{\pi\omega^2} \simeq 2TS(\omega) \tag{11-111}$$

because $\mathbf{N}_T(\omega)$ is the transform of the segment $p_T(t)\mathbf{n}(t)$ of the noise.

Thus, to reduce the variance of $\mathbf{X}_T(\omega)$, we must choose a small value for T. However, a small T results in a large bias $F_T(\omega) - F(\omega)$. The optimum T is determined as in Sec. 10-2.

Windows The estimation can be improved if we use as our estimator the transform

$$\mathbf{X}_w(\omega) = \int_{-\infty}^{\infty} w(t)\mathbf{x}(t)e^{-j\omega t}\, dt = F_w(\omega) + \mathbf{N}_w(\omega) \tag{11-112}$$

of the product $w(t)\mathbf{x}(t)$. Since $F_w(\omega)$ is the transform of $w(t)f(t)$ and the mean of $\mathbf{N}_w(\omega)$ is zero, we conclude from Eq. (11-112) that

$$E\{\mathbf{X}_w(\omega)\} = F_w(\omega) = \frac{1}{2\pi} W(\omega) * F(\omega) \tag{11-113}$$

and, as in (11-96),

$$\text{var}[\mathbf{X}_w(\omega)] = E\{|\mathbf{N}_w(\omega)|^2\} = \frac{1}{2\pi} |W(\omega)|^2 * S(\omega) \tag{11-114}$$

Applying Eq. (3-A-1) to the convolution integral in (11-114), we obtain

$$\text{var}[\mathbf{X}_w(\omega)] = \frac{S(\omega_1)}{2\pi} \int_{-\infty}^{\infty} |W(\omega)|^2\, d\omega \tag{11-115}$$

where ω_1 is some number near ω.

From the preceding discussion it follows that, for $\mathbf{X}(\omega)$ to be a satisfactory estimator of $F(\omega)$, the right side of (11-113) must be close to $F(\omega)$, and the right side of (11-115) close to zero. The first requirement is satisfied if $W(\omega)$ is of short duration and its area equals 2π; the second, if the energy of $W(\omega)$ is small. We can also assume that $f(t) = 0$ outside some interval. This assumption is essential if $\mathbf{x}(t)$ is known for $|t| < T$ only; it is desirable if $f(t) \simeq 0$ for $|t| > T$. In any case, it does not impose a restriction of substance because, for a reduction of the effect of noise, $w(t)$ must attenuate rapidly for large T. Furthermore, it simplifies the computations.

Note: In (11-112), the function $w(t)$ is independent of ω. This assumption is convenient because it permits the evaluation of $\mathbf{X}_T(\omega)$ with the economy of the FFT algorithm; however, it is not necessary.

Consider, for example, the average

$$\mathbf{X}_a(\omega) = \frac{1}{2a} \int_{\omega-a}^{\omega+a} \mathbf{X}(y)\, dy$$

of the transform $\mathbf{X}(\omega)$ of $\mathbf{x}(t)$ obtained from (11-112) with $w(t) = \sin at/at$. If a is chosen so as to minimize the resulting mean-square error as in Sec. 10-2, then its optimum value varies with ω; that is, the resulting window $w(t)$ is frequency-dependent.

Periodic Functions with Random Coefficients

Given a discrete process $\mathbf{x}[n]$ with mean $\eta[n]$ and autocorrelation $R[n_1, n_2]$, we form the sum

$$\mathbf{X}(\omega) = \sum_{n=-\infty}^{\infty} \mathbf{x}[n]e^{-jnT\omega} \tag{11-116}$$

This sum is a periodic, stochastic process with period $\omega_0 = 2\pi/T$, mean

$$E\{\mathbf{X}(\omega)\} = \sum_{n=-\infty}^{\infty} \eta[n]e^{-jnT\omega} \tag{11-117}$$

and autocorrelation

$$E\{\mathbf{X}(u)\mathbf{X}^*(v)\} = \sum_{n_1=-\infty}^{\infty} \sum_{n_2=-\infty}^{\infty} E\{\mathbf{x}[n_1]\mathbf{x}^*[n_2]\}e^{-j(n_1u-n_2v)T} \tag{11-118}$$

We shall discuss its properties briefly. With

$$\Gamma(u, v) = \sum_{n_1=-\infty}^{\infty} \sum_{n_2=-\infty}^{\infty} R[n_1, n_2]e^{-j(n_1u+n_2v)T} \tag{11-119}$$

a doubly periodic function with Fourier-series coefficients $R[n_1, n_2]$, it follows from Eq. (11-118) that

$$E\{\mathbf{X}(u)\mathbf{X}^*(v)\} = \Gamma(u, -v) \tag{11-120}$$

White noise If $R[n_1, n_2] = I[n_1]\,\delta[n_1 - n_2]$, then, with

$$J(y) = \sum_{n=-\infty}^{\infty} I[n]e^{-jnTy}$$

it follows from (11-119) that

$$\Gamma(u, v) = \sum_{n_1=-\infty}^{\infty} I[n_1]e^{-jn_1(u+v)T} = J(u+v) \tag{11-121}$$

Hence,

$$E\{\mathbf{X}(u)\mathbf{X}^*(v)\} = J(u-v) \tag{11-122}$$

Stationary processes If the process $\mathbf{x}[n]$ is stationary with autocorrelation $R[n]$ and power spectrum

$$\bar{S}(\omega) = \sum_{n=-\infty}^{\infty} R[n]e^{-jnT\omega} \tag{11-123}$$

then

$$E\{\mathbf{X}(u)\mathbf{X}^*(v)\} = \omega_0\,\bar{S}(u) \sum_{n=-\infty}^{\infty} \delta(u - v + n\omega_0) \tag{11-124}$$

Proof Since

$$\sum_{n_1=-\infty}^{\infty} R[n_1 - n_2]e^{-jn_1Tu} = e^{-jn_2Tu}\bar{S}(u) \tag{11-125}$$

(elaborate), it follows from (11-119) that

$$\Gamma(u, v) = \bar{S}(u) \sum_{n_2=-\infty}^{\infty} e^{-j(u+v)n_2 T} = \omega_0 \bar{S}(u) \sum_{n=-\infty}^{\infty} \delta(u + v + n\omega_0) \quad (11\text{-}126)$$

and (11-124) results from (11-120). The last equality in (11-126) is a consequence of the identity [see Eq. (1-58)]

$$\sum_{n=-\infty}^{\infty} e^{-jnT\omega} = \omega_0 \sum_{n=-\infty}^{\infty} \delta(\omega + n\omega_0)$$

Windows Given a periodic function

$$W(\omega) = \sum_{n=-\infty}^{\infty} w[n] e^{-jnT\omega} \quad (11\text{-}127)$$

and a stationary discrete process $\mathbf{x}[n]$, we form the process

$$\mathbf{x}_w[n] = w[n]\mathbf{x}[n]$$

Clearly, $\mathbf{x}_w[n]$ is a nonstationary process with autocorrelation

$$R_w[n_1, n_2] = w[n_1]w^*[n_2]R[n_1 - n_2] \quad (11\text{-}128)$$

We maintain that the corresponding function $\Gamma_w(u, v)$ is given by

$$\Gamma_w(u, v) = \frac{1}{\omega_0} \int_{\omega_0}^{\omega_0} W(u - y)W^*(-v - y)\bar{S}(y)\, dy \quad (11\text{-}129)$$

Proof From the convolution theorem (3-72), it follows that the Fourier-series coefficients of the function

$$\frac{1}{\omega_0} \int_{\omega_0}^{\omega_0} W(u - y)\bar{S}(y) e^{-jn_2 Ty} w^*[n_2]\, dy$$

equal the right side of (11-128), where n_2 is considered constant [see (11-125)]. Hence,

$$\Gamma_w(u, v) = \frac{1}{\omega_0} \int_{\omega_0}^{\omega_0} W(\omega - y)\bar{S}(y) \sum_{n_2=-\infty}^{\infty} w^*[n_2] e^{-jn_2 T(v+y)}\, dy$$

and (11-129) results because the last sum equals $W^*(-v - y)$.

From (11-129) and (11-120), it follows that, if

$$\mathbf{X}_w(\omega) = \sum_{n=-\infty}^{\infty} w[n]\mathbf{x}[n] e^{-jnT\omega} \quad (11\text{-}130)$$

then

$$E\{|\mathbf{X}_w(\omega)|^2\} = \frac{1}{\omega_0} \int_{\omega_0}^{\omega_0} |W(\omega - y)|^2 \bar{S}(y)\, dy \quad (11\text{-}131)$$

Special case The truncated sum

$$\mathbf{X}_N(\omega) = \sum_{n=-N}^{N} \mathbf{x}[n]e^{-jnT\omega} \tag{11-132}$$

is a special case of Eq. (11-130), where

$$w[n] = \begin{cases} 1 & |n| \le N \\ 0 & |n| > N \end{cases} \qquad W(\omega) = \sum_{n=-N}^{N} e^{-jnT\omega} = \frac{\sin\,(N+\frac{1}{2})T\omega}{\sin\,T\omega/2}$$

The preceding results show that digital and analog processes have similar properties. To avoid further repetition, we limit the discussion in the next chapter to analog signals. The reader can readily derive the corresponding digital version.

APPENDIX 11-A

Normal Processes and Cumulants

The statistical properties of a normal process $\mathbf{z}(t)$ are uniquely determined if its mean $\eta(t)$ and autocorrelation $R(t_1, t_2)$ are known. In particular, a moment M of any order can be expressed in terms of η and R. Suppose that an arbitrary process $\mathbf{w}(t)$ has the same mean η and autocorrelation R as $\mathbf{z}(t)$. In general, a moment M_w of $\mathbf{w}(t)$, of the same order as M, is different from M; the difference, $K = M_w - M$, known as the *cumulant* of $\mathbf{w}(t)$, gives some measure of the deviation of $\mathbf{w}(t)$ from normality.[1] We shall show that the effect of K on various integrals used in spectral estimation can be neglected if the size of the sample is sufficiently large.

Fourth-Order Moments of Normal Processes

It is well known that, if the random variables $\mathbf{z}_1$, $\mathbf{z}_2$, $\mathbf{z}_3$, and $\mathbf{z}_4$ are jointly normal with zero mean, then

$$E\{\mathbf{z}_1\mathbf{z}_2\mathbf{z}_3\mathbf{z}_4\} = E\{\mathbf{z}_1\mathbf{z}_2\}E\{\mathbf{z}_3\mathbf{z}_4\} + E\{\mathbf{z}_1\mathbf{z}_3\}E\{\mathbf{z}_2\mathbf{z}_4\} + E\{\mathbf{z}_1\mathbf{z}_4\}E\{\mathbf{z}_2\mathbf{z}_3\} \qquad (11\text{-A-}1)$$

With $\mathbf{z}_1 = \mathbf{z}_2 = \mathbf{z}$ and $\mathbf{z}_3 = \mathbf{z}_4 = \mathbf{x}$, the above yields

$$E\{\mathbf{z}^2\mathbf{x}^2\} = E\{\mathbf{z}^2\}E\{\mathbf{x}^2\} + 2E^2\{\mathbf{zx}\} \qquad (11\text{-A-}2)$$

Hence, the covariance of the random variables $\mathbf{z}^2$ and $\mathbf{x}^2$ is given by

$$\operatorname{cov}(\mathbf{z}^2, \mathbf{x}^2) = 2E^2\{\mathbf{zx}\} \qquad (11\text{-A-}3)$$

Setting $\mathbf{z}_1 = \cdots = \mathbf{z}_4$ in (11-A-1), we obtain

$$E\{\mathbf{z}^4\} = 3E^2\{\mathbf{z}^2\} \qquad (11\text{-A-}4)$$

We shall express the fourth-order moments of a real normal process $\mathbf{z}(t)$ with zero mean in terms of its correlation

$$E\{\mathbf{z}(t_i)\mathbf{z}(t_k)\} = R(t_i, t_k)$$

From (11-A-1) it follows that

$$\begin{aligned} E\{\mathbf{z}(t_1)\mathbf{z}(t_2)\mathbf{z}(t_3)\mathbf{z}(t_4)\} = {} & R(t_1, t_2)R(t_3, t_4) + R(t_1, t_3)R(t_2, t_4) \\ & + R(t_1, t_4)R(t_2, t_3) \end{aligned} \qquad (11\text{-A-}5)$$

Note: If the mean $\eta(t)$ of $\mathbf{z}(t)$ is not zero, then (11-A-5) still holds, provided that $\mathbf{z}(t)$ is replaced with the centered process $\mathbf{z}(t) - \eta(t)$.

[1] A. Papoulis, Narrow-Band Systems and Gaussianity, *IEEE Trans. Inf. Theory*, vol. IT-10, no. 1, pp. 20–27, 1972.

Cumulants Suppose that the random variables $\mathbf{z}$ and $\mathbf{w}$ have the same mean η and variance σ^2. If $\mathbf{z}$ is normal, then its central moments

$$\mu_n = E\{(\mathbf{z} - \eta)^n\}$$

are uniquely determined in terms of η and σ^2. This is, in general, not the case for $\mathbf{w}$. The difference

$$k_n = E\{(\mathbf{w} - \eta)^n\} - E\{(\mathbf{z} - \eta)^n\} \tag{11-A-6}$$

is the nth cumulant of $\mathbf{w}$. In particular,

$$k_3 = E\{(\mathbf{w} - \eta)^3\} \qquad k_4 = E\{(\mathbf{w} - \eta)^4\} - 3\sigma^4 \tag{11-A-7}$$

because $\mu_3 = 0$ and $\mu_4 = 3\sigma^4$ [see Eq. (11-A-4)]. The cumulants of several random variables can be defined similarly. We shall assume that the mean of all random variables under consideration is zero.

If $\mathbf{z}_i$ and $\mathbf{w}_i$ are two sequences of random variables such that

$$E\{\mathbf{z}_i \mathbf{z}_k\} = E\{\mathbf{w}_i \mathbf{w}_k\}$$

and the sequence $\mathbf{z}_i$ is normal, then the cumulant of $\mathbf{w}_1, \ldots, \mathbf{w}_4$ is defined by

$$k[\mathbf{w}_1, \ldots, \mathbf{w}_4] = E\{\mathbf{w}_1 \mathbf{w}_2 \mathbf{w}_3 \mathbf{w}_4\} - E\{\mathbf{z}_1 \mathbf{z}_2 \mathbf{z}_3 \mathbf{z}_4\} \tag{11-A-8}$$

The last term can be expressed in terms of $E\{\mathbf{w}_i \mathbf{w}_k\}$ [see (11-A-1)].

Suppose, finally, that $\mathbf{w}(t)$ is a random process with zero mean and autocorrelation $R(t_1, t_2)$. Denoting by $k(t_1, \ldots, t_4)$ the cumulant of the random variables $\mathbf{w}(t_1), \ldots, \mathbf{w}(t_4)$, we conclude from (11-A-8) that

$$\begin{aligned} k(t_1, t_2, t_3, t_4) = E\{\mathbf{w}(t_1)\mathbf{w}(t_2)\mathbf{w}(t_3)\mathbf{w}(t_4)\} - R(t_1, t_2)R(t_3, t_4) - R(t_1, t_3)R(t_2, t_4) \\ - R(t_1, t_4)R(t_2, t_3) \end{aligned} \tag{11-A-9}$$

Input-Output Moments and Cumulants

We have seen that, if $\mathbf{y}(t)$ is the output of a linear system with input $\mathbf{x}(t)$:

$$\mathbf{y}(t) = \mathbf{x}(t) * h(t)$$

then

$$E\{\mathbf{x}(t_1)\mathbf{y}(t_2)\} = E\{\mathbf{x}(t_1)\mathbf{x}(t_2)\} * h(t_2) \tag{11-A-10}$$

$$E\{\mathbf{y}(t_1)\mathbf{y}(t_2)\} = E\{\mathbf{x}(t_1)\mathbf{y}(t_2)\} * h(t_1)$$

This result can be readily extended to higher-order moments. For example, taking expected values of both sides of the identity

$$\mathbf{y}(t_1)\mathbf{x}(t_2)\mathbf{y}(t_3)\mathbf{x}(t_4) = \int_{-\infty}^{\infty} \mathbf{y}(t_1)\mathbf{x}(t_2)\mathbf{x}(t_3 - \tau)h(\tau)\mathbf{x}(t_4)\, d\tau$$

we obtain

$$E\{\mathbf{y}(t_1)\mathbf{x}(t_2)\mathbf{y}(t_3)\mathbf{x}(t_4)\} = E\{\mathbf{y}(t_1)\mathbf{x}(t_2)\mathbf{x}(t_3)\mathbf{x}(t_4)\} * h(t_3) \tag{11-A-11}$$

where the convolution is in the variable t_3.

In general, if an input-output joint moment contains the factor $\mathbf{x}(t_i)$ and is convolved with $h(t_i)$, the result equals the same order of moment with $\mathbf{x}(t_i)$ replaced by $\mathbf{y}(t_i)$.

Repeated application of the above leads to the determination of the moments of $\mathbf{y}(t)$ of any order in terms of the corresponding moments of $\mathbf{x}(t)$.

It can easily be seen that joint cumulants are determined similarly. We shall prove the cumulant version of (11-A-11):

$$k[\mathbf{y}(t_1), \mathbf{x}(t_2), \mathbf{y}(t_3), \mathbf{x}(t_4)] = k[\mathbf{y}(t_1), \mathbf{x}(t_2), \mathbf{x}(t_3), \mathbf{x}(t_4)] * h(t_3) \quad \text{(11-A-12)}$$

As we see from Eq. (11-A-9), the cumulants can be expressed in terms of second- and fourth-order moments. Furthermore, the variable t_3 appears once in each term. Therefore, (11-A-12) follows from (11-A-10) and (11-A-11) (elaborate).

Covariance of Sample Spectrum

For the determination of the covariance of the sample spectrum

$$\mathbf{S}_T(\omega) = \frac{1}{2T}\left|\int_{-T}^{T} \mathbf{w}(t)e^{-j\omega t}\,dt\right|^2$$

of a real process $\mathbf{w}(t)$, the integral

$$E\{\mathbf{S}_T(u)\mathbf{S}_T(v)\} = \frac{1}{4T^2}\int_V E\{\mathbf{w}(t_1)\mathbf{w}(t_2)\mathbf{w}(t_3)\mathbf{w}(t_4)\}e^{-j(ut_1 - ut_2 - vt_3 + vt_4)}\,dV \quad \text{(11-A-13)}$$

needs to be evaluated, where V is a four-dimensional cube with volume $16T^4$.

If the process $\mathbf{w}(t)$ is normal, then [see (11-A-5)] the integral in (11-A-13) can be expressed in terms of the autocorrelation $R(t_1, t_2)$ of $\mathbf{w}(t)$. Otherwise, we must add to the result of the normal case the integral

$$I = \frac{1}{4T^2}\int_V k(t_1, t_2, t_3, t_4)e^{-j(ut_1 - ut_2 - vt_3 + vt_4)}\,dV \quad \text{(11-A-14)}$$

[See (11-A-9).]

We shall show that if the process $\mathbf{w}(t)$ is a-dependent (see the footnote on page 302), then

$$I \to 0 \qquad \text{with } T \to \infty \quad \text{(11-A-15)}$$

It will then follow that if T is sufficiently large, then for the evaluation of the covariance of $\mathbf{S}_T(\omega)$ we can assume that the process $\mathbf{w}(t)$ is normal.

Lemma If the distance between any two neighboring points in the group t_1, t_2, t_3, t_4 is greater than a, then

$$k(t_1, t_2, t_3, t_4) = 0 \quad \text{(11-A-16)}$$

Proof Without loss of generality, we can assume that $t_1 \le t_2 \le t_3 \le t_4$. Suppose, first, that $t_2 > t_1 + a$. It then follows that the random variable $\mathbf{w}(t_1)$ is independent of the product $\mathbf{w}(t_2)\mathbf{w}(t_3)\mathbf{w}(t_4)$; hence,

$$E\{\mathbf{w}(t_1)\mathbf{w}(t_2)\mathbf{w}(t_3)\mathbf{w}(t_4)\} = E\{\mathbf{w}(t_1)\}E\{\mathbf{w}(t_2)\mathbf{w}(t_3)\mathbf{w}(t_4)\} = 0$$

$$R(t_1, t_2) = E\{\mathbf{w}(t_1)\mathbf{w}(t_2)\} = E\{\mathbf{w}(t_1)\}E\{\mathbf{w}(t_2)\} = 0$$

because $E\{\mathbf{w}(t)\} = 0$. Similarly, $R(t_1, t_3) = R(t_1, t_4) = 0$ and (11-A-16) results [see (11-A-9)].

If $t_3 > t_2 + a$, then the products $\mathbf{w}(t_1)\mathbf{w}(t_2)$ and $\mathbf{w}(t_3)\mathbf{w}(t_4)$ are independent; hence,

$$E\{\mathbf{w}(t_1)\mathbf{w}(t_2)\mathbf{w}(t_3)\mathbf{w}(t_4)\} = R(t_1, t_2)R(t_3, t_4)$$

Furthermore, $R(t_2, t_3) = 0$ and $R(t_1, t_4) = 0$; hence, (11-A-16) is again true. The last case, $t_4 > t_3 + a$, is provided similarly.

It can be seen that the points in the region V such that $|t_i - t_j| \le a$ form a set whose volume is less than $216a^3T$. Since $k = 0$ outside this set, it follows from (11-A-14) that

$$|I| < \frac{54a^3k_{\max}}{T} \xrightarrow[T\to\infty]{} 0$$

where $k_{\max}$ is the maximum of $|k|$ in V.

CHAPTER

TWELVE

SPECTRAL ESTIMATION

The most widely used concept in applications of random signals is the power spectrum. This concept is central in system theory, optics, spectroscopy, oil exploration, earthquake analysis, and many other areas. In such applications, a common problem is the estimation of the spectrum in terms of given data. In this chapter, we examine various aspects of the problem.[1,2] *We assume that the data consist of a single sample of finite length and that the estimator is a linear functional of the data. The resulting theory includes, we believe, the aspects of the problem that are basic and of general interest. Methods involving multisample averaging or extrapolation techniques (maximum entropy) are either direct extensions or of rather special interest.*

In Sec. 12-1, we define the problem. In Sec. 12-2, we explain the need for smoothing the sample spectrum, and we discuss the properties of the various estimators. The proofs of the results are given in Sec. 12-3.

---

[1] R. B. Blackman and J. W. Tukey, "The Measurement of Power Spectra," Dover Publications, Inc., New York, 1959.

[2] G. M. Jenkins and D. G. Watts, "Spectral Analysis and Its Applications," Holden-Day, Inc., Publisher, San Francisco, 1968.

12-1 SAMPLE SPECTRUM

We shall consider the problem of estimating the power spectrum $S(\omega)$ of a real, stationary process $\mathbf{x}(t)$ in terms of a single realization of $\mathbf{x}(t)$ available over a finite interval $(-T, T)$. Since $S(\omega)$ is not defined directly in terms of $\mathbf{x}(t)$ but is the transform of the autocorrelation $R(\tau)$ of $\mathbf{x}(t)$, it can be estimated with the transform of the estimate of $R(\tau)$. We shall start with this approach.

In Sec. 11-2, we used, as estimates of $R(\tau)$, the time averages

$$\mathbf{R}^T(\tau) = \frac{1}{2T - |\tau|} \int_{-T+|\tau|/2}^{T-|\tau|/2} \mathbf{x}\left(t + \frac{\tau}{2}\right)\mathbf{x}\left(t - \frac{\tau}{2}\right) dt \tag{12-1}$$

and

$$\mathbf{R}_T(\tau) = \frac{1}{2T} \int_{-T+|\tau|/2}^{T-|\tau|/2} \mathbf{x}\left(t + \frac{\tau}{2}\right)\mathbf{x}\left(t - \frac{\tau}{2}\right) dt \tag{12-2}$$

These functions are defined as above for $|\tau| < 2T$; for $|\tau| > 2T$, it is assumed that they are zero. Their transforms are thus given by

$$\mathbf{S}^T(\omega) = \int_{-2T}^{2T} \mathbf{R}^T(\tau) e^{-j\omega\tau} \, d\tau \tag{12-3}$$

$$\mathbf{S}_T(\omega) = \int_{-2T}^{2T} \mathbf{R}_T(\tau) e^{-j\omega\tau} \, d\tau \tag{12-4}$$

respectively.

With $p_{2T}(t)$ and $q_{2T}(t)$ a pulse and a triangle, as in Fig. 12-1, it follows from Eqs. (12-1) and (12-2) that

$$E\{\mathbf{R}^T(\tau)\} = R(\tau) p_{2T}(\tau) \tag{12-5}$$

$$E\{\mathbf{R}_T(\tau)\} = R(\tau)\left(1 - \frac{|\tau|}{2T}\right) p_{2T}(\tau) = R(\tau) q_{2T}(\tau) \tag{12-6}$$

Hence,

$$E\{\mathbf{S}^T(\omega)\} = \int_{-2T}^{2T} R(\tau) e^{-j\omega\tau} \, d\tau = S(\omega) * \frac{\sin 2T\omega}{\pi\omega} \tag{12-7}$$

$$E\{\mathbf{S}_T(\omega)\} = \int_{-2T}^{2T} R(\tau) q_{2T}(\tau) e^{-j\omega\tau} \, d\tau = S(\omega) * \frac{\sin^2 T\omega}{\pi T\omega^2} \tag{12-8}$$

Thus, the mean of $\mathbf{S}^T(\omega)$ equals the convolution of $S(\omega)$ with the Fourier kernel $\sin 2T\omega/\pi\omega$ (rectangular window), and the mean of $\mathbf{S}_T(\omega)$ equals the convolution of $S(\omega)$ with the Fejér kernel $\sin^2 T\omega/\pi T\omega^2$ (Bartlett window). Both estimators are, therefore, biased. However,

$$E\{\mathbf{S}^T(\omega)\} \underset{T\to\infty}{\to} S(\omega) \qquad \text{and} \qquad E\{\mathbf{S}_T(\omega)\} \underset{T\to\infty}{\to} S(\omega) \tag{12-9}$$

as is easy to see. Clearly,

$$\mathbf{S}_T(\omega) = \mathbf{S}^T(\omega) * \frac{\sin^2 T\omega}{\pi T\omega^2} \tag{12-10}$$

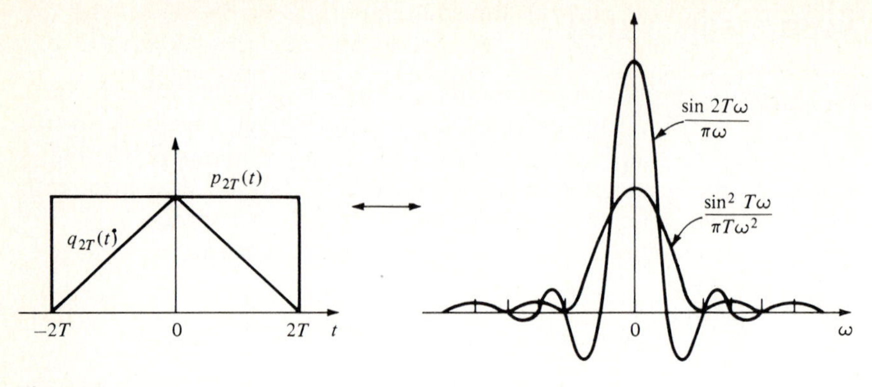

Figure 12-1

because

$$\mathbf{R}_T(\tau) = \mathbf{R}^T(\tau)q_{2T}(\tau) \tag{12-11}$$

Hence, for sufficiently large T, the sample spectra $\mathbf{S}^T(\omega)$ and $\mathbf{S}_T(\omega)$ are approximately equal. We shall continue the analysis in terms of $\mathbf{S}_T(\omega)$ only. We do so because, unlike $\mathbf{S}^T(\omega)$, the spectrum $\mathbf{S}_T(\omega)$ can be determined directly from the given sample of $\mathbf{x}(t)$ [see Eq. (11-63)]:

$$\mathbf{S}_T(\omega) = \frac{1}{2T}\left|\int_{-T}^{T} \mathbf{x}(t)e^{-j\omega t}\,dt\right|^2 \tag{12-12}$$

The variance problem From (12-8), it follows that if T is sufficiently large, then

$$E\{\mathbf{S}_T(\omega)\} \simeq S(\omega)$$

However, as we show later, the variance of $\mathbf{S}_T(\omega)$ is not small. In fact, for any T,

$$\text{var}[\mathbf{S}_T(\omega)] \geq E^2\{\mathbf{S}_T(\omega)\}$$

[See (12-40).] Therefore, $\mathbf{S}_T(\omega)$ is not a good estimator of $S(\omega)$, no matter how large T is. As we noted on page 362, the reason is that the values of the integrand $\mathbf{R}_T(\tau)$ in (12-4) are not reliable (have large variance) for τ close to $\pm 2T$. Thus, the power spectrum $S(\omega)$ of a process $\mathbf{x}(t)$ cannot be determined from a single sample, no matter how large the sample is. To reduce the variance of the estimate, we must accept only a smoothed version of $S(\omega)$; that is, we must sacrifice resolution.

12-2 SMOOTHED SPECTRUM

To reduce the variance of the integral in Eq. (12-4), we must deemphasize the contribution of $\mathbf{R}_T(\tau)$ for τ near $\pm 2T$. For this purpose, we form the estimator

$$\mathbf{S}_w(\omega) = \int_{-2T}^{2T} \mathbf{R}_T(\tau)w(\tau)e^{-j\omega\tau}\,d\tau \tag{12-13}$$

where $w(\tau)$ is a window as in Sec. 7-3, vanishing for $|\tau| > 2T$. In this section, we examine the properties of $\mathbf{S}_w(\omega)$ and the factors affecting the selection of the window $w(\tau)$. Various statements are made without proofs; the proofs are given in the next section.

The estimator $\mathbf{S}_w(\omega)$ is the Fourier transform of the product $\mathbf{R}_T(\tau)w(\tau)$; hence,

$$\mathbf{S}_w(\omega) = \frac{1}{2\pi}\int_{-\infty}^{\infty} \mathbf{S}_T(\omega - y)W(y)\,dy = \frac{1}{2\pi}\mathbf{S}_T(\omega) * W(\omega) \tag{12-14}$$

From the above and (12-8) it follows that

$$E\{\mathbf{S}_w(\omega)\} = \frac{1}{2\pi}E\{\mathbf{S}_T(\omega)\} * W(\omega) = \frac{1}{2\pi}S(\omega) * \frac{\sin^2 T\omega}{\pi T\omega^2} * W(\omega) \tag{12-15}$$

We shall show that, for a reliable estimation, the duration of $W(\omega)$ must be large compared to $1/T$. This leads to the approximation

$$\frac{\sin^2 T\omega}{\pi T\omega^2} * W(\omega) \simeq W(\omega)$$

Inserting into (12-15), we obtain

$$E\{\mathbf{S}_w(\omega)\} \simeq \frac{1}{2\pi}S(\omega) * W(\omega) \tag{12-16}$$

In the next section, we show that, under certain general conditions, the variance of $\mathbf{S}_w(\omega)$ is given by

$$\operatorname{var}[\mathbf{S}_w(\omega)] \simeq \frac{E_w}{2T}S^2(\omega) \qquad \omega \neq 0 \tag{12-17}$$

where
$$E_w = \int_{-2T}^{2T} w^2(t)\,dt = \frac{1}{2\pi}\int_{-\infty}^{\infty} W^2(\omega)\,d\omega$$

is the energy of $w(t)$ [see (12-66)].

Equations (12-16) and (12-17) dictate the factors affecting the selection of the window pair $w(t) \leftrightarrow W(\omega)$: For the bias

$$b = E\{\mathbf{S}_w(\omega)\} - S(\omega)$$

to be small, $W(\omega)$ must be of short duration. For the variance to be small, E_w must be small. We shall presently see that if T is sufficiently large, then both requirements can be satisfied.

Window Selection

For a satisfactory estimation of $S(\omega)$, the variance of the estimator $\mathbf{S}_w(\omega)$ must be small compared to $S^2(\omega)$ or, equivalently, the *variance ratio*

$$\beta = \frac{\operatorname{var}[\mathbf{S}_w(\omega)]}{S^2(\omega)}$$

must be small compared to 1:

$$\beta \ll 1 \tag{12-18}$$

This is the case if [see (12-17)] the energy of $w(t)$ is small compared to $2T$:

$$E_w \simeq 2T\beta \ll 2T$$

The above requirement leads to the conclusion that $w(t)$ must take significant values only in an interval $(-M, M)$ such that $M \ll 2T$. We shall assume that $|w(t)| \le 1$ for all t and that, for $|t| > M$, it is not just small but it vanishes:

$$w(t) = 0 \qquad \text{for } |t| > M \tag{12-19}$$

From these assumptions, it follows that

$$E_w \le 2M \qquad \text{so that } \beta \le \frac{M}{T} \tag{12-20}$$

Thus, to satisfy the variance requirement (12-18), we must choose M such that

$$M \ll T \tag{12-21}$$

With M so determined, the shape of the window is selected so as to minimize the bias

$$b = \frac{1}{2\pi}\int_{-\infty}^{\infty} S(\omega - y)W(y)\,dy - S(\omega) \tag{12-22}$$

Notes:

1. We have seen that it is impossible to determine the value of $S(\omega)$ precisely from a single sample, no matter how large. However, the average of $S(\omega)$ over a small interval can be reliably estimated by suitable smoothing. If, therefore, $S(\omega)$ is continuous at $\omega = \omega_1$ and T is large, then $S(\omega_1)$ can be evaluated with any desired accuracy. However, the presence of an impulse in $S(\omega)$ (line spectrum) cannot be determined directly. We mention without elaboration that it can be inferred from the form of $\mathbf{S}_w(\omega)$ obtained with several windows of different widths.
2. The assumption that $w(t) = 0$ for $|t| > M$ is not necessary but merely convenient. It permits us to satisfy the variance requirement (12-18) in a simple manner and to apply the considerations of Sec. 7-3 to the selection of $W(\omega)$. This assumption restricts the class of signals from which $w(t)$ is selected. However, the restriction is not severe, because for a reliable estimate, E_w must be small compared to T. Hence, $w(t)$ must take significant values near the origin only. If, therefore, it is truncated for $|t| > M$, then the effect on the variance will be small.

   In fact, if the smoothing is performed directly in terms of $W(\omega)$ by the frequency-domain convolution (12-14), it is not even necessary to assume that its inverse $w(t)$ is zero for $|t| > T$. It suffices to find a function $W(\omega)$ of specified energy E_w such as to yield a minimum bias. The bias b

depends not only on $W(\omega)$ but also on the shape of $S(\omega)$. Therefore, there is no well-defined optimum window. However, if T is sufficiently large and

$$W(\omega) \geq 0 \tag{12-23}$$

then [see Eq. (7-66)]

$$b \simeq \frac{S''(\omega)}{4\pi} \int_{-\infty}^{\infty} \omega^2 W(\omega)\, d\omega \tag{12-24}$$

and the problem is to find a positive function $W(\omega)$ of specified energy E_w and minimum second moment m_2. As we noted, the condition that $w(t)$ is time limited need not be imposed, although it is convenient.

3. As we have shown in Sec. 7-3, the optimum $W(\omega)$ minimizing the bias b is the function in (7-64). For this reason, we have called this function the *minimum-bias window*.
4. The integral (12-14) might yield an estimator $\mathbf{S}_w(\omega)$ that takes negative values for some ω. However, if $W(\omega) \geq 0$, then $\mathbf{S}_w(\omega) \geq 0$ for all ω.

Measurement constraints In certain applications, the known quantity is not the signal $\mathbf{x}(t)$ itself, but its sample autocorrelation

$$\mathbf{R}_a(\tau) = \frac{1}{a} \int_0^a \mathbf{x}(t)\mathbf{x}(t-\tau)\, dt \tag{12-25}$$

measured with some analog device. For example, if $\mathbf{x}(t)$ is an optical signal, then its instantaneous values cannot be measured easily with ordinary instruments; however, the time average $\mathbf{R}_a(\tau)$ can be readily determined with an interferometer.

In such applications, the time of observation a is sufficiently large so that the variance of $\mathbf{R}_a(\tau)$ can be neglected. This leads to the conclusion that the measured quantity $\mathbf{R}_a(\tau)$ equals its expected value:

$$\mathbf{R}_a(\tau) \simeq E\{\mathbf{R}_a(\tau)\} = R(\tau) \tag{12-26}$$

for every observable τ. The variance problem is thus eliminated. However, owing to measurement restrictions, the sample autocorrelation $\mathbf{R}_a(\tau)$ can be determined for $|\tau| \leq M$ only, where M depends on the measuring device. In the Michelson interferometer, for example, the constant M is proportional to the maximum displacement of the moving mirror; in the stellar interferometer, its value depends on the size of the viewing lens.

In such cases, we are faced with the deterministic problem of computing the transform $S(\omega)$ of $R(\tau)$ in terms of the finite segment $\mathbf{R}_a(\tau)p_M(\tau) \simeq R(\tau)p_M(\tau)$. This problem was considered in Sec. 7-3.

Numerical methods in spectral estimation The discussion of this section leads to the following numerical methods for estimating the power spectrum $S(\omega)$ in terms of the given segment

$$\mathbf{x}_T(t) = \mathbf{x}(t)p_T(t)$$

These methods are statistically equivalent, differing only in the nature of the computations.

1. We determine the sample autocorrelation $\mathbf{R}_T(\tau)$ by convolving $\mathbf{x}_T(t)$ with $\mathbf{x}_T(-t)$ as in (12-2):

$$\mathbf{R}_T(\tau) = \frac{1}{2T}\mathbf{x}_T(\tau) * \mathbf{x}_T(-\tau)$$

   We multiply $\mathbf{R}_T(\tau)$ by the window $w(\tau)$ and compute the Fourier transform of the product as in Eq. (12-13):

$$\mathbf{R}_T(\tau)w(\tau) \leftrightarrow \mathbf{S}_w(\omega)$$

   Thus, for the determination of the estimator $\mathbf{S}_w(\omega)$ of $S(\omega)$ by this method, the required operations are one convolution, one multiplication, and one Fourier transform.
2. We compute the Fourier transform of $\mathbf{x}_T(t)$:

$$\mathbf{x}_T(t) \leftrightarrow \mathbf{X}_T(\omega)$$

   We multiply $\mathbf{X}_T(\omega)$ by its conjugate and form the sample spectrum $\mathbf{S}_T(\omega)$ as in (12-12):

$$\mathbf{S}_T(\omega) = \frac{1}{2T}|\mathbf{X}_T(\omega)|^2$$

   We convolve $\mathbf{S}_T(\omega)$ with the window $W(\omega)$ as in (12-14):

$$\mathbf{S}_w(\omega) = \frac{1}{2\pi}\mathbf{S}_T(\omega) * W(\omega)$$

   The required operations are one Fourier transform, one multiplication, and one convolution.
3. We compute $\mathbf{X}_T(\omega)$ and $\mathbf{S}_T(\omega)$ as in method 2. We find the inverse Fourier transform $\mathbf{R}_T(\tau)$ of $\mathbf{S}_T(\omega)$. We form the product $\mathbf{R}_T(\tau)w(\tau)$ and compute its Fourier transform $\mathbf{S}_w(\omega)$ as in method 1. The required operations are two multiplications, one Fourier transform, and one inverse Fourier transform.

   Note: The third method involves only multiplications and Fourier transforms. It therefore avoids convolutions and can be carried out with the economy of FFT's.

Data Windows

As we have seen, the sample spectrum $\mathbf{S}_T(\omega)$ is a biased estimator of $S(\omega)$ because its mean equals the convolution $S(\omega)$ with the Bartlett window [see

(12-10)]. To reduce its bias, we introduce the modified sample

$$\mathbf{S}_c(\omega) = \left| \int_{-T}^{T} c(t)\mathbf{x}(t)e^{-j\omega t}\, dt \right|^2 \tag{12-27}$$

The factor $c(t)$ we shall call a *data window*. It will be assumed that $c(t)$ is real and even, and $c(t) = 0$ for $|t| > T$. If $c(t)$ is a rectangular pulse of unit energy,

$$c(t) = \frac{1}{\sqrt{2T}}\, p_T(t) \tag{12-28}$$

then [see (12-12)] $\mathbf{S}_c(\omega) = \mathbf{S}_T(\omega)$.

We show in the next section that the mean of $\mathbf{S}_c(\omega)$ is given by [see (12-37)]

$$E\{\mathbf{S}_c(\omega)\} = \frac{1}{2\pi}\, S(\omega) * C^2(\omega) \tag{12-29}$$

The window pair $c(t) \leftrightarrow C(\omega)$ must therefore be chosen so as to minimize the resulting bias

$$b_c = E\{\mathbf{S}_c(\omega)\} - S(\omega) = \frac{1}{2\pi} \int_{-\infty}^{\infty} S(\omega - y)C^2(y)\, dy - S(\omega) \tag{12-30}$$

From the above it follows that $c(t)$ must be such that

$$\int_{-T}^{T} c^2(t)\, dt = \frac{1}{2\pi} \int_{-\infty}^{\infty} C^2(\omega)\, d\omega = 1 \tag{12-31}$$

because only then does $b_c \to 0$ as $T \to \infty$.

Truncated cosine Since $C^2(\omega) \geq 0$, we conclude as in Eq. (7-66) that

$$b_c = \frac{M_2}{2}\, S''(\omega_1) \qquad \text{where} \qquad M_2 = \frac{1}{2\pi} \int_{-\infty}^{\infty} \omega^2 C^2(\omega)\, d\omega \tag{12-32}$$

If the term $\omega^2 C^2(\omega)$ takes significant values only in a small interval, and if $S''(\omega)$ is sufficiently smooth, then $S''(\omega_1) \simeq S''(\omega)$. This assumption leads to the conclusion that to minimize b_c, it suffices to choose $c(t)$ such that the energy moment M_2 is minimum. As we noted [see (7-63)], the optimum $c(t)$ is a truncated cosine of unit area (Fig. 12-2):

$$c(t) = \frac{1}{\sqrt{T}}\, p_T(t) \cos \frac{\pi}{2T}\, t \tag{12-33}$$

The resulting minimum bias is given by

$$b_c \simeq \frac{\pi^2}{2T^2}\, S''(\omega)$$

as is easy to see from (12-32).

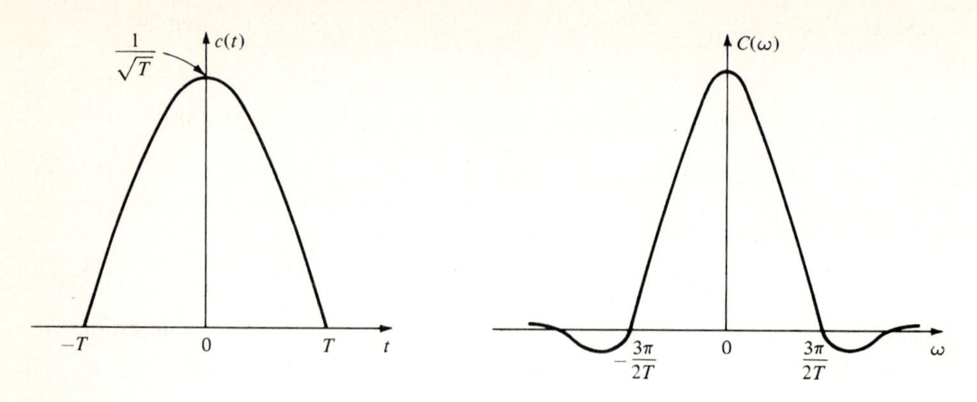

Figure 12-2

12-3 THEORY

We shall determine the second-order moments of the sample spectrum $\mathbf{S}_T(\omega)$ and smoothed spectrum $\mathbf{S}_w(\omega)$ in terms of the autocorrelation $R(\tau)$ of the given process $\mathbf{x}(t)$. We shall assume, for simplicity, subtracting a constant if necessary, that the mean of $\mathbf{x}(t)$ is zero.

In general, the variance of $\mathbf{S}_T(\omega)$ cannot be evaluated in terms of $R(\tau)$ only because it involves fourth-order moments of $\mathbf{x}(t)$. However, if the process $\mathbf{x}(t)$ is normal, then knowledge of $R(\tau)$ is sufficient. We shall carry out the analysis with this simplifying assumption, noting that if the size T of the sample is sufficiently large, then the results also hold for arbitrary processes satisfying only certain mild conditions (see Appendix 11-A).

We start with the modified sample spectrum $\mathbf{S}_c(\omega)$ as defined in Eq. (12-27). This slight generalization does not complicate the analysis. The corresponding results for $\mathbf{S}_T(\omega)$ are presented at the end of the section as a special case.

As we see from (12-27),

$$\mathbf{S}_c(\omega) = |\mathbf{X}_c(\omega)|^2 \tag{12-34}$$

where

$$\mathbf{X}_c(\omega) = \int_{-T}^{T} c(t)\mathbf{x}(t)e^{-j\omega t}\,dt \tag{12-35}$$

Thus, $\mathbf{S}_c(\omega)$ is the energy spectrum of the process $c(t)\mathbf{x}(t)$; hence, we can apply the results of Sec. 11-3. With

$$\Gamma_c(u, v) = \frac{1}{2\pi}\int_{-\infty}^{\infty} C(u + v - \alpha)C(\alpha)S(v - \alpha)\,d\alpha \tag{12-36}$$

as in (11-83), we conclude from (11-96) that

$$E\{\mathbf{S}_c(\omega)\} = E\{|\mathbf{X}_c(\omega)|^2\} = \Gamma_c(\omega, -\omega) = \frac{1}{2\pi}C^2(\omega) * S(\omega) \tag{12-37}$$

because $C(\omega)$ is even.

Covariance As we see from (11-105), the covariance of $\mathbf{S}_c(\omega)$ is given by

$$\text{cov}[\mathbf{S}_c(u), \mathbf{S}_c(v)] = \Gamma_c^2(u, -v) + \Gamma_c^2(u, v) \tag{12-38}$$

The variance of $\mathbf{S}_c(\omega)$ is obtained by setting $u = v = \omega$ in Eq. (12-38):

$$\text{var}[\mathbf{S}_c(\omega)] = \Gamma_c^2(\omega, -\omega) + \Gamma_c^2(\omega, \omega) \tag{12-39}$$

Fundamental note From (12-39) and (12-37), it follows that

$$\text{var}[\mathbf{S}_c^2(\omega)] \geq \Gamma_c^2(\omega, -\omega) = E^2\{\mathbf{S}_c(\omega)\} \tag{12-40}$$

for any T and $c(t)$. Hence, the sample spectrum $\mathbf{S}_c(\omega)$ is not a reliable estimator of $S(\omega)$, no matter how large T is.

Smoothing

To reduce the variance of the estimator, we convolve $\mathbf{S}_c(\omega)$ with the window $W(\omega)$. The resulting $\mathbf{S}_w(\omega)$ is given by

$$\mathbf{S}_w(\omega) = \frac{1}{2\pi} \int_{-\infty}^{\infty} \mathbf{S}_c(\omega - y) W(y)\, dy \tag{12-41}$$

Mean From the above and (12-37), it follows that

$$E\{\mathbf{S}_w(\omega)\} = \frac{1}{4\pi^2} S(\omega) * C^2(\omega) * W(\omega) \tag{12-42}$$

Covariance The right side of (12-41) is a convolution integral similar to (9-37). Hence, the autocovariance of $\mathbf{S}_w(\omega)$ is given by (9-55), suitably modified (elaborate):

$$\text{cov}[\mathbf{S}_w(u), \mathbf{S}_w(v)] = \frac{1}{4\pi^2} \int_{-\infty}^{\infty} \int_{-\infty}^{\infty} \text{cov}[\mathbf{S}_c(u - \alpha), \mathbf{S}_c(v - \beta)] W(\alpha) W(\beta)\, d\alpha\, d\beta \tag{12-43}$$

From Eqs. (12-36) and (12-38), it follows that the above integral depends on $C(\omega)$, $W(\omega)$, and $S(\omega)$. As we show next, the resulting expression can be simplified if T is large.

Large T

Consider the window pairs

$$c(t) \leftrightarrow C(\omega) \qquad w(t) \leftrightarrow W(\omega) \tag{12-44}$$

Since $c(t) = 0$ for $|t| > T$, the duration of $C(\omega)$ is of the order of $1/T$. If $w(t) = 0$ for $|t| > M$, then the duration of $W(\omega)$ is of the order of $1/M$. The assumption that $c(t) = 0$ for $|t| > T$ is essential. The assumption that $w(t) = 0$ for $|t| < M$ is not essential because, in the evaluation of $\mathbf{S}_w(\omega)$ from (12-41),

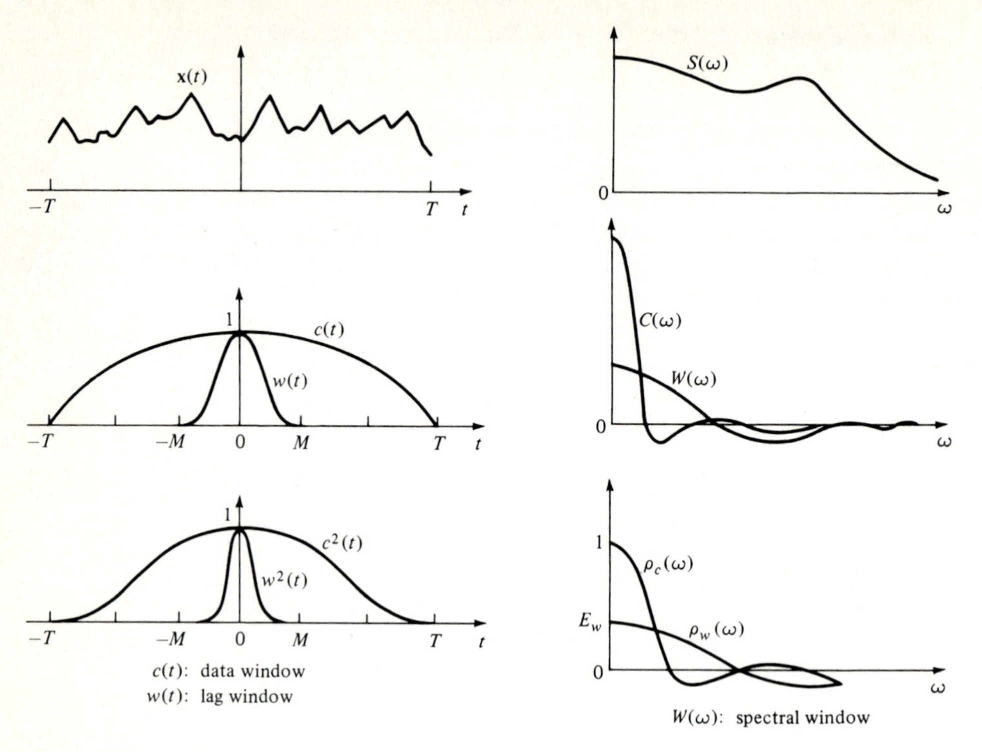

Figure 12-3

no a priori restrictions on $W(\omega)$ are necessary. However, as we pointed out on page 382, this assumption is convenient.

For a reliable estimate of $S(\omega)$, the bias $E\{\mathbf{S}_w(\omega)\} - S(\omega)$ must be small compared to $S(\omega)$, and the variance of $\mathbf{S}_w(\omega)$ must be small compared to $S^2(\omega)$. The first requirement leads to the conclusion that T must be large enough so that, in any interval of the order of $1/T$, the variation of $S(\omega)$ is negligible. The second requirement is satisfied if $M \ll T$ [see (12-21)]. We shall continue the analysis under the assumption that both requirements are satisfied.

In Fig. 12-3 we show the given segment of $\mathbf{x}(t)$, the spectrum $S(\omega)$, and the two window pairs in (12-44). We note that, since $M \ll T$, the function $C(\omega)$ is of short duration relative to $W(\omega)$. In the same figure, we plot also the functions $c^2(t)$ and $w^2(t)$ and their transforms

$$\rho_c(\omega) = \int_{-T}^{T} c^2(t)e^{-j\omega t}\,dt = \frac{1}{2\pi}\int_{-\infty}^{\infty} C(\omega - \alpha)C(\alpha)\,d\alpha \tag{12-45}$$

$$\rho_w(\omega) = \int_{-M}^{M} w^2(t)e^{-j\omega t}\,dt = \frac{1}{2\pi}\int_{-\infty}^{\infty} W(\omega - \alpha)W(\alpha)\,d\alpha \tag{12-46}$$

The assumption that $C(\omega)$ is of short duration relative to $S(\omega)$ leads to the approximation

$$\Gamma_c(u, v) \simeq \frac{S(v)}{2\pi}\int_{-\infty}^{\infty} C(u + v - \alpha)C(\alpha)\,d\alpha = S(v)\rho_c(u + v) \qquad (12\text{-}47)$$

This results from Eq. (12-36) because the integral contains the factor $C(\alpha)$ whose duration is short; hence, the term $S(v - \alpha)$ can be replaced with $S(v)$.

We introduce also the energies E_c and E_w of the signals $c(t)$ and $w(t)$, respectively, [see (12-31)]

$$\rho_c(0) = \int_{-T}^{T} c^2(t)\,dt = E_c = 1 \qquad \rho_w(0) = \int_{-M}^{M} w^2(t)\,dt = E_w \qquad (12\text{-}48)$$

and the integral

$$I_4 = \int_{-T}^{T} c^4(t)\,dt = \frac{1}{2\pi}\int_{-\infty}^{\infty} \rho_c^{\,2}(\omega)\,d\omega \qquad (12\text{-}49)$$

Mean Since $C(\omega)$ is of short duration, we conclude from (12-37) that

$$E\{\mathbf{S}_c(\omega)\} \simeq \frac{S(\omega)}{2\pi}\int_{-\infty}^{\infty} C^2(\omega)\,d\omega = S(\omega) \qquad (12\text{-}50)$$

Furthermore,

$$\frac{1}{2\pi}C^2(\omega) * W(\omega) \simeq W(\omega)$$

Hence [see (12-42)],

$$E\{\mathbf{S}_w(\omega)\} \simeq \frac{1}{2\pi}S(\omega) * W(\omega) \qquad (12\text{-}51)$$

Thus, for large T, $\mathbf{S}_c(\omega)$ is an unbiased estimator, and the bias of $\mathbf{S}_w(\omega)$ is due to smoothing only.

Covariance of sample spectrum Introducing the approximation (12-47) into Eq. (12-38), we obtain

$$\text{cov}[\mathbf{S}_c(u), \mathbf{S}_c(v)] \simeq S^2(v)[\rho_c^{\,2}(u - v) + \rho_c^{\,2}(u + v)] \qquad (12\text{-}52)$$

With $u = v = \omega$, the above yields

$$\text{var}[\mathbf{S}_c(\omega)] \simeq S^2(\omega)[1 + \rho_c^{\,2}(2\omega)] \qquad (12\text{-}53)$$

Hence,

$$S^2(\omega) \le \text{var}[\mathbf{S}_c(\omega)] \le 2S^2(\omega) \qquad (12\text{-}54)$$

because

$$|\rho_c(\omega)| \le \rho_c(0) = 1 \qquad (12\text{-}55)$$

as is easily seen from (12-45).

We note, finally, that since $\rho_c(2\omega)$ is negligible for $|\omega| \gg 1/T$,

$$\operatorname{var}[\mathbf{S}_c(\omega)] = \begin{cases} S^2(\omega) & |\omega| \gg \dfrac{1}{T} \\ 2S^2(\omega) & \omega = 0 \end{cases} \tag{12-56}$$

Covariance of smoothed spectrum We shall show that

$$\operatorname{cov}[\mathbf{S}_w(u), \mathbf{S}_w(v)] \simeq I_4 S^2(v)[\rho_w(u-v) + \rho_w(u+v)] \tag{12-57}$$

Proof From (12-52) and (12-43) it follows that

$$\operatorname{cov}[\mathbf{S}_w(u), \mathbf{S}_w(v)] \simeq \frac{1}{4\pi^2}\int_{-\infty}^{\infty}\int_{-\infty}^{\infty} S^2(v-\beta)[\rho_c{}^2(u-\alpha-v+\beta) + \rho_c{}^2(u-\alpha+v-\beta)]W(\alpha)W(\beta)\,d\alpha\,d\beta \tag{12-58}$$

Since $\rho_c(\omega)$ is of short duration relative to $W(\omega)$, we conclude with $y = u - \alpha - v + \beta$ that

$$\int_{-\infty}^{\infty} \rho_c{}^2(u-\alpha-v+\beta)W(\alpha)\,d\alpha = \int_{-\infty}^{\infty} \rho_c{}^2(y)W(u-v+\beta-y)\,dy \simeq W(u-v+\beta)\int_{-\infty}^{\infty} \rho_c{}^2(y)\,dy \tag{12-59}$$

Similarly,

$$\int_{-\infty}^{\infty} \rho_c{}^2(u-\alpha+v-\beta)W(\alpha)\,d\alpha \simeq W(u+v-\beta)\int_{-\infty}^{\infty} \rho_c{}^2(y)\,dy \tag{12-60}$$

Inserting (12-59) and (12-60) into (12-58), we obtain

$$\operatorname{cov}[\mathbf{S}_w(u), \mathbf{S}_w(v)] \simeq \frac{I_4}{2\pi}\int_{-\infty}^{\infty} S^2(v-\beta)[W(u-v+\beta) + W(u+v-\beta)]W(\beta)\,d\beta \tag{12-61}$$

because

$$\frac{1}{2\pi}\int_{-\infty}^{\infty} \rho_c{}^2(y)\,dy = I_4$$

The factor $W(\beta)$ in (12-61) takes significant values for β of the order of $1/M$ only. Assuming, therefore, that $S(\omega)$ is nearly constant in any interval of length $1/M$, we can use the approximation $S(v-\beta) \simeq S(v)$ in (12-61). The result is (12-57), as we can see from Eq. (12-46) and the fact that $W(-\omega) = W(\omega)$.

Variance Reasoning as in (12-53) and (12-56), we conclude from (12-57) that

$$\operatorname{var}[\mathbf{S}_w(\omega)] \simeq I_4 S^2(\omega)[\rho_w(0) + \rho_w(2\omega)] \simeq \begin{cases} 2I_4 E_w S^2(\omega) & \omega = 0 \\ I_4 E_w S^2(\omega) & |\omega| \gg \dfrac{1}{M} \end{cases} \tag{12-62}$$

where $E_w = \rho_w(0)$ is the energy of the window $w(t)$.

Constant Data Window

The sample spectrum

$$\mathbf{S}_T(\omega) = \frac{1}{2T}\left|\int_{-T}^{T} \mathbf{x}(t)e^{-j\omega t}\,dt\right|^2 \tag{12-63}$$

is a special case of $\mathbf{S}_c(\omega)$. Hence, its moments and the moments of the smoothed spectrum

$$\mathbf{S}_w(\omega) = \frac{1}{2\pi}\mathbf{S}_T(\omega) * W(\omega) \tag{12-64}$$

can be obtained from the preceding results, with

$$c(t) = \frac{1}{\sqrt{2T}}\,p_T(t) \qquad C(\omega) = \sqrt{\frac{2}{T}}\frac{\sin T\omega}{\omega}$$

$$c^2(t) = \frac{1}{2T}\,p_T(t) \qquad \rho_c(\omega) = \frac{\sin T\omega}{T\omega}$$

$$I_4 = \int_{-T}^{T} c^4(t)\,dt = \frac{1}{2T}$$

Thus, from Eq. (12-37) it follows that

$$E\{\mathbf{S}_T(\omega)\} = S(\omega) * \frac{\sin^2 T\omega}{\pi T\omega^2} \tag{12-65}$$

The autocovariances of $\mathbf{S}_T(\omega)$ and $\mathbf{S}_w(\omega)$ are given by (12-38) and (12-43), respectively, where now [see (12-36)]

$$\Gamma_c(u, v) = \int_{-\infty}^{\infty} \frac{\sin T(u+v-\alpha)}{\pi T(u+v-\alpha)}\frac{\sin\alpha}{\alpha}S(v-\alpha)\,d\alpha$$

If T is large, then

$$\operatorname{var}[\mathbf{S}_w(\omega)] \simeq \begin{cases} \dfrac{E_w}{T}S^2(\omega) & \omega = 0 \\[2ex] \dfrac{E_w}{2T}S^2(\omega) & |\omega| \gg \dfrac{1}{M} \end{cases} \tag{12-66}$$

because $I_4 = 1/2T$.

Notes:

1. With

$$\mathbf{X}_T(\omega) = \int_{-T}^{T} \mathbf{x}(t)e^{-j\omega t}\,dt$$

the transform of the given segment of $\mathbf{x}(t)$, Eq. (12-64) yields

$$\mathbf{S}_w(\omega) = \frac{1}{4\pi T} |\mathbf{X}_T(\omega)|^2 * W(\omega)$$

From (12-35) and the convolution theorem, it follows that

$$\mathbf{S}_c(\omega) = \frac{1}{4\pi^2} |\mathbf{X}_T(\omega) * C(\omega)|^2$$

As we see from (12-66), the variance of $\mathbf{S}_w(\omega)$ tends to zero as $T \to \infty$. However, this is not the case for the sample spectrum $\mathbf{S}_c(\omega)$ [see (12-56)]. Thus, for a satisfactory estimation of the power spectrum $S(\omega)$, we must smooth not the Fourier transform $\mathbf{X}_T(\omega)$ but the energy spectrum $|\mathbf{X}_T(\omega)|^2$ of the given segment of $\mathbf{x}(t)$.

2. The function $\mathbf{S}_c(\omega)$ is the Fourier transform of the modified sample autocorrelation

$$\mathbf{R}_c(\tau) = \int_{-T+|\tau|/2}^{T-|\tau|/2} c\left(t+\frac{\tau}{2}\right)c\left(t-\frac{\tau}{2}\right)\mathbf{x}\left(t+\frac{\tau}{2}\right)\mathbf{x}\left(t-\frac{\tau}{2}\right) dt$$

because $\mathbf{S}_c(\omega)$ is the energy spectrum of the process

$$\mathbf{x}_c(t) = c(t)\mathbf{x}(t)$$

[see (12-35)] and

$$\mathbf{R}_c(\tau) = \mathbf{x}_c(\tau) * \mathbf{x}_c(-\tau)$$

as is easily seen.

The smoothed spectrum $\mathbf{S}_w(\omega)$ can thus be expressed as a Fourier integral [see (12-41)]

$$\mathbf{S}_w(\omega) = \int_{-2T}^{2T} \mathbf{R}_c(\tau)w(\tau)e^{-j\omega\tau}\, d\tau$$

Hence, its mean and variance can be derived with the techniques of Sec. 11-3. The approach used in this chapter, however, is more direct.

3. We note, finally, that $w(t)$ is often called the *lag window*, and its transform $W(\omega)$ the *spectral window*.

PROBLEMS

Chapters 1 and 2

1. Show that the discrete convolution of the sequences $f_1[n] = U[n] - U[n-6]$ and $f_2[n] = U[n] - U[n-11]$ is the sequence $g[n]$ in Fig. P-1.

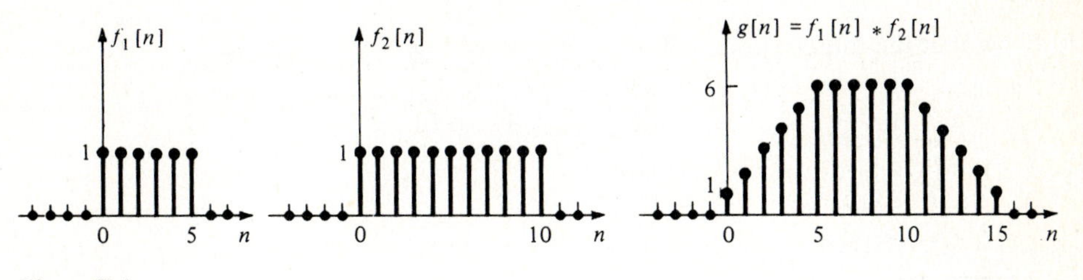

Figure P-1

2. (*a*) The response of a linear system to the input $f[n]$ equals

$$\sum_{k=n-3}^{n+2} f[k]$$

Find $h[n]$ and $H(z)$.

(*b*) Find $g[n] = f[n] * h[n]$ if

$$f[n] = \sum_{k=-\infty}^{\infty} \delta[n+kN]$$

$h[n] = a^n U[n]$, $|a| < 1$.

3. Show that

(*a*) $f_1(t) * f_2(t) = \int_{-\infty}^{\infty} f_1\left(\frac{t}{2} + \tau\right) f_2\left(\frac{t}{2} - \tau\right) d\tau$

(*b*) If $f_1(t) = 0$ and $f_2(t) = 0$ for $|t| > T$

then
$$f_1(t) * f_2(t) = \int_{-T+|t|/2}^{T-|t|/2} f_1\left(\frac{t}{2} + \tau\right) f_2\left(\frac{t}{2} - \tau\right) d\tau$$

4. Show that, if $f(t) = p_T(t) \cos(\pi t/2T)$, then

$$f(t) * f(t) = T p_{2T}(t)\left[\frac{1}{\pi}\left|\sin\frac{\pi t}{2T}\right| + \left(1 - \frac{|t|}{2T}\right)\cos\frac{\pi t}{2T}\right]$$

5. If $f(t) = e^t$ and $h(t) = (3e^{-2t} - 1)U(t)$, find $g(t) = f(t) * h(t)$.

6. The impulse response of a linear system equals $\delta(t) + 2e^{-3t}U(t)$. Find its system function $H(\omega)$. Find the response to $4\cos^2 2t$.

7. The input to the ideal filter $H(\omega) = 2p_a(\omega)e^{-jt_0\omega}$ is an impulse train (Fig. P-7). Find the resulting response $g(t)$ if $a = 10^4$ and $T = 10^{-3}$.

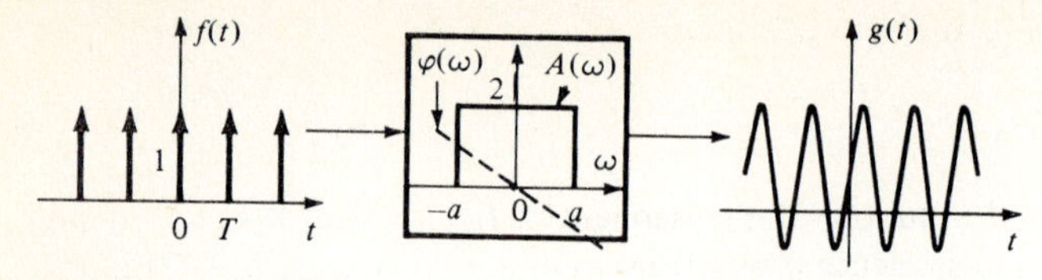

Figure P-7

8. Show that the impulse response of the system of Fig. P-8 is given by

$$h(t) = \sum_{n=0}^{\infty} a^n \, \delta(t - nT)$$

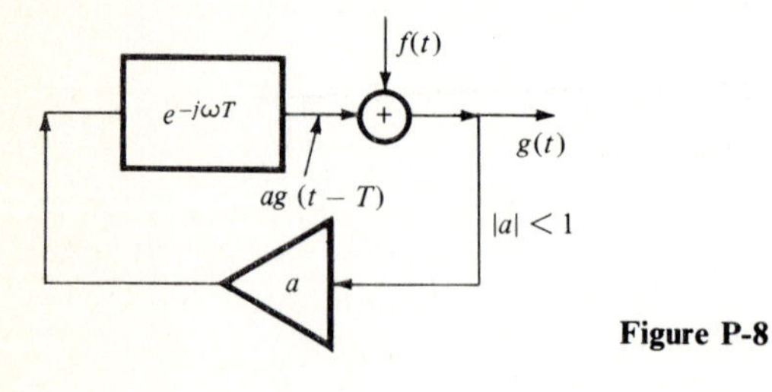

Figure P-8

9. Find the digital simulator of the analog system $H_a(\omega) = (1 + e^{-jc\omega})^2$ for bandlimited inputs with $\sigma = 2\pi/c$.

10. For the simulation of a differentiator $H_a(\omega) = j\omega$, the following systems are often used:

$$g_1[n] = \frac{1}{T} f[n] - \frac{1}{T} f[n-1] \qquad g_2[n] + g_2[n-1] = \frac{2}{T} f[n] - \frac{2}{T} f[n-1]$$

Find the corresponding system functions $H_1(z)$ and $H_2(z)$; plot their frequency responses.

11. Find the causal inverse transforms of the functions

$$F_1(z) = \frac{z^3}{z-1} \qquad F_2(z) = \frac{4z^2 + 8z}{4z^2 - 5z + 1} \qquad F_3(z) = \frac{4}{z^3(2z-1)}$$

12. Find all inverse transforms of the function

$$F(z) = \frac{z}{(z-1)^2(z-2)}$$

13. (*a*) Show that, if $f[n] \leftrightarrow F(z)$, then $nf[n] \leftrightarrow -zF'(z)$.
(*b*) Apply the above to the pair $a^n U[n] \leftrightarrow z/(z-a)$.

14. Find the z transforms of the sequences

$$f_1[n] = \begin{cases} 1 & 0 \le n \le N-1 \\ 0 & \text{otherwise} \end{cases} \qquad f_2[n] = \begin{cases} n & 0 \le n \le N-1 \\ 0 & \text{otherwise} \end{cases}$$

15. (*a*) Show that, if $|a| < 1$ and $f[n] = a^{|n|}$, then

$$F(z) = \frac{z(1-a^2)}{-az^2 + (1+a^2)z - a} \qquad |a| < |z| < \frac{1}{|a|}$$

(*b*) If $f(t) = e^{-\alpha|t|}$ and $f[n] = f(nT) \leftrightarrow F(z)$, find $F(e^{j\omega T})$.

16. (*a*) If $f[n] \leftrightarrow F(z)$, $f_1[n] \leftrightarrow F_1(z)$, and $F_1(z) = F(z^N)$, express $f_1[n]$ in terms of $f[n]$.
(*b*) Use part *a* to find the causal inverse of the function $F_1(z) = z^N/(z^N - 1)$.

17. Show that, if $g[n] = 0$ for $n < 0$ and $g[n + N] = g[n]$ for $n \geq 0$, then

$$G(z) = \frac{z^N}{z^N - 1} \sum_{n=0}^{N-1} g[n]z^{-n} \qquad |z| > 1$$

18. Show that, if $F(e^{j\theta}) = \sum_{n=0}^{\infty} f[n]e^{-jn\theta}$, then

$$F(z) = \frac{1}{2\pi}\int_{-\pi}^{\pi} F(e^{j\theta})\,\frac{z}{z - e^{j\theta}}\,d\theta \qquad |z| > 1$$

19. Show that if $f(t)$ is a function with Fourier transform $F(\omega)$, then, for any T,

$$\sum_{n=0}^{\infty} f(nT)z^{-n} = \frac{1}{2\pi}\int_{-\infty}^{\infty} F(\omega)\,\frac{z}{z - e^{j\omega T}}\,d\omega \qquad |z| > 1$$

20. The equation $g[n] - \frac{1}{2}g[n-1] = f[n] - 2f[n-1]$ defines a causal system with input $f[n]$ and output $g[n]$. Find its system function $H(z)$. Find $g[n]$ if $f[n] = U[n]$. Realize the system.

21. Find $g[n]$ such that $g[0] = \lambda$ and $g[n] - (1 - \lambda)g[n-1] = \lambda$ for $n > 0$.

22. (*a*) Given a function $f(t)$ with Fourier transform $F(\omega)$ such that $F(\omega) = 0$ for $|\omega| > \sigma = \pi/T$, we form the sequence $w[n] = f(nT)$ and its z transform $W(z)$. Show that for $|\omega| < \sigma$, $F(\omega) = TW(e^{j\omega T})$.
(*b*) The input to a discrete system $H(z)$ is the above sequence $w[n]$. With $g[n]$ the resulting response, show that

$$g[n] = \frac{1}{2\pi}\int_{-\sigma}^{\sigma} F(\omega)H(e^{j\omega T})e^{jnT\omega}\,d\omega$$

23. (Hanning windows) Find the delta response $h[n]$, the system function $H(z)$, and the frequency response $H(e^{j\omega T})$ of each of the following systems:
(*a*) $g[n] = \frac{1}{2}f[n+1] + f[n] + \frac{1}{2}f[n-1]$
(*b*) $g[n] = -\frac{1}{2}f[n+2] + f[n] - \frac{1}{2}f[n-2]$

24. Write a recursion equation for the system of Fig. P-24.

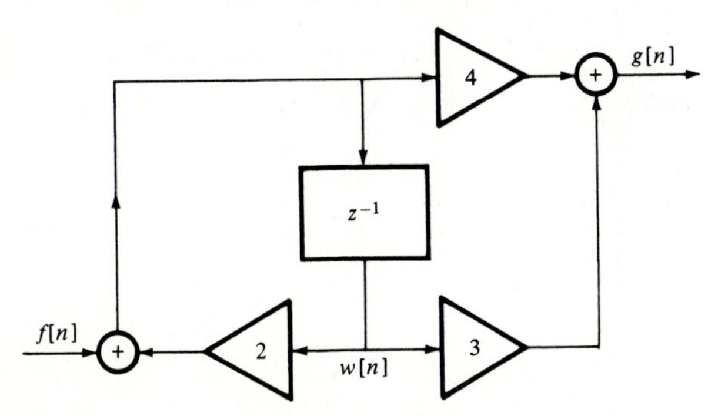

Figure P-24

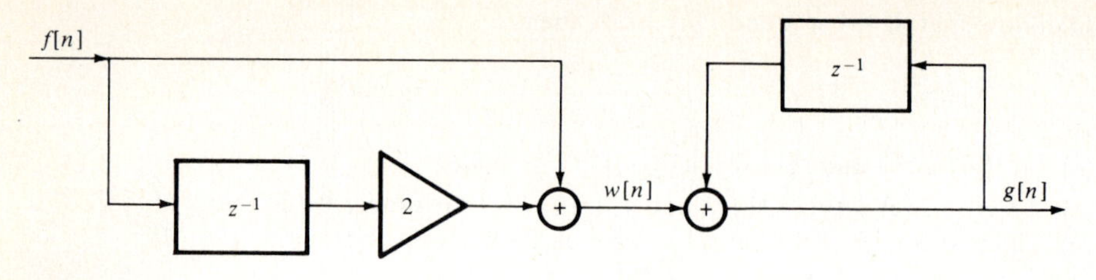

Figure P-25

25. Realize the system of Fig. P-25 using only one delay element. Find $g[n]$ if $f[n] = U[n+2]$.

26. (Continued-fraction expansion) Show that the system functions of the two systems in Figs. P-26*a* and *b*

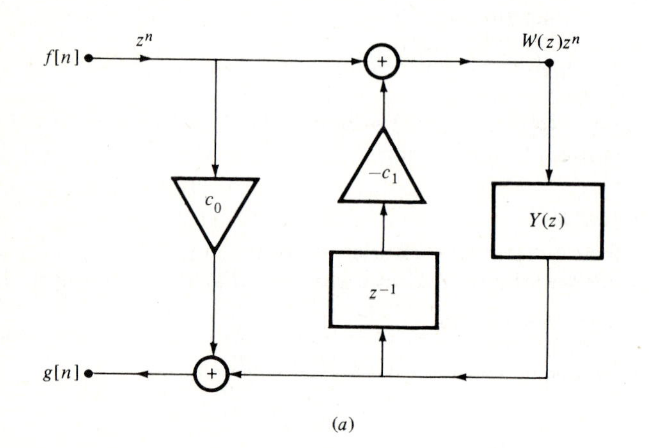

(*a*)

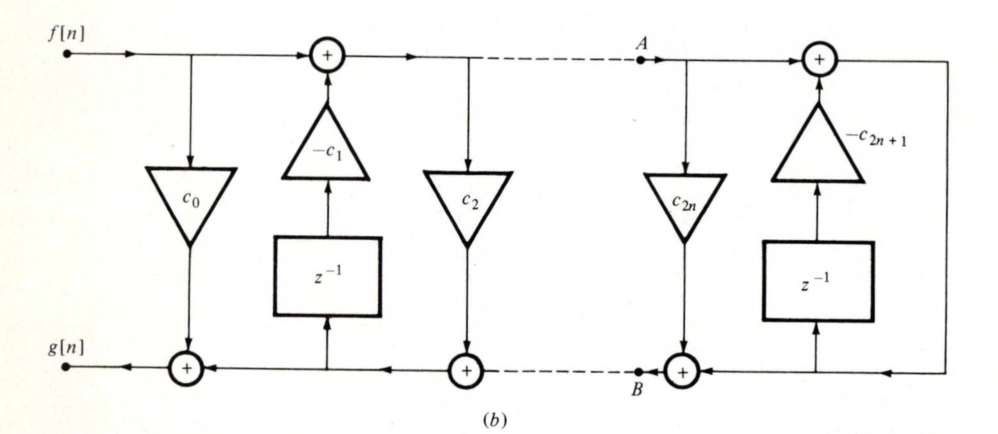

(*b*)

Figure P-26

are

$$H_a(z) = c_0 + \cfrac{1}{c_1 z^{-1} + \cfrac{1}{Y(z)}} \qquad H_b(z) = c_0 + \cfrac{1}{c_1 z^{-1} + \cfrac{1}{c_2 + \cfrac{1}{c_3 z^{-1} + \cdots + \cfrac{1}{c_{2n+1} z^{-1} + 1}}}}$$

27. Find a minimum-phase function $H(z)$ such that

$$|H(e^{j\omega T})| = \frac{\sqrt{\epsilon}}{\sqrt{\epsilon + \sin^4 (\omega T/2)}}$$

28. The stability of a system is often defined as follows: If the input $f[n]$ is bounded, then the output $g[n] = f[n] * h[n]$ is also bounded. Show that a system is stable in this sense iff the sum $I = \sum_{n=-\infty}^{\infty} |h[n]|$ is finite.

29. (*a*) Show that $f[n] * z^n = F(z)z^n$.
(*b*) Using part *a*, prove the convolution theorem for z transforms.

Chapter 3

30. (*a*) Show that $f(t) * e^{j\omega t} = F(\omega)e^{j\omega t}$.
(*b*) Using part *a*, prove the convolution theorem for Fourier transforms.

31. Show that, if α is small enough, then

$$f(t) * e^{j\alpha t^2} \simeq [F(2\alpha t) - j\alpha F''(2\alpha t)]e^{j\alpha t^2} \tag{1}$$

32. Show that

$$e^{-\alpha|t|}(1 + \alpha|t|) \leftrightarrow \frac{4\alpha^3}{(\alpha^2 + \omega^2)^2}$$

33. Using Parseval's formula, show that

$$\int_{-\infty}^{\infty} \frac{dt}{(t^2 + a^2)(t^2 + b^2)} = \frac{\pi}{ab(a + b)}$$

34. Show that

(*a*) If $f(t) \leftrightarrow F(\omega)$ then $f'(t) * \dfrac{1}{\pi t} \leftrightarrow |\omega| F(\omega)$

(*b*) $-\dfrac{1}{\pi t^2} \leftrightarrow |\omega|$

35. Show that, if $\rho(t) = f(t) * f^*(-t) \leftrightarrow |F(\omega)|^2$, then

(*a*) $\displaystyle\int_{-\infty}^{\infty} |f(t + a) - f(t)|^2 \, dt = 2 \text{ Re } [\rho(0) - \rho(a)]$

(*b*) (Holography principle) For any A and a,

$$|A|^2 \, \delta(t) + A^* f(t + a) + Af^*(-t + a) + \rho(t) \leftrightarrow |Ae^{-ja\omega} + F(\omega)|^2$$

36. Show that if $f(t)$ is a real, causal function and $f(t) \leftrightarrow R(\omega) + jX(\omega)$, then

$$\int_0^\infty f^2(t)\,dt = \frac{2}{\pi}\int_0^\infty R^2(\omega)\,d\omega = \frac{2}{\pi}\int_0^\infty X^2(\omega)\,d\omega$$

37. The functions $a(t)$ and $\varphi(t)$ are real, and $a(t)e^{j\varphi(t)} \leftrightarrow F(\omega)$. Express the transform of $a(t)\cos\varphi(t)$ in terms of $F(\omega)$.

38. Prove that, if $f_1(t) = f(t)U(t)$, $f_2(t) = f(-t)U(t)$, and $g(t) = f_1(t) + jf_2(t)$, then $2F(\omega) = G(\omega) - jG(-\omega) + G^*(-\omega) + jG^*(\omega)$. This shows that the transform $F(\omega)$ of a *real* function $f(t)$ can be expressed in terms of the transform $G(\omega)$ of a *causal* function $g(t)$.

39. Show that

$$|\cos\omega_0 t| = -\frac{2}{\pi}\sum_{n=-\infty}^{\infty}\frac{(-1)^n e^{j2n\omega_0 t}}{4n^2-1} \qquad |\sin\omega_0 t| = -\frac{2}{\pi}\sum_{n=-\infty}^{\infty}\frac{e^{j2n\omega_0 t}}{4n^2-1}$$

40. Show that

$$\cot\frac{\omega_0 t}{2} = 2\sum_{n=1}^{\infty}\sin n\omega_0 t \tag{2}$$

41. If

$$f(t) = \sum_{n=-\infty}^{\infty} a_n e^{jn\omega_0 t}$$

and $a_0 = 0$, then the function $g(t) = \int_{-T/2}^{t} f(\tau)\,d\tau$ is periodic. Show that

$$g(t) = \sum_{n=-\infty}^{\infty} b_n e^{jn\omega_0 t} \qquad b_0 = \int_{-T/2}^{T/2} g(t)\,dt \qquad b_n = \frac{a_n}{jn\omega_0} \qquad n \neq 0 \tag{3}$$

Special case If

$$f(t) = T\sum_{n=-\infty}^{\infty}\delta(t+nT) - 1$$

then $g(t)$ is a saw tooth as in Fig. P-41, and Eq. (3) yields $b_0 = 0$ and $b_n = 1/jn\omega_0$ for $n \neq 0$.

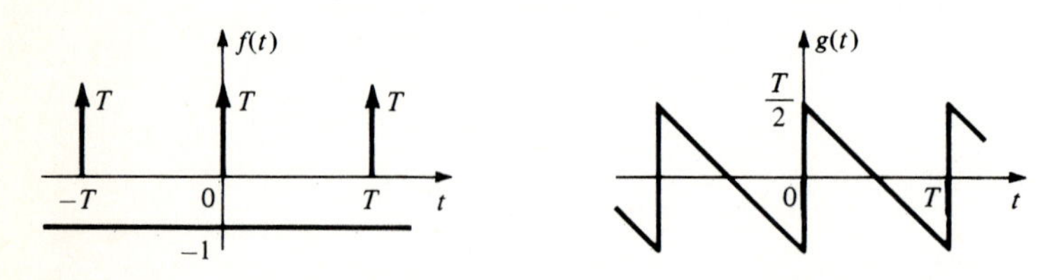

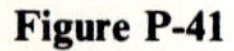

Figure P-41

42. Using Parseval's formula and Prob. 41, show that

$$I_1 = \sum_{n=1}^{\infty}\frac{1}{n^2} = \frac{\pi^2}{6} \qquad I_2 = \sum_{n=1}^{\infty}\frac{1}{(2n+1)^2} = \frac{\pi^2}{8}$$

43. Using the shifting theorem (3-20), show that if $f(t) \leftrightarrow F(\omega)$ and $f(t+T) = f(t)$, then $F(\omega) = 0$ for $\omega \neq 2\pi n/T$.

44. Find the transform $F(\omega)$ and the first three moments of the Poisson density

$$f(t) = e^{-a} \sum_{n=0}^{\infty} \frac{a^n}{n!} \delta(t - n)$$

45. Show that

$$\sum_{n=-\infty}^{\infty} e^{-\alpha T|n|} \delta(t - nT) \leftrightarrow \frac{1 - e^{-2\alpha T}}{1 - 2e^{-\alpha T} \cos \omega T + e^{-2\alpha T}} \tag{4}$$

46. Find the transforms of the functions

$$f_1(t) = \bar{\delta}(t) * [p_a(t) \cos \omega_1 t] \qquad f_2(t) = [\bar{\delta}(t) * p_a(t)] \cos \omega_1 t$$

where $\bar{\delta}(t)$ is an impulse train with period T as in (3-60).

47. Show that if $f(t) = \int_{t-a}^{t+a} \varphi(\tau)\, d\tau$ then

$$\sum_{n=-\infty}^{\infty} f(t + 2na) = \text{constant}$$

48. Given $M + 1$ real numbers, a_n, show that if $N = 2M + 1$ and

$$A_m = a_0 + 2 \sum_{n=1}^{M} a_k \cos \frac{2\pi}{N} mn$$

then

$$a_n = \frac{A_0}{N} + \frac{2}{N} \sum_{m=1}^{M} A_m \cos \frac{2\pi}{N} mn \tag{5}$$

49. Show that (Fig. P-49)

$$a_n = \begin{cases} 1 & |n| \le k \\ 0 & |n| > k \end{cases} \underset{N}{\leftrightarrow} A_m = \frac{\sin [(2k + 1)\pi m/N]}{\sin (\pi m/N)}$$

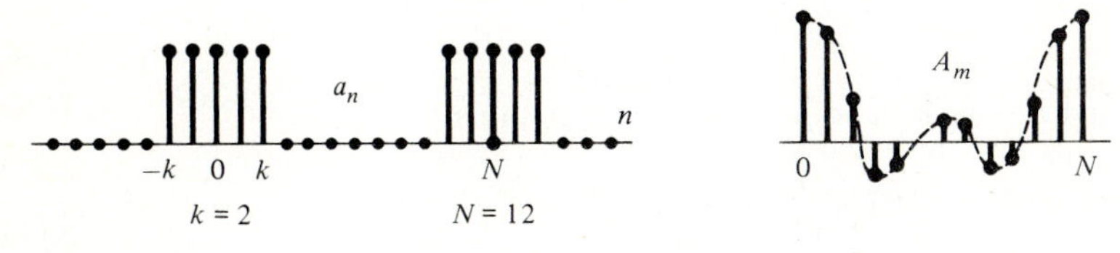

Figure P-49

50. Show that, if $a_n \underset{3N}{\leftrightarrow} A_m$ and

$$b_n = a_{3n} \underset{N}{\leftrightarrow} B_m \qquad c_n = a_{3n+1} \underset{N}{\leftrightarrow} C_m \qquad d_n = a_{3n+2} \underset{N}{\leftrightarrow} D_m$$

then

$$A_m = B_m + w_{3N}{}^m C_m + w_{3N}{}^{2m} D_m \tag{6}$$

51. Show that if $a_n \leftrightarrow A_m$, then $A_{-n} \leftrightarrow N a_m$.

52. From Eq. (3-103), it follows that, if

$$Y(z) = \sum_{n=0}^{N-1} a_n z^{-n}$$

is the z transform of the truncated sequence a_n and $a_n \underset{N}{\leftrightarrow} A_m$, then $A_m = Y(w^{-m})$. Using this and Prob. 14, show that if $a_n = n$, then $A_m = N/(w^m - 1)$ for $m > 0$.

53. Find the DFS a_n of order $N = 20$ of the samples $y(mT_1)$ of $y(t) = 5 + 4 \cos \omega_0 t + 2 \sin \omega_0 t$, where $T_1 = 2\pi/20\omega_0$.

54. (*a*) Find the coefficients $\bar{c}_n$ [see (3-92)] of the function $y(t) = \cos k\omega_0 t$, where k is an integer between 0 and $N/2$.

(*b*) Find a_n such that $a_n \leftrightarrow \cos(2\pi km/N)$.

(*c*) Using (3-125), show that

$$\sum_{m=0}^{N-1} \cos^2 \frac{2\pi k}{N} m = \frac{N}{2}$$

55. Using the pair $\delta^{(n)}(t) \leftrightarrow (j\omega)^n$, show that

$$f(t)\,\delta^{(n)}(t) = \sum_{k=0}^{n} (-1)^k \binom{n}{k} f^{(k)}(0)\,\delta^{(n-k)}(t)$$

Special case $f(t)\,\delta'(t) = f(0)\,\delta'(t) - f'(0)\,\delta(t)$.

Chapters 4 and 5

56. Show that, if $h(t)$ is the impulse response of the filter $H(\omega) = p_a(\omega)e^{j\varphi(\omega)}$, then

$$\int_{-\infty}^{\infty} th(t)\,dt = -\varphi'(0) \qquad \int_{-\infty}^{\infty} th^2(t)\,dt = -\frac{1}{2\pi}\int_{-a}^{a} \varphi'(\omega)\,d\omega$$

57.[1] (Deconvolution) We wish to express the impulse response $h(t)$ of a linear system in terms of its output $g(t)$ and the moments m_n of the input $f(t)$. We assume that $f(t)$ is of short duration relative to $h(t)$. Using the expansion

$$F(\omega) = m_0 - j\omega m_1 - \frac{\omega^2}{2} m_2 + \frac{j\omega^3}{6} m_3 + \cdots$$

show that

(*a*) If $m_0 \neq 0$, and $\eta = m_1/m_0$ and $\mu_2 = m_2 - m_1{}^2/m_0$, then

$$h(t) \simeq \frac{1}{m_0} g(t + \eta) - \frac{\mu_2}{2m_0{}^2} g''(t + \eta) \tag{7}$$

(*b*) If $m_0 = 0$, and $a = m_2/2m_1$ and $b = m_3 - 3m_2{}^2/4m_1$, then

$$h(t) \simeq \frac{1}{m_1}\int_{-\infty}^{t+a} g(\tau)\,d\tau + \frac{b}{6m_1{}^2} g'(t + a) \tag{8}$$

58. Show that if $g(t) = f(t) * h(t)$, $|f'(t)| < M$, and $\int_{-\infty}^{\infty} h(t)\,dt = 1$, then

$$|g(t) - f(t)| \leq M \int_{-\infty}^{\infty} |th(t)|\,dt \tag{9}$$

[1] A. Papoulis, Approximations of Point Spreads for Deconvolution, *J. Opt. Soc. Am.*, vol. 62, no. 1, pp. 77–80, January, 1972.

59. (*a*) Show that the Hilbert transform $\hat{f}(t)$ of an even function $f(t)$ is odd.
(*b*) Show that if $f(t) = \sin at/\pi t$, then $\hat{f}(t) = (1 - \cos at)/\pi t$.

60. Using the pairs

$$e^{s_i t}U(t) \leftrightarrow \frac{1}{s - s_i} \qquad \text{Re } s > \text{Re } s_i$$

$$-e^{s_i t}U(-t) \leftrightarrow \frac{1}{s - s_i} \qquad \text{Re } s < \text{Re } s_i$$

find all the inverse transforms of the function $2\alpha/(\alpha^2 - s^2)$.

61. Prove Schwarz' inequality (4-A-1) for complex signals.

62. Show that if

$$f_i(t) = \sin \sigma t \sum_{n=-\infty}^{\infty} \frac{(-1)^n f(nT)}{\sigma(t - nT)}$$

as in Eq. (5-21), then

$$|f_i(t)| \leq \frac{1}{2\pi} \int_{-\infty}^{\infty} |F(\omega)| \, d\omega \tag{10}$$

$$|f(t) - f_i(t)| \leq \frac{1}{\pi} \int_{|\omega| > \sigma} |F(\omega)| \, d\omega \tag{11}$$

63. Given n arbitrary numbers t_k and a real σ-BL function $f(t)$ with energy E, we determine the constants a_k such that the sum

$$\hat{f}(t) = \sum_{k=1}^{n} a_k y_k(t) \qquad y_k(t) = \frac{\sin \sigma(t - t_k)}{\pi(t - t_k)}$$

is the optimum mean-square approximation of $f(t)$. Show that $\hat{f}(t)$ is also an interpolator of $f(t)$, that is, $\hat{f}(t_k) = f(t_k)$.

64. The input to an ideal low-pass filter $H_a(\omega) = p_{\omega_c}(\omega)$ is a σ-BL class of functions where $\sigma = 3\omega_c$.
(*a*) Find a discrete system $y[n]$ simulating $H_a(\omega)$.
(*b*) Find a nonrecursive filter

$$H(z) = \sum_{m=-12}^{12} b_m z^{-m}$$

interpolating $H_a(\omega)$ at the points $n\omega_0$, where $\omega_0 = 2\sigma/25$.

Chapters 6 and 7

65. We have shown that a function cannot be BL and TL; the following is the DFS version of this result: Show that if $a_n \underset{N}{\leftrightarrow} A_m$ and $a_n = 0$ for $n_1 < n \leq N - 1$, then $A_m = 0$ for at most n_1 values of m.

66. (Interpolation with trigonometric and Tchebycheff polynomials)[1]
(*a*) Show that, if $f(t)$ is an arbitrary real function defined in the interval $(0, \pi)$ and

[1] C. Lanczos, "Discourse on Fourier Series," Hafner Publishing Company, Inc., New York, 1966.

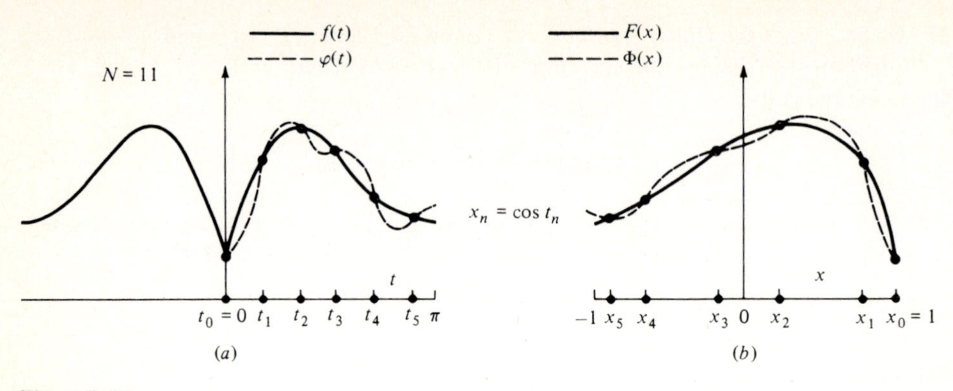

Figure P-66

$\varphi(t)$ is a trigonometric polynomial interpolating $f(t)$ at the points $t_m = 2\pi m/N$, where $N = 2M + 1$ and $m = 0, \ldots, M$:

$$\varphi(t) = b_0 + 2\sum_{n=1}^{M} b_n \cos nt \qquad \varphi(t_m) = f(t_m)$$

(Fig. P-66*a*), then

$$b_n = \frac{1}{N} f(0) + \frac{2}{N}\sum_{m=1}^{M} f\left(\frac{2\pi}{N} m\right) \cos \frac{2\pi}{N} mn \tag{12}$$

(*b*) Show that, if $F(x)$ is a real function defined in the interval $(-1, 1)$ and $\Phi(x)$ is a polynomial of degree M such that

$$\Phi(x_m) = F(x_m) \qquad x_m = \cos \frac{2\pi}{N} m \qquad 0 \le m \le M$$

(Fig. P-66*b*), then

$$\Phi(x) = b_0 + 2\sum_{n=1}^{M} b_n C_n(x) \qquad b_n = \frac{1}{N} F(0) + \frac{2}{N}\sum_{n=1}^{M} F(x_m) \cos \frac{2\pi}{N} mn \tag{13}$$

where $C_n(x) = \cos(n \arccos x)$ are the *Tchebycheff* polynomials.

67. Show that, if

$$w(t) = a_0 + 2\sum_{n=1}^{\infty} a_n \cos 2\pi nt$$

and $w(t) \ge 0$, then $|a_1| \le a_0$.

68. Show that if $f(t)$ is a σ-BL function with $\sigma = \pi/T$, then for any t and τ,

$$\sigma \cos \sigma\tau\, f(t) - \sin \sigma\tau\, f'(t) = \sum_{n=-\infty}^{\infty} \frac{(-1)^n \sin^2 \sigma\tau}{\sigma(\tau - nT)^2} f(t + nT - \tau) \tag{14}$$

If, also, $|f'(t)| \le M$, then for any α,

$$|\sigma \cos \alpha\, f(t) - \sin \alpha\, f'(t)| \le \sigma M \tag{15}$$

69. Show that, if $f(t)$ is a τ-TL function satisfying the Lipschitz condition

$$|f(t+x)-f(t)| < L|x|^{\alpha},$$

then

$$|F(\omega)| < L\left|\frac{\pi}{\omega}\right|^{\alpha}\left(\tau+\frac{\pi}{2|\omega|}\right) \tag{16}$$

70.[1] Find a σ-BL function $f(t)$ with given energy E, crossing the t axis at specified points $t_1 = a$ and $t_2 = -a$ and such that its value $f(0)$ at the origin is maximum.

Answer
$$f(t) = k\left[\frac{\sin \sigma t}{t} + \alpha_1 \frac{\sin \sigma(t-a)}{t-a} + \alpha_2 \frac{\sin \sigma(t+a)}{t+a}\right] \tag{17}$$

where $\alpha_1 = \alpha_2 = -2 \sin \sigma a/(2\sigma a + \sin 2\sigma a)$.

71. Show that if $f(t)$ is a σ-BL function and $f(t) \geq 0$, then $f(t) \leq (\sigma/2\pi)\int_{-\infty}^{\infty} f(t)\,dt$.

72. Given a τ-TL function

$$y(t) = \sum_{n=0}^{\infty} b_n \varphi_n(t) p_\tau(t)$$

[see Eq. (6-99)], find a σ-BL function

$$f(t) = \sum_{n=0}^{\infty} a_n \varphi_n(t)$$

of specified energy E such that if $y(t)$ is approximated by $f(t)$, then the mean-square error $e = \int_{-\tau}^{\tau} |f(t) - y(t)|^2\,dt$ is minimum.

73. Show that, if $v(t)$ and $y(t)$ are two periodic functions with Fourier-series coefficients v_n and y_n, respectively, then

$$\frac{1}{T}\int_{-T/2}^{T/2} y(t)v(t)y^*(t)\,dt = \sum_{k,\,n=-\infty}^{\infty} y_k v_{n-k} y_n^* \tag{18}$$

74. Show that, if $F(s)$ is a minimum-phase function and $|F(j\omega)| = e^{\alpha(\omega)}$ belongs to L^2, then

$$\int_{-\infty}^{\infty} \frac{|\alpha(\omega)|}{1+\omega^2}\,d\omega < \infty \tag{19}$$

75. Given a *real* function $Y(\omega)$, find a *causal* function $f(t)$ with transform $R(\omega) + jX(\omega)$ such that $R(\omega) = Y(\omega)$.

76. (*a*) Show that, if $w(0) < \infty$, $w(t) \leftrightarrow W(\omega)$, and $W(\omega) \geq 0$, then $w(t)$ is continuous.
(*b*) Show that if $f(t) \leftrightarrow F(\omega)$ and $\int_{-\infty}^{\infty} |F(\omega)|\,d\omega < \infty$, then $f(t)$ is continuous.

77. Show that, if

$$\frac{P(x)}{Q(x)} = \sum_{n=0}^{m} \frac{A_n}{x - x_n}$$

and $x_n = (n\pi/a)^2$, then

$$p_a(t) \sum_{n=0}^{m} (-1)^n A_n \cos \frac{n\pi}{a} t \leftrightarrow 2\omega \sin a\omega \frac{P(\omega^2)}{Q(\omega^2)}$$

[1] R. Barakat, Solution of the Luneberg Apodization Problems, *J. Opt. Soc. Am.*, vol. 52, pp. 264–284, 1962.

78. We assign to a real function $y(x)$ the numbers

$$E = \int_{-\infty}^{\infty} y^2(x)\,dx \qquad I = \int_{-\infty}^{\infty} y^4(x)\,dx \qquad M = \int_{-\infty}^{\infty} x^2 y^2(x)\,dx$$

Show that, for given values of E and I, M is minimum if $y^2(x) = A(x^2 - a^2)p_a(x)$.

79. Show that, if $H(z)$ is the system function of a real, discrete, minimum-phase system and $H(e^{j\omega T}) = e^{\alpha(\omega) + j\varphi(\omega)}$, then with $\sigma = \pi/T$,

$$\varphi(\omega) = -\frac{1}{2\sigma}\int_{-\sigma}^{\sigma} \alpha(\omega - y)\cot\frac{Ty}{2}\,dy \qquad \alpha(\omega) = \alpha(0) + \frac{1}{2\sigma}\int_{-\sigma}^{\sigma} \varphi(\omega - y)\cot\frac{Ty}{2}\,dy$$

Chapter 8

80. Show that $(1/\sqrt{t})U(t) \leftrightarrow \sqrt{\pi/j\omega}$.

81. Show that

$$e^{-\alpha t^2} * e^{-\beta t^2} = \sqrt{\frac{\pi}{\alpha + \beta}}\,e^{-\gamma t^2} \qquad \frac{1}{\alpha} + \frac{1}{\beta} = \frac{1}{\gamma}$$

82. Show that

$$\cos \alpha t^2 \leftrightarrow \sqrt{\frac{\pi}{\alpha}}\cos\left(\frac{\omega^2}{4\alpha} - \frac{\pi}{4}\right) \qquad \sin \alpha t^2 \leftrightarrow -\sqrt{\frac{\pi}{\alpha}}\sin\left(\frac{\omega^2}{4\alpha} - \frac{\pi}{4}\right)$$

83. (Theta function) Show that

$$\sqrt{\frac{\alpha}{\pi}}\sum_{n=-\infty}^{\infty} e^{-\alpha(t+n)^2} = 1 + 2\sum_{n=1}^{\infty} e^{-\pi^2 n^2/\alpha}\cos 2\pi n t$$

84. Show that, if

$$h_n(t) = (-1)^n e^{t^2/2}\frac{d^n(e^{-t^2})}{dt^n} \leftrightarrow H_n(\omega)$$

then $\sqrt{2\pi}\,h_n(t) = j^n H_n(t)$. Thus, the *hermite* functions $h_n(t)$, properly scaled, equal their Fourier transforms.

85. Given E and α, find a σ-BL function $f(t)$ such that

$$\int_{-\infty}^{\infty} f^2(t)\,dt = E \qquad \frac{1}{E}\int_{-\tau}^{\tau} f^2(t)\,dt = \alpha$$

86. Given a function $f(t)$ with projections $f_\tau(t)$ and $f_\sigma(t)$ as in (6-118), we form the sum $\hat{f}(t) = a_1 f_\tau(t) + a_2 f_\sigma(t)$, where a_1 and a_2 are such that the energy e of the difference $y(t) = f(t) - \hat{f}(t)$ is minimum. With α, β and $\hat{\alpha}$, $\hat{\beta}$ the energy concentrations of $f(t)$ and $\hat{f}(t)$, respectively [see (6-109)], show that $\hat{\alpha} \geq \alpha$ and $\hat{\beta} \geq \beta$.

87. Show that

$$\frac{1}{\pi}\int_{-\pi/2}^{\pi/2} e^{jt\sin\theta}\,d\theta \leftrightarrow \begin{cases} \dfrac{2}{\sqrt{1-\omega^2}} & |\omega| < 1 \\ 0 & |\omega| > 1 \end{cases} \tag{20}$$

88. If $f_1(r) \overset{h}{\leftrightarrow} \bar{F}_1(w)$, $f_2(r) \overset{h}{\leftrightarrow} \bar{F}_2(w)$, and $f(r) \overset{h}{\leftrightarrow} \bar{F}(w)$ are three Hankel transform pairs and $\bar{F}(w) = \bar{F}_1(w)\bar{F}_2(w)$, express $f(r)$ in terms of $f_1(r)$ and $f_2(r)$.

89.[1] (*a*) Show that if

$$g(x) = \int_{-\infty}^{\infty} f(\sqrt{x^2 + y^2})\, dy \qquad h(x) = \int_{-\infty}^{\infty} g(\sqrt{x^2 + t^2})\, dt$$

then
$$f(x) = -\frac{1}{2\pi x}\frac{dh(x)}{dx} \tag{21}$$

(*b*) With $g(x)$ as in part *a*, show that if $g(x) = \lambda f(x)$, then $f(x) = Ae^{-\pi^2 x^2/\lambda^2}$. (22)

Chapters 9 to 12

90. Show that $E\{|\mathbf{x}(t + \tau) \pm \mathbf{x}(t)|^2\} = 2\,\mathrm{Re}\,[R(0) \pm R(\tau)]$.

91. Show that, if $\mathbf{x}(t)$ is a stationary process with autocorrelation $R(\tau)$ and $R(\tau_1) = R(\tau_2) = R(0)$, where τ_1 and τ_2 are two noncommensurate numbers, then $R(\tau) = R(0) =$ constant.

92. Show that, if $\mathbf{n}(t)$ is white noise with autocorrelation $I\,\delta(\tau)$, and $\mathbf{y}(t) = \int_0^t \mathbf{n}(\alpha)\, d\alpha$ and $\mathbf{z}(t) = \mathbf{y}(t) - \mathbf{y}(T)t/T$, then $E\{\mathbf{z}^2(t)\} = I(t - t^2/T)$.
Application Suppose that we wish to estimate the integral $g(t) = \int_0^t f(\alpha)\, d\alpha$ of a signal $f(t)$ in terms of the data $\mathbf{x}(t) = f(t) + \mathbf{n}(t)$, *knowing that* $g(T) = 0$. If we use, as the estimate of $g(t)$, the integral $\mathbf{w}(t) = \int_0^t \mathbf{x}(\alpha)\, d\alpha$, then the variance of the estimate equals $E\{\mathbf{y}^2(t)\} = It$ (see Example 9-5). If, however, we use the difference $\mathbf{w}(t) - \mathbf{w}(T)t/T$ as the estimate, then the variance equals $I(t - t^2/T)$.

93. A process $\mathbf{x}(t)$ has constant mean and bounded autocovariance $C(t_1, t_2)$. Show that it is mean-ergodic if $C(t + \tau, t)$ tends to zero uniformly as $\tau \to \infty$; that is, given $\epsilon > 0$, there exists a constant α such that $|C(t + \tau, t)| < \epsilon$ for $|\tau| > \alpha$.

94. (*a*) Show that, if $R_n(\tau) = I\,\delta(\tau)$ and

$$\mathbf{N}(\omega) = \int_{-T}^{T} \mathbf{n}(t)e^{-j\omega t}\, dt$$

then $E\{|\mathbf{N}(\omega)|^2\} = 2IT$.

(*b*) Show that, if $E\{\mathbf{a}_n \mathbf{a}_k^*\} = I\,\delta[n - k]$ and

$$\mathbf{A}_m = \sum_{n=0}^{N-1} \mathbf{a}_n w^{mn} \qquad w = e^{j2\pi/N}$$

then $E\{|\mathbf{A}_m|^2\} = IN$.

[1] A. Papoulis, Joint Densities with Circular Symmetry, *IEEE Trans. Inf. Theory*, vol. 1T-14, pp. 164–165, 1968.

SOLUTIONS

Chapters 1 and 2

1. From Eq. (1-11) it follows that $g[n]$ equals $\sum_{k=0}^{n} 1 = n + 1$ for $0 \le n \le 5$; $\sum_{k=n-5}^{n} 1 = 6$ for $5 < n \le 10$; $\sum_{k=n-5}^{10} 1 = 16 - n$ for $10 < n \le 15$; and zero otherwise.

2. (*a*) $g[n] = \sum_{k=n-3}^{n+2} f[k] = \sum_{k=-\infty}^{\infty} h[n-k]f[k]$; hence, $h[n-k] = 1$ if $n - 3 \le k \le n + 2$ and $h[n-k] = 0$ otherwise. Therefore,

$$h[n] = \begin{cases} 1, & -2 \le n \le 3 \\ 0 & \text{otherwise} \end{cases} \qquad H(z) = z^2 + z + 1 + z^{-1} + z^{-2} + z^{-3}$$

(*b*) Clearly, $g[n] = \sum_{k=-\infty}^{\infty} a^{n+kN} U[n + kN] = g[n + N]$ is a periodic sequence with period N and for $0 \le n < N$ it is given by

$$g[n] = \sum_{k=0}^{\infty} a^{n+kN} = \frac{a^n}{1 - a^N}$$

3. (*a*) With $\alpha = t/2 - \tau$,

$$\int_{-\infty}^{\infty} f_1(t - \alpha) f_2(\alpha)\, d\alpha = \int_{-\infty}^{\infty} f_1\left(\frac{t}{2} + \tau\right) f_2\left(\frac{t}{2} - \tau\right) d\tau$$

(*b*) If $t > 0$, then $f_1(t/2 + \tau) = 0$ for $\tau > T - t/2$, and $f_2(t/2 - \tau) = 0$ for $\tau < -T + t/2$; similarly for $t < 0$.

4. For $|t| < 2T$,

$$f(t) * f(t) = \int_{-T+|t|/2}^{T-|t|/2} \cos\frac{\pi}{2T}\left(\frac{t}{2} + \tau\right) \cos\frac{\pi}{2T}\left(\frac{t}{2} - \tau\right) d\tau$$

5. $g(t) = \int_{0}^{\infty} e^{t-\tau}(3e^{-2\tau} - 1)\, d\tau = 0$

6. $H(\omega) = 1 + 2/(j\omega + 3)$. If $f(t) = 4\cos^2 2t = 2 + 2\cos 4t$, then

$$g(t) = \tfrac{10}{3} + 2\sqrt{\tfrac{41}{25}} \cos\left(4t + \tan^{-1}\tfrac{4}{5} - \tan^{-1}\tfrac{4}{3}\right)$$

7. $\sum_{n=-\infty}^{\infty} \delta(t - nT) = \frac{1}{T} \sum_{n=-\infty}^{\infty} e^{jn\omega_0 t} \qquad \omega_0 = \frac{2\pi}{T} = 2\pi \times 10^3$

Hence,
$$g(t) = \frac{1}{T} \sum_{n=-\infty}^{\infty} H(n\omega_0) e^{jn\omega_0(t - t_0)} = \frac{2}{T}[1 + 2\cos\omega_0(t - t_0)]$$

because $H(n\omega_0) = 0$ for $|n| > 1$.

8. $g(t) - ag(t - T) = f(t) \qquad G(\omega)(1 - ae^{-j\omega T}) = F(\omega) \qquad H(\omega) = \frac{1}{1 - ae^{-j\omega T}}$

$$= \sum_{n=0}^{\infty} a^n e^{-jnT\omega}$$

9. With $T = \pi/\sigma$ and $c = 2T$, $H_a(\omega) = 1 + 2e^{-j2T\omega} + e^{-j4T\omega}$; hence, $H(z) = 1 + 2z^{-2} + z^{-4}$ and $h[n] = \delta[n] + 2\,\delta[n-2] + \delta[n-4]$.

10. As in Fig. S-10,

$$H_1(z) = \frac{1}{T}(1 - z^{-1}) \qquad H_1(e^{j\omega T}) = e^{-j\omega T/2}\frac{2j}{T}\sin\frac{\omega T}{2}$$

$$H_2(z) = \frac{2}{T}\frac{1 - z^{-1}}{1 + z^{-1}} \qquad H_2(e^{j\omega T}) = \frac{2j}{T}\tan\frac{\omega T}{2}$$

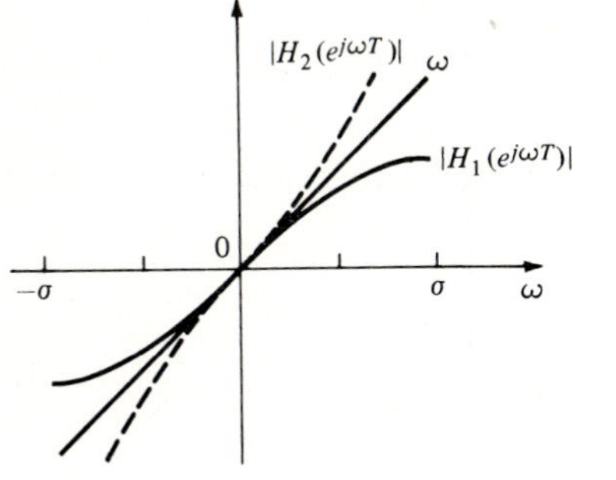

Figure S-10

11. $F_1(z) = z^2\dfrac{z}{z-1} \qquad F_2(z) = \dfrac{4z}{z-1} - \dfrac{3z}{z-\frac{1}{4}} \qquad F_3(z) = z^{-4}\dfrac{2z}{z-\frac{1}{2}}$

$f_1[n] = U[n+2] \qquad f_2[n] = 4U[n] - 3(\tfrac{1}{4})^n U[n] \qquad f_3[n] = 2(\tfrac{1}{2})^{n-4}U[n-4]$

12. $F(z) = \dfrac{z}{z-2} - \dfrac{z}{(z-1)^2} - \dfrac{z}{z-1}$

For $|z| > 2$: $\quad f[n] = 2^n U[n] - (n+1)U[n]$

For $1 < |z| < 2$: $\quad f[n] = -2^n U[-n-1] - (n+1)U[n]$

For $|z| < 1$: $\quad f[n] = -2^n U[-n-1] + (n+1)U[-n-1]$

13. (*a*) $F'(z) = \displaystyle\sum_{n=-\infty}^{\infty}(-n)f[n]z^{-n-1} = -z^{-1}\sum_{n=-\infty}^{\infty} nf[n]z^{-n}$

(*b*) $na^n U[n] \leftrightarrow -z\dfrac{d}{dz}\left(\dfrac{z}{z-a}\right) = \dfrac{az}{(z-a)^2}$

14. $f_1[n] = U[n] - U[n-N] \qquad F_1(z) = \dfrac{z}{z-1} - z^{-N}\dfrac{z}{z-1}$

$f_2[n] = nf_1[n] \qquad F_2(z) = -zF_1'(z) = \dfrac{1 - z^{-N}}{(z-1)^2}z - \dfrac{Nz^{-N}}{z-1}z$

15. (*a*) $F(z) = \sum_{n=-\infty}^{\infty} a^{|n|}z^{-n} = \sum_{n=-\infty}^{-1} a^{-n}z^{-n} + \sum_{n=0}^{\infty} a^{n}z^{-n} = \frac{az}{1-az} + \frac{z}{z-a}$

(*b*) $f[n] = e^{-\alpha T|n|} = a^{|n|}$, where $a = e^{-\alpha T}$. Hence,

$$F(e^{j\omega T}) = \frac{1 - e^{-2\alpha T}}{1 - 2e^{-\alpha T}\cos \omega T + e^{-2\alpha T}}$$

16. (*a*) $F_1(z) = \sum_{n=-\infty}^{\infty} f[n]z^{-nN}$. Hence (Fig. S-16), $f_1[n] = \begin{cases} f[k] & n = kN \\ 0 & \text{otherwise} \end{cases}$

(*b*) $F(z) = \frac{z}{z-1} \qquad f[n] = U[n] \qquad f_1[n] = \begin{cases} U[k] & n = kN \\ 0 & \text{otherwise} \end{cases}$

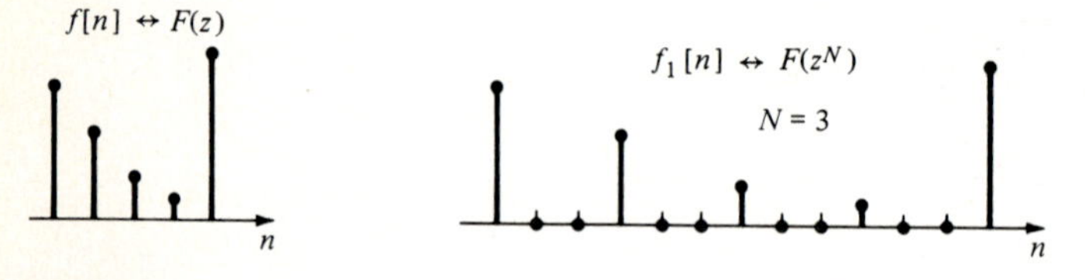

Figure S-16

17. The truncated, periodic sequence $g[n]$ can be written as a convolution $g[n] = f_1[n] * h[n]$, where $f_1[n] = U[k]$ as in Prob. 16*b*, and $h[n] = g[n]$ for $0 \le n < N$ and $h[n] = 0$ otherwise. Hence, $G(z) = F_1(z)H(z)$, where

$$F_1(z) = \frac{z^N}{z^N - 1} \qquad H(z) = \sum_{n=0}^{N-1} g[n]z^{-n}$$

18. Since [see Eq. (2-12)]

$$f[n] = \frac{1}{2\pi}\int_{-\pi}^{\pi} F(e^{j\theta})e^{jn\theta}\, d\theta \qquad \text{and} \qquad \sum_{n=0}^{\infty} (e^{j\theta}z^{-1})^n = \frac{1}{1 - e^{j\theta}z^{-1}}$$

for $|z| > 1$, we conclude that

$$\sum_{n=0}^{\infty} f[n]z^{-n} = \frac{1}{2\pi}\int_{-\pi}^{\pi} F(e^{j\theta})\left(\sum_{n=0}^{\infty} e^{jn\theta}z^{-n}\right) d\theta = \frac{1}{2\pi}\int_{-\pi}^{\pi} F(e^{j\theta})\,\frac{z}{z - e^{j\theta}}\, d\theta$$

19. $\sum_{n=0}^{\infty} f(nT)z^{-n} = \frac{1}{2\pi}\sum_{n=0}^{\infty}\left[\int_{-\infty}^{\infty} F(\omega)e^{jnT\omega}\, d\omega\right]z^{-n} = \frac{1}{2\pi}\int_{-\infty}^{\infty} F(\omega)\sum_{n=0}^{\infty}(e^{jT\omega}z^{-1})^n\, d\omega$

20. $H(z) = (1 - 2z^{-1})/(1 - z^{-1}/2)$; if $f[n] = U[n]$, then

$$G(z) = \frac{z(2z-4)}{(z-1)(2z-1)} = \frac{-2z}{z-1} + \frac{3z}{z - \frac{1}{2}} \qquad g[n] = -2U[n] + 3(\tfrac{1}{2})^n U[n]$$

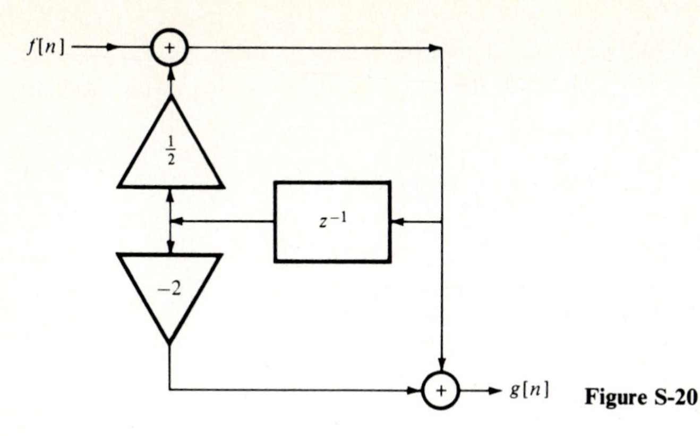

Figure S-20

21. $g[-1] = 0$ because $g[0] = \lambda$ and $g[0] - (1 - \lambda)g[-1] = \lambda$. Hence [see (2-24)],

$$G_1(z) = \frac{\lambda z}{z-1}\frac{1}{1-(1-\lambda)z^{-1}} = \frac{z}{z-1} - \frac{(1-\lambda)z}{z-(1-\lambda)} \qquad g[n] = 1 - (1-\lambda)^{n+1}$$

22. (*a*) From the Poisson formula (3-87), it follows that

$$W(e^{j\omega T}) = \sum_{n=-\infty}^{\infty} f(nT)e^{-jnT\omega} = \frac{1}{T}\sum_{n=-\infty}^{\infty} F(\omega + 2n\sigma)$$

The last sum equals $F(\omega)$ for $|\omega| < \sigma$ because $F(\omega) = 0$ for $|\omega| > \sigma$ by assumption.

(*b*) Since $G(z) = W(z)H(z)$, the inversion formula (2-12) yields

$$g[n] = \frac{1}{2\sigma}\int_{-\sigma}^{\sigma} W(e^{j\omega T})H(e^{j\omega T})e^{jnT\omega}\,d\omega = \frac{1}{2\sigma T}\int_{-\sigma}^{\sigma} F(\omega)H(e^{j\omega T})e^{jnT\omega}\,d\omega$$

23. (*a*) $h[n] = \frac{1}{2}\,\delta[n+1] + \delta[n] + \frac{1}{2}\,\delta[n-1]$

$$H(z) = \frac{z}{2} + 1 + \frac{z^{-1}}{2} \qquad H(e^{j\omega T}) = 1 + \cos \omega T \qquad \text{(Fig. S-23}a\text{)}$$

(*b*) $h[n] = -\frac{1}{2}\,\delta[n+2] + \delta[n] - \frac{1}{2}\,\delta[n-2]$

$$H(z) = -\frac{z^2}{2} + 1 - \frac{z^{-2}}{2} \qquad H(e^{j\omega T}) = 1 - \cos 2\omega T \text{ (Fig. S-23}b\text{)}$$

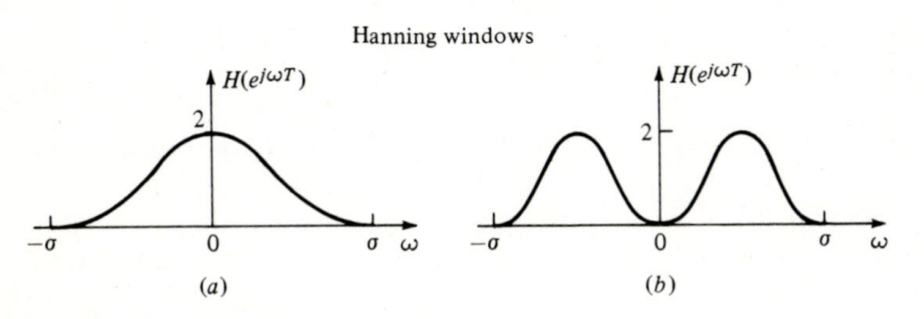

Figure S-23

24. As we see from Fig. P-24, $g[n] = 4f[n] + 8w[n] + 3w[n]$, where $w[n] = f[n-1] + 2w[n-1]$. Hence, $G(z) = 4F(z) + 11\,W(z)$ and $W(z) = z^{-1}F(z) + 2z^{-1}W(z)$. Solving for $G(z)$, we obtain

$$G(z) = \frac{4 + 3z^{-1}}{1 - 2z^{-1}} F(z)$$

Hence, $g[n] - 2g[n-1] = 4f[n] + 3f[n-1]$.

25. $g[n] - g[n-1] = w[n] = f[n] + 2f[n-1]$. Hence, $H(z) = (1 + 2z^{-1})/(1 - z^{-1})$ as in Fig. S-25. If $f[n] = U[n+2]$, then $F(z) = z^3/(z-1)$ and

$$G(z) = \frac{z^3(z+2)}{(z-1)^2} \qquad g[n] = (n+3)U[n+3] + 2(n+2)U[n+2]$$

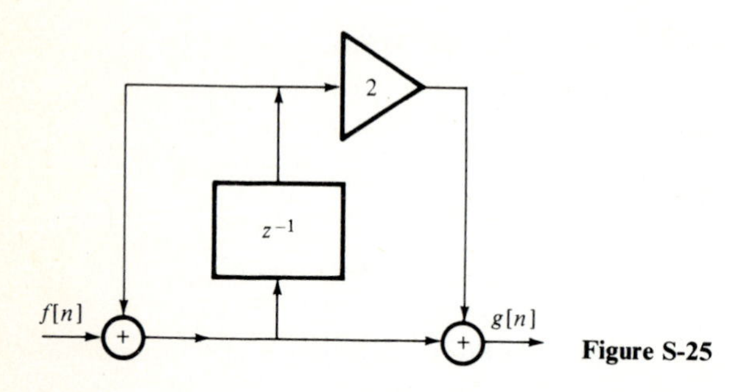

Figure S-25

26. From Fig. P-26*a*, it follows that if $f[n] = z^n$, then

$$W(z)z^n = z^n - c_1 z^{-1} W(z) z^n Y(z) \text{ and } H_a(z)z^n = c_0\, z^n + W(z)z^n Y(z)$$

Eliminating $W(z)$, we find $H_a(z)$. The formula for $H_b(z)$ follows by induction because the system function of the system with input A and output B equals $c_{2n} + 1/(c_{2n+1}z^{-1} + 1)$.

27. Since $4 \sin^4 (\omega T/2) = 1 - 2 \cos \omega T + \cos^2 \omega T$, it follows that [see (2-46)]

$$V(w) = \frac{4\epsilon}{4\epsilon + 1 - 2w + w^2} \qquad \text{so that } H(z) = \frac{1 + a_1 + a_2}{1 + a_1 z^{-1} + a_2 z^{-2}}$$

The poles of $V(w)$ equal $1 \pm j2\sqrt{\epsilon}$; hence, the poles of $H(z)$ are the roots of the quadratic $z^2 - 2(1 \pm j2\sqrt{\epsilon})z + 1$ that are inside the unit circle.

28. *Sufficiency* If $I < \infty$ and $|f[n]| < M$, then

$$|g[n]| = \left|\sum_{k=-\infty}^{\infty} f[n-k]h[k]\right| \le M \sum_{k=-\infty}^{\infty} |h[k]| < \infty$$

Necessity We shall show that if $I = \infty$, then the response to some bounded input can be infinite. Indeed, if $f[n] = |h[-n]|/h[-n]$, then $f[n]$ is bounded because it equals 1 or -1. However,

$$g[0] = \sum_{k=-\infty}^{\infty} f[-k]h[k] = \sum_{k=-\infty}^{\infty} |h[k]| = \infty$$

29. (*a*) $f[n] * z^n = \sum_{k=-\infty}^{\infty} f[k]z^{n-k} = z^n \sum_{k=-\infty}^{\infty} f[k]z^{-k}$

(*b*) Suppose that $f[n] = f_1[n] * f_2[n]$. From part *a* it follows that

$$F(z)z^n = f[n] * z^n = f_1[n] * f_2[n] * z^n = f_1[n] * F_2(z)z^n$$
$$= (f_1[n] * z^n)F_2(z) = F_1(z)z^n F_2(z)$$

Hence, $F(z) = F_1(z)F_2(z)$.

Chapter 3

30. (*a*) $f(t) * e^{j\omega t} = \int_{-\infty}^{\infty} f(\tau)e^{j\omega(t-\tau)}\,d\tau = e^{j\omega t}\int_{-\infty}^{\infty} f(\tau)e^{-j\omega\tau}\,d\tau$

(*b*) Suppose that $f(t) = f_1(t) * f_2(t)$. From part *a* it follows that

$$F(\omega)e^{j\omega t} = f(t) * e^{j\omega t} = f_1(t) * f_2(t) * e^{j\omega t} = f_1(t) * F_2(\omega)e^{j\omega t}$$
$$= [f_1(t) * e^{j\omega t}]F_2(\omega) = F_1(\omega)e^{j\omega t}F_2(\omega)$$

Hence, $F(\omega) = F_1(\omega)F_2(\omega)$.

31. $\int_{-\infty}^{\infty} f(\tau)e^{j\alpha(t-\tau)^2}\,d\tau = e^{j\alpha t^2}\int_{-\infty}^{\infty} f(\tau)e^{j\alpha\tau^2}e^{-j2\alpha t\tau}\,d\tau$

and (1) follows with $e^{j\alpha\tau^2} \simeq 1 + j\alpha\tau^2$ because

$$j\alpha t^2 f(t) \leftrightarrow -j\alpha F''(\omega)$$

32. From (3-14) and the pairs

$$e^{-\alpha t}U(t) \leftrightarrow \frac{1}{\alpha + j\omega} \qquad te^{-\alpha t}U(t) \leftrightarrow \frac{1}{(\alpha + j\omega)^2}$$

it follows that

$$e^{-\alpha|t|}(1 + \alpha|t|) \leftrightarrow 2\mathrm{Re}\left[\frac{1}{\alpha + j\omega} + \frac{\alpha}{(\alpha + j\omega)^2}\right]$$

33. $\dfrac{2\alpha}{t^2 + \alpha^2} \leftrightarrow 2\pi e^{-\alpha|\omega|}$

Hence [see (3-35)],

$$\int_{-\infty}^{\infty} \frac{2a}{t^2 + a^2}\frac{2b}{t^2 + b^2}\,dt = \frac{4\pi^2}{2\pi}\int_{-\infty}^{\infty} e^{-a|\omega|}e^{-b|\omega|}\,d\omega = 4\pi\int_0^{\infty} e^{-(a+b)\omega}\,d\omega$$

34. (*a*) It follows from (3-28) because

$$f'(t) \leftrightarrow j\omega F(\omega) \qquad \frac{1}{\pi t} \leftrightarrow -j\,\mathrm{sgn}\,\omega \qquad j\omega(-j\,\mathrm{sgn}\,\omega) = |\omega|$$

(*b*) This is a special case of part *a* if $f(t) = \delta(t)$, because $\delta'(t) * 1/\pi t = -1/\pi t^2$.

35. (*a*) $f(t+a) - f(t) \leftrightarrow (e^{ja\omega} - 1)F(\omega)$. Hence [see (3-37)],

$$\int_{-\infty}^{\infty} |f(t+a) - f(t)|^2\, dt = \frac{1}{2\pi}\int_{-\infty}^{\infty} |e^{ja\omega} - 1|^2 |F(\omega)|^2\, d\omega$$

$$= \frac{2}{2\pi}\int_{-\infty}^{\infty} |F(\omega)|^2(1 - \cos a\omega)\, d\omega = 2\rho(0) - 2\,\text{Re}\,\rho(a)$$

(*b*) This follows from the identity

$$|Ae^{-ja\omega} + F(\omega)|^2 = [Ae^{-ja\omega} + F(\omega)][A^*e^{ja\omega} + F^*(\omega)]$$

$$= |A|^2 + A^*F(\omega)e^{ja\omega} + AF^*(\omega)e^{-ja\omega} + |F(\omega)|^2$$

36. With $f_e(t)$ as in (3-14), $f(t) = 2f_e(t)U(t)$. Hence,

$$\int_0^{\infty} f^2(t)\, dt = 4\int_0^{\infty} f_e^2(t)\, dt = 2\int_{-\infty}^{\infty} f_e^2(t)\, dt = \frac{2}{2\pi}\int_{-\infty}^{\infty} R^2(\omega)\, d\omega$$

The last equality follows because $f_o(t) \leftrightarrow jX(\omega)$ and $|f_e(t)| = |f_o(t)|$.

37. $2a(t)\cos\varphi(t) = a(t)e^{j\varphi(t)} + a(t)e^{-j\varphi(t)} \leftrightarrow F(\omega) + F^*(-\omega)$

38. $G(\omega) = F_1(\omega) + jF_2(\omega)$, and since $f_1(t)$ and $f_2(t)$ are real,

$$G^*(-\omega) = F_1^*(-\omega) - jF_2^*(-\omega) = F_1(\omega) - jF_2(\omega)$$

Hence, $2F_1(\omega) = G(\omega) + G^*(-\omega)$ and $2jF_2(\omega) = G(\omega) - G^*(-\omega)$, and the proof follows because $f(t) = f_1(t) + f_2(-t)$; that is, $F(\omega) = F_1(\omega) + F_2(-\omega)$.

39. The function $|\cos\omega_0 t|$ is periodic with period $T/2 = \pi/\omega_0$. Since [see Eq. (3-65)]

$$f_0(t) = p_{T/4}(t)\cos\omega_0 t \leftrightarrow F_0(\omega) = \frac{2\omega_0}{\omega_0^2 - \omega^2}\cos\frac{T}{4}\omega$$

it follows from (3-69) that

$$a_n = \frac{2}{T}F_0(2n\omega_0) = \frac{(-1)^n 2}{\pi(1 - 4n^2)}$$

The second equality results when t is changed to $t - T/4$, because $\sin\omega_0 t = \cos\omega_0(t - T/4)$.

40.
$$\cot\frac{\omega_0 t}{2} = 2\sum_{n=1}^{\infty}\gamma_n \sin n\omega_0 t \qquad \gamma_n = \frac{1}{T}\int_{-T/2}^{T/2}\cot\frac{\omega_0 t}{2}\sin n\omega_0 t\, dt$$

because $\cot(\omega_0 t/2)$ is odd. Inserting in the last integral the identity

$$\cot\frac{\omega_0 t}{2}\sin n\omega_0 t = \frac{\sin(n+\frac{1}{2})\omega_0 t}{2\sin(\omega_0 t/2)} + \frac{\sin(n-\frac{1}{2})\omega_0 t}{2\sin(\omega_0 t/2)}$$

we conclude with (3-57) that $\gamma_n = \text{sgn}\, n$ and (2) follows readily.

41. Integrating the series expansion of $f(t)$ termwise from $-T/2$ to t, we obtain

$$g(t) = c + \sum_{n=-\infty}^{\infty}\frac{a_n}{jn\omega_0}e^{jn\omega_0 t}$$

and (3) results.

42. Applying Eq. (3-77) to the function $g(t) = T/2 - t$ of Fig. P-41, we obtain

$$2\sum_{n=1}^{\infty} \frac{1}{n^2\omega_0^2} = \frac{1}{T}\int_0^T \left(\frac{T}{2} - t\right)^2 dt = \frac{T^2}{12}$$

Hence, $I_1 = \pi^2/6$. To find I_2, we express I_1 in terms of even and odd components:

$$I_1 = \sum_{n=1}^{\infty} \frac{1}{(2n)^2} + \sum_{n=1}^{\infty} \frac{1}{(2n+1)^2} = \frac{I_1}{4} + I_2$$

Hence, $I_2 = 3I_1/4 = \pi^2/8$.

43. From $f(t + T) = f(t)$, it follows that $e^{j\omega T}F(\omega) = F(\omega)$. This is possible only if $F(\omega) = 0$ for every ω such that $e^{j\omega T} \neq 1$; that is, for $\omega \neq 2\pi n/T$.

44. $F(\omega) = e^{-a}\sum_{n=0}^{\infty}(a^n/n!)e^{-jn\omega} = e^{a(e^{-j\omega}-1)}$ because $\delta(t - n) \leftrightarrow e^{-jn\omega}$. Applying (3-43), we obtain $m_0 = F(0) = 1$, $m_1 = jF'(0) = a$, and $m_2 = -F''(0) = a^2 + a$.

45. $F(\omega) = \sum_{n=-\infty}^{\infty} e^{-\alpha T|n|}e^{-jnT\omega} = G(e^{j\omega T})$, where $G(z) = \sum_{n=-\infty}^{\infty} e^{\alpha T|n|}z^{-n}$, is the z transform of the sequence $f[n] = e^{-\alpha T|n|}$. Hence, (4) follows from Prob. 15.

46. Using the pair $\bar{\delta}(t) \leftrightarrow \omega_0\,\bar{\delta}(\omega)$ and the convolution theorem, we obtain (Fig. S-46)

$$F_1(\omega) = \omega_0\,\bar{\delta}(\omega)\left[\frac{\sin a(\omega - \omega_1)}{\omega - \omega_1} + \frac{\sin a(\omega + \omega_1)}{\omega + \omega_1}\right]$$

$$F_2(\omega) = \omega_0\,\bar{\delta}(\omega - \omega_1)\frac{\sin a(\omega - \omega_1)}{\omega - \omega_1} + \omega_0\,\bar{\delta}(\omega + \omega_1)\frac{\sin a(\omega + \omega_1)}{\omega + \omega_1}$$

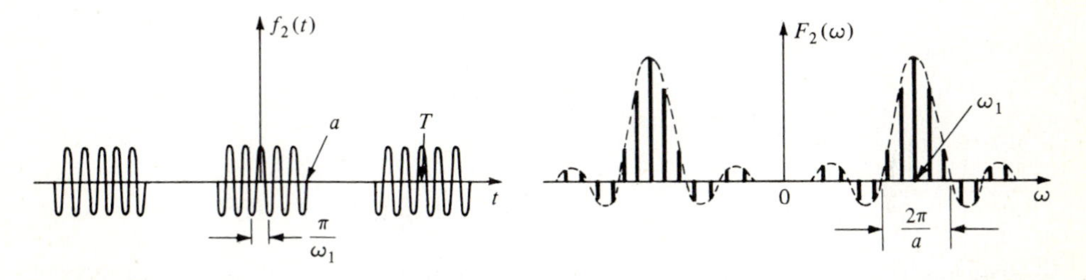

Figure S-46

47. $F(\omega) = \Phi(\omega)\dfrac{2\sin a\omega}{\omega}$ because $f(t) = \varphi(t) * p_a(t)$

With $T = 2a$, (3-86) yields

$$\sum_{n=-\infty}^{\infty} f(t + 2na) = \frac{1}{2a}\sum_{n=-\infty}^{\infty} F\left(\frac{\pi n}{a}\right)e^{jnt\pi/a} = \frac{F(0)}{2a}$$

48. Extending the definition of a_n for $n < 0$ such that $a_{-n} = a_n$, we see from (3-107) that its DFS of order N equals

$$\sum_{n=-M}^{M} a_n w^{mn} = a_0 + \sum_{n=1}^{M} a_n(w^{mn} + w^{-mn}) = A_m$$

Since $A_{-m} = A_m$, (5) follows similarly from the second equation in (3-107).

49. $A_m = \sum_{n=-k}^{k} w^{mn} = \dfrac{w^{(k+1)m} - w^{-km}}{w^m - 1} = \dfrac{w^{(k+1/2)m} - w^{-(k+1/2)m}}{w^{m/2} - w^{-m/2}}$

50. $A_m = \sum_{n=0}^{3N-1} a_n w_{3N}{}^{mn} = \sum_{k=0}^{N-1} [a_{3k} w_{3N}{}^{3km} + a_{3k+1} w_{3N}{}^{(3k+1)m} + a_{3k+2} w_{3N}{}^{(3k+2)m}]$

and (6) follows because $w_{3N}{}^{3km} = w_N{}^{km}$.

51. $\sum_{n=0}^{N-1} A_{-n} w^{mn} = \sum_{k=0}^{N-1} A_k w^{-mk} = Na_m$

52. The sum $\sum_{n=0}^{N-1} nz^{-n}$ is the z transform of the sequence $nU[n] - nU[n-N]$. Hence (Prob. 14),

$$Y(z) = z\frac{1 - z^{-N}}{(z-1)^2} - \frac{Nz^{-N+1}}{z-1} \qquad A_m = Y(w^{-m}) = \frac{-N}{w^{-m} - 1} w^{-m}$$

53. $y(t) = 5 + (2 - j)e^{j\omega_0 t} + (2 + j)e^{-j\omega_0 t}$; hence [see (3-135)], $c_0 = \bar{c}_0 = a_0 = 5$; $c_1 = \bar{c}_1 = a_1 = 2 - j$; $c_{-1} = \bar{c}_{-1} = \bar{c}_{19} = a_{19} = 2 + j$; and $a_n = 0$ for $1 < n < 19$.

54. (*a*) $\cos k\omega_0 t = \frac{1}{2}(e^{jk\omega_0 t} + e^{-jk\omega_0 t})$; hence, $c_n = \frac{1}{2}$ for $n = \pm k$ and 0 otherwise; $\bar{c}_n = c_n$ for $|n| < N/2$.

(*b*) $a_k = \bar{c}_k = \frac{1}{2}$; $a_{N-k} = \bar{c}_{N-k} = \frac{1}{2}$; $a_n = 0$ otherwise.

(*c*) From Eq. (3-125) it follows that

$$\sum_{m=0}^{N-1} \cos^2 \frac{2\pi k}{N} m = N \sum_{n=0}^{N-1} a_n{}^2 = \frac{N}{2}$$

55. From (3-30) it follows that

$$f(t)\,\delta^{(n)}(t) \leftrightarrow \frac{1}{2\pi} F(\omega) * (j\omega)^n = \frac{j^n}{2\pi}\int_{-\infty}^{\infty} F(y)(\omega - y)^n\,dy$$

$$= \frac{1}{2\pi}\sum_{k=0}^{n}(-1)^k \binom{n}{k} (j\omega)^{n-k} \int_{-\infty}^{\infty} (jy)^k F(y)\,dy$$

and the given equation results because the last integral equals $2\pi f^{(k)}(0)$ [see (3-43)].

Chapters 4 and 5

56. The first equality follows from Eq. (4-20) with $m_0 = H(0) = 1$; the second follows from (3-40) with $A(\omega) = p_a(\omega)$.

57. (*a*) We can show, as in (4-22), that $F(\omega)e^{j\eta\omega} = m_0 - \mu_2\omega^2/2 + \cdots$. But

$$(m_0 - \mu_2\omega^2/2)^{-1} \simeq 1/m_0 + \mu_2\omega^2/2m_0{}^2.$$

Hence,
$$H(\omega) = \frac{G(\omega)e^{j\eta\omega}}{F(\omega)e^{j\eta\omega}} \simeq \left(\frac{1}{m_0} + \frac{\mu_2\omega^2}{2m_0{}^2}\right) G(\omega)e^{j\eta\omega}$$

and (7) results.

(*b*) Similarly, if $m_0 = 0$, then $F(\omega)e^{ja\omega} \simeq -j\omega m_1 + jb\omega^3/6$. Hence,

$$H(\omega) = \frac{G(\omega)e^{ja\omega}}{F(\omega)e^{ja\omega}} \simeq -\frac{1}{m_1}\frac{G(\omega)}{j\omega} e^{ja\omega} + \frac{jb\omega}{6m_1{}^2} G(\omega)e^{ja\omega}$$

58. As in (4-12),

$$g(t) - f(t) = \int_{-\infty}^{\infty} [f(t-\tau) - f(t)]h(\tau)\, d\tau = -\int_{-\infty}^{\infty} \tau f'(t-\tau_1)h(\tau)\, d\tau$$

and (9) follows because $|f'(t-\tau_1)| \le M$.

59. (*a*) If $f(t)$ is even, then $F(\omega)$ is real; hence, the transform $-j \operatorname{sgn} \omega F(\omega)$ of $\hat{f}(t)$ is imaginary.

(*b*) The transform of $\sin at/\pi t$ equals $p_a(\omega)$; hence,

$$z_f(t) = \frac{1}{\pi}\int_0^a e^{j\omega T}\, d\omega = \frac{e^{ja\omega} - 1}{j\pi t} = \frac{\sin at}{\pi t} + j\frac{1 - \cos at}{\pi t}$$

60. $\dfrac{2\alpha}{\alpha^2 - s^2} = \dfrac{1}{\alpha - s} + \dfrac{1}{\alpha + s}$

Hence,

$$f_1(t) = (e^{-\alpha t} - e^{\alpha t})U(t) \qquad \alpha < \operatorname{Re} s$$

$$f_2(t) = e^{\alpha t}U(-t) + e^{-\alpha t}U(t) = e^{-\alpha|t|} \qquad -\alpha < \operatorname{Re} s < \alpha$$

$$f_3(t) = (e^{\alpha t} - e^{-\alpha t})U(-t) \qquad \operatorname{Re} s < -\alpha$$

61. With $\int_a^b z(x)w(x)\, dx \equiv Be^{j\theta}$ and y any real number,

$$\int_a^b |z(x) - ye^{j\theta}w^*(x)|^2\, dx = \int_a^b |z(x)|^2\, dx - 2y\left|\int_a^b z(x)w(x)\, dx\right|^2 + y^2\int_a^b |w(x)|^2\, dx$$

The above is a nonnegative quadratic; hence, Eq. (4-A-1) follows.

62. Since $F_i(\omega) = F_*(\omega)p_\sigma(\omega)$, it follows that

$$|f_i(t)| \le \frac{1}{2\pi}\int_{-\sigma}^{\sigma} \sum_{n=-\infty}^{\infty} |F(\omega + 2n\sigma)|\, d\omega = \frac{1}{2\pi}\int_{-\infty}^{\infty} |F(\omega)|\, d\omega$$

To prove (11), we form the functions $f_1(t) \leftrightarrow F(\omega)p_\sigma(\omega)$ and $f_2(t) \leftrightarrow F(\omega)[1 - p_\sigma(\omega)]$ and their interpolators $f_{1i}(t)$ and $f_{2i}(t)$ as in (5-21). Clearly, $f_{1i}(t) = f_1(t)$ (sampling theorem); hence, $f(t) - f_i(t) = f_2(t) - f_{2i}(t)$, and (11) results because

$$|f_2(t) - f_{2i}(t)| \le |f_2(t)| + |f_{2i}(t)| \le \frac{1}{2\pi}\int_{|\omega|>\sigma} |F(\omega)|\, d\omega + \frac{1}{2\pi}\int_{|\omega|>\sigma} |F(\omega)|\, d\omega$$

63. For any σ-BL function $w(t)$, $\langle w, y_k\rangle = w(t_k)$. Since the functions $f(t)$ and $\hat{f}(t)$ are σ-BL, it follows that [see (5-32)] $0 = \langle f - \hat{f}, y_k\rangle = f(t_k) - \hat{f}(t_k)$.

64. (*a*) $h_a(t) = \dfrac{\sin \omega_c t}{\pi t} \qquad y[n] = \dfrac{T \sin \omega_c nT}{\pi n T} = \dfrac{\sin(\pi n/3)}{\pi n}$

(*b*) As we know, the numbers b_m are the DFS coefficients of $H_a(n\omega_0)$ of order 25. Since $H_a(n\omega_0) = 1$ for $|n|\omega_0 \le \omega_c$ (that is, for $|n| \le 4$) and 0 otherwise, we conclude from Prob. 49 with $k = 4$ that $b_m = \sin(9\pi m/25)/\sin(\pi m/25)$.

Chapters 6 and 7

65. We form the polynomial $F(z) = \sum_{n=0}^{n_1} a_n z^n$. Clearly, $F(w^m) = A_m$. Hence, if $A_m = 0$ for some m, then w^m is a root of $F(z)$. Therefore, there are at most n_1 such m's because $F(z)$ has n_1 roots.

66. (a) We extend the definition of $f(t)$ in the interval $(-\pi, \pi)$ such that $f(-t) = f(t)$. It then follows that $b_n \underset{N}{\leftrightarrow} f(2\pi m/N)$, and (12) results (see Prob. 48).

(b) With $x = \cos t$, $F(x) = F(\cos t) = f(t)$, $\Phi(x) = \Phi(\cos t) = \varphi(t)$, and $C_n(x) = \cos nt$. If $x = x_m$, then $t = 2\pi m/N$. Hence, the interpolation of $F(x)$ by the polynomial $\Phi(x)$ is equivalent to the interpolation of $f(t)$ by the trigonometric polynomial $\varphi(t)$. Therefore, (13) follows from (12).

67. From Eq. (3-80) it follows that $a_0 \pm a_1 = \int_0^1 (1 \pm \cos 2\pi t)w(t)\, dt \geq 0$. Hence, $-a_0 \leq a_1 \leq a_0$.

68. The left side of (14) is the output of the system $H(\omega) = \sigma \cos \sigma\tau - j\omega \sin \sigma\tau$ with input $f(t)$. Since

$$h_\sigma(t) = \frac{1}{2\pi}\int_{-\sigma}^{\sigma} H(\omega)e^{j\omega t}\, d\omega = \frac{\sigma t \sin \sigma(t-\tau) + \sin \sigma\tau \sin \sigma t}{\pi t^2}$$

(14) follows from (5-18) mutatis mutandis. Applying (5-12) to (14), we obtain (15).

69. $$F(\omega) = \int_{-\infty}^{\infty} f(t)e^{-j\omega t}\, dt = \int_{-\infty}^{\infty} f\left(t + \frac{\pi}{\omega}\right)e^{-j\omega(t+\pi/\omega)}\, dt$$

and since $f(t) = 0$ for $|t| > \tau$, it follows that for every $\omega > 0$,

$$|F(\omega)| = \frac{1}{2}\left|\int_{-\tau-\pi/\omega}^{\tau}\left[f(t) - f\left(t + \frac{\pi}{\omega}\right)\right]e^{-j\omega t}\, dt\right| < \frac{L}{2}\left(\frac{\pi}{\omega}\right)^{\alpha}\int_{-\tau-\pi/\omega}^{\tau} dt$$

and (15) results. The proof is similar for $\omega < 0$.

70. Since $f(0) = (1/2\pi)\int_{-\sigma}^{\sigma} F(\omega)\, d\omega$ and $f(\pm a) = (1/2\pi)\int_{-\sigma}^{\sigma} F(\omega)e^{\pm ja\omega}\, d\omega = 0$ by assumption, it follows that for any α_1 and α_2,

$$f(0) = \frac{1}{2\pi}\int_{-\sigma}^{\sigma} F(\omega)(1 + \alpha_1 e^{ja\omega} + \alpha_2 e^{-ja\omega})\, d\omega$$

Applying Schwarz' inequality to the above integral, we conclude that $f(0)$ is maximum if $F(\omega)$ is proportional to $1 + \alpha_1^* e^{-ja\omega} + \alpha_2^* e^{ja\omega}$. Determining the constants α_1 and α_2 from the condition $f(a) = f(-a) = 0$, we obtain (17).

71. From the Akhiezer theorem (6-A-33), it follows that $f(t) = |y(t)|^2$, where $y(t)$ is $\sigma/2$ – BL. Hence [see (6-53)],

$$f(t) = |y(t)|^2 \leq \frac{\sigma}{2\pi}\int_{-\infty}^{\infty} |y(t)|^2\, dt$$

72. We wish to minimize the error $e = \sum_{n=0}^{\infty} (a_n - b_n)^2 \lambda_n$ subject to the constraint $E = \sum_{n=0}^{\infty} a_n^2$. With μ a Lagrange multiplier, we must have

$$\frac{\partial}{\partial a_n}(e + \mu E) = 0 = 2(a_n - b_n)\lambda_n + 2\mu a_n$$

Hence, $a_n = b_n \lambda_n/(\mu + \lambda_n)$. The constant μ is determined from the energy constraint.

73. With $y(t)v(t) = \sum_{n=-\infty}^{\infty} b_n e^{jn\omega_0 t}$, it follows from Eq. (3-73) that $b_n = \sum_{k=-\infty}^{\infty} y_k v_{n-k}$. Applying (3-76) to the functions $f_1(t) = y(t)v(t)$ and $f_2(t) = y(t)$, we obtain (18).

74. With $\alpha^+(\omega) = \alpha(\omega)$ if $\alpha(\omega) > 0$ and $\alpha^+(\omega) = 0$ otherwise, we have $e^{2\alpha(\omega)} \geq 2\alpha^+(\omega)$. Hence,

$$\infty > \int_{-\infty}^{\infty} e^{2\alpha(\omega)}\, d\omega > 2\int_{-\infty}^{\infty} \alpha^+(\omega)\, d\omega > 2\int_{-\infty}^{\infty} \frac{\alpha^+(\omega)}{1+\omega^2}\, d\omega$$

It suffices, therefore, to prove that

$$\int_{-\infty}^{\infty} \frac{\alpha(\omega)}{1+\omega^2}\, d\omega < \infty$$

We form the function $\ln F(s)/(1-s^2)$. This function is analytic everywhere in the region $\mathrm{Re}\, s > 0$ except at $s = 1$, where it has a pole with residue $\ln F(1)/2$. Therefore, its integral along the closed curve C_1 of Fig. 7-1*a* equals $j\pi \ln F(1)$. The contribution of the semicircle Γ_1 to this integral tends to zero as $r \to \infty$. Hence,

$$\int_{-\infty}^{\infty} \frac{\ln F(j\omega)}{1+\omega^2}\, d\omega = \pi \ln F(1) < \infty$$

75. With $y(t)$ the inverse of $Y(\omega)$, we maintain that $f(t) = 2y(t)U(t)$. Indeed, since $Y(\omega)$ is real, $y^*(-t) = y(t)$. Hence, $f(t) + f^*(-t) = 2y(t)U(t) + 2y(t)U(-t) = 2y(t)$. But $f^*(-t) \leftrightarrow F^*(\omega) = R(\omega) - jX(\omega)$. Therefore, $R(\omega) + jX(\omega) + R(\omega) - jX(\omega) = 2Y(\omega)$.

76. (*a*) Given $\epsilon > 0$, we can find a number ω_1 such that $\int_{|\omega|>\omega_1} W(\omega)\, d\omega < \pi\epsilon$

$$\begin{aligned}|w(t+\tau) - w(t-\tau)| &= \frac{1}{\pi}\left|\int_{-\infty}^{\infty} W(\omega) \sin \omega\tau e^{j\omega t}\, d\omega\right| \\ &\le \frac{1}{\pi}\int_{|\omega|>\omega_1} W(\omega)\, d\omega + \frac{1}{\pi}\int_{-\omega 1}^{\omega_1} |\omega|\tau W(\omega)\, d\omega \le \epsilon + \frac{\tau\omega_1}{\pi}\int_{-\omega_1}^{\omega_1} W(\omega)\, d\omega \\ &\le \epsilon + 2\tau\omega_1 w(0)\end{aligned}$$

If $\tau < \epsilon/2\omega_1 w(0)$, then $|w(t+\tau) - w(t-\tau)| < 2\epsilon$ and, since ϵ is arbitrary, $w(t+\tau) - w(t-\tau) \to 0$ as $\tau \to 0$.

(*b*) We substitute $|F(\omega)|$ with $W(\omega)$ and reason as in part *a*.

77. It follows from the pair

$$p_a(t) \cos \frac{n\pi}{a} t \leftrightarrow \frac{2(-1)^n \omega \sin a\omega}{\omega^2 - (n\pi/a)^2}$$

78. With λ and μ two Lagrange multipliers, it suffices to minimize the integral $\int_{-\infty}^{\infty} F(x, y)\, dx$, where $F(x, y) = x^2 y^2(x) - \lambda y^2(x) - \mu y^4(x)$. Since $F(x, y)$ does not depend on y', Euler's condition $dF_{y'}/dx = F_y$ yields $0 = 2x^2 y(x) - 2\lambda y(x) - 4\mu y^3(x)$, from which it follows that $2\mu y^2(x) = x^2 - \lambda$ or $y(x) = 0$. The rest of the proof is simple.

79. Since $H(z)$ is analytic and has no zeros outside the unit circle, its logarithm can be expanded into a power series $\ln H(z) = \sum_{n=0}^{\infty} a_n z^{-n}$ converging for every $|z| \ge 1$. With $z = e^{j\omega T}$, the above yields

$$\alpha(\omega) + j\varphi(\omega) = \sum_{n=0}^{\infty} a_n \cos n\omega T - j \sum_{n=1}^{\infty} a_n \sin n\omega t$$

Thus, the functions $\alpha(\omega) - \alpha(0)$ and $\varphi(\omega)$ are cosine and sine series with the same coefficients as in (7-137); hence, they satisfy (8-103), properly modified.

Chapter 8

80. $$\int_0^{\infty} \frac{1}{\sqrt{t}} e^{-j\omega t}\, dt = 2\int_0^{\infty} e^{-j\omega y^2}\, dy = \int_{-\infty}^{\infty} e^{-j\omega y^2}\, dy = \sqrt{\frac{\pi}{j\omega}}$$

81. $$\int_{-\infty}^{\infty} e^{-\alpha(t-\tau)^2} e^{-\beta\tau^2}\, d\tau = e^{-\gamma t^2}\int_{-\infty}^{\infty} e^{-(\alpha+\beta)[\tau - \alpha t/(\alpha+\beta)]^2}\, d\tau = \sqrt{\frac{\pi}{\alpha+\beta}}\, e^{-\gamma t^2}$$

82. It follows, as in Prob. 37, because [see (8-6)]

$$\cos \alpha t^2 + j \sin \alpha t^2 = e^{j\alpha t^2} \leftrightarrow \sqrt{\frac{\pi}{\alpha}}\, e^{-j(\omega^2/4\alpha - \pi/4)} = \sqrt{\frac{\pi}{\alpha}} \left[\cos\left(\frac{\omega^2}{4\alpha} - \frac{\pi}{4}\right) - j \sin\left(\frac{\omega^2}{4\alpha} - \frac{\pi}{4}\right)\right]$$

83. It follows from the pair $e^{-\alpha t^2} \leftrightarrow \sqrt{\pi/\alpha}\, e^{-\omega^2/4\alpha}$ and the Poisson sum formula (3-86), with $T = 1$.

84. It can easily be seen, by induction, that $h_n(t) = th_{n-1}(t) - h'_{n-1}(t)$. Transforming both sides, we obtain $H_n(\omega) = jH'_{n-1}(\omega) - j\omega H_{n-1}(\omega)$. With $y_n(t) = j^n H_n(t)$, the above yields $y_n(t) = ty_{n-1}(t) - y'_{n-1}(t)$. Thus, the functions $h_n(t)$ and $y_n(t)$ satisfy the same recursion equation. Furthermore,

$$h_0(t) = e^{-t^2/2} \leftrightarrow \sqrt{2\pi}\, e^{-\omega^2/2} = H_0(\omega)$$

That is, $y_0(t) = \sqrt{2\pi}\, h_0(t)$. Hence, $y_n(t) = \sqrt{2\pi}\, h_n(t)$ for all n.

85. With $f(t) = \sum_{n=0}^{\infty} a_n \varphi_n(t)$ as in Eq. (6-83), we must find a_n such that $\sum_{n=0}^{\infty} a_n{}^2 = E$ and $\sum_{n=0}^{\infty} \lambda_n a_n{}^2 = \alpha E$. It follows that if $\alpha > \lambda_0$, then there are no solutions; if $\alpha = \lambda_0$, there is one solution $f(t) = \sqrt{E}\, \varphi_0(t)$; if $\alpha < \lambda_0$, then there are infinitely many solutions.

86. We can assume that the energy E of $f(t)$ equals 1. Defining inner products as in Eq. (5-26), we have $1 = \langle f, f \rangle$; $\alpha = \langle f_\tau, f_\tau \rangle$; $\beta = \langle f_\sigma, f_\sigma \rangle$. From the orthogonality principle (5-32), it follows that $\langle f_\tau, y \rangle = 0$ and $\langle f_\sigma, y \rangle = 0$; hence, $\langle \hat{f}, y \rangle = 0$. Finally, the energy $\hat{E}$ of $\hat{f}$ is given by $\hat{E} = \langle \hat{f}, \hat{f} \rangle = 1 - e$. Denoting by e_τ and e_σ the energy of the projections $y_\tau(t)$ and $y_\sigma(t)$, respectively, we conclude that $\hat{E}\hat{\alpha} = \alpha + e_\tau$ and $\hat{E}\hat{\beta} = \beta + e_\sigma$ because $\hat{f}(t) = f(t) - y(t)$. Hence, $\hat{\alpha} \ge \alpha$ and $\hat{\beta} \ge \beta$.

87. With $\omega = \sin \theta$ and $d\omega = \cos \theta\, d\theta = \sqrt{1 - \omega^2}\, d\theta$, we have

$$\frac{1}{\pi}\int_{-\pi/2}^{\pi/2} e^{jt \sin \theta}\, d\theta = \frac{1}{\pi}\int_{-1}^{1} \frac{1}{\sqrt{1-\omega^2}}\, e^{j\omega t}\, d\omega = \frac{1}{2\pi}\int_{-\infty}^{\infty} \frac{2p_1(\omega)}{\sqrt{1-\omega^2}}\, e^{j\omega t}\, d\omega$$

and (20) follows [see (3-2)].

88. From (8-69) and (8-60), it follows that

$$f(\sqrt{x^2+y^2}) = \frac{1}{2\pi}\int_{-\infty}^{\infty}\int_{-\infty}^{\infty} f_1(\sqrt{(x-\xi)^2+(y-\eta)^2})\, f_2(\sqrt{\xi^2+\eta^2})\, d\xi\, d\eta$$

89. (*a*) With $t = r \cos \theta$, $y = r \sin \theta$, and $z = \sqrt{x^2 + r^2}$,

$$h(x) = \int_{-\infty}^{\infty}\int_{-\infty}^{\infty} f(\sqrt{x^2+y^2+t^2})\, dy\, dt = 2\pi \int_{0}^{\infty} rf(\sqrt{x^2+r^2})\, dr = 2\pi \int_{x}^{\infty} zf(z)\, dz$$

and (21) follows by differentiation.

(*b*) From $g(x) = \lambda f(x)$, it follows that $h(x) = \lambda^2 f(x)$, and (21) yields $2\pi x f(x) = -\lambda^2 f'(x)$; solving, we obtain (22).

Chapters 9 to 12

90. $$E\{|\mathbf{x}(t+\tau) \pm \mathbf{x}(t)|^2\} = E\{[\mathbf{x}(t+\tau) \pm \mathbf{x}(t)][\mathbf{x}^*(t+\tau) \pm \mathbf{x}^*(t)]\}$$
$$= R(0) \pm R(\tau) \pm R^*(\tau) + R(0)$$

91. From (9-80), it follows that $R(\tau) = w(\tau)$ is a periodic function with periods τ_1 and τ_2. Hence, $w(\tau)$ must be constant because the ratio τ_1/τ_2 is irrational by assumption.

92. $E\{\mathbf{n}(t_1)\mathbf{y}(t)\} = 0$ for $t_1 > t$. Hence, $E\{\mathbf{y}(t)\mathbf{y}(T)\} = E\{\mathbf{y}^2(t)\} = It$ and

$$E\{\mathbf{z}^2(t)\} = E\{\mathbf{y}^2(t)\} - 2E\{\mathbf{y}(t)\mathbf{y}(T)\}\frac{t}{T} + E\{\mathbf{y}^2(T)\}\frac{t^2}{T^2}$$

93. It suffices to show that [see (11-30)]

$$\sigma_\eta{}^2 = \frac{1}{4T^2}\iint_R C(t_1, t_2)\, dt_1\, dt_2 \underset{T\to\infty}{\to} 0 \tag{23}$$

where R is a square as in Fig. S-93. Defining by R_1 the part of R where $|t_1 - t_2| > \alpha$, we conclude that

$$\sigma_{\eta T}^2 \le \frac{1}{4T^2}(4T^2\epsilon + 4\alpha TM)\epsilon + \frac{\alpha M}{T} \underset{T\to\infty}{\to} \epsilon$$

because the area of R_1 is less than $4T^2$, the area of $R_2 = R - R_1$ is less than $4\alpha T$, $|C(t_1, t_2)| < M$ everywhere, and $|C(t_1, t_2)| < \epsilon$ in R_1. And, since ϵ is arbitrary, (23) follows.

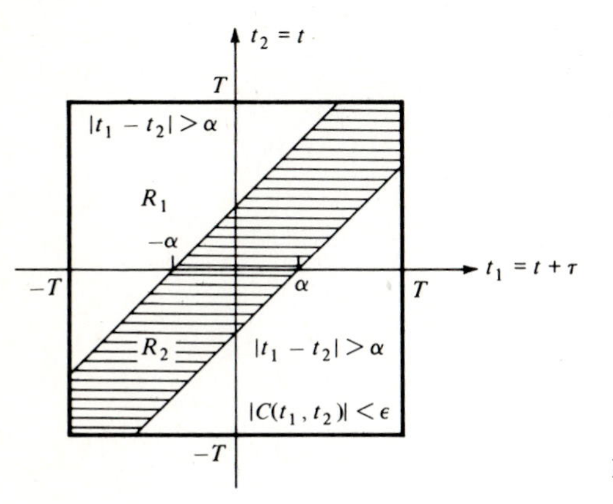

Figure S-93

94. (*a*) $E\{|\mathbf{N}(\omega)|^2\} = I\int_{-T}^{T}\int_{-T}^{T}\delta(t_1 - t_2)e^{-j\omega(t_1 - t_2)}\, dt_1\, dt_2 = I\int_{-T}^{T} dt_2$

(*b*) $E\{|\mathbf{A}_m|^2\} = I\sum_{n=0}^{N-1}\sum_{k=0}^{N-1}\delta[n-k]w^{m(n-k)} = I\sum_{n=0}^{N-1} 1$

INDEX

INDEX